CW01197375

To
My Parents

Preface

The purpose of this book is to provide a mature approach to the basic principles of classical thermodynamics which is one of the few core subjects for the undergraduate students of almost all branches of engineering. The system-surroundings interactions involving work and heat transfer with associated property changes, and the system-control volume approaches of the first law have been emphasized. The second law has been elaborated upon in considerable detail. Except for some physical explanations, a statistical or microscopic analysis of the subject has not been made.

The first eight chapters of the book are devoted to a thorough treatment of the basic principles and concepts of classical thermodynamics. The second law and entropy have been introduced using the concept of heat engine. Chapters 9 and 10 present the properties of substances. Chapter 11 gives the general thermodynamic relationships among properties. A detailed analysis of power and refrigeration cycles is given in Chapters 12 to 14. Chapter 15 deals with air-water vapour mixtures, and reactive systems are analyzed in Chapter 16. To increase the utility of the book, Chapters 17 and 18 dealing with compressible fluid flow and heat transfer respectively, have been added.

Many illustrative examples are solved and many problems are provided in each chapter to aid comprehension and to stimulate the interest of the students.

Throughout the text SI units have been used. Tables and charts given in the Appendix are also in SI units.

This book is based mainly on the lecture notes prepared for classes on the subject at IIT, Kharagpur. I am grateful to all the authors of the books I had used in preparing the notes, a list of which is given in the bibliography. I am thankful to my colleagues in the mechanical engineering department of IIT, Kharagpur, for many stimulating discussions, and for encouragement. I am indebted to all those who have helped in the preparation of the book.

I would very much appreciate criticisms, suggestions for improvement, and detection of errors from my readers, which will be gratefully acknowledged.

P K NAG

It is a remarkable illustration of the ranging power of the human intellect that a principle first detected in connection with the clumsy puffing of the early steam engines should be found to apply to the whole world, and possibly, even to the whole cosmic universe.

A.R. Ubbelohde, *Man and Energy*
(Hutchinson, London, 1954)

Contents

List of Symbols

A	Area, availability
AE	Available energy
a	Acceleration, specific availability
B	Virial coefficient, Keenan function
b	Specific Keenan function
C	Heat capacity, number of components, virial coefficient
c	Specific heat, velocity of sound
°C	Degree Celsius
D	Diameter, virial coefficient
d	An infinitesimal increase in a point function
E	Energy
e	Specific energy
F	Helmholtz function, force, impulse function, configuration factor for gray bodies, tension
f	Degree of freedom or variance, fugacity, specific Helmholtz function
G	Gibbs function, mass of dry air
ΔG	Gibbs function change
g	Acceleration due to gravity, specific Gibbs function
g_0	Constant of proportionality in Newton's law
Gr	Grashof number
H	Enthalpy, magnetic field strength
ΔH	Heat of reaction
h	Specific enthalpy, heat transfer coefficient
$\bar{h}_{RP}$	Enthalpy of combustion
I	Electric current, irreversibility, moment of inertia, impulse pressure
J	Joule's equivalent
K	Boltzmann constant, equilibrium constant, thermal conductivity,
K	Degree Kelvin
K.E.	Kinetic energy
kgf	Kilogram force (in MKS)
kgm	Kilogram mass (in MKS)
kg	Kilogram (in SI)
L	Length, amount of liquid
l	Latent heat
M	Mach number
μ	Molecular weight
m	Mass, mass of water vapour
$\dot{m}$	Mass flow rate

N	RPM, number of molecules
Nu	Nusselt number
n	Number of moles, number of cylinders, polytropic exponent
P.E.	Potential energy
Pr	Prandtl number
p	Pressure, partial pressure
Q	Heat transfer, rate of heat transfer
q	Rate of heat flux
R	Characteristic gas constant, thermal resistance, electrical resistance
$\bar{R}$	Universal gas constant
Re	Reynolds number
r	Radius
r_c	Cut-off ratio
r_k	Compression ratio
r_e	Expansion ratio
S	Entropy
s	Specific entropy
T	Absolute temperature, torque
t	Temperature
t_{db}	Dry bulb temperature
t_{dp}	Dew point temperature
t_{wb}	Wet bulb temperature
U	Internal energy, overall heat transfer coefficient
U.E.	Unavailable energy
u	Specific internal energy
$\bar{u}_{RP}$	Internal energy of combustion
V	Volume
$\mathbf{V}$	Velocity
v	Specific volume
W	Work transfer, specific humidity
w	Mass flow rate, angular velocity
X	Value of thermometric property
x	Quality, mole fraction, thickness, Cartesian co-ordinate
y	Cartesian co-ordinate
Z	Compressibility factor, elevation, height
z	Cartesian co-ordinate

Greek symbols and special notations

α	Absorptivity
β	Coefficient of volume expansion
γ	Specific heat ratio
Δ	A finite increase in a point function or property
$đ$	An infinitesimal amount of work or heat transfer (a path function)

δ	A small change in a property
ϵ	Degree of reaction, energy of a particle
ε	Thermal emf
η	Efficiency
θ	Angle, temperature
μ	Joule-Kelvin coefficient, chemical potential, degree of saturation, coefficient of viscosity
ν	Kinematic viscosity, stoichiometric coefficient
ρ	Density, reflectivity
σ	Control surface, surface tension, Stefan-Boltzmann constant
τ	Time, transmissivity
ϕ	Availability function for a closed system, number of phases, relative humidity
Ψ	Availability function for a steady flow system

Subscripts and superscripts

c	Critical state, cold fluid
f	Saturated liquid
fg	Change in property between saturated vapour and saturated liquid phases
g	Saturated vapour
i	Fluid at initial state, initial phase
f	Fluid at final state; final phase
h	Hot fluid
m	Mixture value
o	Surrounding state
p	Constant pressure
R	Reversible
r	Reduced value
s	Constant entropy
t	Triple point
v	Constant volume

1

Introduction : Basic Concepts, Definitions and Units

Thermodynamics is the science of energy transfer and its effect on the physical properties of substances. It is based upon observations of common experience which have been formulated into thermodynamic laws. These laws govern the principles of energy conversion. The applications of the thermodynamic laws and principles are found in all fields of energy technology, notably in steam and nuclear power plants, internal combustion engines, gas turbines, air conditioning, refrigeration, gas dynamics, jet propulsion, compressors, chemical process plants, and direct energy conversion devices.

1.1 MACROSCOPIC vs MICROSCOPIC

There are two points of view from which the behaviour of matter can be studied: the macroscopic and the microscopic. In the macroscopic approach, a certain quantity of matter is considered, without the events occurring at the molecular level being taken into account. From the microscopic point of view, matter is composed of myriads of molecules. If it is a gas, each molecule at a given instant has a certain position, velocity, and energy, and for each molecule these change very frequently as a result of collisions. The behaviour of the gas is described by summing up the behaviour of each molecule. Such a study is made in *microscopic or statistical thermodynamics*. *Macroscopic thermodynamics* is only concerned with the effects of the action of many molecules, and these effects can be perceived by human senses. For example, the macroscopic quantity, pressure, is the average rate of change of momentum due to all the molecular collisions made on a unit area. The effects of pressure can be felt. The macroscopic point of view is not concerned with the action of individual molecules, and the force on a given unit area can be measured by using, e.g., a pressure gauge. These macroscopic observations are completely independent of the assumptions regarding the nature of matter. All the results of classical or macroscopic thermodynamics can, however, be derived from the microscopic and statistical study of matter.

1.2 THERMODYNAMIC SYSTEM AND CONTROL VOLUME

A thermodynamic *system* is defined as a quantity of matter or a region in space upon which attention is concentrated in the analysis of a problem. Everything external to the system is called the *surroundings* or the *environment*. The system is separated from the surroundings by the system boundary (Fig. 1.1). The boundary may be either *fixed* or *moving*. A system and its surroundings together comprise a *universe*.

There are three classes of systems: (a) closed system, (b) open system, and (c) isolated system. The *closed system* (Fig. 1.2) is a system of fixed

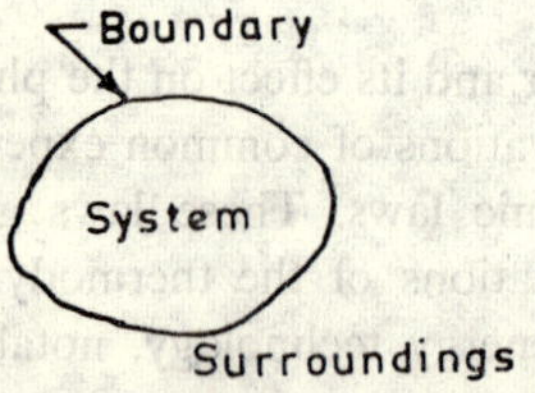

Fig. 1.1 A thermodynamic system

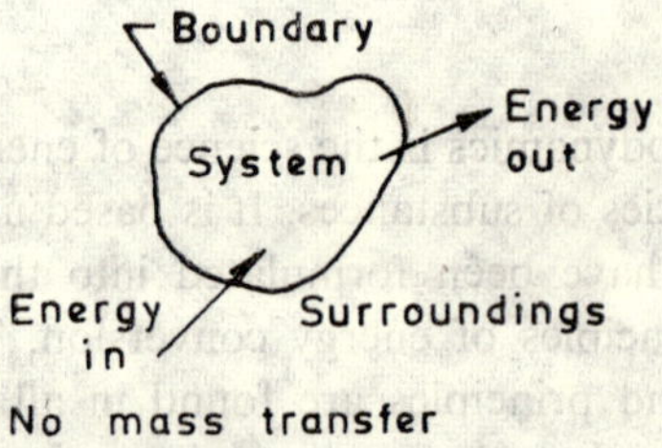

Fig. 1.2 A closed system

mass. There is no mass transfer across the system boundary. There may be energy transfer into or out of the system. A certain quantity of fluid in a cylinder bounded by a piston constitutes a closed system. The *open system* (Fig. 1.3) is one in which matter crosses the boundary of the system. There may be energy transfer also. Most of the engineering devices are generally open systems, e.g., an air compressor in which air enters at low pressure and leaves at high pressure and there are energy transfers across the system boundary. The *isolated system* (Fig. 1.4) is one in which there is no in-

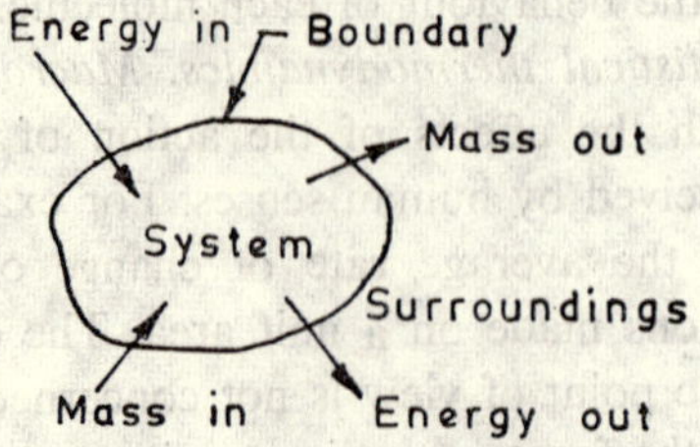

Fig. 1.3 An open system

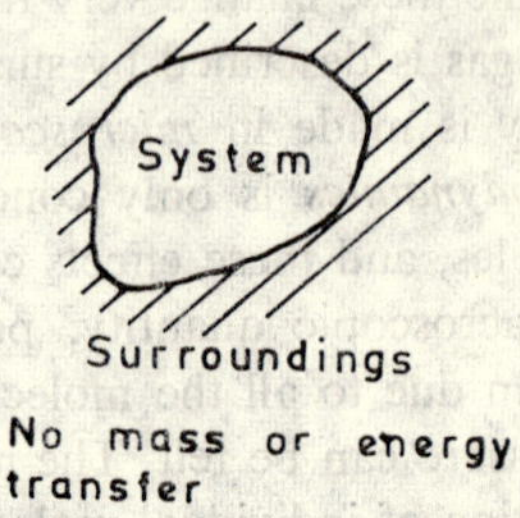

Fig. 1.4 An isolated system

teraction between the system and the surroundings. It is of fixed mass and energy, and there is no mass or energy transfer across the system boundary.

For thermodynamic analysis of an open system, such as an air compressor, in which there is a flow of mass into and out of the system (Fig. 1.5), attention is focussed on a certain volume in space surrounding the compressor, known as the *control volume,* bounded by a surface called the *control surface.* Matter as well as energy can cross the control surface.

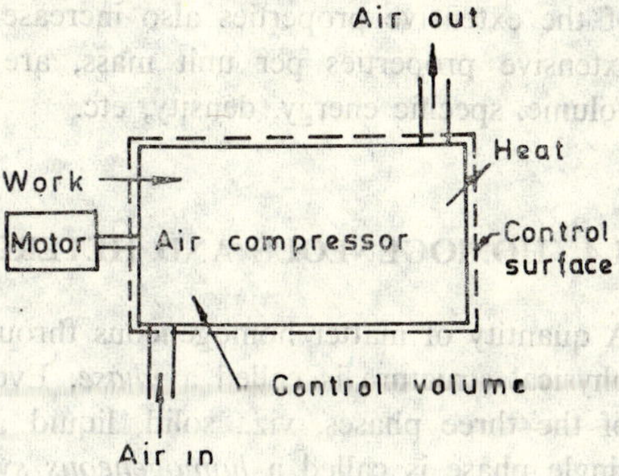

Fig. 1.5 Control volume and control surface

While dealing with a fixed quantity of mass, a system is defined, and in the case of a device involving a flow of mass a control volume is specified. The system approach concentrates on a fixed mass and the control volume approach focusses on the mass (and energy) flowing across the control surface. These are equivalent to the closed system and the open system respectively. The difference in these two approaches is elaborated in Chapter 5.

1.3 THERMODYNAMIC PROPERTIES, PROCESSES, AND CYCLES

Every system has certain characteristics by which its physical condition may be described, e.g., volume, temperature, pressure, etc. Such characteristics are called *properties* of the system. These are all *macroscopic* in nature. When all the properties of a system have definite values, the system is said to exist at a definite *state.* Properties are the coordinates to describe the state of a system. They are the state variables of the system. Any operation in which one or more of the properties of a system changes is called a *change of state.* The succession of states passed through during a change of state is called the *path* of the change of state. When the path is completely specified, the change of state is called a *process,* e.g., a constant pressure process. A thermodynamic *cycle* is defined as a series of state changes such that the final state is identical with the initial state (Fig. 1.6).

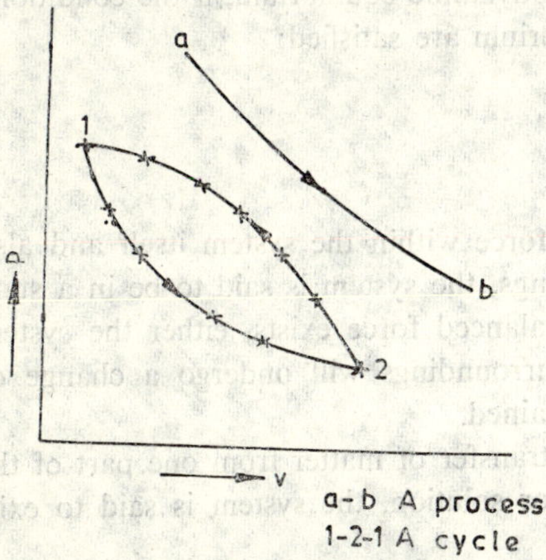

Fig. 1.6 A process and a cycle

Properties may be of two types. *Intensive properties* are independent of the mass in the system, e.g., pressure, temperature, etc. *Extensive properties* are related to mass, e.g., volume, energy, etc. If mass is increased, the values of the extensive properties also increase. Specific extensive properties, i.e., extensive properties per unit mass, are intensive properties, e.g., specific volume, specific energy, density, etc.

1.4 HOMOGENEOUS AND HETEROGENEOUS SYSTEM

A quantity of matter homogeneous throughout in chemical composition and physical structure is called a *phase*. Every substance can exist in any one of the three phases, viz., solid, liquid and gas. A system consisting of a single phase is called a *homogeneous system,* while a system consisting of more than one phase is known as a *heterogeneous system.*

1.5 THERMODYNAMIC EQUILIBRIUM

A system is said to exist in a state of *thermodynamic equilibrium* when no change in any macroscopic property is registered, if the system is isolated from its surroundings.

An isolated system always reaches in course of time a state of thermodynamic equilibrium and *can never depart from it spontaneously.*

Therefore, there can be no *spontaneous change in any macroscopic property* if the system exists in an equilibrium state. Thermodynamics studies mainly the properties of physical systems that are found in equilibrium states.

A system will be in a state of thermodynamic equilibrium, if the conditions for the following three types of equilibrium are satisfied:

(a) Mechanical equilibrium
(b) Chemical equilibrium
(c) Thermal equilibrium

In the absence of any unbalanced force within the system itself and also between the system and the surroundings, the system is said to be in a state of *mechanical equilibrium.* If an unbalanced force exists, either the system alone or both the system and the surroundings will undergo a change of state till mechanical equilibrium is attained.

If there is no chemical reaction or transfer of matter from one part of the system to another, such as diffusion or solution, the system is said to exist in a state of *chemical equilibrium.*

When a system existing in mechanical and chemical equilibrium is separated from its surroundings by a diathermic wall (diathermic means 'which allows heat to flow') and if there is no spontaneous change in any property of the system, the system is said to exist in a state of *thermal equilibrium.* When this is not satisfied, the system will undergo a change of state till thermal equilibrium is restored.

When the conditions for any one of the three types of equilibrium are not satisfied, a system is said to be in a *nonequilibrium state.* If the nonequilibrium of the state is due to an unbalanced force in the interior of a system or between the system and the surroundings, the pressure varies from one part of the system to another. There is no single pressure that refers to the system as a whole. Similarly, if the nonequilibrium is because of the temperature of the system being different from that of its surroundings, there is a non-uniform temperature distribution set up within the system and there is no single temperature that stands for the system as a whole. It can thus be inferred that when the conditions for thermodynamic equilibrium are not satisfied, the states passed through by a system cannot be described by thermodynamic properties which represent the system as a whole.

Thermodynamic properties are the macroscopic co-ordinates defined for, and significant to, only thermodynamic equilibrium states. Both classical and statistical thermodynamics study mainly the equilibrium states of a system.

1.6 QUASI-STATIC PROCESS

Let us consider a system of gas contained in a cylinder (Fig system initially is in equilibrium state, represented by the properties t_1. The weight on the piston just balances the upward force exerted by the gas. If the weight is removed, there will be an unbalanced force between the system and the surroundings, and under gas pressure, the piston will move up till it hits the stops. The system again comes to an equilibrium state, being described by the properties p_2, v_2, t_2. But the intermediate states passed through by the system are nonequilibrium states which cannot be described by thermodynamic coordinates. Figure 1.8 shows points 1 and 2

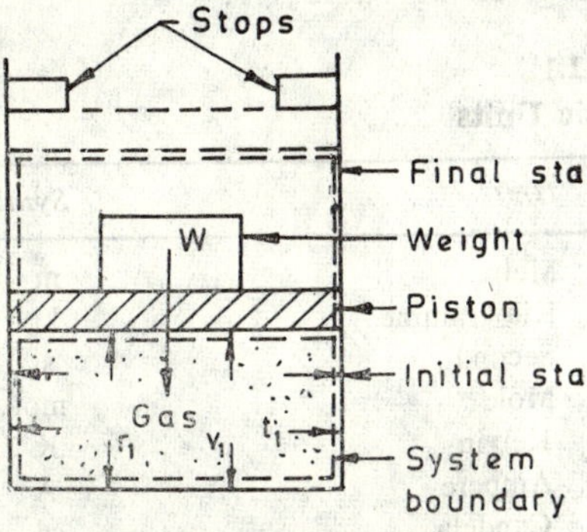

Fig. 1.7 Transition between two equilibrium states by an unbalanced force

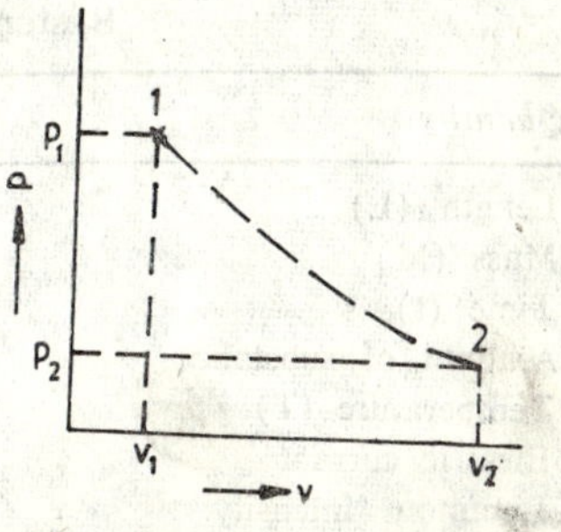

Fig. 1.8 Plot representing the transition between two equilibrium states

as the initial and final equilibrium states joined by a dotted line, which has got no meaning otherwise. Now if the single weight on the piston is made up of many very small pieces of weights (Fig. 1.9), and these weights are removed one by one very slowly from the top of the piston, at any instant of the upward travel of the piston, if the gas system is isolated, *the departure of the state of the system from the thermodynamic equilibrium state will be infinitesimally small.* So every state passed through by the system will be an equilibrium state. Such a process, which is but a locus of all the equilibrium points passed through by the system, is known as a *quasi-static process* (Fig. 1.10), 'quasi' meaning 'almost'. *Infinite showness is the characteristic*

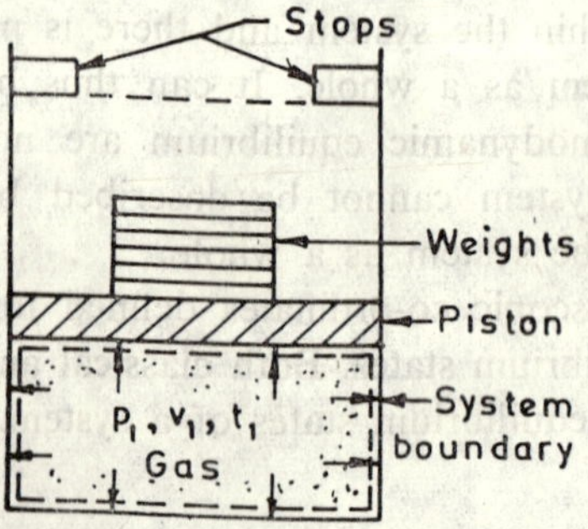

Fi[illegible]itely slow transition [illegible] by infinitesimal

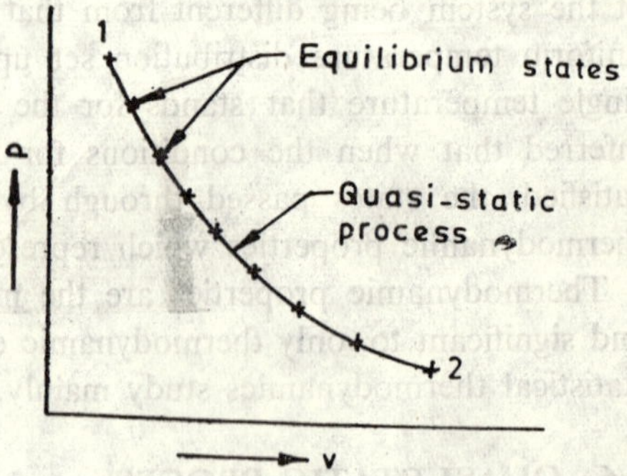

Fig. 1.10 A quasi-static process

feature of a quasi-static process. A quasi-static process is thus a succession of equilibrium states. A quasi-static process is also called a *'reversible process'.*

1.7 UNITS AND DIMENSIONS

In the present text, the SI (System International) system of units has been used. The basic units in this system are given in Table 1.1.

Table 1.1
System: Basic Units

Quantity	*Unit*	*Symbol*
Length (L)	Metre	m
Mass (M)	Kilogramme	kg
Time (t)	Second	s
Amount of substance	Mole	mol
Temperature (T)	Kelvin	K
Electric current	Ampere	A
Luminous intensity	Candela	cd
Plane angle	Radian	rad
Solid angle	Steradian	sr

The dimensions of all other quantities are derived from these basic units, which are given in Table 1.2.

Table 1.2
SI System: Derived Units

Quantity	*Unit*	*Symbol*	*Alternative unit*	*In Basic units*
Force (F)	Newton	N		kg m/s^2
Energy (E)	Joule	J	Nm	kg m^2/s^2
Power	Watt	W	J/s	kg m^2/s^3
Pressure	Pascal	Pa	N/m^2	kg/(ms^2)
Frequency	Hertz	Hz		s^{-1}
Electric charge	Coulomb	C		As
Electric potential	Volt	V	W/A=J/C	kg m^2/(s^{3}A)
Capacitance	Farad	F	C/V	s^{4}A^2/(kg m^2)
Electrical resistance	Ohm		V/A	kg m^2/(s^{3}A^2)
Magnetic flux	Weber	Wb	Vs	kg m^2/(s^{2}A)
Magnetic flux density	Tesla	T	Wb/m^2	kg/(s^{2}A)
Inductance	Henry	H	Wb/A	kg m^2/(s^{2}A^2)

The unit of force is the newton (N). A force of one newton produces an acceleration of one metre per sec^2 when applied to a mass of 1 kg.

In the MKS system (metre-kilogramme-second system), the standard unit of force is the kilogramme-force (kgf), which is defined as that force which, when applied to a mass of 1 kg (the unit of mass is kg) will produce the standard gravitational acceleration $g=9.80665$ m/s^2. The MKS system is not coherent, because the product (or quotient) of any two quantities is not the unit of the resultant quantity and a constant has to be introduced. By Newton's second law of motion

$$F \times C = m \times a$$

where F is the force, m the mass, a the acceleration, and C a constant.

$$\therefore \qquad 1 \text{ kgf} \times C = 1 \text{ kg} \times 9.806 \text{ m/s}^2$$

where

$$C = g_0 = 9.80665 \frac{\text{kg}}{\text{kgf}} \cdot \frac{\text{m}}{\text{s}^2}$$

So

$$F = \frac{ma}{g_0}$$

The introduction of the constant g_0 is not necessary and often leads to errors and undesirable complications.

Pressure is the normal force exerted by a system against unit area of its bounding surface. In SI units the unit of pressure is the *pascal* (N/m^2), which is very small. Instead, the unit bar, kPa or MPa (kilo- or mega-pascal) is used.

$$1 \text{ bar} = 10^5 \text{ N/m}^2 = 100 \text{ kN/m}^2 = 100 \text{ kPa}$$

Pressures are also stated in mm or cm of mercury and, in MKS system, in kgf/m^2. The unit kgf/m^2 is very large by comparison with many engineering values, and the unit kgf/cm^2, sometimes called *ata* (atmosphere technical absolute), is used.

1 standard atmosphere

$$= 760 \text{ mm Hg} = 1.01325 \text{ bar} = 101.325 \text{ kN/m}^2$$

$$= 1.0332 \text{ ata} = 1.0332 \text{ kgf/cm}^2$$

Most instruments indicate pressure relative to the atmospheric pressure, whereas the pressure of a system is its pressure above zero, or relative to a perfect vacuum. The pressure relative to the atmosphere is called *gauge pressure*. The pressure relative to a perfect vacuum is called *absolute pressure*.

Absolute pressure = Gauge pressure + Atmospheric pressure

When the pressure in a system is less than atmospheric pressure, the gauge pressure becomes negative, but is frequently designated by a positive number and called *vacuum*. For example, 16 cm vacuum will be

$$\frac{76 - 16}{76} \times 1.013 = 0.80 \text{ bar}$$

Figure 1.11 shows a few pressure measuring devices. Figure (a) shows the Bourdon gauge which measures the difference between the system pressure inside the tube and atmospheric pressure. It relies on the deformation of a bent hollow tube of suitable material which, when subjected to the pressure to be measured on the inside (and atmospheric pressure on the outside), tends to unbend. This moves a pointer through a suitable gear-and-lever mechanism against a calibrated scale. Figure (b) shows an open U-tube indicating gauge pressure, and Fig. (c) shows an open U-tube indicating vacuum. Figure (d) shows a closed U-tube indicating absolute pressure. If p is atmospheric pressure, this is a *barometer*. These are called U-tube manometers.

If Z is the difference in the heights of the fluid columns in the two limbs of the U-tube [Fig. (b) and Fig. (c)], ρ the density of the fluid and g the

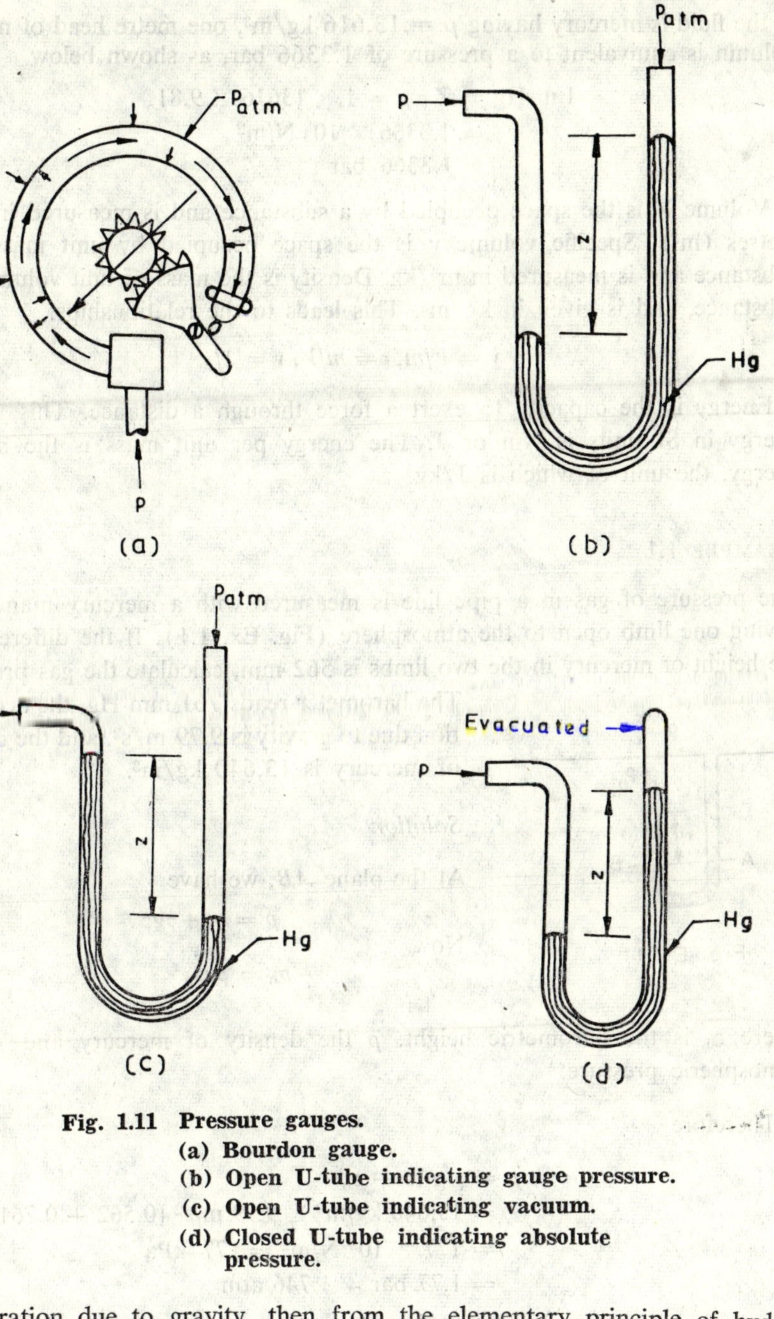

Fig. 1.11 Pressure gauges.
(a) Bourdon gauge.
(b) Open U-tube indicating gauge pressure.
(c) Open U-tube indicating vacuum.
(d) Closed U-tube indicating absolute pressure.

acceleration due to gravity, then from the elementary principle of hydrostatics, the gauge pressure p_g is given by

$$p_g = Z \rho g \left[\text{m.} \frac{\text{kg}}{\text{m}^3} \cdot \frac{\text{m}}{\text{s}^2} \right]$$
$$= Z \rho g \quad \text{N/m}^2$$

If the fluid is mercury having $\rho = 13{,}616$ kg/m³, one metre head of mercury column is equivalent to a pressure of 1.3366 bar, as shown below

$$\begin{aligned} 1 \text{ m Hg} &= Z \rho g = 1 \times 13616 \times 9.81 \\ &= 1.3366 \times 10^5 \text{ N/m}^2 \\ &= 1.3366 \text{ bar} \end{aligned}$$

Volume V is the space occupied by a substance and is measured in cubic metres (m³). Specific volume v is the space occupied by unit mass of a substance and is measured in m³/kg. Density is the mass of unit volume of a substance, and is given in kg/m³. This leads to the relationships.

$$v = V/m, \rho = m/V, v = 1/\rho$$

Energy is the capacity to exert a force through a distance. This unit of energy in SI units is Nm or J. The energy per unit mass is the specific energy, the unit of which is J/kg.

EXAMPLE 1.1

The pressure of gas in a pipe line is measured with a mercury manometer having one limb open to the atmosphere (Fig. Ex. 1.1). If the difference in the height of mercury in the two limbs is 562 mm, calculate the gas pressure. The barometer reads 761 mm Hg, the acceleration due to gravity is 9.79 m/s², and the density of mercury is 13,640 kg/m³.

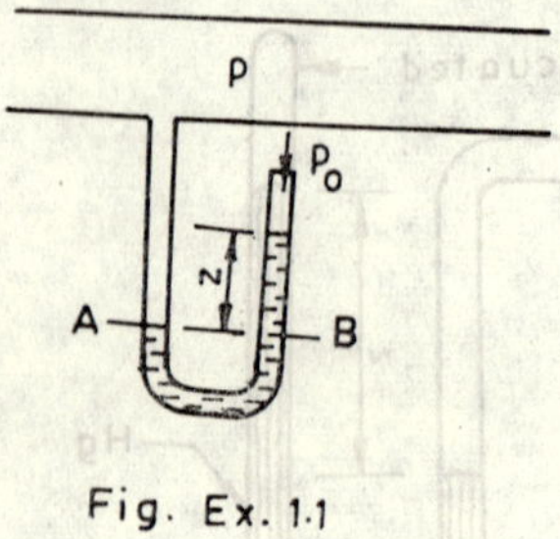

Fig. Ex. 1.1

Solution

At the plane AB, we have

$$p = p_0 + \rho g z$$

Now

$$p_0 = \rho g z_0$$

where z_0 is the barometric height, ρ the density of mercury and p_0 the atmospheric pressure.

Therefore

$$\begin{aligned} p &= \rho\, g\, (z + z_0) \\ &= 13{,}640 \text{ kg/m}^3 \times 9.79 \text{ m/s}^2\, (0.562 + 0.761) \text{ m} \\ &= 177 \times 10^3 \text{ N/m}^2 = 177 \text{ kPa} \\ &= 1.77 \text{ bar} = 1.746 \text{ atm} \end{aligned}$$

Ans.

EXAMPLE 1.2

A turbine is supplied with steam at a gauge pressure of 1.4 MPa. After expansion in the turbine the steam flows into a condenser which is maintained at a vacuum of 710 mm Hg. The barometric pressure is 772 mm Hg. Express the inlet and exhaust steam pressures in pascals (absolute). Take the density of mercury as 13.6×10^3 kg/m³.

Solution

The atmospheric pressure p_0

$$= \rho g z_0 = 13.6 \times 10^3 \text{ kg/m}^3 \times 9.81 \text{ m/s}^2 \times 0.772 \text{ m}$$

$$= 1.03 \times 10^5 \text{ Pa}$$

Inlet steam pressure

$$= [(1.4 \times 10^6) + (1.03 \times 10^5)] \text{ Pa}$$

$$= 15.03 \times 10^5 \text{ Pa}$$

$$= 1.503 \text{ MPa} \qquad \text{Ans.}$$

Condenser pressure

$$= (0.772 - 0.710) \text{ m} \times 9.81 \text{ m/s}^2 \times 13.6 \times 10^3 \text{kg/m}^3$$

$$= 0.827 \times 10^4 \text{ Pa}$$

$$= 8.27 \text{ kPa} \qquad \text{Ans.}$$

EXAMPLE 1.3

A lunar excursion module (*LEM*) weighs 1500 kgf at sea-level on earth. What will be its weight on the surface of the moon where $g = 1.7$ m/s²? On the surface of the moon, what will be the force in kgf required to accelerate the module at 10 m/s²?

Solution

The mass m of the *LEM* is given by

$$W = \frac{mg}{g_0}$$

$$1500 \text{ kgf} = m \frac{9.806 \dfrac{\text{m}}{\text{s}^2}}{9.806 \dfrac{\text{kg}}{\text{kgf}} \cdot \dfrac{\text{m}}{\text{s}^2}}$$

$$\therefore \quad m = 1500 \text{ kg}$$

The weight of the *LEM* on the moon would be

$$W = 1500 \text{ kg} \times \frac{1.7 \dfrac{\text{m}}{\text{s}^2}}{9.806 \dfrac{\text{kg}}{\text{kgf}} \cdot \dfrac{\text{m}}{\text{s}^2}} \text{ kgf} \qquad \text{Ans.}$$

The force required to accelerate the module at 10 m/s²

$$= \frac{1500 \text{ kg}}{9.806 \dfrac{\text{kg}}{\text{kgf}} \cdot \dfrac{\text{m}}{\text{s}^2}} \times 10 \frac{\text{m}}{\text{s}^2} = 1530 \text{ kgf} \qquad \text{Ans.}$$

PROBLEMS

1.1 A pump discharges a liquid into a drum at the rate of 0.032 m^3/. The drum, 1.50 m in diameter and 4.20 m in length, can hold 3000 kg of the liquid. Find the density of the liquid and the mass flow rate of the liquid handled by the pump.

1.2 The acceleration of gravity is given as a function of elevation above sea level by

$$g = 980.6 - 3.086 \times 10^{-6}\ H$$

where g is in cm/s^2, and H is in cm. If an aeroplane weighs 10,000 kgf at sea level, what is the gravity force upon it at 10,000 m elevation? What is the percentage difference from the sea-level weight?

1.3 Prove that the weight of a body at an elevation H above sea-level is given by

$$W = \frac{mg}{g_0}\left(\frac{d}{d + 2H}\right)^2$$

where d is the diameter of the earth..

1.4 The first artificial earth satellite is reported to have encircled the earth at a speed of 28,840 km/hr and its maximum height above the earth's surface was stated to be 916 km. Taking the mean diameter of the earth to be 12,680 km, and assuming the orbit to be circular, evaluate the value of the gravitational acceleration at this height.

The mass of the satellite is reported to have been 86 kg at sea-level. Estimate the gravitational force acting on the satellite at the operational altitude.

Ans. 8.9 m/s^2; 78 kgf.

1.5 Convert the following readings of pressure to kPa, assuming that the barometer reads 760 mm Hg:

(a) 90 cm Hg gauge, (b) 40 cm Hg vacuum, (c) 2.1 atg, (d) 1.2 m H_2O gauge, (e) 100 psig, (f) 76 inches Hg gauge, (g) 3.1 bar, and (h) 8 in, Hg vacuum.

1.6 Assume that the pressure p and the specific volume v of the atmosphere are related according to the equation $pv^{1.4} = 2.3 \times 10^4$, where p is in kgf/m^2 abs and v, is in m^3/kg. The acceleration due to gravity is constant at 9.81 m/s^2. What is the depth of atmosphere necessary to produce a pressure of 1.0332 ata at the earth's surface? Consider the atmosphere as a fluid column.

Ans. 64.8 km.

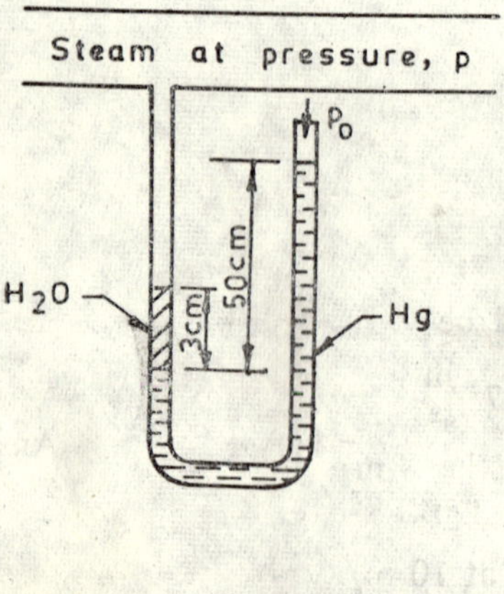

Fig P.1-7

1.7 The pressure of steam flowing in a pipe line is measured with a mercury manometer, as shown in Fig. P. 1.7. Some steam condenses into water. Estimate the steam pressure in kPa. Take the density of mercury as 13.6×10^3 kg/m^3, density of water as 10^3 kg/m^3, the barometer reading as 76.1 cm Hg, and g as 9.806 m/s^2.

1.8 A vacuum gauge mounted on a condenser reads 0.66 m Hg. What is the absolute pressure in the condenser in kPa when the atmospheric pressure is 101.3 kPa?

Ans. 8.8 kPa.

2

Temperature

2.1 ZEROTH LAW OF THERMODYNAMICS

The property which distinguishes thermodynamics from other sciences is temperature. One might say that temperature bears as important a relation to thermodynamics as force does to statics or velocity does to dynamics. Temperature is associated with the ability to distinguish hot from cold. When two bodies at different temperatures are brought into contact, after some time they attain a common temperature and are then said to exist in thermal equilibrium.

When a body A is in thermal equilibrium with a body B, and also separately with a body C, then B and C will be in thermal equilibrium with each other.

This is known as the *zeroth law of thermodynamics*. It is the basis of temperature measurement.

In order to obtain a quantitative measure of temperature, a reference body is used, and a certain physical characteristic of this body which changes with temperature is selected. The changes in the selected characteristic may be taken as an indication of change in temperature. The selected characteristic is called the *thermometric property,* and the reference body which is used in the determination of temperature is called the *thermometer*. A very common thermometer consists of a small amount of mercury in an evacuated capillary tube. In this case the extension of the mercury in the tube is used as the thermometric property.

There are five different kinds of thermometer, each with its own thermometric property, as shown in Table 2.1.

Table 2.1
Thermometers and Thermometric Properties

Thermometer	*Thermometric Property*	*Symbol*
1. Constant volume gas thermometer	Pressure	p
2. Constant pressure gas thermometer	Volume	V
3. Electrical resistance thermometer	Resistance	R
4. Thermocouple	Thermal e.m.f.	ε
5. Mercury-in-glass thermometer	Length	L

2.2 MEASUREMENT OF TEMPERATURE—THE REFERENCE POINTS

The temperature of a system is a property that determines whether or not a system is in thermal equilibrium with other systems. If a body is at, say, 70°C, it will be 70°C, whether measured by a mercury-in-glass thermometer, resistance thermometer or constant volume gas thermometer. If X is the thermometric property, let us arbitrarily choose for the temperature common to the thermometer and *to all systems in thermal equilibrium with it* the following *linear* function of X:

$\theta(X) = aX$, where a is an arbitrary constant.

If X_1 corresponds to $\theta(X_1)$ and then X_2 will correspond to

$$\frac{\theta(X_1)}{X_1} \cdot X_2$$

that is

$$\theta(X_2) = \frac{\theta(X_1)}{X_1} \cdot X_2$$

Two temperatures on the linear X scale are to each other as the ratio of the corresponding X's.

2.2.1 Method in Use Before 1954

The thermometer is first placed in contact with the system whose temperature $\theta(X)$ is to be measured, and then in contact with an arbitrarily chosen standard system in an easily reproducible state where the temperature is $\theta(X_1)$. Thus

$$\frac{\theta(X_1)}{\theta(X)} = \frac{X_1}{X} \tag{2.1}$$

Then the thermometer at the temperature $\theta(X)$ is placed in contact with another arbitrarily chosen standard system in another easily reproducible state where the temperature is $\theta(X_2)$. It gives

$$\frac{\theta(X_2)}{\theta(X)} = \frac{X_2}{X} \tag{2.2}$$

From equations (2.1) and (2.2)

$$\frac{\theta(X_1) - \theta(X_2)}{\theta(X)} = \frac{X_1 - X_2}{X}$$

or

$$\theta(X) = \frac{\theta(X_1) - \theta(X_2)}{X_1 - X_2} \cdot X$$

If we assign an arbitrary number of degrees to the temperature interval $\theta(X_1) - \theta(X_2)$, then $\theta(X)$ can be calculated from the measurements of X, X_1 and X_2.

An easily reproducible state of an arbitrarily chosen standard system is called a fixed point. Before 1954, there were two fixed points: (a) *the ice point,* the temperature at which pure ice coexisted in equilibrium with air-saturated water at one atmosphere pressure, and (b) *the steam point,* the temperature of equilibrium between pure water and pure steam at one atmosphere pressure. The temperature interval, $\theta(X_1) - \theta(X_2)$, between these two fixed points was chosen to be 100 degrees.

The use of two fixed points was found unsatisfactory and later abandoned, because of (a) the difficulty of achieving equilibrium between pure ice and air-saturated water (since when ice melts, it surrounds itself only with pure water and prevents intimate contact with air-saturated water), and (b) extreme sensitiveness of the steam point to the change in pressure.

2.2.2 Method in Use After 1954

Since 1954 only one fixed point has been in use, viz, *the triple point of water,* the state at which ice, liquid water and water vapour coexist in equilibrium. The temperature at which this state exists is arbitrarily assigned the value of 273.16 degrees Kelvin, or 273.16 K (the reason for using Kelvin's name will be explained later). Designating the triple point of water by θ_t, and with X_t being the value of the thermometric property when the body, whose temperature θ is to be measured, is placed in contact with water at its triple point, it follows that

$$\theta_t = a X_t$$

$$\therefore \quad a = \frac{\theta_t}{X_t} = \frac{273.16}{X_t}$$

Therefore

$$\theta = aX$$

$$= \frac{273.16}{X_t} . X$$

or

$$\theta = 273.16 \frac{X}{X_t} \tag{2.3}$$

The temperature of the triple point of water, which is an easily reproducible state, is now the *standard fixed point of thermometry.*

2.3 COMPARISON OF THERMOMETERS

Applying the above principle to the five thermometers listed in Table 2.1, the temperatures are given as

(a) Constant volume gas thermometer $\theta(P) = 273.16\frac{p}{p_t}$

(b) Constant pressure gas thermometer $\theta(V) = 273.16\frac{V}{V_t}$

(c) Electric resistance thermometer $\theta(R) = 273.16\,\frac{R}{R_t}$

(d) Thermocouple $\theta(\varepsilon) = 273.26\frac{\epsilon}{\epsilon_t}$

(e) Liquid-in-glass thermometer $\theta(L) = 273.16\frac{L}{L_t}$

If the temperature of a given system is measured simultaneously with each of the five thermometers, it is found that there is considerable difference among the readings. The smallest variation is, however, observed among different gas thermometers. That is why a gas is chosen as the standard thermometric substance.

2.4 GAS THERMOMETERS

A schematic diagram of a constant volume gas thermometer is given in Fig. 2.1. A small amount of gas is enclosed in bulb B which is in communication via the capillary tube C with one limb of the mercury manometer M. The other limb of the mercury manometer is open to the atmosphere and can be moved vertically to adjust the mercury levels so that the mercury just touches lip L of the capillary. The pressure in the bulb is used as a thermometric property and is given by

$$p = p_0 + \rho_M mg$$

where p_0 is the atmospheric pressure and ρ_M is the density of mercury.

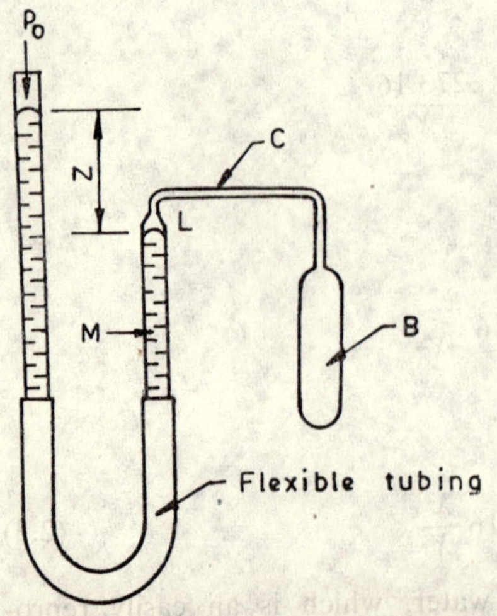

Fig. 2.1 Constant volume gas thermometer

In a constant pressure gas thermometer, the mercury levels have to be adjusted to keep Z constant, and the volume of gas V, which would vary with the temperature of the system, becomes the thermometric property. The constant volume gas thermometer is, however, mostly in use, since it is simpler in construction and easier to operate.

2.5 IDEAL GAS TEMPERATURE

Let us suppose that the bulb of a constant volume gas thermometer contains an amount of gas such that when the bulb is surrounded by water at its triple point, the pressure p_t is 1000 mm Hg. Keeping the volume V constant, let the following procedure be conducted:

(a) Surround the bulb with steam condensing at 1 atm, determine the gas pressure p and calculate

$$\theta = 273.16 \frac{p}{1000}$$

(b) Remove some gas from the bulb so that when it is surrounded by water at its triple point, the pressure p_t is 500 mm Hg. Determine the new values of p and then θ for steam condensing at 1 atm.

$$\theta = 273.16 \frac{p}{500}$$

(c) Continue reducing the amount of gas in the bulb so that p_t and p have smaller and smaller values, e.g., p_t having, say, 250 mm Hg, 100 mm Hg, and so on. At each value of p_t calculate the corresponding θ.

(d) Plot θ vs. p_t and extrapolate the curve to the axis where $p_t = 0$. Read from the graph

$$\lim_{p_t \to 0} \theta$$

The graph, as shown in Fig. 2.2, indicates that although the readings of a constant volume gas thermometer depend upon the nature of the gas, *all*

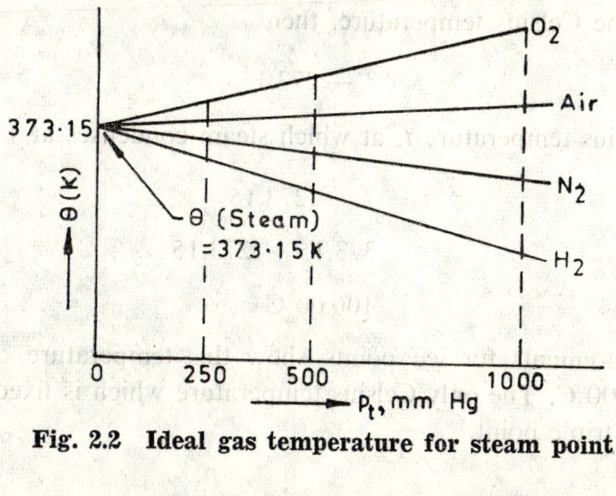

Fig. 2.2 Ideal gas temperature for steam point

gases indicate the same temperature as p_t is lowered and made to approach zero.

A similar series of tests may be conducted with a constant pressure gas thermometer. The constant pressure may first be taken to be 1000 mm Hg, then 500 mm Hg, etc. and at each value of p, the volumes of gas V and V_t may be recorded when the bulb is surrounded by steam condensing at 1 atm and the triple point of water, respectively. The corresponding value of θ may be calculated from

$$\theta = 273.16 \frac{V}{V_t}$$

and θ vs. p may be plotted, similar to Fig. 2.2. It is found from the experiments that all gases indicate the same value of θ as p approaches zero.

Since a real gas, as used in the bulb, behaves as an ideal gas as pressure approaches zero (which would be explained later in Chapter 10), the *ideal gas temperature T* is defined by either of the two equations

$$T = 273.16 \lim_{p_t \to 0} \frac{p}{p_t}$$

$$= 273.16 \lim_{p \to 0} \frac{V}{V_t}$$

where θ has been replaced by T to denote this particular temperature scale, the *ideal gas temperature scale.*

2.6 CELSIUS TEMPERATURE SCALE

The Celsius temperature scale employs a degree of the same magnitude as that of the ideal gas scale, but its zero point is shifted, so that the Celsius temperature of the triple point of water is 0.01 degree Celsius or 0.01°C. If t denotes the Celsius temperature, then

$$t = T - 273.15^\circ$$

Thus the Celsius temperature t_s at which steam condenses at 1 atm. pressure

$$t_s = T_s - 273.15^\circ$$

$$= 373.15 - 273.15$$

$$= 100.00^\circ\text{C}$$

Similar measurements for ice points show this temperature on the Celsius scale to be 0.00°C. The only Celsius temperature which is fixed by definition is that of the triple point.

2.7 ELECTRICAL RESISTANCE THERMOMETER

In the resistance thermometer (Fig. 2.3) the change in resistance of a metal wire due to its change in temperature is the thermometric property. The wire,

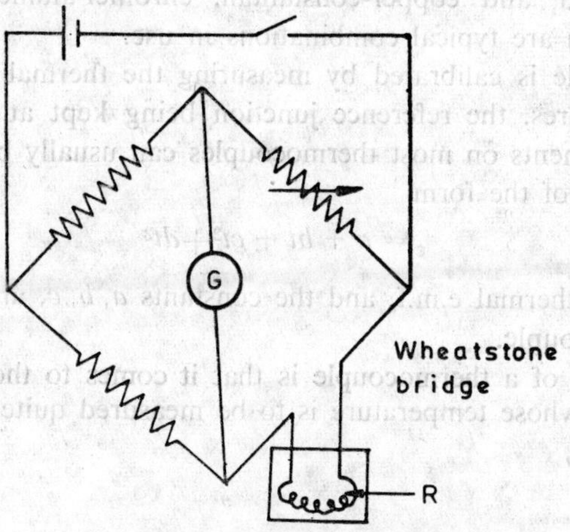

Fig. 2.3 Resistance thermometer

frequently platinum, may be incorporated in a Wheatstone bridge circuit. The platinum resistance thermometer measures temperature to a high degree of accuracy and sensitivity, which makes it suitable as a standard for the calibration of other thermometers.

In a restricted range, the following quadratic equation is often used

$$R = R_0 (1 + At + Bt^2)$$

where R_0 is the resistance of the platinum wire when it is surrounded by melting ice and A and B are constants.

2.8 THERMOCOUPLE

A thermocouple circuit made up from joining two wires A and B made of dissimilar metals is shown in Fig. 2.4. Due to the Seebeck effect, a net e.m.f. is

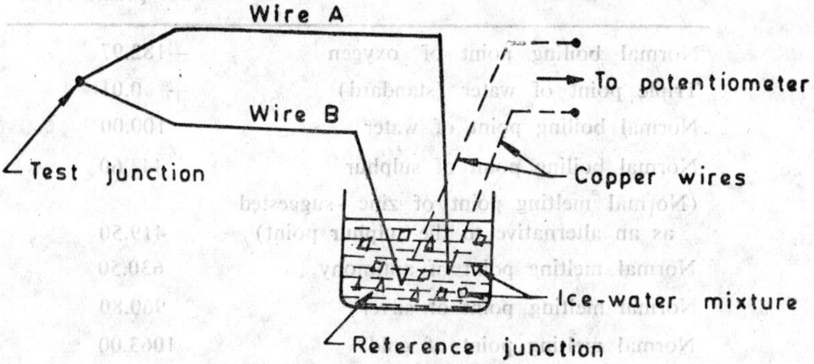

Fig. 2.4 Thermocouple

generated in the circuit which depends on the difference in temperature between the hot and cold junctions and is, therefore, a thermometric property of the circuit. This e.m.f. can be measured by a microvoltmeter to a high degree of accuracy. The choice of metals depends largely on the temperature range to be investigated, and copper-constantan, chromel-alumel and platinum-platinum-rhodium are typical combinations in use.

A thermocouple is calibrated by measuring the thermal e.m.f. at various known temperatures, the reference junction being kept at 0°C. The results of such measurements on most thermocouples can usually be represented by a cubic equation of the form

$$\varepsilon = a + bt + ct^2 + dt^3$$

where ε is the thermal e.m.f. and the constants a, b, c, and d are different for each thermocouple.

The advantage of a thermocouple is that it comes to thermal equilibrium with the system whose temperature is to be measured quite rapidly, because its mass is small.

2.9 INTERNATIONAL PRACTICAL TEMPERATURE SCALE

An international temperature scale was adopted at the Seventh General Conference on Weights and Measures held in 1927. It was not to replace the Celsius or ideal gas scales, but to provide a scale that could be easily and rapidly used to calibrate scientific and industrial instruments. Slight refinements were incorporated into the scale in revisions adopted in 1948, 1954, 1960 and 1968. The international practical scale agrees with the Celsius scale at the defining fixed points listed in Table 2.2. The temperature interval from the oxygen point to the gold point is divided into three main parts, as given below.

Table 2.2
Temperatures of Fixed Points

	Temperature °C
Normal boiling point of oxygen	−182.97
Triple point of water (standard)	+ 0.01
Normal boiling point of water	100.00
Normal boiling point of sulphur	444.60
(Normal melting point of zinc—suggested as an alternative to the sulphur point)	419.50
Normal melting point of antimony	630.50
Normal melting point of silver	960.80
Normal melting point of gold	1063.00

(a) From 0 to 660°C

A platinum resistance thermometer with a platinum wire whose diameter must lie between 0.05 and 0.20 mm is used, and the temperature is given by the equation

$$R = R_0\ (1 + At + Bt^2)$$

where the constants R_0, A, and B are computed by measurements at the ice point, steam point, and sulphur point.

(b) From −190 to 0°C

The same platinum resistance thermometer is used, and the temperature is given by

$$R = R_0\ [1 + At + Bt^2 + C\ (t - 100)\ t^3]$$

where R_0, A, and B are the same as before, and C is determined from a measurement at the oxygen point.

(c) From 660 to 1063°C

A thermocouple, one wire of which is made of platinum and the other of an alloy of 90% platinum and 10% rhodium, is used with one junction at 0°C. The temperature is given by the formula

$$\varepsilon = a + bt + ct^2$$

where a, b, and c are computed from measurements at the antimony point, silver point, and gold point. The diameter of each wire of the thermocouple must lie between 0.35 and 0.65 mm.

An optical method is adopted for measuring temperatures higher than the gold point. The intensity of radiation of any convenient wavelength is compared with the intensity of radiation of the same wavelength emitted by a black body at the gold point. The temperature is then determined with the help of Planck's law of thermal radiation.

EXAMPLE 2.1

Two mercury-in-glass thermometers are made of identical materials and are accurately calibrated at 0°C and 100°C. One has a tube of constant diameter, while the other has a tube of conical bore, ten per cent greater in diameter at 100°C than at 0°C. Both thermometers have the length between 0 and 100 subdivided uniformly. What will be the straight bore thermometer read in a place where the conical bore thermometer reads 50°C?

Solution

The volume of mercury in the tube at t°C, V_t, is given by

$$V_t = V_0\ [1 + \beta\ (t - t_0)]$$

where V_0 is the volume of mercury at 0°C, β is the coefficient of volume expansion of mercury, and t_0 is the ice point temperature which is 0°C. The volume change of glass is neglected.

Therefore $$V_t - V_0 = \beta V_0 t$$

The temperature t is thus a linear function of volume change of mercury $(V_t - V_0)$.

Therefore $$\Delta V_{0-100} = \beta V_0 \cdot 100$$

$$\Delta V_{0-50} = \beta V_0 \cdot 50$$

$$\therefore \quad \frac{\Delta V_{0-50}}{\Delta V_{0-100}} = \frac{1}{2}$$

i.e., at 50°C, the volume of mercury will be half of that at 100°C, for the straight bore thermometer (Fig. Ex. 2.1a).

But if the bore is conical (Fig. Ex. 2.1b), mercury will fill up the volume *ACDB*, which is less than half of the mercury volume at 100°C,

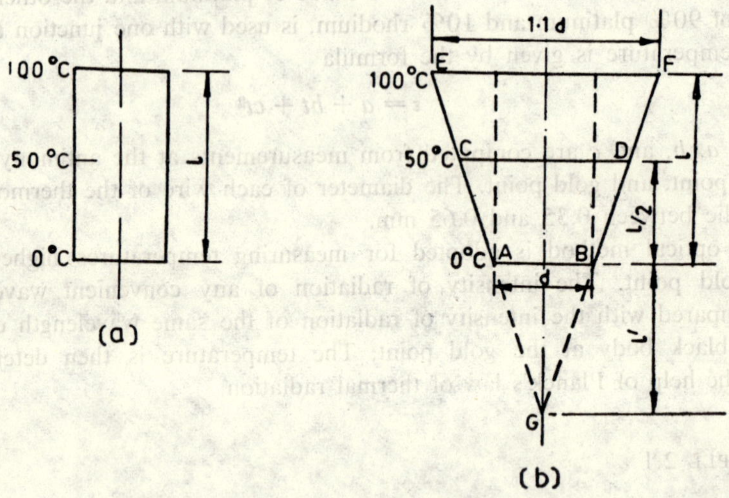

Fig. Ex. 2·1

i.e., volume *AEFB*. Let t be the true temperature when mercury rises half the length of the conical tube (the apparent temperature being 50°C). Let *EA* and *FB* be extended to meet at *G*. Let l represent the length of the thermometers and l' the vertical height of the cone *ABG*, as shown in the figure. Now,

$$\frac{l'}{l + l'} = \frac{d}{1 \cdot 1d} = \frac{1}{1 \cdot 1}$$

$$\therefore \quad l' = 10$$

and $$\frac{l'}{l'+l/2} = \frac{d}{CD}$$

$\therefore$ $$CD = \frac{10.5}{10}\, d = 1.05\, d$$

Again

$$\Delta V_{0-100} = V_0 \cdot \beta \cdot 100$$

$$\Delta V_{0-t} = V_0\, \beta\, t$$

$$\frac{\Delta\, V_{0-t}}{\Delta\, V_{0-100}} = \frac{t}{100}$$

or $$\frac{\text{Volume } ACDB}{\text{Volume } AEFB} = \frac{t}{100}$$

or $$\frac{\frac{1}{3}\frac{\pi}{4}\,(1.05\, d)^2 \times 10.5\, l - \frac{1}{3}\frac{\pi}{4}\, d^2 \cdot 10\, l}{\frac{1}{3}\frac{\pi}{4}(1.1 d)^2 \times 11\, l - \frac{1}{3}\frac{\pi}{4}\, d^2 \cdot 10\, l} = \frac{t}{100}$$

or $$\frac{1.05 \times 1.05 \times 10.5 - 10}{1.1 \times 1.1 \times 11 - 10} = \frac{t}{100}$$

$\therefore$ $$t = \frac{1.58}{3.31} \times 100 = 47.7°\text{C}$$ Ans.

EXAMPLE 2.2

The e.m.f. in a thermocouple with the test junction at t°C on gas thermometer scale and reference junction at ice point is given by

$$\varepsilon = 0.20\ t - 5 \times 10^{-4}\ t^2 \text{ mV}$$

The millivoltmeter is calibrated at ice and steam points. What will this thermometer read in a place where the gas thermometer reads 50°C?

Solution

At ice point, when $t = 0$°C, $\varepsilon = 0$ mV

At steam point, when $t = 100$°C, $\varepsilon = 0.20 \times 100 - 5 \times 10^{-4} \times (100)^2$

$= 15$ mV

At $t = 50$°C, $\varepsilon = 0.20 \times 50 - 5 \times 10^{-4}\ (50)^2 = 8.75$ mV

When the gas thermometer reads 50°C, the thermocouple will read

$$\frac{100}{15} \times 8.75, \text{ or } 58.33°\text{C}$$ Ans.

PROBLEMS

2.1 The limiting value of the ratio of the pressure of gas at the steam point and at the triple point of water when the gas is kept at constant volume is found to be 1.36605. What is the ideal gas temperature of the steam point?

2.2 In a constant volume gas thermometer the following pairs of pressure readings were taken at the boiling point of water and the boiling point of sulphur, respectively:

Water b.p.	50.0	100	200	300
Sulphur b.p.	96.4	193	387	582

The numbers are the gas pressures, mm Hg, each pair being taken with the same amount of gas in the thermometer, but the successive pairs being taken with different amounts of gas in the thermometer. Plot the ratio of $S_{b.p.}$: $H_2O_{b.p.}$ against the reading at the water boiling point, and extrapolate the plot to zero pressure at the water boiling point. This gives the ratio of $S_{b.p.}$: $H_2O_{b.p.}$ on a gas thermometer operating at zero gas pressure, i.e., an ideal gas thermometer. What is the boiling point of sulphur on the gas scale, from your plot? Ans. 445°C.

2.3 The resistance of a platinum wire is found to be 11,000 ohms at the ice point, 15.247 ohms at the steam point, and 28.887 ohms at the sulphur point. Find the constants A and B in the equation

$$R = R_0 (1 + At + Bt^2)$$

and plot R against t in the range 0 to 660°C.

2.4 When the reference junction of a thermocouple is kept at the ice point and the test junction is at the Celsius temperature t, and e.m.f. ε of the thermocouple is given by the equation

$$\varepsilon = a t + b t^2$$

where $a = 0.20$ mV/deg, and $b = -5.0 \times 10^{-4}$ mV/deg²

(a) Compute the e.m.f. when $t = -100$°C, 200°C, 400°C, and 500°C, and draw graph of ε against t in this range.

(b) Suppose the e.m.f. ε is taken as a thermometric property and that a temperature scale t^* is defined by the linear equation.

$$t^* = a' \varepsilon + b'$$

and that $t^* = 0$ at the ice point and $t^* = 100$ at the steam point. Find the numerical values of a' and b' and draw a graph of ε against t^*.

(c) Find the values of t^* when $t = -100$°C, 200°C, 400°C, and 500°C, and draw a graph of t^* against t.

(d) Compare the Celsius scale with the t^* scale.

2.5 The temperature t on a thermometric scale is defined in terms of a property K by the relation

$$t = a \ln K + b$$

where a and b are constants.

The values of K are found to be 1.83 and 6.78 at the ice point and the steam point, the temperatures of which are assigned the numbers 0 and 100 respectively. Determine the temperature corresponding to a reading of K equal to 2.42 on the thermometer.

2.6 The resistance of the windings in a certain motor is found to be 80 ohms at room temperature (25°C). When operating at full load under steady state conditions, the motor is switched off and the resistance of the windings, immediately measured again, is found to be 93 ohms. The windings are made of copper whose resistance at temperature t°C is given by

$$R_t = R_0 \,[1 + 0.00393\, t]$$

where R_0 is the resistance at 0°C. Find the temperature attained by the coil during full load.

3

Energy Interactions

A closed system and its surroundings can interact in two ways: (a) by work transfer, and (b) by heat transfer. These are called *energy interactions* and these bring about changes in the properties of the system. Thermodynamics mainly studies these energy interactions and the associated property changes of the system.

3.1 WORK TRANSFER

Work is one of the basic modes of energy transfer. In mechanics the action of a force on a moving body is identified as work. A force is a means of transmitting an effect from one body to another. But a force itself never produces a physical effect except when coupled with motion and hence it is not a form of energy. An effect such as the raising of a weight through a certain distance can be performed by using a small force through a large distance or a large force through a small distance. The product of force and distance is the same to accomplish the same effect. In mechanics work is defined as:

The work is done by a force as it acts upon a body moving in the direction of the force.

The action of a force through a distance (or of a torque through an angle) is called *mechanical work* since other forms of work can be identified, as discussed later. The product of the force and the distance moved parallel to the force is the magnitude of mechanical work.

In thermodynamics, work transfer is considered as occurring between the system and the surroundings. *Work is said to be done by a system if the sole effect on things external to the system can be reduced to the raising of a weight.* The weight may not actually be raised, but the net effect external to the system would be the raising of a weight. Let us consider the battery and the motor in Fig. 3.1 as a system. The motor is driving a fan. The system is doing work upon the surroundings. When the fan is replaced by a pulley and a weight, as shown in Fig. 3.2, the weight may be raised with the pulley driven by the motor. The sole effect on things external to the system is then the raising of a weight.

When work is done by a system, it is arbitrarily taken to be positive, and when work is done on a system, it is taken to be negative (Fig. 3.3). The symbol W is used for work transfer.

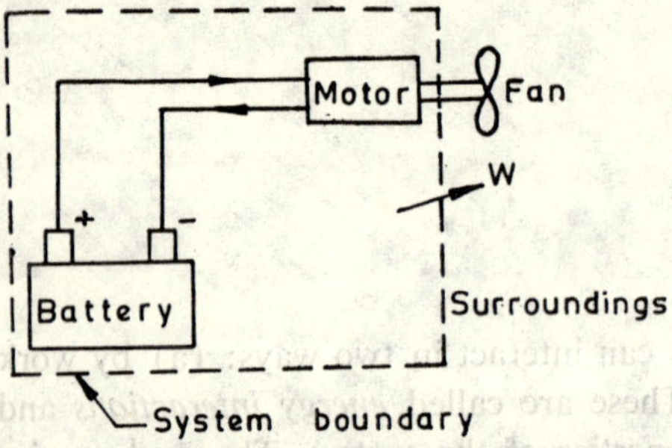

Fig. 3.1 Battery-motor system driving a fan

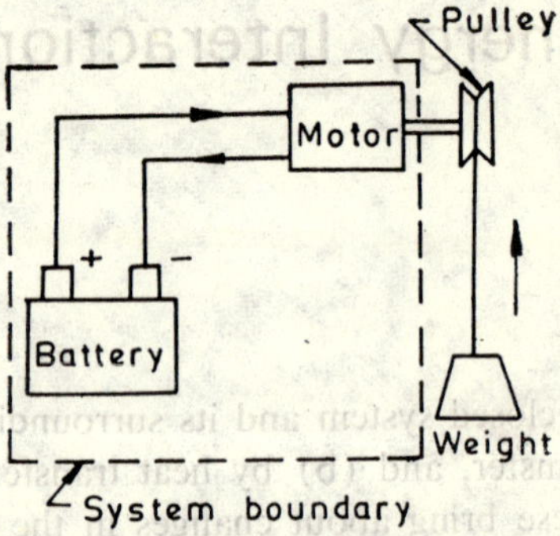

Fig. 3.2 Work transfer from a system

(a) W **is positive** **(b)** W **is negative**

Fig. 3.3 Work interaction between a system and the surroundings

The unit of work is N.m or Joule [1 Nm = 1 Joule]. In MKS units it is kgf.m. The rate at which work is done by, or upon, the system is known as *power*. The units of power are J/s or watt and kgf.m/s. In MKS units, there is a unit of power known as horse power (HP). It is also, sometimes, referred to as P.S. in the abbreviated form*. One horse power is defined as the rate of work equivalent to 75 kgf.m/s.

$$1 \text{ kgf.m} = 9.806 \text{ Nm}$$

$$1 \text{ HP (metric) or } 1 \text{ PS} = 75 \times 9.806 \text{ Nm/s}$$

$$= 735.5 \text{ W}$$

$$= 0.7355 \text{ kW}$$

Work is one of the forms in which a system and its surroundings can interact with each other. There are various types of work transfers which can get involved between them.

---

*From the German words, *pferd stärke*, meaning horse strength.

3.2 *pdV*-WORK OR DISPLACEMENT WORK

Let the gas in the cylinder (Fig. 3.4) be a system having initially the pressure p_1 and volume V_1. The system is in thermodynamic equilibrium, the state of which is described by the coordinates p_1, V_1. The piston is the only boundary which moves due to gas pressure. Let the piston move out to a new final position 2, which is also a thermodynamic equilibrium state specified by pressure p_2 and volume V_2. At any intermediate point in the travel of the piston, let the pressure be p and the volume V. This must also be an equilibrium state, since macroscopic properties p and V are significant only for equilibrium states. When the piston moves an infinitesimal distance dl, and if 'a' be the area of the piston, the force F acting on the piston $F = p.a$ and the infinitesimal amount of work done by the gas on the piston

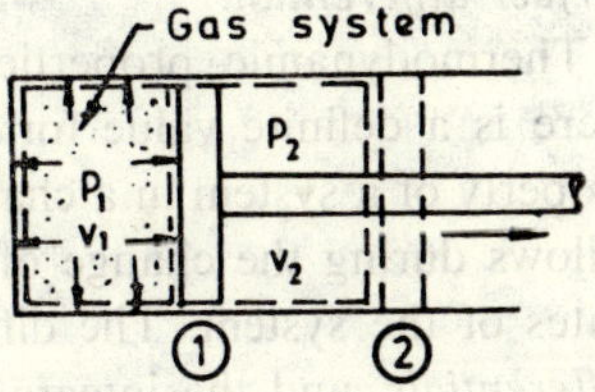

Fig. 3.4 ***pdV* work**

$$đW = F \cdot dl = padl = pdV$$

where $dV = adl$ = infinitesimal displacement volume. The differential sign in $đW$ with the line drawn at the top of it will be explained later.

When the piston moves out from position 1 to position 2 with the volume changing from V_1 to V_2, the amount of work W done by the system will be

$$W_{1-2} = \int_{V_1}^{V_2} pdV$$

The magnitude of the work done is given by the area under the path 1-2, as shown in Fig. 3.5. Since p is at all times a thermodynamic coordinate, all the states passed through by the system as the volume changes from V_1 *to* V_2 must be equilibrium states, and the path 1-2 *must be quasi-static.* The piston moves infinitely slowly so that every state passed through is an equilibrium state. *The integration* $\int pdV$ *can be performed only on a quasi-static path.*

With reference to Fig. 3.6, it is possible to take a system from state 1 to state 2 along many quasi-static paths, such as *A, B* or *C*. Since the area

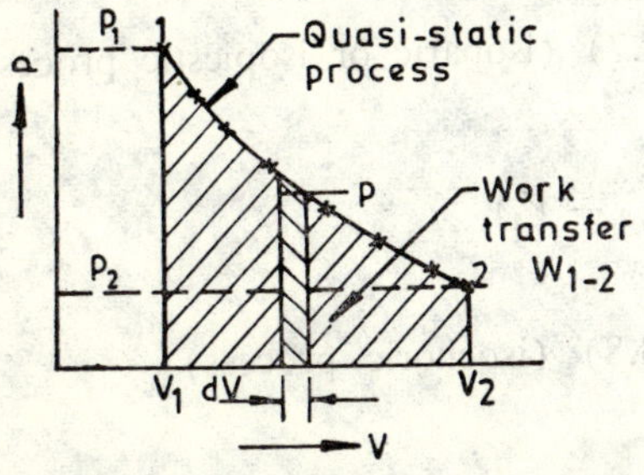

Fig. 3.5 Quasi-static *pdV* work

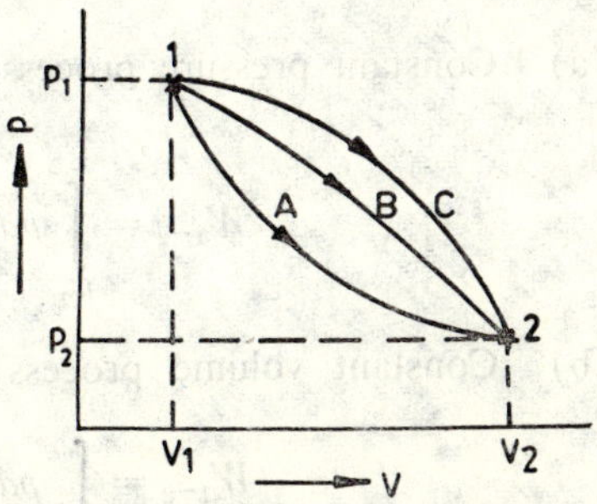

Fig. 3.6 Work—a path function

under each curve represents the work for each process, the amount of work involved in each case is not a function of the end states of the process, and it depends on the path the system follows in going from state 1 to state 2. For this reason, work is called a *path function, and* $đW$ is an *inexact or imperfect differential.*

Thermodynamic properties are *point functions,* since for a given state, there is a definite value for each property. The change in a thermodynamic property of a system in a change of state is independent of the path the system follows during the change of state, and depends only on the initial and final states of the system. The differentials of point functions are *exact or perfect differentials,* and the integration is simply

$$\int_{V_1}^{V_2} dV = V_2 - V_1$$

The change in volume thus depends only on the end states of the system irrespective of the path the system follows.

On the other hand, work done in a quasi-static process between two given states depends on the path followed.

$$\int_1^2 đW \neq W_2 - W_1$$

Rather,

$$\int_1^2 đW = W_{1-2} \text{ or } {}_1W_2$$

To distinguish an inexact differential $đW$ from an exact differential dV or dp the differential sign is being cut by a line at its top.

3.2.1 pdV-work in various quasi-static processes

(a) Constant pressure process (Fig. 3.7) (isobaric or isopiestic process)

$$W_{1-2} = \int_{V_1}^{V_2} pdV = p\,(V_2 - V_1)$$

(b) Constant volume process (Fig. 3.8) (isochoric process)

$$W_{1-2} = \int pdV = 0$$

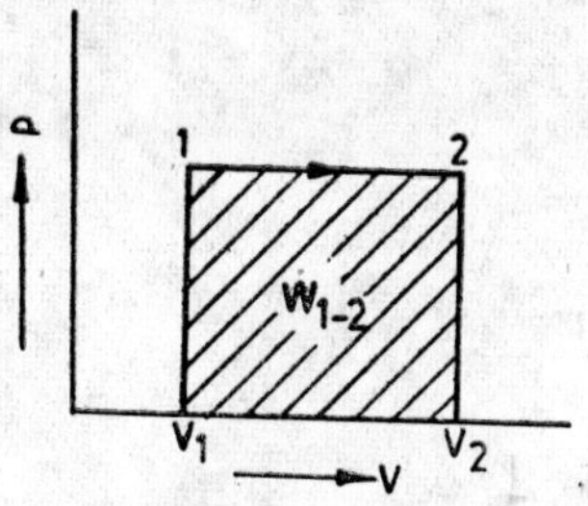

Fig. 3.7 Constant pressure process

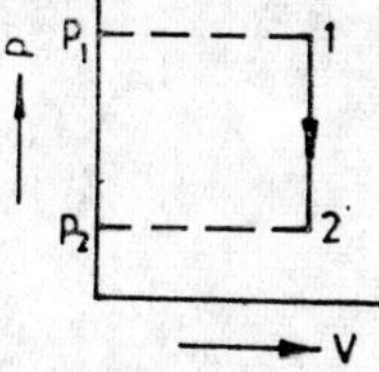

Fig. 3.8 Constant volume process

(c) Process in which $pV = C$ (Fig. 3.9)

$$\therefore \quad W_{1-2} = \int_{V_1}^{V_2} pdV \qquad pV = p_1V_1 = C$$

$$p = \frac{(p_1V_1)}{V}$$

$$W_{1-2} = p_1V_1 \int_{V_1}^{V_2} \frac{dV}{V} = p_1V_1 \ln \frac{V_2}{V_1}$$

$$= p_1V_1 \ln \frac{p_1}{p_2}$$

(d) Process in which $pV^n = C$, where n is a constant (Fig. 3.10)

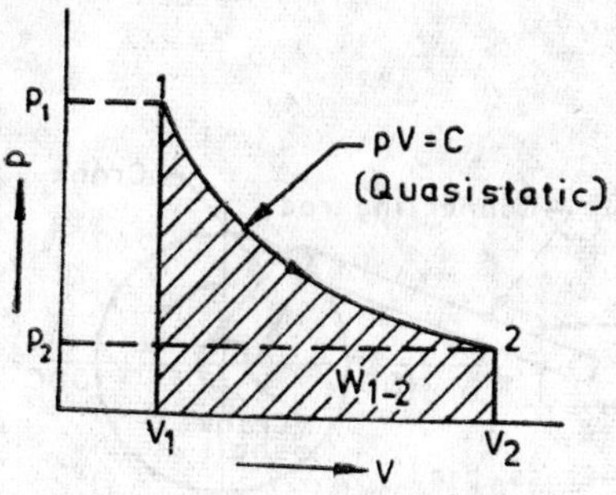

Fig. 3.9 Process in which pV = constant

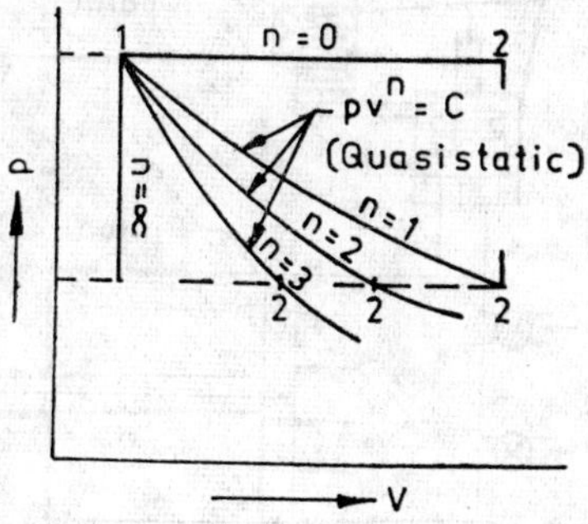

Fig. 3.10 Process in which pV^n= constant

$$pV^n = p_1V_1^n = p_2V_2^n = C$$

$$\therefore \quad p = \frac{(p_1V_1^n)}{V^n}$$

$$\therefore \qquad W_{1-2} = \int_{V_1}^{V_2} pdV$$

$$= \int_{V_1}^{V_2} \frac{p_1 V_1^n}{V^n} \cdot dV$$

$$= (p_1 V_1^n) \left[\frac{V^{-n+1}}{-n+1} \right]_{V_1}^{V_2}$$

$$= \frac{p_1 V_1^n}{1-n} (V_2^{1-n} - V_1^{1-n})$$

$$= \frac{p_2 V_2^n \times V_2^{1-n} - p_1 V_1^n \times V_1^{1-n}}{1-n}$$

$$= \frac{p_1 V_1 - p_2 V_2}{n-1} = \frac{p_1 V_1}{n-1} \left[1 - \left(\frac{p_2}{p_1} \right)^{n-1/n} \right]$$

3.3 INDICATOR DIAGRAM

An indicator diagram is a trace made by a recording pressure gauge, called the indicator, attached to the cylinder of a reciprocating engine. This represents the work done in one engine cycle. Figure 3.11 shows a typical engine indicator.

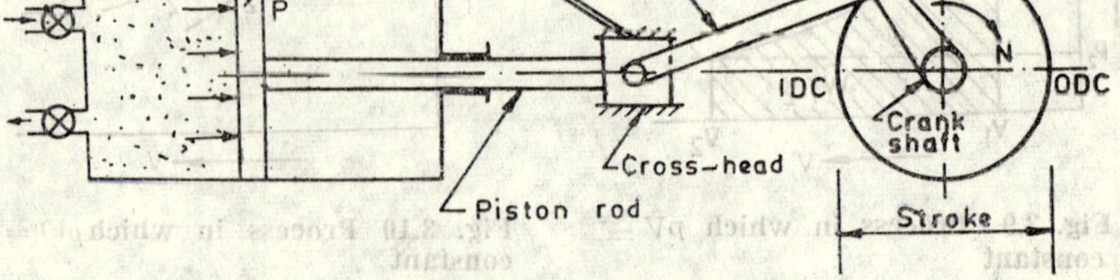

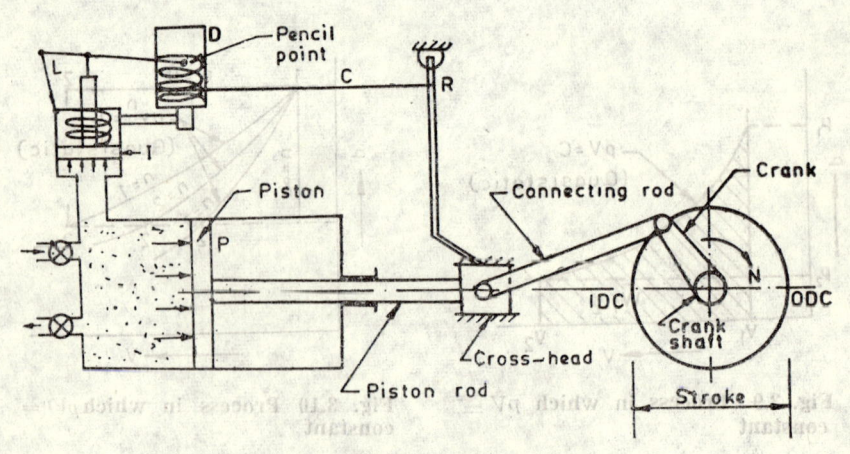

Fig. 3.11 Engine indicator

The same gas pressure acts on both the engine piston *P* and the indicator piston *I*. The indicator piston is loaded by a spring and it moves in direct proportion to the change in pressure. The motion of the indicator piston causes a pencil held at the end of the linkage L to move upon a strip of paper wrapped around drum D. The drum is rotated about its axis by cord C, which is connected through a reducing motion R to the piston P of the engine. The surface of drum D moves horizontally under the pencil while the pencil moves vertically over the surface and a plot of pressure upon the piston vs. piston travel is obtained.

Before tracing the final indicator diagram, a pressure reference line is recorded by subjecting the indicator to the atmosphere and tracing a line at a constant pressure of one atmosphere.

The area of the indicator diagram represents the magnitude of the net work done by the system in one engine cycle. The area under the path 1-2 represents work done by the system and the area under the path 2-1 represents work done upon the system (Fig. 3.12). The area of the diagram, a_d, is measured by means of a planimeter, and the length of the diagram,

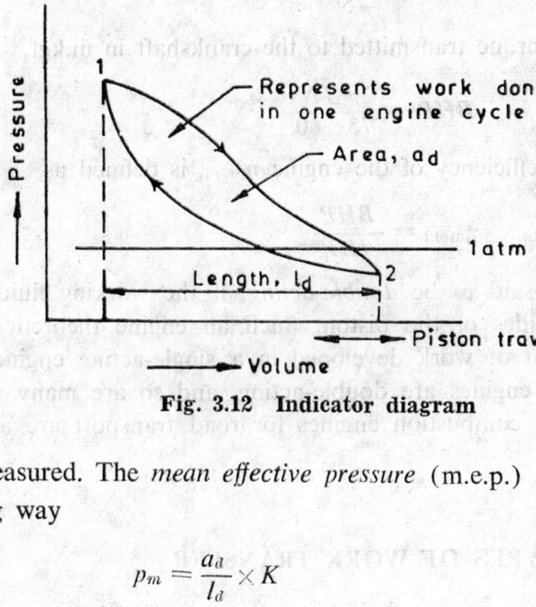

Fig. 3.12 Indicator diagram

l_d, is also measured. The *mean effective pressure* (m.e.p.) p_m is defined in the following way

$$p_m = \frac{a_d}{l_d} \times K$$

where K is the indicator spring constant (N/cm^2 × cm travel). Work done in one engine cycle

$$= (p_m . A) \, L$$

where A = cross-sectional area of the cylinder

$$= \frac{\pi}{4} D^2, \text{ where } D \text{ is the cylinder diameter}$$

and L = stroke of piston, or length of cylinder.

Let N be the revolutions per minute (r.p.m.) of the crankshaft. In a two-stroke cycle, the engine cycle is completed in two strokes of the piston or in one revolution of the crankshaft. In a four-stroke cycle, the engine cycle is completed in four strokes of the piston or two revolutions of the crankshaft.

For a two-stroke engine, work done in one minute = $p_m ALN$, and for a four-stroke engine, work done in one minute = $p_m ALN/2$.

The power that is developed inside the cylinder of the engine is called *Indicated Horse Power* (*I.H.P.*).

$$\therefore \qquad IHP = \frac{p_m AL\,(N \text{ or } N/2)\,n}{75\times 60}$$

where n is the number of cylinders in the engine.

The power that is available at the crankshaft is always less than this value (I.H.P.), and it is called the *Brake Horse Power* (*B.H.P.*), If w is the angular velocity of the crankshaft in radian/sec, then

$$BHP = \frac{Tw}{75}$$

where T is the torque transmitted to the crankshaft in m.kgf,

$$\therefore \qquad BHP = \frac{2\pi\, TN}{75\times 60}$$

The mechanical efficiency of the engine η_{mech} is defined as

$$\eta_{mech} = \frac{BHP}{IHP}$$

An engine is said to be *double-acting,* if the working fluid is made to work on both sides of the piston. Such an engine theoretically develops twice the amount of work developed in a single-acting engine. Most reciprocating steam engines are double-acting, and so are many marine diesel engines. Internal combustion engines for road transport are always single-acting.

3.4 OTHER TYPES OF WORK TRANSFER

(a) Electrical work

When a current flows through a resistor (Fig. 3.13), taken as a system, there is work transfer into the system. This is because the current can drive a motor, the motor can drive a pulley and the pulley can raise a weight.

The rate of work transfer = $I.E$ where E is the potential difference across the wire and I the current flowing.

(b) Shaft Work

When a shaft, taken as the system, is rotated (Fig. 3.14), there is work transfer into the system. This is because the shaft can rotate a pulley which can raise a weight. If T is the torque applied to the shaft rotating at N r.p.m., then the rate of work done would be $2\pi TN$ (N.m/min). To

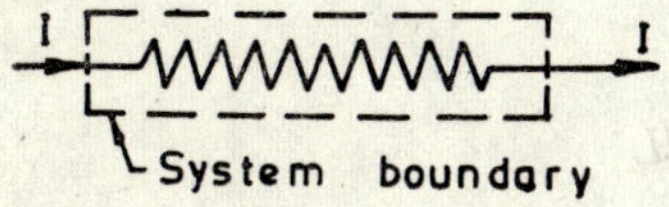

Fig. 3.13 Electrical work

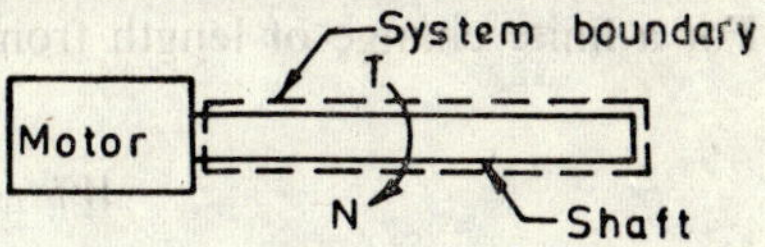

Fig. 3.14 Shaft work

differentiate between torque and work the units of the former may be written as m-N, while those of the latter are in N-m.

(c) Paddle-wheel Work or Stirring Work

As the weight is lowered, and the paddle wheel turns (Fig. 3.15), there is work transfer into the fluid system which gets stirred. Since the volume of the system remains constant, $\int pdv = 0$ although $đW \neq 0$.

(d) Flow Work

The flow work represents the amount of work that must be done on a system to introduce a unit mass of material into it. In Fig. 3.16, the fluid

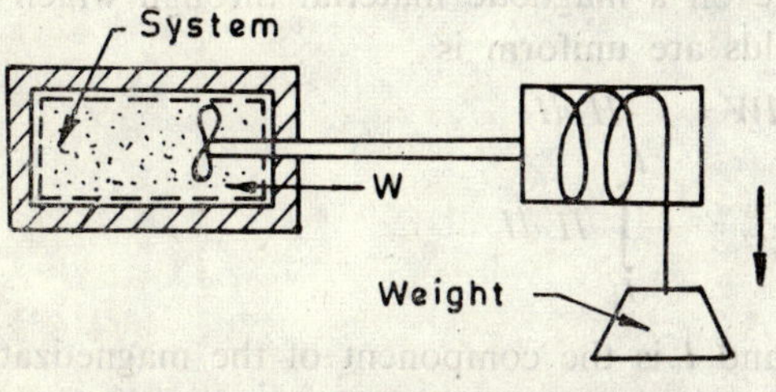

Fig. 3.15 Paddle-wheel work

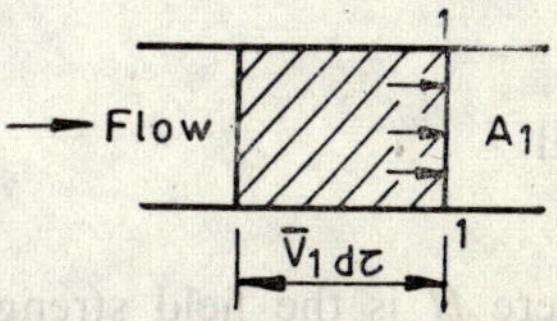

Fig. 3.16 Flow work

crossing section 1 exerts a normal pressure p_1 against the area A_1, giving a total force p_1A_1. In time $d\tau$, this force moves in its own direction a distance $\mathbf{V}_1 d\tau$ where $\mathbf{V}_1$ is the velocity of flow. The work in time $d\tau$ is $p_1A_1\mathbf{V}_1\, d\tau$, or the work per unit time is $p_1A_1\mathbf{V}_1$.

Now, the flow rate, $\quad w_1 = \dfrac{A_1\mathbf{V}_1}{v_1}$

where v_1 is the specific volume.

$$\text{So} \qquad A_1\mathbf{V}_1 = w_1v_1$$

$$\therefore \text{ work done/time} = p_1w_1v_1$$

$$\therefore \text{ work done per unit mass} = p_1v_1$$

This is the *flow work* or the flow energy associated with unit mass of a fluid.

(e) Work Done in Stretching a Wire

If the length of a wire in which there is a tension $\mathcal{J}$ is changed from L to $L+dL$, the infinitesimal amount of work that is done is equal to

$$đW = -\mathcal{J}\, dL$$

The minus sign is used because a positive value of dL means an expansion of the wire, for which work must be done on the wire, i.e., negative work. For a finite change of length from L_i to L_f

$$W_{i-f} = -\int_{L_i}^{L_f} \mathcal{J}\, dL$$

(f) Work Done in Changing the Area of a Surface Film

The work done on a homogeneous liquid film in changing its surface area by an infinitesimal amount dA is

$$đW = \sigma\, dA$$

and

$$W_{1-2} = \int_{A_1}^{A_2} \sigma\, dA$$

where σ is the interfacial tension per unit length.

(g) Magnetization of a Paramagnetic Solid

The work done per unit volume on a magnetic material through which the magnetic and magnetization fields are uniform is

$$đW = -H\, dI$$

and

$$W_{1-2} = -\int_{I_1}^{I_2} H\, dI$$

where H is the field strength, and I is the component of the magnetization field in the direction of the field. The minus sign provides that an increase in magnetization (positive dI) involves negative work.

3.5 FREE EXPANSION

The expansion of a gas against vacuum is called *free or unrestrained expansion.* The system consists of an insulated box which is divided into a number of compartments by means of ideal frictionless sliding partitions (Fig. 3.17). The largest partition of volume V is filled with gas, the others, whose volumes are dV, having been evacuated. If the partitions are removed one by one, the gas expands and its volume is increased by dV each time. But the gas performs no work against the surroundings, and hence $đW = 0$ although $pdv \neq 0$.

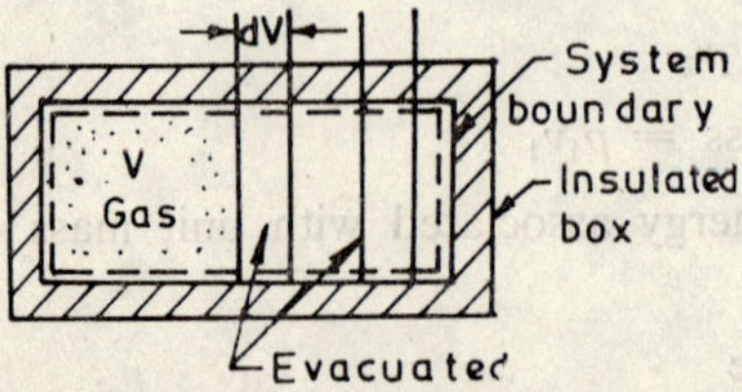

Fig. 3.17 Free expansion

3.6 NET WORK DONE BY A SYSTEM

Often different forms of work transfer occur simultaneously during a process executed by a system. When all these work interactions have been evaluated, the total or net work done by the system would be equal to the algebraic sum of these as given below

$$W_{\text{total}} = W_{\text{displacement}} + W_{\text{shear}} + W_{\text{electrical}} + W_{\text{strring}} + \cdots\cdots$$

3.7 HEAT TRANSFER

Heat is defined as the form of energy that is transferred across a boundary *by virtue of a temperature difference.* The temperature difference is the 'potential' and heat transfer is the 'flux'.

The transfer of heat between two bodies in direct contact is called *conduction.* Heat may be transferred between two bodies separated by empty space or gases by the mechanism of *radiation* through electromagnetic waves. A third method of heat transfer is *convection* which refers to the transfer of heat between a wall and a fluid system in motion.

The direction of heat transfer is taken from the high temperature system to the low temperature system. *Heat flow into a system is taken to be positive, and heat flow out of a system is taken as negative* (Fig 3.18). The symbol Q is used for heat transfer, i.e., the quantity of heat transferred within a certain time.

A proceses in which no heat crosses the boundary of the system is called an *adiabatic process.*

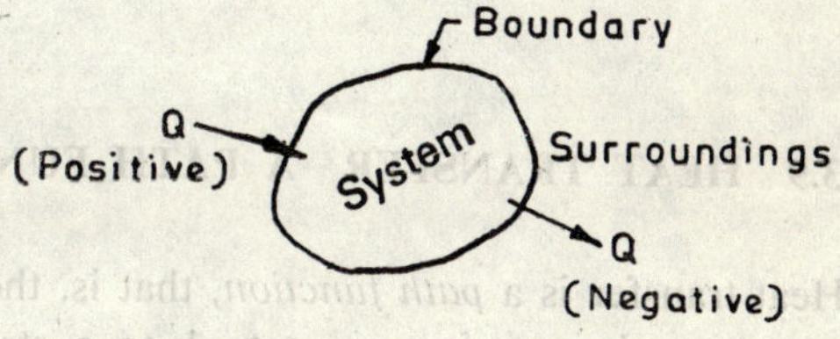

Fig. 3.18 Direction of heat transfer

A wall which is impermeable to the flow of heat is an *adiabatic wall,* whereas a wall which permits the flow of heat is a *diathermic wall.*

The unit of heat is Joule in S.I. units and cal in MKS units. One calorie is the energy required to raise the temperature of 1 g of water by 1 °C.

$$1 \text{ cal} = 4.187 \text{ J}$$

The rate of heat transfer or work transfer is given in kW

$$1 \text{ kW} = 860 \text{ kcal/hr}$$

3.8 SPECIFIC HEAT AND LATENT HEAT

The *specific heat* of a substance is defined as the amount of heat required to raise a unit mass of the substance through a unit rise in temperature. The symbol c will be used for specific heat.

$$\therefore \qquad c = \frac{Q}{m \cdot \Delta t} \text{ J/kg-K}$$

where Q is the amount of heat transfer (J), m, the mass of the substance (kg), and Δt, the rise in temperature (K).

Since heat is not a property, as explained later, so the specific heat is qualified with the process through which exchange of heat is made. For gases, if the process is at constant pressure, it is c_p, and if the process is at constant volume, it is c_v. For solids and liquids, however, the specific heat does not depend on the process. It will be shown in Chapters 10 and 11 that the specific heat of a substance is a function of temperature.

The product of mass and specific heat (mc) is called the *heat capacity* of the substance. The capital letter C, C_p or C_v, is used for heat capacity.

The latent heat is the amount of heat transfer required to cause a phase change in unit mass of a substance *at a constant pressure and temperature.* There are three phases in which matter can exist : solid, liquid, and vapour or gas. The *latent heat of fusion* (l_{fu}) is the amount of heat transferred to melt unit mass of solid into liquid, or to freeze unit mass of liquid to solid. The *latent heat of vaporization* (l_{vap}) is the quantity of heat required to vaporize unit mass of liquid into vapour, or condense unit mass of vapour into liquid. The *latent heat of sublimation* (l_{sub}) is the amount of heat transferred to convert unit mass of solid to vapour or vice versa. l_{fu} is not much affected by pressure, whereas l_{vap} is highly sensitive to pressure.

3.9 HEAT TRANSFER—A PATH FUNCTION

Heat transfer is a *path function,* that is, the amount of heat transferred when a system changes from a state 1 to a state 2 depends on the intermediate states through which the system passes, i.e., its path. Therefore $đQ$ is an inexact differential, and we write

$$\int_1^2 đQ = Q_{1-2} \text{ or } {}_1Q_2$$

3.10 POINTS TO REMEMBER REGARDING HEAT TRANSFER AND WORK TRANSFER

(a) Heat transfer and work transfer are the *energy interactions.* A closed system and its surroundings can interact in two ways: by heat transfer and by work transfer. Thermodynamics studies how these interactions bring about property changes in a system.

(b) The same effect in a closed system can be brought about either by heat transfer or by work transfer. Whether heat transfer or work transfer has taken place depends on what constitutes the system.

(c) Both heat transfer and work transfer are boundary phenomena. Both are observed at the boundaries of the system, and both represent energy crossing the boundaries of the system.

(d) *It is wrong to say 'total heat' or 'heat content' of a closed system,* because heat or work is not a property of the system. Heat, like work, cannot be stored by the system. Both heat and work are the energy in transit.

(e) Heat transfer is the energy interaction due to temperature difference only. All other energy interactions may be termed as work transfer.

(f) Both heat and work are path functions and inexact differentials. The magnitude of heat transfer or work transfer depends upon the path the system follows during the change of state.

EXAMPLE 3.1

Gas from a bottle of compressed helium is used to inflate an inelastic flexible balloon, originally folded completely flat to a volume of 0.5 m^3. If the barometer reads 760 mm Hg, what is the amount of work done upon the atmosphere by the balloon? Sketch the system before and after the process.

Solution

The firm line P_1 (Fig. Ex. 3.1) shows the boundary of the system before the process, and the dotted line P_2 shows the boundary after the process. The

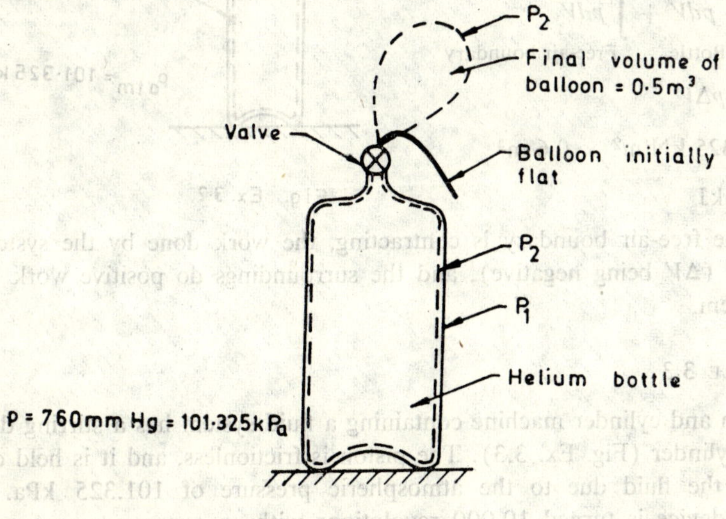

Fig. Ex. 3.1

displacement work

$$W_d = \int_{\text{Balloon}} pdV + \int_{\text{Bottle}} pdV = p\Delta V + 0$$

$$= 101.325 \frac{\text{kN}}{\text{m}^2} \times 0.5 \text{ m}^3$$

$$= 50.66 \text{ kJ}$$

This is positive, because work is done by the system. Work done by the atmosphere is -50.66 kJ. Since the wall of the bottle is rigid, there is no pdV-work involved in it.

It is assumed that the pressure in the balloon is atmospheric at all times, since the balloon fabric is light, inelastic and unstressed. If the balloon were elastic and stressed during the filling process, the work done by the gas would be greater than 50.66 kJ by an amount equal to the work done in stretching the balloon, although the displacement work done by the atmosphere is still -50.66 kJ. However, if the system includes both the gas and the balloon, the displacement work would be 50.66 kJ, as estimated above.

Example 3.2

When the valve of the evacuated bottle (Fig. Ex. 3.2) is opened, atmospheric air rushes into it. If the atmospheric pressure is 101.325 kPa, and 0.6 m³ of air (measured at atmospheric conditions) enters into the bottle, calculate the work done by air.

Solution

The displacement work done by air

$$W_d = \int_{\text{Bottle}} pdV + \int_{\text{Free-air boundary}} pdV$$

$$= 0 + p\Delta V$$

$$= 101.325 \text{ kN/m}^2 \times 0.6 \text{ m}^3$$

$$= 60.8 \text{ kJ}$$

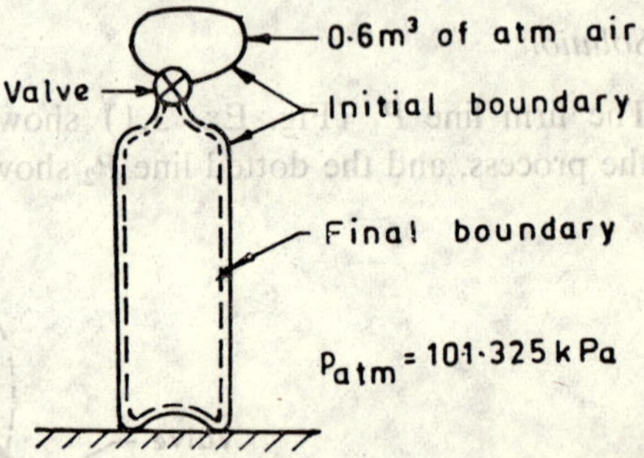

Fig. Ex. 3·2

Since the free-air boundary is contracting, the work done by the system is negative (ΔV being negative), and the surroundings do positive work upon the system.

Example 3.3

A piston and cylinder machine containing a fluid system has a stirring device in the cylinder (Fig. Ex. 3.3). The piston is frictionless, and it is held down against the fluid due to the atmospheric pressure of 101.325 kPa. The stirring device is turned 10,000 revolutions with an average torque against the fluid of 1.275 mN. Meanwhile the piston of 0.6 m diameter moves out 0.8 m. Find the net work transfer for the system.

Solution

Work done by the stirring device upon the system (Fig. Ex. 3.3)

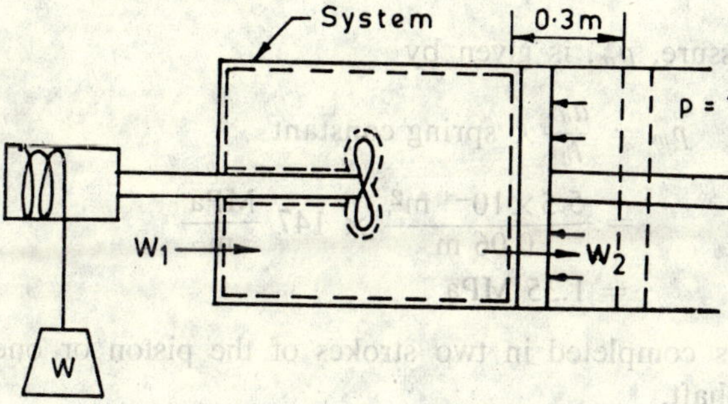

Fig. Ex. 3·3

$$W_1 = 2\pi\, TN$$
$$= 2\pi \times 1.275 \times 10{,}000 \text{ Nm}$$
$$= \mathbf{80\ kJ}$$

This is negative work for the system.

Work done by the system upon the surroundings

$$W_2 = (pA) \cdot L$$

$$= 101.325 \frac{\text{kN}}{\text{m}^2} \times \frac{\pi}{4} (0.6)^2 \text{ m}^2 \times 0.80 \text{ m}$$

$$= 22.9 \text{ kJ}$$

This is positive work for the system. Hence, the net work transfer for the system

$$W = W_1 + W_2 = -80 + 22.9 = -\ 57.1 \text{ kJ}$$

EXAMPLE 3.4

The following data refer to a 12-cylinder, single-acting, two-stroke marine diesel engine:

Speed—150 rpm

Cylinder diameter—0.8 m

Stroke of piston—1.2 m

Area of indicator diagram—5.5×10^{-4} m²

Length of diagram—0.06 m

Spring value—147 MPa per m

Find the net rate of work transfer from the gas to the pistons in kW.

Solution

Mean effective pressure, p_m, is given by

$$p_m = \frac{a_d}{l_d} \times \text{spring constant}$$

$$= \frac{5.5 \times 10^{-4}\ \text{m}^2}{0.06\ \text{m}} \times 147\ \frac{\text{MPa}}{\text{m}}$$

$$= 1.35\ \text{MPa}$$

One engine cycle is completed in two strokes of the piston or one revolution of the crank-shaft.

∴ Work done in one minute

$$= p_m\, L\, A\, N$$

$$= 1.35 \times \frac{\pi}{4}(0.8)^2 \times 1.2 \times 150 = 122\ \text{MJ}$$

Since the engine is single-acting, and it has 12 cylinders, each contributing an equal power, the rate of work transfer from the gas to the piston is given by

$$\dot{W} = 122 \times 12\ \text{MJ/min}$$

$$= 24.4\ \text{MJ/sec}$$

$$= 24.4\ \text{MW} = 24{,}400\ \text{kW}$$ Ans.

Example 3.5

It is required to melt 5 tonnes/hr of iron from a charge at 15°C to molten metal at 1650°C. The melting point is 1535°C, and the latent heat is 270 kJ/kg. The specific heat in solid state is 0.502 and in liquid state (29.93/atomic weight) kJ/kg°C. If an electric furnace has 70% efficiency, find the kW rating needed. If the density in molten state is 6900 kg/m³ and the bath volume is three times the hourly melting rate, find the dimensions of the cylindrical furnace if the length to diameter ratio is 2. The atomic weight of iron is 56.

Solution

Heat required to melt 1 kg of iron at 15°C to molten metal at 1650°C

= Heat required to raise the temperature from 15°C to 1535°C + Latent heat + Heat required to raise the temperature from 1535°C to 1650°C

$= 0.502\ (1535 - 15) + 270 + 29.93\ (1650 - 1535)/56$
$= 763 + 270 + 61.5$
$= 1094.5$ kJ/kg

Melting rate $= 5 \times 10^3$ kg/hr

So, the rate of heat supply required

$= (5 \times 10^3 \times 1094.5)$ kJ/hr

Since the furnace has 70% efficiency, the rating of the furnace would be

$$= \frac{\text{Rate of heat supply per sec}}{\text{Furnace efficiency}}$$

$$= \frac{5 \times 10^3 \times 1094.5}{0.7 \times 3600} = 2.17 \times 10^3 \text{ kW} \qquad \text{Ans.}$$

$$\text{Volume needed} = \frac{3 \times 5 \times 10^3}{6900} \text{m}^3 = 2.18 \text{ m}^3$$

If d is the diameter and l the length of the furnace

$$\frac{\pi}{4} d^2 l = 2.18 \text{ m}^3$$

or $$\frac{\pi}{4} d^2 \times 2d = 2.18 \text{ m}^3$$

$\therefore$ $$d = 1.15 \text{ m}$$

and $$l = 2.30 \text{ m} \qquad \text{Ans.}$$

EXAMPLE 3.6

If it is desired to melt aluminium with solid state specific heat 0.9 kJ/kg °C, latent heat 390 kJ/kg, atomic weight 27, density in molten state 2400 kg/m³ and final temperature 700°C, find out how much metal can be melted per hour with the above kW rating. Other data are as in the above example. Also, find the mass of aluminium that the above furnace will hold. The melting point of aluminium is 660°C.

Solution

Heat required per kg of aluminium

$$= 0.9(660 - 15) + 390 + \frac{29.93}{27}(700 - 660)$$

$$= 580.5 + 390 + 44.3$$

$$= 1014.8 \text{ kJ}$$

$$\text{Heat to be supplied} = \frac{1014.8}{0.7} = 1449.7 \text{ kJ/kg}$$

With the given power, the rate at which aluminium can be melted

$$= \frac{2.17 \times 10^3 \times 3600}{1449.7} \text{ kg/hr}$$

$= 5.39$ tonnes/hr **Ans.**

Mass of aluminium that can be held in the above furnance

$= 2.18 \times 2400$ kg

$= 5.23$ tonnes **Ans.**

PROBLEMS

3.1 (a) A pump forces 1 m³/min of water horizontally from an open well to a closed tank where the pressure is 0.9 MPa. Compute the work the pump must do upon the water in an hour just to force the water into the tank against the pressure. Sketch the system upon which the work is done before and after the process.

(b) If the work done as above upon the water had been used solely to raise the same amount of water vertically against gravity without change of pressure, how many metres would the water have been elevated?

(c) If the work done in (a) upon the water had been used solely to accelerate the water from zero velocity without change of pressure or elevation, what velocity would the water have reached? If the work had been used to accelerate the water from an initial velocity of 10 m/sec, what would the final velocity have been?

3.2 The piston of an oil engine, of area 0.0045 m², moves downwards 75 mm, drawing in 0.00028 m³ of fresh air from the atmosphere. The pressure in the cylinder is uniform during the process at 80 kPa, while the atmospheric pressure is 101.325 kPa, the difference being due to the flow resistance in the induction pipe and the inlet valve. Estimate the displacement work done by the air finally in the cylinder.

3.3 An engine cylinder has a piston of area 0.12 m³ and contains gas at a pressure of 1.5 MPa. The gas expands according to a process which is represented by a straight line on a pressure-volume diagram. The final pressure is 0.15 MPa. Calculate the work done by the gas on the piston if the stroke is 0.30 m.

3.4 A mass of 1.5 kg of air is compressed in a quasi-static process from 0.1 MPa to 0.7 MPa for which *pv*-constant. The initial density of air is 1.16 kg/m³. Find the work done by the piston to compress the air. Ans. 251.62 kJ

3.5 A mass of gas is compressed in a quasi-static process from 80 kPa, 0.1 m³ to 0.4 MPa, 0.03 m³. Assuming that the pressure and volume are related by pv^n = constant, find the work done by the gas system. Ans. −11.83 kJ

3.6 A single-cylinder, double-acting, reciprocating water pump has an indicator diagram which is a rectangle 0.075 m long and 0.03 m high. The indicator spring constant is 147 MPa per m. The pump runs at 50 rpm. The pump cylinder diameter is 0.15 m and the piston stroke is 0.20 m. Find the rate in kW at which the piston does work on the water. Ans. 30.29 kW.

3.7 A single-cylinder, single-acting, 4-stroke engine of 0.15 m bore develops an indicated power of 4 kW when running at 216 rpm. Calculate the area of the indicator diagram that would be obtained with an indicator having a spring constant of

25×10^6 N/m³. The length of the indicator diagram is 0.1 times the length of the stroke of the engine.

Ans. 505 mm²

3.8 A six-cylinder, 4-stroke gasoline engine is run at a speed of 2520 RPM. The area of the indicator card of one cylinder is 2.45×10^3 mm² and its length is 58.5 mm. The spring constant is 20×16^6 N/m³. The bore of the cylinders is 140 mm and the piston stroke is 150 mm. Determine the indicated power, assuming that each cylinder contributes an equal power.

3.9 A closed cylinder of 0.25 m diameter is fitted with a light frictionless piston. The piston is retained in position by a catch in the cylinder wall and the volume on one side of the piston contains air at a pressure of 750 kN/m². The volume on the other side of the piston is evacuated. A helical spring is mounted coaxially with the cylinder in this evacuated space to give a force of 120 N on the piston in this position. The catch is released and the piston travels along the cylinder until it comes to rest after a stroke of 1.2 m. The piston is then held in its position of maximum travel by a ratchet mechanism. The spring force increases linearly with the piston displacement to a final value of 5 kN. Calculate the work done by the compressed air on the piston.

3.10 A steam turbine drives a ship's propeller through an 8:1 reduction gear. The average resisting torque imposed by the water on the propeller is 750×10^3 mN and the shaft power delivered by the turbine to the reduction gear is 14 MW. The turbine speed is 1450 rpm. Determine (a) the torque developed by the turbine, (b) the power delivered to the propeller shaft, and (c) the net rate of working of the reduction gear.

3.11 A fluid, contained in a horizontal cylinder fitted with a frictionless leakproof piston, is continuously agitated by means of a stirrer passing through the cylinder cover. The cylinder diameter is 0.40 m. During the stirring process lasting 10 minutes, the piston slowly moves out a distance of 0.485 m against the atmosphere. The net work done by the fluid during the process is 2 kJ. The speed of the electric motor driving the stirrer is 840 rpm. Determine the torque in the shaft and the power output of the motor.

Ans. 0.08 mN, 6.92 W.

3.12 At the beginning of the compression stroke of a two-cylinder internal combustion engine the air is at a pressure of 101.325 kPa. Compression reduces the volume to 1/5 of its original volume, and the law of compression is given by $pv^{1.2}$ = constant. If the bore and stroke of each cylinder is 0.15 m and 0.25 m, respectively, determine the power absorbed in kW by compression strokes when the engine speed is such that each cylinder undergoes 500 compression strokes per minute.

3.13 Determine the total work done by a gas system following an expansion process as shown in Fig. P. 3.13.

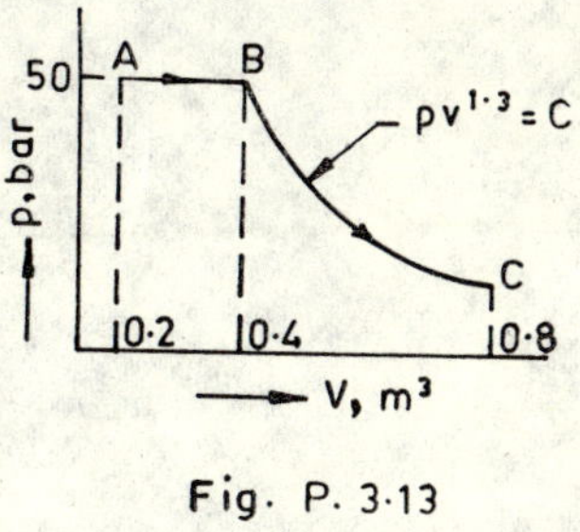

Fig. P. 3·13

3.14 A system of volume V contains a mass m of gas at pressure p and temperature T. The macroscopic properties of the system obey the following relationship:

$$\left(P + \frac{a}{V^2}\right)(V - b) = mRT$$

where a, b, and R are constants.

Obtain an expression for the displacement work done by the system during a constant-temperature expansion from volume V_1 to volume V_2. Calculate the work done by a system which contains 10 kg of this gas expanding from 1 m^3 to 10 m^3 at a temperature of 293 K. Use the values $a = 15.7 \times 10^4$, $b = 1.07 \times 10^{-2}$ m^3, and $R = 0.278$ kJ/kg–K.

Ans. 1758.7 kJ

3.15 A milk chilling unit can remove heat from the milk at the rate of 41.87 MJ/hr. Heat leaks into the milk from the surroundings at an average rate of 4.187 MJ/hr. Find the time required for cooling a batch of 500 kg of milk from 45°C to 5°C. Take the c_p of milk to be 4.187 kJ/kg °C.

3.16 680 kg of fish at 5°C are to be frozen and stored at —12°C. The specific heat of fish above freezing point is 3.182, and below freezing point is 1.717 kJ/kg °C. The freezing point is —2°C, and the latent heat of fusion is 234.5 kJ/kg. How much heat must be removed to cool the fish, and what per cent of this is latent heat?

Ans. 186.28 MJ, 85.6%

4

First Law of Thermodynamics

4.1 FIRST LAW FOR A CLOSED SYSTEM UNDERGOING A CYCLE

The transfer of heat and the performance of work may both cause the same effect in a system. Heat and work are different forms of the same entity, called energy, which is conserved. Energy which enters a system as heat may leave the system as work, or energy which enters the system as work may leave as heat.

Let us consider a closed system which consists of a known mass of water contained in an adiabatic vessel having a themometer and a paddle wheel, as shown in Fig. 4.1. Let a certain amount of work W_{1-2} be done

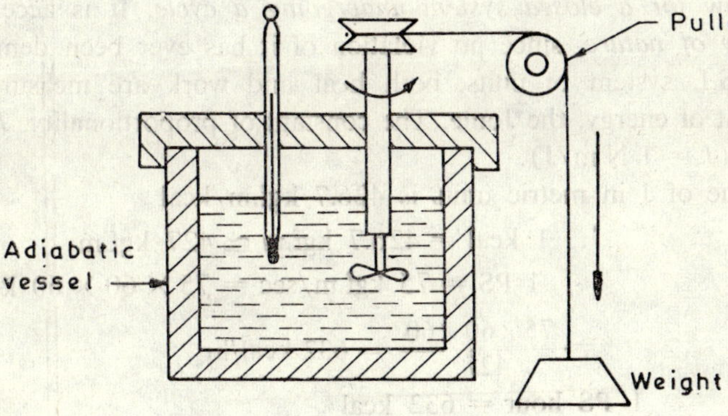

Fig. 4.1 Adiabatic work

upon the system by the paddle wheel. The quantity of work can be measured by the fall of weight which drives the paddle wheel through a pulley. The system was initially at temperature t_1, the same as that of atmosphere, and after work transfer let the temperature rise to t_2. The pressure is always 1 atm. The process 1-2 undergone by the system is shown in Fig. 4.2 in generalized thermodynamic coordinates X, Y. Let the insulation now be removed. The system and the surroundings interact by heat transfer till the system returns to the original temperature t_1, attaining the condition of thermal equilibrium with the atmosphere. The amount of heat transfer Q_{2-1} from the system during this process, 2-1, shown in Fig. 4.2, can be estimated. The system thus executes a cycle, which consists of a definite amount of work input W_{1-2} to the system followed by the transfer of an amount of heat Q_{2-1} from the system. It has been found that this W_{1-2} is always proportional to the heat Q_{2-1}, and the constant of proportionality is called the Joule's equivalent or the *mechanical equivalent of heat*. In the simple example given here, there are only two energy transfer quantities as the system performs a thermodynamic cycle. If the cycle involves many more heat and work quantities, the same result will be found. Expressed algebraically.

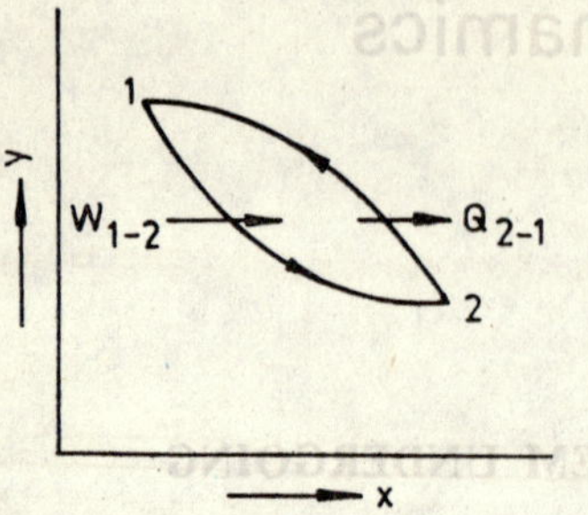

Fig. 4.2 Cycle completed by a system with two energy interactions: adiabatic work transfer W_{1-2} followed by heat transfer Q_{2-1}

$$(\Sigma W)_{\text{cycle}} = J\,(\Sigma Q)_{\text{cycle}} \tag{4.1}$$

where J is the Joule's equivalent. This is also expressed in the form

$$\oint dW = J \oint dQ$$

where the symbol $\oint$ denotes the cyclic integral for the closed path. This is the *first law for a closed system undergoing a cycle*. It is accepted as a *general law of nature*, since no violation of it has ever been demonstrated.

In the S.I. system of units, both heat and work are measured in the derived unit of energy, the Joule. The constant of proportionality, J, is therefore unity ($J = 1$ N m/J).

The value of J in metric units is 426.7 kgf.m/kcal

$$\therefore \quad 1 \text{ kcal} = 426.7 \text{ kgf.m} \cong 427 \text{ kgf.m}$$

$$1 \text{ PS} = 75 \text{ kgf.m/sec} = 75 \times 60 \times 60 \text{ kgf.m/hr}$$

$$= \frac{75 \times 60 \times 60}{427} = 632 \text{ kcal/hr}$$

$$1 \text{ PS hour} = 632 \text{ kcal}$$

$$1 \text{ kW hour} = 860 \text{ kcal}$$

$$1 \text{ kcal} = 4.187 \text{ kJ}$$

The first law of thermodynamics owes much to J.P. Joule who, during the period 1840-1849, carried out a series of experiments to investigate the equivalence of work and heat. In one of these experiments, Joule used an apparatus similar to the one shown in Fig. 4.1. Work was transferred to the measured mass of water by means of a paddle wheel driven by the falling weight. The rise in the temperature of water was recorded. Joule also used mercury as the fluid system, and later a solid system of metal blocks which absorbed work by friction when rubbed against each other. Other experiments involved the supplying of work in an electric current. In every case, he found the same ratio (J) between the amount of work and the quantity of heat that would produce identical effects in the system.

Prior to Joule, heat was considered to be an invisible fluid flowing from a body of higher calorie to a body of lower calorie, and this was known as the *caloric theory of heat*. It was Joule who first established that heat is a form of energy, and thus laid the foundation of the first law of thermodynamics.

4.2 FIRST LAW FOR A CLOSED SYSTEM UNDERGOING A CHANGE OF STATE

The expression $(\Sigma W)_{\text{cycle}} = (\Sigma Q)_{\text{cycle}}$ applies only to systems undergoing cycles, and the algebraic summation of all energy transfer across system boundaries is zero. But if a system undergoes a change of state during which both heat transfer and work transfer are involved, the *net* energy transfer will be stored or accumulated within the system. If Q is the amount of heat transferred to the system and W is the amount of work transferred from the system during the process (Fig. 4.3), the net energy transfer $(Q-W)$ will be stored in the system. Energy in storage is neither heat nor work, and is given the name *internal energy* or simply, the *energy* of the system.

Therefore
$$Q - W = \Delta E$$
where ΔE is the increase in the energy of the system

or
$$Q = \Delta E + W \tag{4.2}$$
Here Q, W, and ΔE are all expressed in the same units (in Joules). Energy may be stored by a system in different modes, as explained in Article 4.4.

If there are more energy transfer quantities involved in the process, as shown in Fig. 4.4, the first law gives
$$(Q_2 + Q_3 - Q_1) = \Delta E + (W_2 + W_3 - W_1 - W_4)$$
Energy is thus conserved in the operation. The first law is a particular formulation of the principle of the conservation of energy. Equation (4.2)

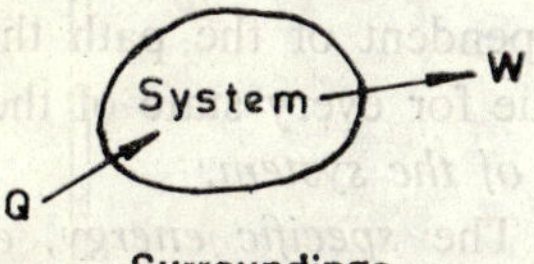

Fig. 4.3 Heat and work interactions of a system with its surroundings in a process

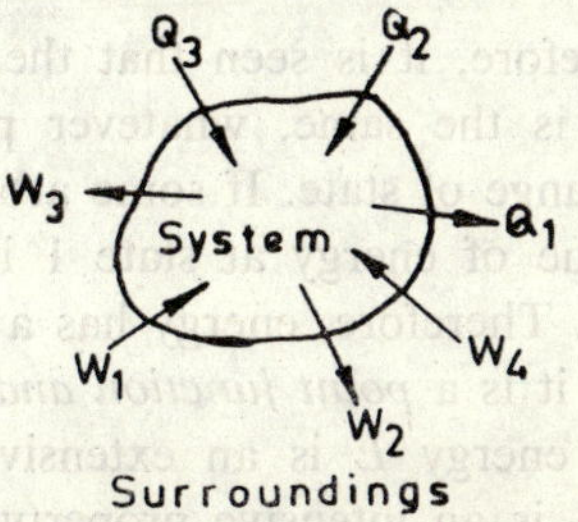

Fig. 4.4 System-surroundings interaction in a process involving many energy fluxes

may also be considered as the definition of energy. This definition does not give an absolute value of energy E, but only the change of energy ΔE for the process. It can, however, be shown that the energy has a definite value at every state of a system and is, therefore, a property of the system.

4.3 ENERGY—A PROPERTY OF THE SYSTEM

Consider a system which changes its state from state 1 to state 2 by following the path A, and returns from state 2 to state 1 by following the path B (Fig. 4.5). So the system undergoes a cycle. Writing the first law for path A

$$Q_A = \Delta E_A + W_A \tag{4.3}$$

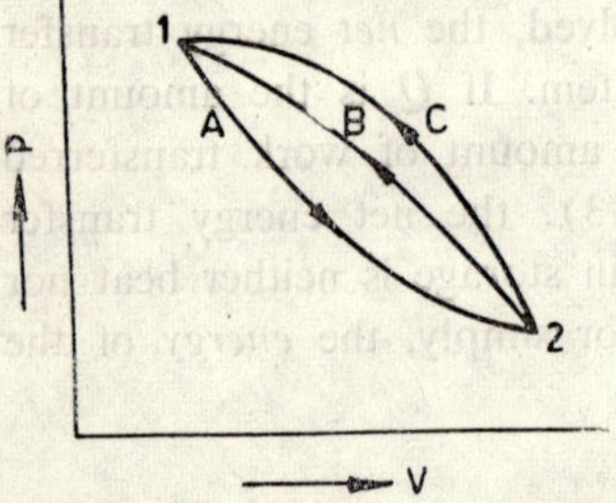

Fig. 4.5 Energy—a property of a system

and for path B

$$Q_B = \Delta E_B + W_B \tag{4.4}$$

The processes A and B together constitute a cycle, for which

$$(\Sigma W)_{\text{cycle}} = (\Sigma Q)_{\text{cycle}}$$

or $$W_A + W_B = Q_A + Q_B$$

or $$Q_A - W_A = W_B - Q_B \tag{4.5}$$

From equations (4.3), (4.4), and (4.5), it yields

$$\Delta E_A = -\Delta E_B \tag{4.6}$$

Similarly, had the system returned from state 2 to state 1 by following the path C instead of path B

$$\Delta E_A = -\Delta E_C \tag{4.7}$$

From equations (4.6) and (4.7)

$$\Delta E_B = \Delta E_C \tag{4.8}$$

Therefore, it is seen that the change in energy between two states of a system is the same, whatever path the system may follow in undergoing that change of state. If some arbitrary value of energy is assigned to state 2, the value of energy at state 1 is fixed independent of the path the system follows. Therefore, energy has a definite value for every state of the system. Hence, it is a *point function and a property of the system.*

The energy E is an extensive property. The *specific energy,* $e = E/m$ (J/kg), is an intensive property.

The cyclic integral of any property is zero, because the final state is identical with the initial state. $dE = 0$, $\oint dV = 0$, etc. So for a cycle, the equation (4.2) reduces to equation (4.1).

4.4 DIFFERENT FORMS OF STORED ENERGY

The symbol E refers to the total energy of the system. Basically there are three forms in which energy may be stored in a system, which are:

(a) Kinetic energy

(b) Potential energy

(c) Molecular internal energy

The kinetic energy refers to the energy of a system in motion. The potential energy is associated with external conservative forces, such as the force of gravity or internal forces like the elastic force. The kinetic energy and potential energy of a system are sometimes referred to as the *bulk energy modes*, and these are measurable. The part of the total energy which is stored in the molecular and atomic structure of the system is known as the *molecular internal energy or simply, internal energy*, customarily denoted by U. Matter is composed of molecules. Each molecule can possess energy in many forms, sometimes called *microscopic energy modes*. Molecules, which are in random thermal motion (for a gas), can have translatory kinetic energy, rotational kinetic energy, vibrational energy, electronic energy, nuclear binding energy, etc. If ε is the energy of one molecule, which is usually a function of temperature, and N is the total number of molecules in the system, then the total internal energy $U = N\varepsilon$. The internal energy cannot be measured directly, but determined by inference. U is a function of the mass and state of a system composed of a pure substance and is different for different substances. A *pure substance* is defined as *a substance of constant chemical composition throughout its mass.*

The total energy E is, therefore, given by

$$E = E_K + E_P + U \tag{4.9}$$

where E_K, E_P, and U refer to the kinetic, potential, and internal energy, respectively. In the absence of motion and gravity

$$E_K = 0, \ E_P = 0$$

$$E = U$$

and equation (4.2) becomes

$$Q = \Delta U + W \tag{4.10}$$

U is an extensive property of the system. The specific internal energy u is equal to U/m and its unit is J/kg.

In the differential forms, equations (4.2) and (4.10) become

$$đQ = dE + đW \tag{4.11}$$

$$đQ = dU + đW \tag{4.11a}$$

where $đW = đW_{pdv} + đW_{\text{shaft}} + đW_{\text{electrical}} + \ldots,$

considering the different forms of work transfer which may be present. When only pdV work is present, the equations become

$$đQ = dE + pdV \tag{4.11b}$$

$$đQ = dU + pdV \tag{4.11c}$$

or, in the integral form

$$Q = \Delta E + \int pdV \tag{4.12a}$$

$$Q = \Delta U + \int pdV \tag{4.12b}$$

4.5 SPECIFIC HEAT AT CONSTANT VOLUME

The specific heat of a substance at constant volume c_v is defined as the rate of change of specific internal energy with respect to temperature when the volume is held constant, i.e.

$$c_v = \left(\frac{\partial u}{\partial T}\right)_v \tag{4.13}$$

For a constant-volume process

$$(\Delta u)_v = \int_{T_1}^{T_2} c_v \cdot dT \tag{4.14}$$

The first law may be written for a closed stationary system composed of a unit mass of a pure substance

$$Q = \Delta u + W$$

or

$$đQ = du + đW$$

For a process in the absence of work other than pdV work

$$đW = pdV$$

$$\therefore \quad đQ = du + pdV \tag{4.15}$$

$\therefore$ When the volume is held constant

$$(Q)_v = (\Delta u)_v$$

$$\therefore \quad (Q)_v = \int_{T_1}^{T_2} c_v \, dT \tag{4.16}$$

Since u, T, and v are properties, c_v is a property of the system. The product $mc_v = C_v$ is called the *heat capacity at constant volume* (J/K).

4.6 ENTHALPY

The enthalpy of a substance, h, is defined as

$$h = vu + pv$$

It is an intensive property of a system (kJ/kg).

Internal energy change is equal to the heat transferred in a constant volume process involving no work other than pdV work. From equation (4.15), it is possible to derive an expression for the heat transfer in a constant pressure process involving no work other than pdV work. In such a process in a closed stationary system of unit mass of a pure substance

$$\bar{d}Q = du + pdV$$

At constant pressure

$$pdV = d(pv)$$

$$\therefore \quad (\bar{d}Q)_p = du + d(pv)$$

or $$(\bar{d}Q)_p = d(u + pv)$$

or $$(\bar{d}Q)_p = dh \tag{4.17}$$

where $h = u + pv$ is the *specific enthalpy,* a property of the system.

Total enthalpy $H = mh$

Also $$H = U + pV$$

and $$h = H/m \text{ (J/kg)}$$

4.7 SPECIFIC HEAT AT CONSTANT PRESSURE

The specific heat at constant pressure c_p is defined as the rate of change of enthalpy with respect to temperature when the pressure is held constant.

$$c_p = \left(\frac{\partial h}{\partial T}\right)_p \tag{4.18}$$

For a constant pressure process

$$(\Delta h)_p = \int_{T_1}^{T_2} c_p \, dT \tag{4.19}$$

The first law for a closed stationary system of unit mass

$$\bar{d}Q = du + pdV$$

Again, $$h = u + pv$$

$$\therefore \quad dh = du + pdV + vdp$$

$$= dQ + vdp$$

$$\therefore \quad \bar{d}Q = dh - vdp \tag{4.20}$$

$$\therefore \qquad (đQ)_p = dh$$

or $$(Q)_p = (\Delta h)_p$$

$\therefore$ From equations (4.19) and (4.20)

$$(Q)_p = \int_{T_1}^{T_2} c_p \, dT$$

c_p is a property of the system, just like c_v. The *heat capacity at constant pressure* C_p is equal to mc_p (J/K).

4.8 ENERGY OF AN ISOLATED SYSTEM

An isolated system is one in which there is no interaction of the system with the surroundings. For an isolated system, $đQ = 0$, $đW = 0$.

The first law gives

$$dE = 0$$

or $$E = \text{constant}$$

The energy of an isolated system is always constant.

4.9 PERPETUAL MOTION MACHINE OF THE FIRST KIND—PMM1

The first law states the general principle of the conservation of energy. *Energy is neither created nor destroyed, but only gets transformed from one form to another.* There can be no machine which would continuously supply mechanical work without some other form of energy disappearing simultaneously (Fig. 4.6). Such a *fictitious machine* is called a *perpetual motion machine of first kind,* or in brief, PMM1. *A PMM1 is thus impossible.*

The converse of the above statement is also true, i.e., there can be no machine which would continuously consume work without some other form of energy appearing simultaneously (Fig. 4.7).

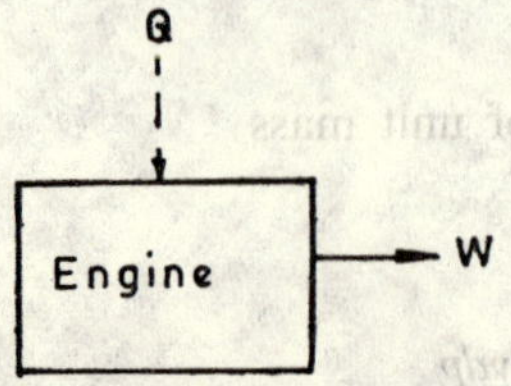

Fig. 4.6 A PMM1

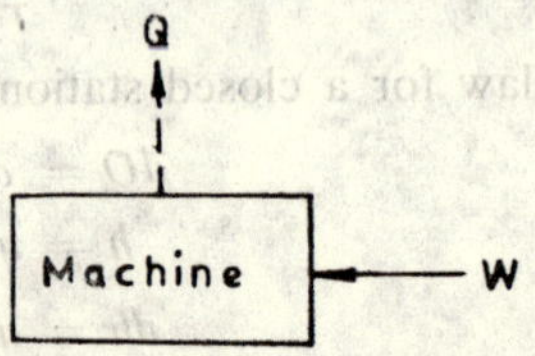

Fig. 4.7 The converse of PMM1

EXAMPLE 4.1

A stationary mass of gas is compressed without friction from an initial state of 0.3 m³ and 0.105 MPa to a final state of 0.15 m³ and 0.105 MPa, the pressure remaining constant during the process. There is a transfer of 37.6 kJ of heat from the gas during the process. How much does the internal energy of the gas change?

Solution

First law for a stationary system in a process gives

$$Q = \Delta E + W$$

or

$$Q_{1-2} = E_2 - E_1 + W_{1-2} \qquad (1)$$

Here

$$W_{1-2} = \int_{V_1}^{V_2} p dV = p(V_2 - V_1)$$

$$= 0.105\ (0.15 - 0.30)\ \text{MJ}$$

$$= -15.75\ \text{kJ}$$

$$Q_{1-2} = -37.6\ \text{kJ}$$

$\therefore$ Substituting in equation (1)

$$-37.6\ \text{kJ} = E_2 - E_1 - 15.75\ \text{kJ}$$

$$\therefore \quad E_2 - E_1 = -21.85\ \text{kJ} \qquad \text{Ans.}$$

The internal energy of the gas decreases by 21.85 kJ in the process.

EXAMPLE 4.2

When a system is taken from state a to state b, in Fig. Ex. 4.2, along path acb, 84 kJ of heat flow into the system, and the system does 32 kJ of work. (a) How much will the heat that flows into the system along path adb be, if the work done is 10.5 kJ? (b) When the system is returned from b to a along the curved path, the work done on the system is 21 kJ. Does the system absorb or liberate heat, and how much of the heat is absorbed or liberated? (c) If $U_a = 0$ and $U_d = 42$ kJ, find the heat absorbed in the processes ad and db.

Solution

$$Q_{acb} = 84\ \text{kJ}$$

$$W_{acb} = 32\ \text{kJ}$$

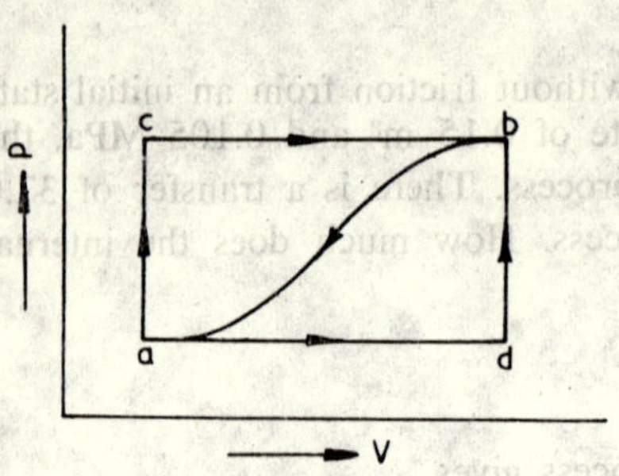

Fig. Ex. 4·2

We have

$$Q_{acb} = U_b - U_a + W_{acb}$$

$$\therefore \quad U_b - U_a = 84 - 32 = 52 \text{ kJ Ans.}$$

(a) $$Q_{adb} = U_b - U_a + W_{adb}$$
$$= 52 + 10.5$$
$$= 62.5 \text{ kJ} \qquad \text{Ans.}$$

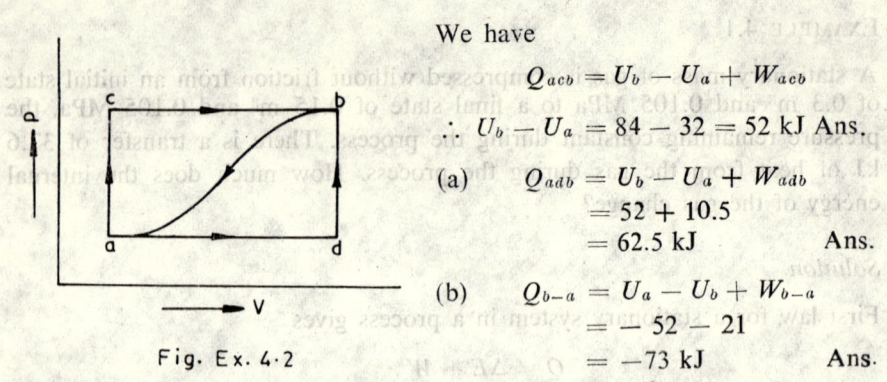

(b) $$Q_{b-a} = U_a - U_b + W_{b-a}$$
$$= -52 - 21$$
$$= -73 \text{ kJ} \qquad \text{Ans.}$$

The system liberates 73 kJ of heat.

(c) $$W_{adb} = W_{ad} + W_{db} = W_{ad} = 10.5 \text{ kJ.}$$

$\therefore$ $$Q_{ad} = U_d - U_a + W_{ad}$$
$$= 42 - 0 + 10.5 = 52.5 \text{ kJ}$$

Now $$Q_{adb} = 62.5 \text{ kJ} = Q_{ad} + Q_{ad}$$

$\therefore$ $$Q_{db} = 62.5 - 52.5 = 10 \text{ kJ.} \qquad \text{Ans.}$$

Example 4.3

A piston and cylinder machine contains a fluid system which passes through a complete cycle of four processes. During a cycle, the sum of all heat transfers is − 170 kJ. The system completes 100 cycles per min. Complete the following table showing the method for each item, and compute the net rate of work output in kW.

Process	Q (kJ/min)	W (kJ/min)	ΔE (kJ/min)
a–b	0	2,170	–
b–c	21,000	0	–
c–d	−2,100	–	−36,600
d–a	–	–	–

Solution

Process *a–b*:

$$Q = \Delta E + W$$
$$0 = \Delta E + 2170$$
$$\therefore \quad \Delta E = -2170 \text{ kJ/min}$$

Process *b–c*:

$$Q = \Delta E + W$$
$$21{,}000 = \Delta E + 0$$
$$\therefore \quad \Delta E = 21{,}000 \text{ kJ/min}$$

Process *c–d*:

$$Q = \Delta E + W$$
$$-2100 = -36{,}600 + W$$
$$\therefore \quad W = 34{,}500 \text{ kJ/min}$$

Process *d–a*:

$$\sum_{\text{cycle}} Q = -170 \text{ kJ}$$

The system completes 100 cycles/min.

$$\because \quad Q_{ab} + Q_{bc} + Q_{cd} + Q_{da} = -17000 \text{ kJ/min}$$
$$0 + 21{,}000 - 2100 + Q_{da} = -17{,}000$$
$$\therefore \quad Q_{da} = -35{,}900 \text{ kJ/min}$$

Now $\oint dE = 0$, since cyclic integral of any property is zero.

$$\therefore \quad \Delta E_{a-b} + \Delta E_{b-c} + \Delta E_{c-d} + \Delta E_{d-a} = 0$$
$$-2170 + 21{,}000 - 36{,}600 + \Delta E_{d-a} = 0$$
$$\therefore \quad \Delta E_{d-a} = 17{,}770 \text{ kJ/min}$$
$$\therefore \quad W_{d-a} = Q_{d-a} - \Delta E_{d-a}$$
$$= -35{,}900 - 17{,}770$$
$$= -53{,}670 \text{ kJ/min}$$

The table becomes

Process	Q (kJ/min)	W (kJ/min)	ΔE (kJ/min)
a–b	0	2170	– 2170
b–c	21,000	0	21,000
c–d	−2100	34,500,	−36,600
d–a	−35,900	−53,670	17,770

Since

$$\sum_{\text{cycle}} Q = \sum_{\text{cycle}} W$$

Rate of work output

$$= -17{,}000 \text{ kJ/min}$$
$$= -283.3 \text{ kW}$$ Ans.

EXAMPLE 4.4

The internal energy of a certain substance is given by the following equation

$$u = 3.56\, pv + 84$$

where u is given in kJ/kg, p is in kPa, and v is in m³/kg.

A system composed of 3 kg of this substance expands from an initial pressure of 500 kPa and a volume of 0.22 m³ to a final pressure 100 kPa in a process in which pressure and volume are related by $pv^{1.2}$ = constant.

(a) If the expansion is quasi-static, find Q, ΔU, and W for the process.

(b) In another process the same system expands according to the same pressure-volume relationship as in part (a), and from the same initial state to the same final state as in part (a), but the heat transfer in this case is 30 kJ. Find the work transfer for this process.

(c) Explain the difference in work transfer in parts (a) and (b).

Solution

(a)
$$u = 3.56\,pv + 84$$
$$\Delta u = u_2 - u_1 = 3.56\,(p_2v_2 - p_1v_1)$$
$$\therefore \quad \Delta U = 3.56\,(p_2V_2 - p_1V_1)$$

Now
$$p_1V_1^{1.2} = p_2V_2^{1.2}$$
$$\therefore \quad V_2 = V_1\left(\frac{p_1}{p_2}\right)^{1/1.2} = 0.22\left(\frac{5}{1}\right)^{1/1.2}$$
$$= 0.22 \times 3.83 = 0.845 \text{ m}^3$$
$$\therefore \quad \Delta U = 356(1 \times 0.845 - 5 \times 0.22) \text{ kJ}$$
$$= -356 \times 0.255 = -91 \text{ kJ}$$
Ans. (a)

For a quasi-static process

$$W = \int pdV = \frac{p_2V_2 - p_1V_1}{1 - n}$$
$$= \frac{(1 \times 0.845 - 5 \times 0.22)\,100}{1 - 1.2}$$
$$= 127.5 \text{ kJ}$$
$$\therefore \quad Q = \Delta U + W$$
$$= -91 + 127.5 = 36.5 \text{ kJ}$$
Ans. (a)

(b) Here Q = 30 kJ

Since the end states are the same, ΔU would remain the same as in (a).

$$\therefore \quad W = Q - \Delta U$$
$$= 30 - (-91)$$
$$= 121 \text{ kJ}$$
Ans. (b)

(c) The work in (b) is not equal to $\int pdV$ since the process is not quasi-static.

EXAMPLE 4.5

A fluid is confined in a cylinder by a spring-loaded, frictionless piston so that the pressure in the fluid is a linear function of the volume ($p = a + bV$). The internal energy of the fluid is given by the following equation

$$U = 34 + 3.15\ pV$$

where U is in kJ, p in kPa, and V in cubic metre. If the fluid changes from an initial state of 170 kPa, 0.03 m³ to a final state of 400 kPa, 0.06 m³, with no work other than that done on the piston, find the direction and magnitude of the work and heat transfer.

Solution

The change in the internal energy of the fluid during the process

$$U_2 - U_1 = 3.15\ (p_2V_2 - p_1V_1)$$
$$= 315\ (4 \times 0.06 - 1.7 \times 0.03)$$
$$= 315 \times 0.189 = 59.5 \text{ kJ}$$

Now

$$p = a + bV$$
$$170 = a + b \times 0.03$$
$$400 = a + b \times 0.06$$

From these two equations

$$a = -60 \text{ kN/m}^2$$
$$b = 7667 \text{ kN/m}^2$$

Work transfer involved during the process

$$W_{1-2} = \int_{V_1}^{V_2} pdV = \int_{V_1}^{V_2} (a + bV)\ dV$$
$$= a\ (V_2 - V_1) + b\ \frac{V_2^2 - V_1^2}{2}$$
$$= (V_2 - V_1)\ [a + \frac{b}{2}\ (V_1 + V_2)]$$
$$= 0.03 \text{ m}^3\ [- 60 \text{ kN/m}^2 + \frac{7667}{2}\ \frac{\text{kN}}{\text{m}^5} \times 0.09 \text{ m}^3]$$
$$= 10.35 \text{ kJ}$$

Work is done by the system, the magnitude being 10.35 kJ.

∴ Heat transfer involved is given by

$$Q_{1-2} = U_2 - U_1 + W_{1-2}$$
$$= 59.5 + 10.35$$
$$= 69.85 \text{ kJ}$$

69.85 kJ of heat flow into the system during the process.

PROBLEMS

4.1 An engine is tested by means of a water brake at 1000 rpm. The measured torque of the engine is 10000 mN and the water consumption of the brake is 0.5 m^3/s, its inlet temperature being 20°C. Calculate the water temperature at exit, assuming that the whole of the engine power is ultimately transformed into heat which is absorbed by the cooling water. Ans. 20.5°C

4.2 In a cyclic process, heat transfers are +14.7 kJ, −25.2 kJ, −3.56 kJ and +31.5 kJ. What is the net work for this cyclic process? Ans. 17.34 kJ

4.3 A slow chemical reaction takes place in a fluid at the constant pressure of 0.1 MPa. The fluid is surrounded by a perfect heat insulator during the reaction which begins at state 1 and ends at state 2. The insulation is then removed and 105 kJ of heat flow to the surroundings as the fluid goes to state 3. The following data are observed for the fluid at states 1, 2, and 3.

State	$V(m^3)$	$t(°C)$
1	0.003	20
2	0.3	370
3	0.06	20

For the fluid system, calculate E_2 and E_3, if $E_1=0$.

Ans. $E_2 = -29.7$ kJ, $E_3 = -110.7$ kJ

4.4 During one cycle the working fluid in an engine engages in two work interactions: 15 kJ to the fluid and 44 kJ from the fluid, and three heat interactions, two of which are known: 75 kJ to the fluid and 40 kJ from the fluid. Evaluate the magnitude and direction of the third heat transfer. Ans. −6 kJ

4.5 A domestic refrigerator is loaded with food and the door closed. During a certain period the machine consumes 1 kW hr of energy and the internal energy of the system drops by 5000 kJ. Find the net heat transfer for the system. Ans. −8.6 MJ

4.6 1.5 kg of liquid having a constant specific heat of 2.5 kJ/kg °C is stirred in a well-insulated chamber causing the temperature to rise by 15°C. Find ΔE and W for the process. Ans. ΔE=56.25 kJ, W= −56.25 kJ.

4.7 The same liquid as in Problem 4.6 is stirred in a conducting chamber. During the process 1.7 kJ of heat are transferred from the liquid to the surroundings, while the temperature of the liquid is rising to 15°C. Find ΔE and W for the process.

Ans. ΔE=54.55 kJ, W=56.25 kJ.

4.8 The properties of a certain fluid are related as follows

$$u = 196 + 0.718\ t$$

$$pv = 0.287\ (t+273)$$

where u is the specific internal energy (kJ/kg), t is in °C, p is pressure (kN/m^2), and v is specific volume (m^3/kg).

For this fluid, find c_v and c_p. Ans. 0.718, 1.005 kJ/kgK.

4.9 A system composed of 2 kg of the above fluid expands in a frictionless piston and cylinder machine from an initial state of 1 MPa, 100°C to a final temperature of 30°C. If there is no heat transfer, find the net work for the process. Ans. 100.52 kJ

4.10 If all the work in the expansion of Problem 4.9 is done on the moving piston, show that the equation representing the path of the expansion in the pv-plane is given by $pv^{1.4} = \text{constant}$.

4.11 A stationary system consisting of 2 kg of the fluid of Problem 4.8 expands in an adiabatic process according to $pv^{1.2} = \text{constant}$. The initial conditions are 1 MPa and 200°C, and the final pressure is 0.1 MPa. Find W and ΔE for the process. Why is the work transfer not equal to $\int pdV$?

Ans. $W = 217.35$, $\Delta E = -217.35$ kJ, $\int pdV = 434.4$ kJ.

4.12 A mixture of gases expands at constant pressure from 1 MPa, 0.03 m^3 to 0.06 m^3 with 84 kJ positive heat transfer. There is no work other than that done on a piston. Find ΔE for the gaseous mixture.

The same mixture expands through the same state path while a stirring device does 21 kJ of work on the system. Find ΔE, W, and Q for the process.

5

First Law Applied to Flow Processes

5.1 CONTROL VOLUME

For any system and in any process, the first law can be written as

$$Q = \Delta E + W$$

where E represents all forms of energy stored in the system.

For a pure substance

$$E = E_K + E_P + U$$

where E_K is the K.E., E_P the P.E., and U the residual energy stored in the molecular structure of the substance.

$$\therefore \qquad Q = \Delta E_K + \Delta E_P + \Delta U + W \qquad (5.1)$$

When there is mass transfer across the system boundary, the system is called an open system. Most of the engineering equipments are open systems involving the flow of fluids through them.

Equation (5.1) refers to a system having a particular mass of substance, and it is free to move from place to place.

Consider a steam turbine (Fig. 5.1) in which steam enters at a high pressure, does work upon the turbine rotor, and then leaves the turbine at low pressure through the exhaust pipe.

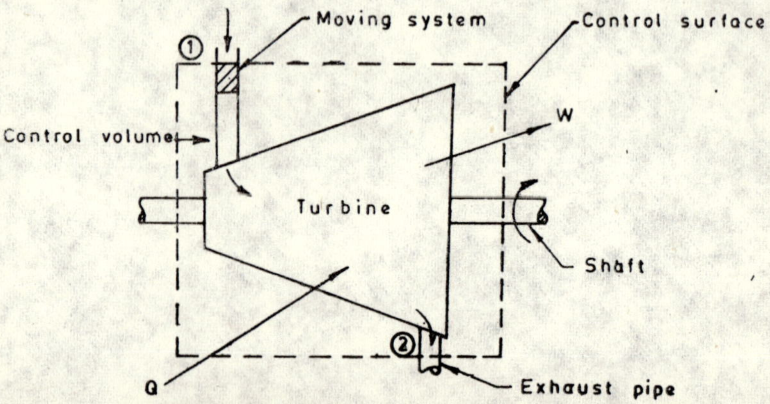

Fig. 5.1 Flow process involving work and heat interactions

If a certain mass of steam is considered as the thermodynamic system, then the energy equation becomes

$$Q = \Delta E_K + \Delta E_P + \Delta U + W$$

and in order to analyze the expansion process in turbine the moving system is to be followed as it travels through the turbine, taking into account the work and heat interactions all the way through. This method of analysis is similar to that of Lagrange in fluid mechanics.

Although the system approach is quite valid, there is another approach which is found to be highly convenient. Instead of concentrating attention upon a certain quantity of fluid, which constitutes a moving system in flow processes, attention is focussed upon a certain fixed region in space called a *control volume* through which the moving substance flows. This is similar to the analysis of Euler in fluid mechanics.

The broken line in Fig. 5.1 represents the surface of the control volume which is known as the *control surface*. This is the same as the system boundary of the open system. The method of analysis is to inspect the control surface and account for all energy quantities transferred through this surface. Since there is mass transfer across the control surface, a mass balance also has to be made. Sections 1 and 2 allow mass transfer to take place, and Q and W are the heat and work interaction respectively.

5.2 STEADY FLOW PROCESS

As a fluid flows through a certain control volume, its thermodynamic properties may vary along the space coordinates as well as with time. If the rates of flow of mass and energy through the control surface change with time, the mass and energy within the control volume also would change with time.

In most engineering devices, there is a constant rate of flow of mass and energy through the control surface, and the control volume in course of time attains a steady state. *At the steady state of a system, any thermodynamic property will have a fixed value at a particular location, and will not alter with time.* Thermodynamic properties may vary along space coordinates, but do not vary with time. 'Steady state' means that the state is steady or invariant with time.

5.3 MASS BALANCE AND ENERGY BALANCE IN A SIMPLE STEADY FLOW PROCESS

In Fig. 5.2, a steady flow system has been shown in which one stream of fluid enters and one stream leaves the control volume. There is no accumulation of mass or energy within the control volume, and the properties at any location within the control volume are steady with time. Sections 1.1 and

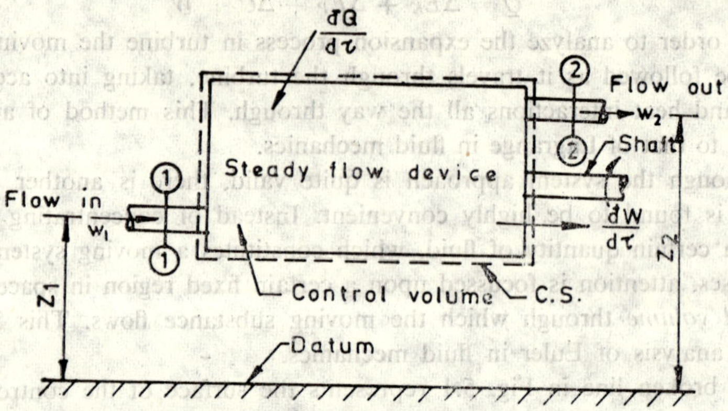

Fig. 5.2 Steady flow process.

2.2 indicate, respectively, the entrance and exit of the fluid across the control surface. The following quantities are defined with reference to Fig. 5.2.

A_1, A_2—cross-section of stream, m²

w_1, w_2—mass flow rate, kg/s

p_1, p_2—pressure, absolute, N/m²

v_1, v_2—specific volume, m³/kg

u_1, u_2—specific internal energy, J/kg

$\mathbf{V}_1$, $\mathbf{V}_2$—velocity, m/s

Z_1, Z_2—elevation above an arbitrary datum, m

$\frac{đQ}{d\tau}$ — net rate of heat transfer through the control surface, J/s

$\frac{đW}{d\tau}$ —net rate of work transfer through the control surface, J/s

τ—time, secs.

Subscripts 1 and 2 refer to the inlet and exit sections.

5.3.1 Mass Balance

By the conservation of mass, if there is no accumulation of mass within the control volume, the mass flow rate entering must equal the mass flow rate leaving, or

$$w_1 = w_2 \tag{5.2}$$

or

$$\frac{A_1 \mathbf{V}_1}{v_1} = \frac{A_2 \mathbf{V}_2}{v_2} \tag{5.3}$$

This equation is known as the *equation of continuity*.

5.3.2 Energy Balance

Since there is no accumulation of energy, by the conservation of energy, the total rate of flow of all energy streams entering the control volume must equal the total rate of flow of all energy streams leaving the control volume. This may be expressed in the following equation

$$w_1 e_1 + w_1 \cdot p_1 v_1 + \frac{đQ}{d\tau} = w_2 e_2 + w_2 p_2 v_2 + \frac{đW}{d\tau} \tag{5.4}$$

where e_1 and e_2 refer to the energy carried into or out of the control volume with unit mass of fluid, and $p_1 v_1$ and $p_2 v_2$ are the 'flow-work' associated with the fluid in sections 1 and 2.

The specific energy e is given by

$$\begin{aligned} e &= e_k + e_p + u \\ &= \frac{\mathbf{V}^2}{2} + Zg + u \end{aligned} \tag{5.5}$$

Substituting the expression for e in equation (5.4)

$$w_1 \left(\frac{\mathbf{V}_1^2}{2} + Z_1 g + u_1 \right) + w_1 p_1 v_1 + \frac{đQ}{d\tau} = w_2 \left(\frac{\mathbf{V}_2^2}{2} + Z_2 g + u_2 \right) + w_2 p_2 v_2 + \frac{đW}{d\tau} \tag{5.6}$$

or

$$w_1 \left(h_1 + \frac{\mathbf{V}_1^2}{2} + Z_1 g \right) + \frac{đQ}{d\tau} = w_2 \left(h_2 + \frac{\mathbf{V}_2^2}{2} + Z_2 g \right) + \frac{đW}{d\tau} \tag{5.7}$$

where $h = u + pv$.

In MKS units, the energy equation at steady flow becomes

$$w_1 \left(h_1 + \frac{\mathbf{V}_1^2}{2 g_0 J} + \frac{Z_1 g}{g_0 J} \right) + \frac{đQ}{d\tau} = w_2 \left(h_2 + \frac{\mathbf{V}_2^2}{2 g_0 J} + \frac{Z_2 g}{g_0 J} \right) + \frac{1}{J} \frac{đW}{d\tau}$$

And, since $w_1 = w_2$, let $w = w_1 = w_2 = \dfrac{dm}{d\tau}$

Dividing equation (5.7) by $\frac{dm}{d\tau}$

$$h_1 + \frac{\mathbf{V}_1^2}{2} + Z_1 g + \frac{đQ}{dm}$$

$$= h_2 + \frac{\mathbf{V}_2^2}{2} + Z_2 g + \frac{đW}{dm} \tag{5.8}$$

Equations (5.7) and (5.8) are known as *steady flow energy equations* (S.F.E.E.), for a single stream of fluid entering and a single stream of fluid leaving the control volume. All the terms in equation (5.8) represent energy flow per unit mass of fluid (J/kg), whereas all the terms in equation (5.7) represent energy flow per unit time (J/s). The basis of energy flow per unit mass is usually more convenient when only a single stream of fluid enters and leaves a control volume. When more than one fluid stream is involved, the basis of energy flow per unit time is more convenient.

When more than one stream of fluid enters or leaves the control volume (Fig. 5.3), the mass balance and energy balance for steady flow are given below.

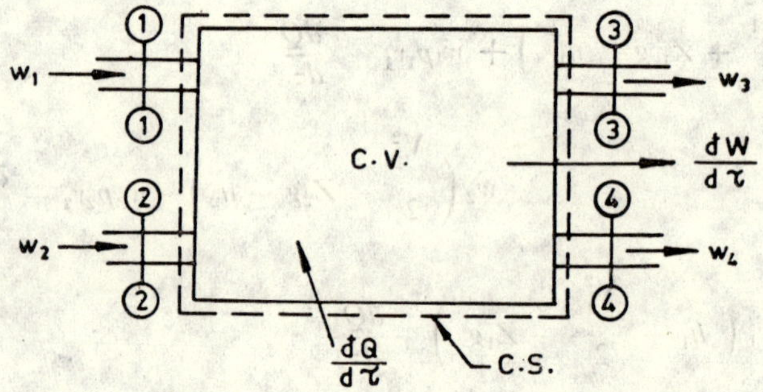

Fig. 5.3 Steady flow process involving two fluid streams at the inlet and exit of the control volume.

Mass balance

$$w_1 + w_2 = w_3 + w_4 \tag{5.9}$$

$$\frac{A_1 \mathbf{V}_1}{v_1} + \frac{A_2 \mathbf{V}_2}{v_2} = \frac{A_3 \mathbf{V}_3}{v_3} + \frac{A_4 \mathbf{V}_4}{v_4} \tag{5.10}$$

Energy balance

$$w_1\left(h_1 + \frac{\mathbf{V}_1^2}{2} + Z_1 g\right) + w_2\left(h_2 + \frac{\mathbf{V}_2^2}{2} + Z_2 g\right) + \frac{đQ}{d\tau}$$

$$= w_3\left(h_3 + \frac{\mathbf{V}_3^2}{2} + Z_3 g\right) + w_4\left(h_4 + \frac{\mathbf{V}_4^2}{2} + Z_4 g\right) + \frac{đW}{d\tau} \tag{5.11}$$

The steady flow energy equation applies to a wide variety of processes like pipe line flows, heat transfer processes, mechanical power generation in engines and turbines, combustion process, and flows through nozzles and diffusors. In certain problems, some of the terms in steady flow energy equation may be negligible or zero. But it is best to write the full equation first, and then eliminate the terms which are unnecessary.

5.4 SOME EXAMPLES OF STEADY FLOW PROCESSES

The following examples illustrate the applications of the steady flow energy equation in some of the engineering systems.

5.4.1 Nozzle and Diffusor

A nozzle is a device which increases the velocity or K.E. of a fluid at the expense of its pressure drop, whereas a diffusor increases the pressure of

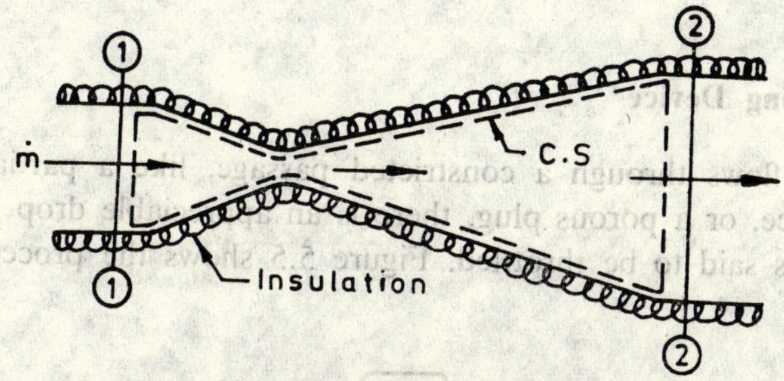

Fig. 5.4 Flow through a nozzle

a fluid at the expense of its K.E. Figure 5.4 shows a nozzle which is insulated. The steady flow energy equation of the control surface gives

$$h_1 + \frac{\mathbf{V}_1^2}{2} + Z_1 g + \frac{đQ}{dm} = h_2 + \frac{\mathbf{V}_2^2}{2} + Z_2 g + \frac{đW}{dm}$$

Here $\frac{đQ}{dm} = 0$, $\frac{đW}{dm} = 0$, and the change in potential energy is zero.

The equation reduces to

$$h_1 + \frac{\mathbf{V}_1^2}{2} = h_2 + \frac{\mathbf{V}_2^2}{2} \tag{5.12}$$

The continuity equation gives

$$w = \frac{A_1 \mathbf{V}_1}{v_1} = \frac{A_2 \mathbf{V}_2}{v_2} \tag{5.13}$$

When the inlet velocity or the 'velocity of approach' $\mathbf{V}_1$ is small compared to the exit velocity $\mathbf{V}_2$, equation (5.12) becomes

$$h_1 = h_2 + \frac{\mathbf{V}_2^2}{2}$$

or

$$\mathbf{V}_2 = \sqrt{2(h_1 - h_2)} \text{ m/s},$$

where $(h_1 - h_2)$ is in J/kg. In MKS units, $\mathbf{V}_2$ will be given by

$$\mathbf{V}_2 = \sqrt{2g_0 J\,(h_1 - h_2)}.$$

Putting

$$g_0 = 9.806 \frac{\text{kgm}}{\text{kgf}} \cdot \frac{\text{m}}{\text{s}^2},$$

and

$$J = 427 \frac{\text{kgf} \cdot \text{m}}{\text{kcal}}$$

$$\mathbf{V}_2 = 91.5\sqrt{h_1 - h_2} \text{ m/s} \tag{5.14}$$

where $(h_1 - h_2)$ is in kcal/kgm.

Equations (5.12) and (5.13) hold good for a diffusor as well.

5.4.2 Throttling Device

When a fluid flows through a constricted passage, like a partially opened valve, an orifice, or a porous plug, there is an appreciable drop in pressure, and the flow is said to be throttled. Figure 5.5 shows the process of throttling by a partially opened valve on a fluid flowing in an insulated pipe.

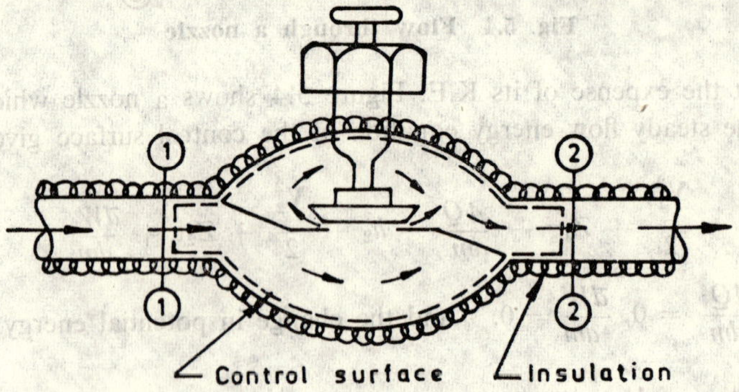

Fig. 5.5 Flow through a valve

In the steady-flow energy equation (5.8),

$$\frac{đQ}{dm} = 0, \quad \frac{đW}{dm} = 0$$

and the changes in P.E. are very small and ignored. Thus the S.F.E.E. reduces to

$$h_1 + \frac{V_2^2}{2} = h_2 + \frac{V_2^2}{2}$$

Often the pipe velocities in throttling are so low that the K.E. terms are also negligible. So

$$h_1 = h_2 \tag{5.15}$$

or *the enthalpy of the fluid before throttling is equal to the enthalpy of the fluid after throttling.*

5.4.3 Turbine and Compressor

Turbines and engines give positive power output, whereas compressors and pumps require power input.

For a turbine (Fig. 5.6) which is well insulated, the flow velocities are often small, and the K.E. terms can be neglected. The S.F.E.E. then becomes

$$h_1 = h_2 + \frac{dW}{dm}$$

or

$$\frac{W}{m} = (h_1 - h_2)$$

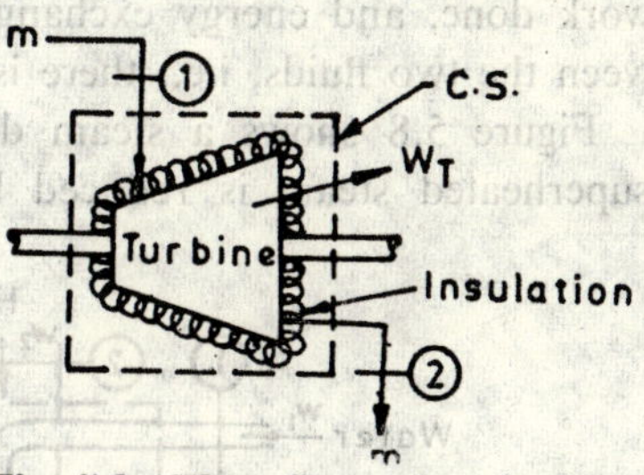

Fig. 5.6 Flow through a turbine

It is seen that work is done by the fluid at the expense of its enthalpy.

Similarly, for an adiabatic pump or compressor, work is done upon the fluid and W is negative. So the S.F.E.E. becomes

$$h_1 = h_2 - \frac{W}{m}$$

or

$$\frac{W}{m} = h_2 - h_1$$

The enthalpy of the fluid increases by the amount of work input.

5.4.4 Heat Exchanger

A heat exchanger is a device in which heat is transferred from one fluid to another. Figure 5.7 shows a steam condenser, where steam condenses outside the tubes and cooling water flows through the tubes. The S.F.E.E. for the C.S. gives

$$w_c h_1 + w_s h_2 = w_c h_3 + w_s h_4$$

or

$$w_s (h_2 - h_4) = w_c (h_3 - h_1)$$

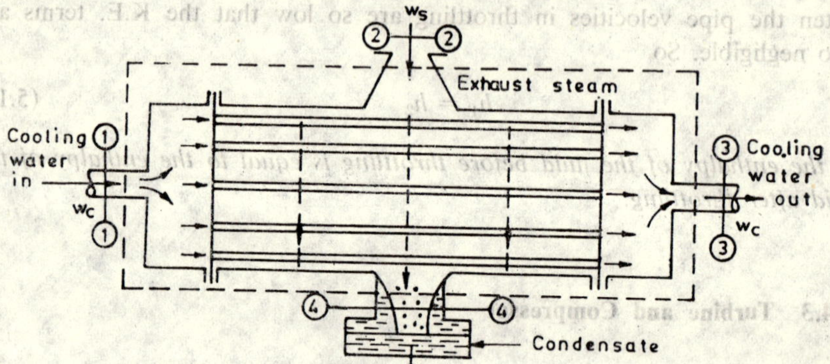

Fig. 5.7 Steam condenser

Here the K.E. and P.E. terms are considered small, there is no external work done, and energy exchange in the form of heat is confined only between the two fluids, i.e., there is no external heat interaction or heat loss.

Figure 5.8 shows a steam desuperheater where the temperature of the superheated steam is reduced by spraying water. If w_1, w_2, and w_3 are

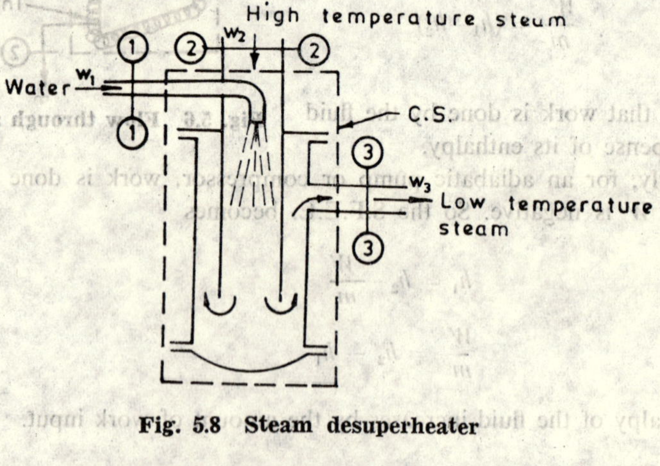

Fig. 5.8 Steam desuperheater

the mass flow rates of the injected water, of the steam entering, and of the steam leaving, respectively, and h_1, h_2, and h_3 are the corresponding enthalpies, and if K.E. and P.E. terms are neglected as before, the S.F.E.E. becomes

$$w_1 h_1 + w_2 h_2 = w_3 h_3$$

and the mass balance gives

$$w_1 + w_2 = w_3$$

5.5 VARIABLE FLOW PROCESSES

Many flow processes, such as filling up and evacuating gas cylinders, are not steady. Such processes can be analyzed by the control volume technique. Consider a device through which a fluid is flowing under non-steady state conditions (Fig. 5.9). The rate at which the mass of fluid within the control

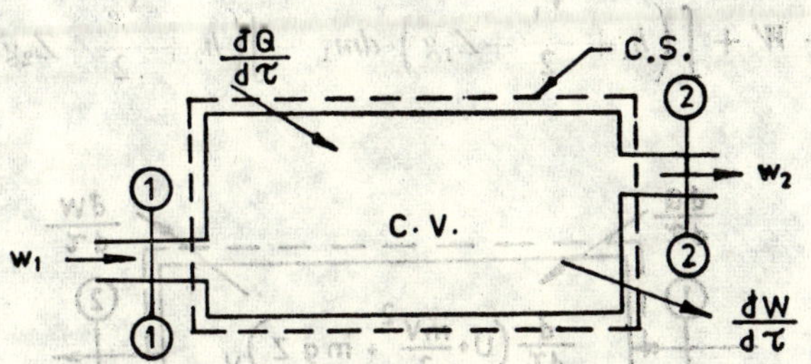

Fig. 5.9 Variable flow process

volume is accumulated is equal to the net rate of mass flow across the control surface, as given below

$$\frac{dm_V}{d\tau} = w_1 - w_2 = \frac{dm_1}{d\tau} - \frac{dm_2}{d\tau} \tag{5.16}$$

where m_V is the mass of fluid within the control volume at any instant.

Over any finite period of time

$$\Delta m_V = \Delta m_1 - \Delta m_2 \tag{5.17}$$

The rate of accumulation of energy within the control volume is equal to the net rate of energy flow across the control surface. If E_v is the energy of fluid within the control volume at any instant

Rate of energy increase = Rate of energy inflow − Rate of energy outflow

$$\frac{dE_V}{d\tau} = w_1\left(h_1 + \frac{\mathbf{V}_1^2}{2} + Z_1 g\right) + \frac{đQ}{d\tau}$$

$$- w_2\left(h_2 + \frac{\mathbf{V}_2^2}{2} + Z_2 g\right) - \frac{đW}{d\tau} \tag{5.18}$$

Now $\quad E_V = \left(U + \dfrac{mV^2}{2} + mgZ\right)_V$

where m is the mass of fluid in the control volume at any instant.

$$\therefore \quad \frac{dE_V}{d\tau} = \frac{d}{d\tau}\left(U + \frac{mV^2}{2} + mgZ\right)_V$$

$$= \left(h_1 + \frac{\mathbf{V}_1^2}{2} + Z_1 g\right)\frac{dm_1}{d\tau} + \frac{đQ}{d\tau} - \left(h_2 + \frac{\mathbf{V}_2^2}{2} + Z_2 g\right)\frac{dm_2}{d\tau} - \frac{đW}{d\tau} \tag{5.19}$$

Figure 5.10 shows all these energy flux quantities. For any finite time interval, equation (5.19) becomes

$$\Delta E_V = Q - W + \int\left(h_1 + \frac{\mathbf{V}_1^2}{2} + Z_1 g\right) dm_1 - \int\left(h_2 + \frac{\mathbf{V}_2^2}{2} + Z_2 g\right) dm_2 \tag{5.20}$$

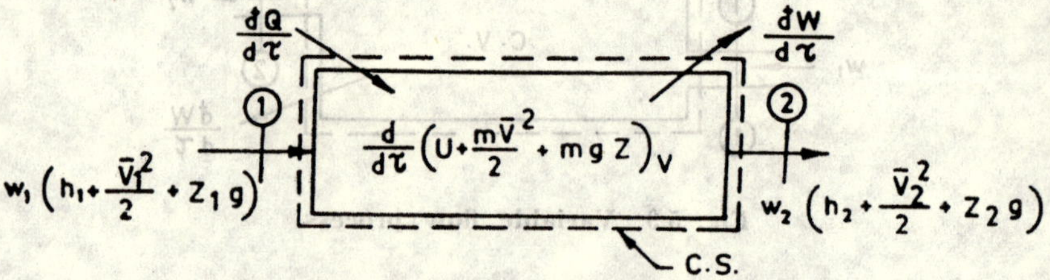

Fig. 5.10 Energy fluxes in an unsteady system

Equation (5.18) is the *general energy equation*. For steady flow,

$$\frac{dE_V}{d\tau} = 0,$$

and the equation reduces to equation (5.7). For a closed system $w_1 = 0$, $w_2 = 0$, then from equation (5.18),

$$\frac{dE_V}{d\tau} = \frac{dQ}{d\tau} - \frac{đW}{d\tau}$$

or $\quad dE_V = đQ - đW \quad$ or, $\quad đQ = dE + đW$

as obtained earlier.

5.6 EXAMPLE OF A VARIABLE FLOW PROBLEM

Variable flow processes may be analyzed either by the system technique or the control volume technique, as illustrated below.

Consider a process in which a gas bottle is filled from a pipeline (Fig. 5.11). In the beginning the bottle contains gas of mass m_1 at state p_1, t_1, v_1, h_1 and u_1. The valve is opened and gas flows into the bottle till the mass of gas in the bottle is m_2 at state p_2, t_2, v_2, h_2 and u_2. The supply to the pipeline is very large so that the state of gas in the pipeline is constant at p_p, t_p, v_p, h_p, u_p, and $\mathbf{V}_p$.

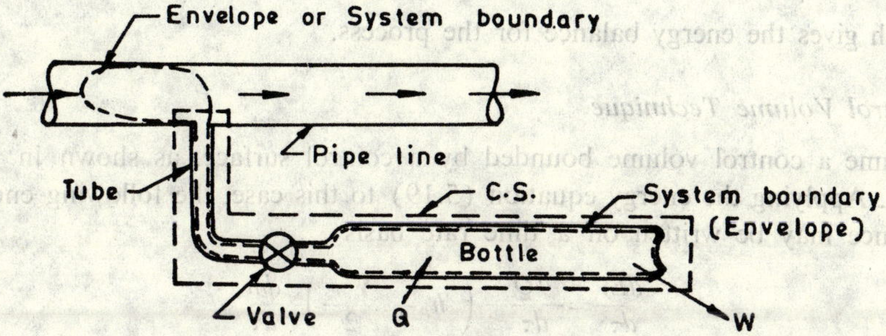

Fig. 5.11 Bottle-filling process

System Technique

Assume an envelope (which is extensible) of gas in the pipeline and the tube which would eventually enter the bottle, as shown in Fig. 5.11.

Energy of the gas before filling

$$E_1 = m_1u_1 + (m_2 - m_1)\left(\frac{\mathbf{V}_p^2}{2} + u_p\right)$$

where $(m_2 - m_1)$ is the mass of gas in the pipeline and tube which would enter the bottle.

Energy of the gas after filling

$$E_2 = m_2u_2$$

$$\therefore \quad \Delta E = E_2 - E_1 = m_2u_2 - \left[m_1u_1 + (m_2 - m_1)\left(\frac{\mathbf{V}_p^2}{2} + u_p\right)\right] \tag{5.21}$$

The P.E. terms are neglected. The gas in the bottle is not in motion, and so the K.E. terms have been omitted.

Now, there is a change in the volume of gas because of the collapse of the envelope to zero volume. Then the work done

$$W = p_p(V_2 - V_1) = p_p[0 - (m_2 - m_1)v_p]$$
$$= -(m_2 - m_1)p_pv_p$$

$\therefore$ Using the first law for the process

$$Q = \Delta E + W$$

$$= m_2u_2 - m_1u_1 - (m_2 - m_1)\left(\frac{\mathbf{V}_p^2}{2} + u_p\right) - (m_2 - m_1)p_pv_p$$

$$=m_2u_2-m_1u_1-(m_2-m_1)\left(\frac{\mathbf{V}_p^2}{2}+h_p\right) \tag{5.23}$$

which gives the energy balance for the process.

Control Volume Technique

Assume a control volume bounded by a control surface, as shown in Fig. 5.11. Applying the energy equation (5.19) to this case, the following energy balance may be written on a time rate basis

$$\frac{dE_V}{d\tau}=\frac{đQ}{d\tau}+\left(h_p+\frac{\mathbf{V}^2_p}{2}\right)\frac{dm}{d\tau}$$

Since h_p and $\mathbf{V}_p$ are constant, the equation is integrated to give for the total process

$$\Delta E_V=Q+\left(h_p+\frac{\mathbf{V}^2_p}{2}\right)(m_2-m_1)$$

Now

$$\Delta E_V=U_2-U_1=m_2u_2-m_1u_1$$

$$\therefore \qquad Q=m_2u_2-m_1u_1-\left(h_p+\frac{\mathbf{V}_p^2}{2}\right)(m^2-m_1)$$

This equation is the same as equation (5.23).

Example 5.1

Air flows steadily at the rate of 0.5 kg/s through an air compressor, entering at 7 m/s velocity, 100 kPa pressure, and 0.95 m³/kg volume, and leaving at 5 m/s, 700 kPa, and 0.19 m³/kg. The internal energy of the air leaving is 90 kJ/kg greater than that of the air entering. Cooling water in the compressor jackets absorbs heat from the air at the rate of 58 kW. (a) Compute the rate of shaft work input to the air in kW. (b) Find the ratio of the inlet pipe diameter to outlet pipe diameter.

Solution

Figure Ex. 5.1 shows the details of the problem.

(a) Writing the steady flow energy equation, we have

$$w\left(u_1+p_1v_1+\frac{\mathbf{V}_1^2}{2}+Z_1g\right)+\frac{đQ}{d\tau}$$

$$=w\left(u_2+p_2v_2+\frac{\mathbf{V}_2^2}{2}+Z_2g\right)+\frac{đW}{d\tau}$$

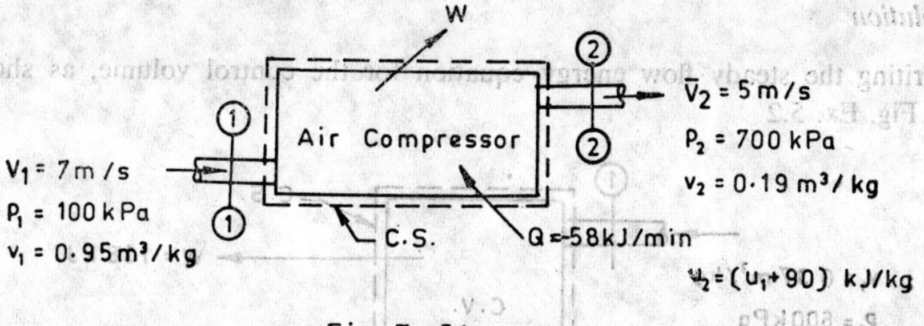

Fig. Ex. 5.1

$$\therefore \quad \frac{dW}{d\tau} = -w\left[(u_2-u_1)+(p_2v_2-p_1v_1)\right.$$

$$\left. + \frac{\mathbf{V}_2^2-\mathbf{V}_1^2}{2} + (Z_2-Z_1)\,g\right] + \frac{dQ}{d\tau}$$

$$\therefore \quad \frac{dW}{d\tau} = -0.5\ \frac{\text{kg}}{\text{s}}\left[90\ \frac{\text{kJ}}{\text{kg}} + (7\times0.19-1\times0.95)\ 100\ \frac{\text{kJ}}{\text{kg}}\right.$$

$$\left. + \frac{(5^2-7^2)\times10^{-3}}{2}\ \frac{\text{kJ}}{\text{kg}} + 0\right] - 58\ \text{kW}$$

$$= -0.5\ [90+38-0.012]\ \text{kJ/s} - 58\ \text{kW}$$

$$= -122\ \text{kW}$$

Rate of work input is 122 kW **Ans. (a)**

(b) From mass balance, we have

$$w = \frac{A_1\,\mathbf{V}_1}{v_1} = \frac{A_2\,\mathbf{V}_2}{v_2}$$

$$\therefore \quad \frac{A_1}{A_2} = \frac{v_1}{v_2}\cdot\frac{\mathbf{V}_2}{\mathbf{V}_1} = \frac{0.95}{0.19}\times\frac{5}{7} = 3.57$$

$$\therefore \quad \frac{d_1}{d_2} = \sqrt{3.57} = 1.89$$

Ans. (b).

EXAMPLE 5.2

In a steady flow apparatus, 135 kJ of work is done by each kg of fluid. The specific volume of the fluid, pressure, and velocity at the inlet are 0.37 m³/kg, 600 kPa, and 16 m/s. The inlet is 32 m above the floor, and the discharge pipe is at floor level. The discharge conditions are 0.62 m³/kg, 100 kPa, and 270 m/s. The total heat loss between the inlet and discharge is 9 kJ/kg of fluid. In flowing through this apparatus, does the specific internal energy increase or decrease, and by how much?

Solution

Writing the steady flow energy equation for the control volume, as shown in Fig. Ex. 5.2

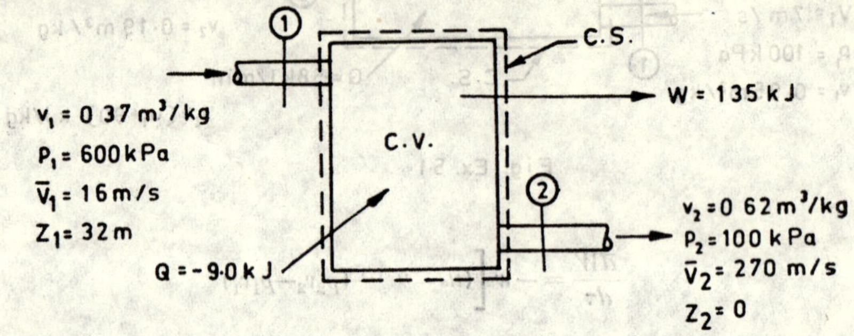

Fig. Ex. 5·2

$$u_1+p_1v_1+\frac{\mathbf{V}_1^2}{2}+Z_1g+\frac{đQ}{dm}=u_2+p_2v_2+\frac{\mathbf{V}_2^2}{2}+Z_2g+\frac{đW}{dm}$$

$$\therefore \quad u_1-u_2=(p_2v_2-p_1v_1)+\frac{\mathbf{V}_2^2-\mathbf{V}_1^2}{2}+(Z_2-Z_1)\,g+\frac{đW}{dm}-\frac{đQ}{dm}$$

$$=(1\times0.62-6\times0.37)\times10^2+\frac{(270^2-16^2)\times10^{-3}}{2}$$

$$+(-32\times9.81\times10^{-3})+135-(-9.0)$$

$$=-160+36.45-0.314+135+9$$

$$=20.136 \text{ kJ/kg}$$

Specific internal energy decreases by 20.136 kJ.

Example 5.3

In a steam power station, steam flows steadily through a 0.2 m diameter pipeline from the boiler to the turbine. At the boiler end, the steam conditions are found to be : $p = 4$ MPa, $t = 400°$C, $h = 3213.6$ kJ/kg, and $v = 0.073$ m³/kg. At the turbine end, the conditions are found to be : $p = 3.5$ MPa, $t = 392°$C, $h = 3202.6$ kJ/kg, and $v = 0.084$ m³/kg. There is a heat loss of 8.5 kJ/kg from the pipeline. Calculate the steam flow rate.

Solution

Writing the steady flow energy equation for the control volume as shown in Fig. Ex. 5.3

$$h_1 + \frac{V_1^2}{2} + Z_1 g + \frac{đQ}{dm} = h_2 + \frac{V_2^2}{2} + Z_2 g + \frac{đW}{dm}$$

Here, there is no change in datum, so change in potential energy will be zero.

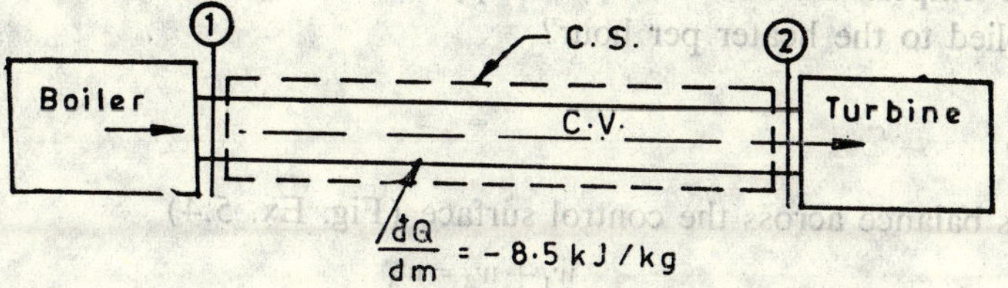

Fig. Ex. 5.3

Now $$\frac{A_1 V_1}{v_1} = \frac{A_2 V_2}{v_2}$$

$\therefore$ $$V_2 = \frac{A_1 V_1}{v_1} \cdot \frac{v_2}{A_2} = \frac{v_2}{v_1} \cdot V_1 = \frac{0.084}{0.073} V_1 = 1.15\ V_1$$

and $$\frac{đW}{dm} = 0$$

$$h_1 + \frac{V_1^2}{2} + \frac{đQ}{dm} = h_2 + \frac{V_2^2}{2}$$

$\therefore$ $$\frac{(V_2^2 - V_1^2) \times 10^{-3}}{2} = h_1 - h_2 + \frac{đQ}{dm}$$

$$= 3213.6 - 3202.6 + (-8.5)$$

$$= 2.5 \text{ kJ/kg}$$

$$V_1^2 (1.15^2 - 1^2) = 5 \times 10^3$$

$$V_1^2 = 15{,}650 \text{ m}^2/\text{s}^2$$

$\therefore$ $$V_1 = 125.1 \text{ m/s}$$

$\therefore$ Mass flow rate $$w = \frac{A_1 V_1}{v_1} = \frac{\frac{\pi}{4} \times (0.2)^2 \text{ m}^2 \times 125.1 \text{ m/s}}{0.073 \text{ m}^3/\text{kg}}$$

$$= 53.8 \text{ kg/s}$$ Ans.

EXAMPLE 5.4

A certain water heater operates under steady flow conditions receiving 4.2 kg/s of water at 75°C temperature, enthalpy 313.93 kJ/kg. The water is heated by mixing with steam which is supplied to the heater at temperature 100.2°C and enthalpy 2676 kJ/kg. The mixture leaves the heater as liquid water at temperature 100°C and enthalpy 419 kJ/kg. How much steam must be supplied to the heater per hour?

Solution

By mass balance across the control surface (Fig. Ex. 5.4)

$$w_1 + w_2 = w_3$$

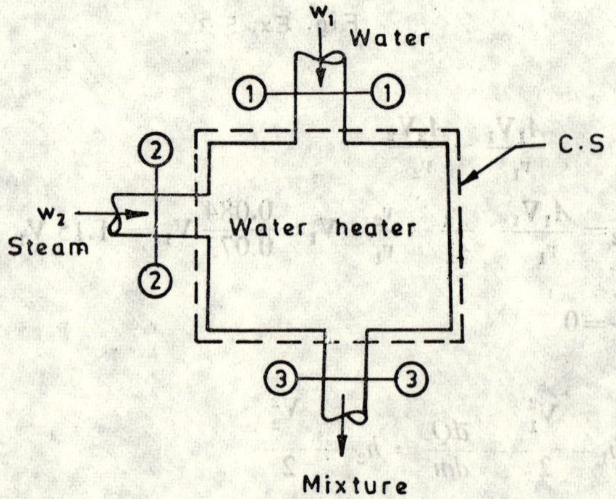

Fig. Ex 5.4

By energy balance

$$w_1\left(h_1 + \frac{\mathbf{V}_1^2}{2} + Z_1 g\right) + \frac{đQ}{d\tau} + w_2\left(h_2 + \frac{\mathbf{V}_2^2}{2} + Z_2 g\right)$$

$$= w_3\left(h_3 + \frac{\mathbf{V}_3^2}{2} + Z_3 g\right) + \frac{đW}{d\tau}$$

By the nature of the process, there is no shaft work. Potential and kinetic energy terms are assumed to balance zero. The heater is assumed to be insulated. So the steady flow energy equation reduces to

$$w_1 h_1 + w_2 h_2 = w_3 h_3$$

$$4.2 \times 313.93 + w_2 \times 2676 = (4.2 + w_2)\ 419$$

$$\therefore \quad w_2 = 0.196 \text{ kg/s}$$

$$= 705 \text{ kg/hr}$$

Ans.

EXAMPLE 5.5

Air at a temperature of 15°C passes through a heat exchanger at a velocity of 30 m/s where its temperature is raised to 800°C. It then enters a turbine with the same velocity of 30 m/s and expands until the temperature falls to 650°C. On leaving the turbine, the air is taken at a velocity of 60 m/s to a nozzle where it expands until the temperature has fallen to 500°C. If the air flow rate is 2 kg/s, calculate (a) the rate of heat transfer to the air in the heat exchanger, (b) the power output from the turbine assuming no heat loss, and (c) the velocity at exit from the nozzle, assuming no heat loss. Take the enthalpy of air as $h = c_p$, where c_p is the specific heat equal to 1.005 kJ/kg°C and t the temperature.

Solution

As shown in Fig. Ex. 5.5, writing the S.F.E.E. for the heat exchanger and eliminating the terms not relevant,

$$w\left(h_1+\frac{V_1^2}{2}+Z_1g\right)+Q_{1-2}=w\left(h_2+\frac{V_2^2}{2}+Z_2g\right)+W_{1-2}$$

$$\therefore \quad wh_1+Q_{1-2}=wh_2$$

$$\therefore \quad Q_{1-2}=w(h_2-h_1)=w\,c_p\,(t_2-t_1)$$

$$=2\times 1.005\;(800-15)$$

$$=2.01\times 785$$

$$=1580\ \text{kJ/s}$$

Ans. (a).

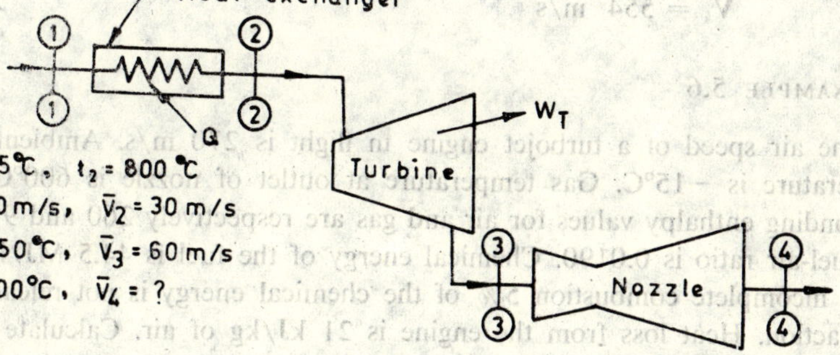

Fig. Ex. 5·5

Energy equation for the turbine gives

$$w\left(\frac{V_2^2}{2}+h_2\right)=wh_3+w\frac{V_3^2}{2}+W_T$$

$$\frac{V_2^2 - V_3^2}{2} + (h_2 - h_3) = W_T/w$$

$$\frac{(30^2 - 60\ ) \times 10^{-3}}{2} + 1.005\ (800 - 650) = W_T/w$$

$$\therefore \quad \frac{W_T}{w} = -1.35 + 150.75$$

$$= 149.4 \text{ kJ/kg}$$

$$\therefore \quad W_T = 149.4 \times 2 \text{ kJ/s}$$

$$= 298.8 \text{ kW}$$

Ans. (b).

Writing the energy equation for the nozzle

$$\frac{V_3^3}{2} + h_3 = \frac{V_4^2}{2} + h_4$$

$$\frac{V_4^2 - V_3^2}{2} = c_p(t_3 - t_4)$$

$$V_4^2 - V_3^2 = 1.005\ (650 - 500) \times 2 \times 10^3$$

$$= 301.50 \times 10^3 \text{ m}^2/\text{s}^2$$

$$V_4^2 = 30.15 \times 10^4 + 0.36 \times 10^4$$

$$= 30.51 \times 10^4 \text{ m}^2/\text{s}^2$$

$\therefore$ Velocity at exit from the nozzle

$$V_4 = 554 \text{ m/s}$$

Ans. (c).

EXAMPLE 5.6

The air speed of a turbojet engine in flight is 270 m/s. Ambient air temperature is −15°C. Gas temperature at outlet of nozzle is 600°C. Corresponding enthalpy values for air and gas are respectively 260 and 912 kJ/kg. Fuel-air ratio is 0.0190. Chemical energy of the fuel is 44.5 MJ/kg. Owing to incomplete combustion 5% of the chemical energy is not released in the reaction. Heat loss from the engine is 21 kJ/kg of air. Calculate the velocity of the exhaust jet.

Solution

Energy equation for the turbojet engine (Fig. Ex. 5.6) gives

$$w_a\left(h_a + \frac{V_a^2}{2}\right) + w_f E_f + Q = w_g\left(h_g + \frac{V_g^2}{2} + E_g\right)$$

$$1\left(260+\frac{270^2\times 10^{-3}}{2}\right)+0.0190\times 44500-21$$

$$=1.0190\left(912+\frac{\mathbf{V}_g^2\times 10^{-3}}{2}+0.05\,\frac{0.019}{1.019}\times 44500\right)$$

$$260+36.5+845-21=1.019\left(912+\frac{\mathbf{V}_g^2\times 10^{-3}}{2}+42\right)$$

$$\therefore \qquad \frac{\mathbf{V}_g^2}{2}=156\times 10^3\ \text{m}^2/\text{s}^2$$

$$\mathbf{V}_g=\sqrt{31.2}\times 100\ \text{m/s}$$

Velocity of exhaust gas, $\mathbf{V}_g$ = 560 m/s Ans.

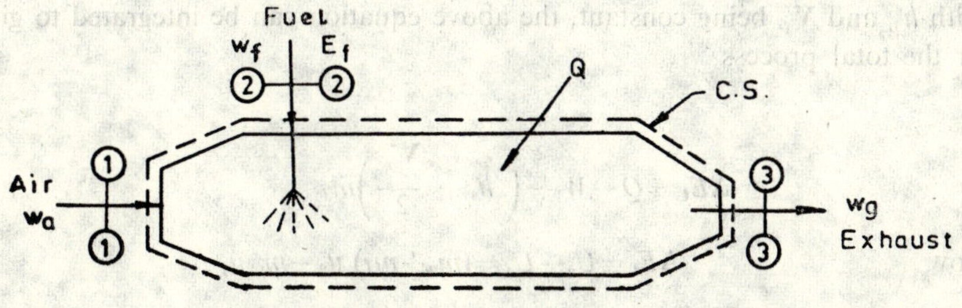

Fig. Ex. 5.6

EXAMPLE 5.7

In a reciprocating engine, the mass of gas occupying the clearance volume is m_c kg at state p_1, u_1, v_1 and h_1. By opening the inlet valve, m_f kg of gas is taken into the cylinder, and at the conclusion of the intake process the state of the gas is given by p_2, u_2, v_2, h_2. The state of the gas in the supply pipe is constant and is given by p_p, u_p, v_p, h_p, $\mathbf{V}_p$. How much heat is transferred between the gas and the cylinder walls during the intake process?

Solution

Let us consider the control volume as shown in Fig. Ex. 5.7. Writing the energy balance on a time rate basis

$$\frac{dE_V}{d\tau} = \frac{đQ}{d\tau} - \frac{đW}{d\tau} + \left(h_p + \frac{\mathbf{V}_p^2}{2} \right) \frac{dm}{d\tau}$$

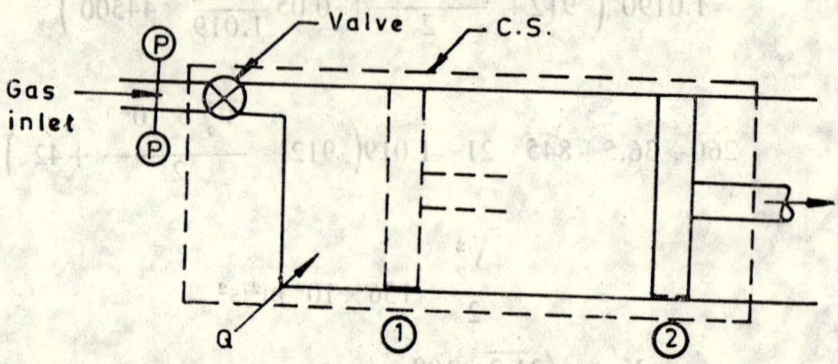

Fig. Ex. 5·7

With h_p and $\mathbf{V}_p$ being constant, the above equation can be integrated to give for the total process

$$\Delta E_V = Q - W + \left(h_p + \frac{\mathbf{V}_p^2}{2} \right) m_f$$

Now $$\Delta E_v = U_2 - U_1 = (m_c + m_f)\, u_2 - m_c\, u_1$$

$$\therefore \quad Q = (m_c + m_f)\, u_2 - m_c\, u_1 + m_f \left(h_p + \frac{\mathbf{V}_p^2}{2} \right) + W \qquad \text{Ans.}$$

EXAMPLE 5.8

The internal energy of air is given by

$$u = u_o + 0.718\, t$$

where u is in kJ/kg, u_o is any arbitrary value of u at 0°C, kJ/kg, and t is the temperature in °C. Also for air, $pv = 0.287\,(t + 273)$, where p is in kPa and v is in m³/kg.

A mass of air is stirred by a paddle wheel in an insulated constant-volume tank. The velocities due to stirring make a negligible contribution to the internal energy of the air. Air flows out through a small valve in the tank at a rate controlled to keep the temperature in the tank constant. At a certain instant the conditions are as follows: tank volume 0.12 m³, pressure 1 MPa, temperature 150°C, and power to paddle wheel 0.1 kW. Find the rate of flow of air out of the tank at this instant.

Solution

Writing the energy balance for the control volume as shown in Fig. Ex. 5.8

$$\frac{dE_V}{d\tau} = \frac{đW}{d\tau} - (h_p)\frac{dm}{d\tau}$$

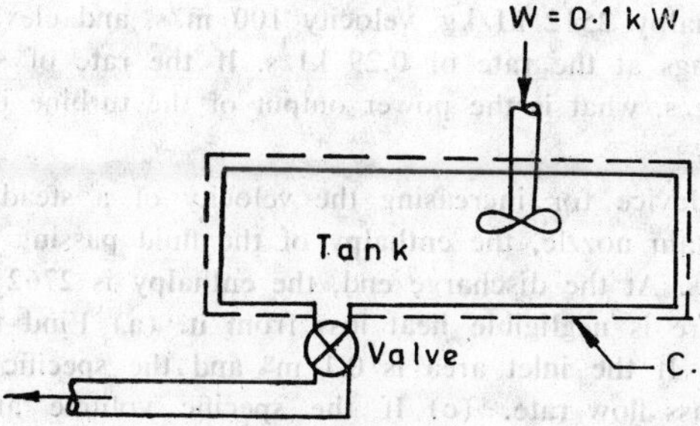

Fig. Ex. 5·8

Since there is no change in internal energy of air in the tank,

$$h_p \cdot \frac{dm}{d\tau} = \frac{đW}{d\tau}$$

where $h_p = u + pv$

Let $u = 0$ at $t = 0\ K = -273°C$

$$u = u_0 + 0.718\, t$$

$$0 = u_0 + 0.718\,(-273)$$

$$u_0 = 0.718 \times 273 \text{ kJ/kg}$$

At t°C

$$u = 0.718 \times 273 + 0.718\, t$$

$$= 0.718\,(t + 273) \text{ kJ/kg}$$

$$h_p = 0.718\,(t + 273) + 0.287\,(t + 273)$$

or $$h_p = 1.005\,(t + 273)$$

At 150°C

$$h_p = 1.005 \times 423$$

$$= 425 \text{ kJ/kg}$$

$$\therefore \quad \frac{dm}{d\tau} = \frac{1}{h_p}\,\frac{đW}{d\tau}$$

$$= \frac{0.1 \text{ kJ/s}}{425 \text{ kJ/kg}} = 0.236 \times 10^{-3} \text{ kg/s}$$

$$= 0.845 \text{ kg/hr}$$

This is the rate at which air flows out of the tank.

PROBLEMS

5.1 A blower handles 1 kg/s of air at 20°C and consumes a power of 15 kW. The inlet and outlet velocities of air are 100 m/s and 150 m/s respectively. Find the exit air temperature, assuming adiabatic conditions. Take c_p of air as 1.005 kJ/kg-K.

Ans. 28.38°C.

5.2 A turbine operates under steady flow conditions, receiving steam at the following state: pressure 1.2 MPa, temperature 188°C, enthalpy 2785 kJ/kg, velocity 33.3 m/s. and elevation 3 m. The steam leaves the turbine at the following state: pressure 20 kPa, enthalpy 2512 kJ/kg, velocity 100 m/s, and elevation 0 m. Heat is lost to the surroundings at the rate of 0.29 kJ/s. If the rate of steam flow through the turbine is 0.42 kg/s, what is the power output of the turbine in kW?

Ans. 112.51 kW.

5.3 A nozzle is a device for increasing the velocity of a steadily flowing stream. At the inlet to a certain nozzle, the enthalpy of the fluid passing is 3000 kJ/kg and the velocity is 60 m/s. At the discharge end, the enthalpy is 2762 kJ/kg. The nozzle is horizontal and there is negligible heat loss from it. (a) Find the velocity at exit from the nozzle. (b) If the inlet area is 0.1 m^2 and the specific volume at inlet is 0.187 m^3/kg, find mass flow rate. (c) If the specific volume at the nozzle exist is 0.498 m^3/kg, find the exit area of the nozzle.

Ans. (a) 692.5 m/s, (b) 32.08 kg/s, (c) 0.023 m^2.

5.4 In an oil cooler, oil flows steadily through a bundle of metal tubes submerged in a steady stream of cooling water. Under steady flow conditions, the oil enters at 90°C and leaves at 30°C, while the water enters at 25°C and leaves at 70°C. The enthalpy of oil at t°C is given by

$$h = 1.68\ t + 10.5 \times 10^{-4}\ t^2 \text{ kJ/kg}$$

What is the cooling water flow required for cooling 2.78 kg/s of oil?

5.5 A thermoelectric generator consists of a series of semiconductor elements (Fig. P. 5.5), heated on one side and cooled on the other. Electric current flow is pro-

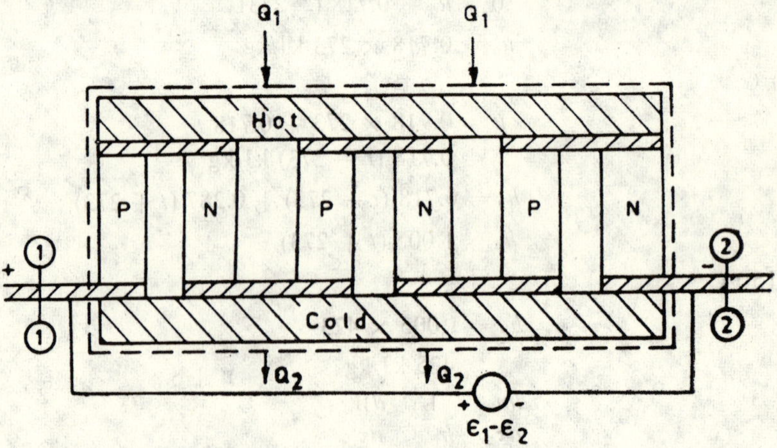

Fig. P. 5·5

duced as a result of energy transfer as heat. In a particular experiment the current was measured to be 0.5 amp and the electrostatic potential at (1) was 0.8 volt above that at (2). Energy transfer as heat to the hot side of the generator was taking place at a rate of 5.5 watts. Determine the rate of energy transfer as heat from the cold side and the energy conversion efficiency. Ans. $Q_2 = 5.1$ watts, $\eta = 0.073$.

5.6 A turbocompressor delivers 2.33 m³/s at 0.276 MPa, 43°C which is heated at this pressure to 430°C and finally expanded in a turbine which delivers 1860 kW. During the expansion, there is a heat transfer of 0.09 MJ/s to the surroundings. Calculate the turbine exhaust temperature if changes in kinetic and potential energy are negligible. Ans. 157°C.

5.7 A reciprocating air compressor takes in 2 m³/min at 0.11 MPa, 20°C which it delivers at 1.5 MPa, 111°C to an aftercooler where the air is cooled at constant pressure to 25°C. The power absorbed by the compressor is 4.15 kW. Determine the heat transfer in (a) the compressor, and (b) the cooler. State your assumptions.

Ans. —0.17 kJ/sec, —3.76 kJ/sec.

5.8 In a water cooling tower air enters at a height of 1 m above the ground level and leaves at a height of 7 m. The inlet and outlet velocities are 20 m/s and 30 m/s respectively. Water enters at a height of 8 m and leaves at a height of 0.8 m. The velocity of water at entry and exit are 3 m/s and 1 m/s respectively. Water temperatures are 80°C and 50°C at the entry and exit respectively. Air temperatures are 30°C and 70°C at the entry and exit respectively. The cooling tower is well insulated and a fan of 2.25 kW drives the air through the cooler. Find the amount of air per sec repuired for 1 kg/s of water flow. c_p of air and water are 1.005 and 4.187 kJ/kg-K respectively.

5.9 Air at 101.325 kPa, 20°C is taken into a gas turbine power plant at a velocity of 140 m/s through an opening of 0.15 m² cross-sectional area. The air is compressed, heated, expanded through a turbine, and exhausted at 0.18 MPa, 150°C through an opening of 0.10 m² cross-sectional area. The power output is 375 kW. Calculate the net amount of heat added to the air in kJ/kg. Assume that air obeys the law $pv = 0.287\ (t + 273)$, where p is the pressure in kPa, v is the specific volume in m³/kg, and t is the temperature in °C. Take $c_p = 1.005$ kJ/kg-K.

Ans. 3825.6 kW.

5.10 A gas flows steadily through a rotary compressor. The gas enters the compressor at a temperature of 16°C, a pressure of 100 kPa, and an enthalpy of 391.2 kJ/kg. The gas leaves the compressor at a temperature of 245°C, a pressure of 0.6 MPa, and an enthalpy of 534.5 kJ/kg. There is no heat transfer to or from the gas as it flows through the compressor. (a) Evaluate the external work done per unit mass of gas assuming the gas velocities at entry and exit to be negligible. (b) Evaluate the external work done per unit mass of gas when the gas velocity at entry is 80 m/s and that at exit is 160 m/s.

Ans. 143.3 kJ/kg, 152.9 kJ/kg.

5.11 The steam supply to an engine comprises two streams which mix before entering the engine. One stream is supplied at the rate of 0.01 kg with an enthalpy of 2952 kJ/kg and a velocity of 20 m/s. The other stream is supplied at the rate of 0.1 kg/s with an enthalpy of 2569 kJ/kg and a velocity of 120 m/s. At the exit from the engine the fluid leaves as two streams, one of water at the rate of 0.001 kg/s with an enthalpy of 420 kJ/kg and the other of steam; the fluid velocities at the exit are negligible. The engine develops a shaft power of 25 kW. The heat transfer is negligible. Evaluate the enthalpy of the second exit stream.

Ans. 2462 kJ/kg.

5.12 The stream of air and gasoline vapour, in the ratio of 14:1 by mass, enters a gasoline engine at a temperature of 30°C and leaves as combustion products at a temperature of 790°C. The engine has a specific fuel consumption of 0.3 kg/kWhr.

The net heat transfer rate from the fuel-air stream to the jacket cooling water and to the surroundings is 35 kW. The shaft power delivered by the engine is 26 kW. Compute the increase in the specific enthalpy of the fuel-air stream, assuming the changes in kinetic energy and in elevation to be negligible.

Ans. −1877 kJ/kg mixture.

5.13 An air turbine forms part of an aircraft refrigerating plant. Air at a pressure of 295 kPa and a temperature of 58°C flows steadily into the turbine with a velocity of 45 m/s. The air leaves the turbine at a pressure of 115 kPa, a temperature of 2°C, and a velocity of 150 m/s. The shaft work delivered by the turbine is 54 kJ/kg of air. Neglecting changes in elevation, determine the magnitude and sign of the heat transfer per unit mass of air flowing. For air, take $c_p = 1{,}005$ kJ/kgK and the enthalpy $h = c_p\ t$.

Ans. +7.96 kJ/kg.

5.14 In a turbomachine handling an incompressible fluid with a density of 1000 kg/m³, the conditions of the fluid at the rotor entry and exit are as given below

	Inlet	Exit
Pressure	1.15 MPa	0.05 MPa
Velocity	30 m/s	15.5 m/s
Height above datum	10 m	2 m

If the volume flow rate of the fluid is 40 m³/s, estimate the net energy transfer from the fluid as work.

Ans. 54.3 MW.

5.15 A room for four persons has two fans, each consuming 0.18 kW power, and three 100 W lamps. Ventilation air at the rate of 80 kg/hr enters with an enthalpy of 84 kJ/kg and leaves with an enthalpy of 59 kJ/kg. If each person puts out heat at the rate of 630 kJ/hr, determine the rate at which heat is to be removed by a room cooler, so that a steady state is maintained in the room.

Ans. 1.92 kW.

5.16 Steam flowing in a pipeline is at a steady state represented by p_p, t_p, u_p, v_p, h_p and V_p. A small amount of the total flow is led through a small tube to an evacuated chamber which is allowed to fill slowly until the pressure is equal to the pipeline pressure. If there is no heat transfer, derive an expression for the final specific internal energy in the chamber, in terms of the properties in the pipeline.

5.17 The internal energy of air is given, at ordinary temperatures, by

$$u = u_0 + 0.718\ t$$

where u is in kJ/kg, u_0 is any arbitrary value of u at 0°C, kJ/kg, and t is temperature in °C.

Also for air, $pv = 0.287\ (t + 273)$

where p is in kPa and v is in m³/kg.

(a) An evacuated bottle is fitted with a valve through which air from the atmosphere, at 760 mm Hg and 25°C, is allowed to flow slowly to fill the bottle. If no heat is transferred to or from the air in the bottle, what will its temperature be when the pressure in the bottle reaches 760 mm Hg?

(b) If the bottle initially contains 0.03 m³ of air at 400 mm Hg and 25°C, what will the temperature be when the pressure in the bottle reaches 760 mm Hg?

6

Second Law of Thermodynamics

6.1 QUALITATIVE DIFFERENCE BETWEEN HEAT AND WORK

The first law of thermodynamics states that a certain energy balance will hold when a system undergoes a change of state or a thermodynamic process. But it does not give any information on whether that change of state or the process is at all feasible or not. The first law cannot indicate whether a metallic bar of uniform temperature can spontaneously become warmer at one end and cooler at the other. All that the law can state is that if this process did occur, the energy gained by one end would be exactly equal to that lost by the other. *It is the second law of thermodynamics which provides the criterion as to the probability of various processes.*

Spontaneous processes in nature occur only in one direction. Heat always flows from a body at a higher temperature to a body at a lower temperature, water always flows downward, time always flows in the forward direction. The reverse of these never happens spontaneously. The spontaneity of the process is due to a finite driving potential, sometimes called the 'force' or the 'cause', and what happens is called the 'flux', the 'current' or the 'effect'. The typical forces like temperature gradient, concentration gradient, and electric potential gradient, have their respective conjugate fluxes of heat transfer, mass transfer, and flow of electric current. These transfer processes can never spontaneously occur from a lower to a higher potential. This directional law puts a limitation on energy transformation other than that imposed by the first law.

Joule's experiments (Article 4.1) amply demonstrate that energy, when supplied to a system in the form of work, can be completely converted into heat (work transfer → internal energy increase → heat transfer). But the complete conversion of heat into work in a cycle is not possible. *So heat and work are not completely interchangeable forms of energy.*

When work is converted into heat, we always have

$$W \underset{\rightarrow}{=} Q$$

but when heat is converted into work in a complete closed cycle process

$$Q \underset{\rightarrow}{\geq} W$$

The arrow indicates the direction of energy transformation. This is illustrated in Fig. 6.1. As shown in Fig. 6.1(a), a system is taken from state 1 to state 2 by work transfer W_{1-2}, and then by heat transfer Q_{2-1} the system is brought back from state 2 to state 1 to complete a cycle. It is always found that $W_{1-2} = Q_{2-1}$. But if the system is taken from state 1 to state 2 by heat transfer Q_{1-2}, as shown in Fig. 6.1(b), and then by work

Fig. 6.1 Qualitative distinction between heat and work

transfer W_{2-1} the system is brought back from state 2 to state 1 to complete a cycle, it is always found that $Q_{2-1} > W_{2-1}$. Hence, heat cannot be converted completely and continuously into work. This underlies the work of Sadi Carnot, a French military engineer, who first studied this aspect of energy transformation (1824). Work is said to be a *high grade energy* and heat a *low grade energy. The complete conversion of low grade energy into high grade energy in a cycle is impossible.*

6.2 CYCLIC HEAT ENGINE

For engineering purposes, the second law is best expressed in terms of the conditions which govern the production of work by a thermodynamic system operating in a cycle.

A heat engine cycle is a thermodynamic cycle in which there is a net heat transfer *to* the system and a net work transfer *from* the system. The system which executes a heat engine cycle is called a *heat engine*.

A heat engine may be in the form of a mass of gas confined in a cylinder and piston machine (Fig. 6.2a) or a mass of water moving in a steady flow through a steam power plant (Fig. 6.2b).

In the cyclic heat engine, as represented in Fig. 6.2(a), heat Q_1 is transferred to the system, work W_E is done by the system, work W_c is done upon the system, and then heat Q_2 is rejected from the system. The system is brought back to the initial state through all these four successive processes which constitute a heat engine cycle. In Fig. 6.2(b) heat Q_1 is transferred from the furnace to the water in the boiler to form steam which then works on the turbine rotor to produce work W_T, then the steam is condensed to water in the condenser in which an amount Q_2 is rejected from the system, and finally work W_p is done on the system (water) to pump it to the boiler. The system repeats the cycle.

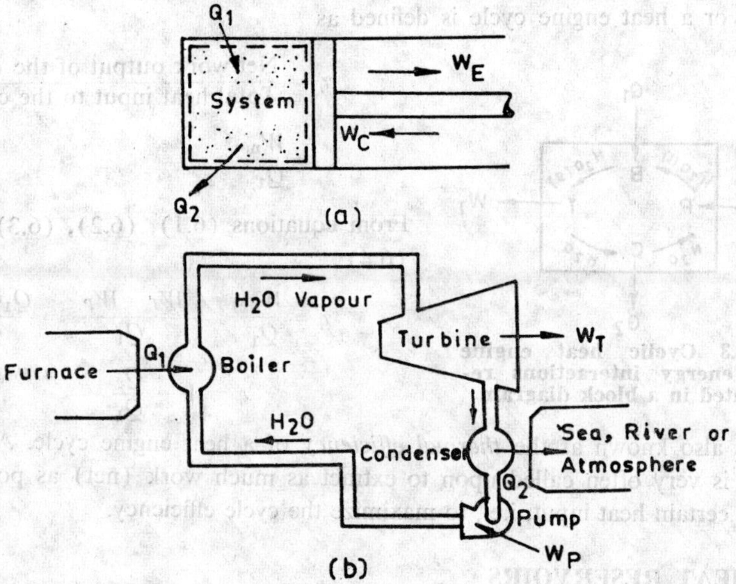

Fig. 6.2 Cyclic heat engine
(a) Heat engine cycle performed by a closed system undergoing four successive energy interactions with the surroundings
(b) Heat engine cycle performed by a steady flow system interacting with the surroundings as shown

The net heat transfer in a cycle to either of the heat engines

$$Q_{\text{net}} = Q_1 - Q_2 \tag{6.1}$$

and the net work transfer in a cycle

$$W_{\text{net}} = W_T - W_P \tag{6.2}$$

(or $\quad W_{\text{net}} = W_E - W_C$)

By the first law of thermodynamics, we have

$$\sum_{\text{cycle}} Q = \sum_{\text{cycle}} W$$

$$\therefore \quad Q_{\text{net}} = W_{\text{net}}$$

or $$Q_1 - Q_2 = W_T - W_P \tag{6.3}$$

Figure 6.3 represents a cyclic heat engine in the form of a block diagram indicating the various energy interactions during a cycle. Boiler (*B*), turbine (*T*), condenser (*C*), and pump (*P*), all four together constitute a heat engine. A heat engine is here a certain quantity of water undergoing the energy interactions, as shown, in cyclic operations to produce net work from a certain heat input.

The function of a heat engine cycle is to produce work continuously at the expense of heat input to the system. So the net work W_{net} and heat input Q_1 referred to the cycle are of primary interest. The efficiency of a heat engine or a heat engine cycle is defined as

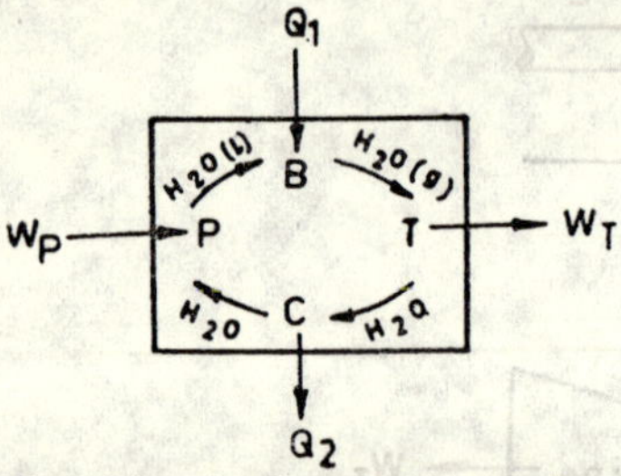

Fig. 6.3 Cyclic heat engine with energy interactions represented in a block diagram

$$\eta = \frac{\text{Net work output of the cycle}}{\text{Total heat input to the cycle}}$$

$$= \frac{W_{net}}{Q_1} \quad (6.4)$$

From equations (6.1), (6.2), (6.3), and (6.4)

$$\eta = \frac{W_{net}}{Q_1} = \frac{W_T - W_P}{Q_1} = \frac{Q_1 - Q}{Q_1}$$

$$\eta = 1 - \frac{Q_2}{Q_1} \quad (6.5)$$

This is also known as the *thermal efficiency* of a heat engine cycle. A heat engine is very often called upon to extract as much work (net) as possible from a certain heat input, i.e., to maximize the cycle efficiency.

6.3 HEAT RESERVOIRS

A heat reservoir is defined as a body of infinite heat capacity, which is capable of absorbing or rejecting an unlimited quantity of heat without suffering any change in any of its thermodynamic properties.

The heat reservoir from which heat Q_1 is transferred to the system operating in a heat engine cycle is called the *source*. The heat reservoir to which heat Q_2 is rejected from the system during a cycle is called the *sink*. A typical source is a constant temperature furnace where fuel is continuously burnt, and a typical sink is a river or sea, or the atmosphere itself.

Figure 6.4 shows a cyclic heat engine exchanging heat with a source and a sink and producing W_{net} in a cycle.

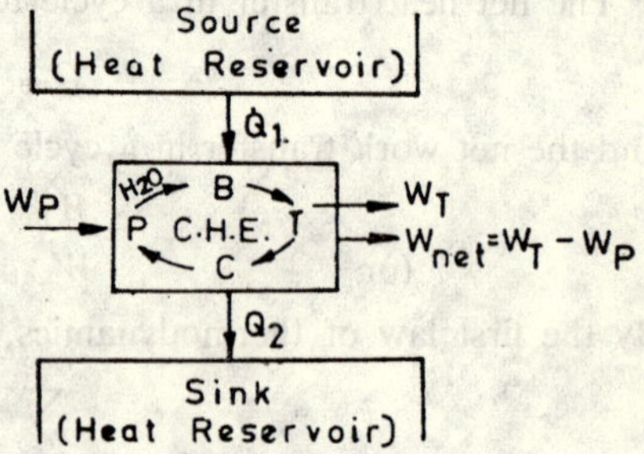

Fig. 6.4 Cyclic heat engine with source and sink

6.4 KELVIN-PLANCK STATEMENT OF SECOND LAW

The efficiency of a heat engine is given by

$$\eta = \frac{W_{net}}{Q_1} = 1 - \frac{Q_2}{Q_1}$$

Experience shows that $W_{net} < Q_1$, since heat Q_1 transferred to a system cannot be completely converted to work in a cycle (Article 6.1). Therefore, η is less than unity. A heat engine can never be 100% efficient. Therefore, $Q_2 > 0$, i.e., there has always to be a heat rejection. To produce net work in a thermodynamic cycle, a heat engine has to exchange heat with two reservoirs, the source and the sink.

The *Kelvin-Planck statement* of the second law states : *It is impossible for a heat engine to produce net work in a complete cycle if it exchanges heat only with bodies at a single fixed temperature.*

If $Q_2 = 0$ (i.e., $W_{net} = Q_1$, or $\eta = 1.00$), the heat engine will produce net work in a complete cycle by exchanging heat with only one reservoir, thus violating the Kelvin-Planck statement (Fig. 6.5). Such a heat engine is called a *perpetual motion machine of the second kind*, abbreviated to PMM2. A PMM2 is impossible.

A heat engine has, therefore, to exchange heat with two heat reservoirs at two different temperatures to produce net work in a complete cycle (Fig. 6.6). So long as there is a difference in temperature, motive power (i.e., work) can be produced. If the bodies with which the heat engine exchanges heat are of finite heat capacities, work will be produced by the heat engine till the temperatures of the two bodies are equalized.

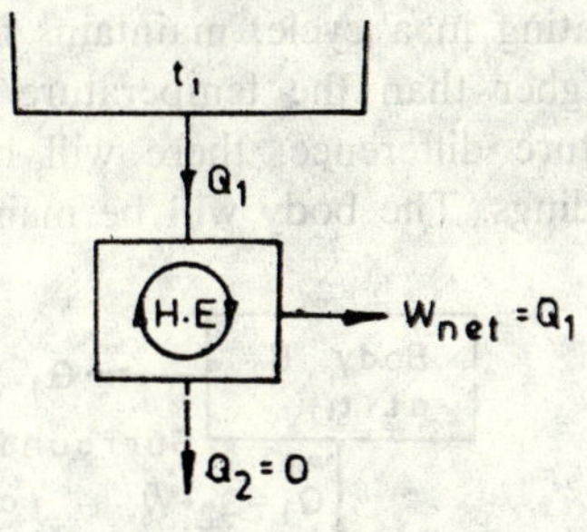

Fig. 6.5 A PMM2

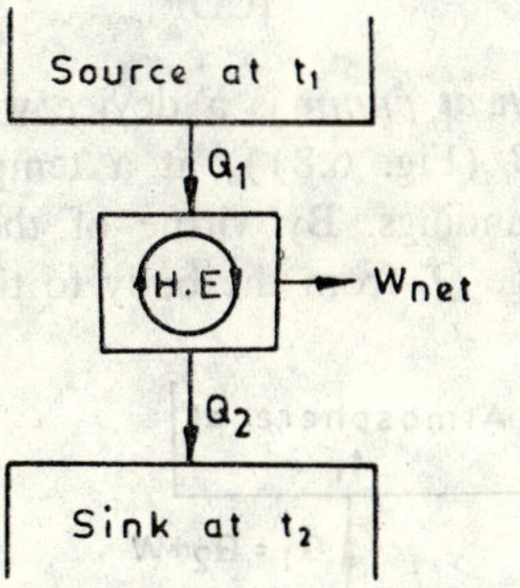

Fig. 6.6 Heat engine producing net work in a cycle by exchanging heat at two different temperatures

6.5 CLAUSIUS' STATEMENT OF THE SECOND LAW

Heat always flows from a body at a higher temperature to a body at a lower temperature. The reverse process never occurs spontaneously.

Clausius' statement of the second law gives : *It is impossible to construct a device which, operating in a cycle, will produce no effect other than the transfer of heat from a cooler to a hotter body.*

Heat cannot flow of itself from a body at a lower temperature to a body at a higher temperature. Some work must be expended to achieve this.

6.6 REFRIGERATOR AND HEAT PUMP

A *refrigerator* is a device which, operating in a cycle, maintains a body at a temperature lower than the temperature of the surroundings. Let body A (Fig. 6.7) be maintained at t_2, which is lower than the ambient temperature t_1. Even though A is insulated, there will always be heat leakage Q_2 into the body from the surroundings by virtue of the temperature difference. In order to maintain body A at the constant temperature t_2, heat has to be removed from the body at the same rate at which heat is leaking into the body. This heat is discharged into the atmosphere which is at t_1, with the expenditure of work W in a device operating in a cycle. The device is then called a refrigerator. In a refrigerator cycle, attention is concentrated on the body the Q_2 and W are of primary interest. Just like efficiency in a heat engine cycle, there is a performance parameter in a refrigerator cycle, called the *coefficient of performance,* abbreviated to COP, which is defined as

$$\text{COP} = \frac{\text{Desired effect}}{\text{Work input}} = \frac{Q_2}{W}$$

$$\therefore \qquad [\text{COP}]_{\text{ref.}} = \frac{Q_2}{Q_1 - Q_2} \qquad (6.6)$$

A *heat pump* is a device which, operating in a cycle, maintains a body, say B (Fig. 6.8)), at a temperature higher than the temperature of the surroundings. By virtue of the temperature difference, there will be heat leakage Q_1 from the body to the surroundings. The body will be maintained

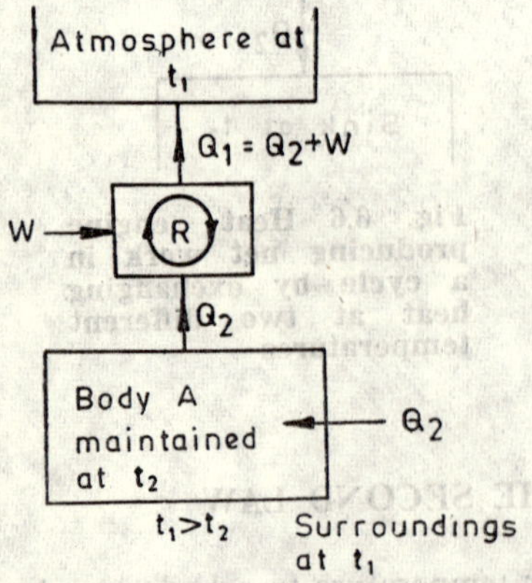

Fig. 6.7 Refrigerator

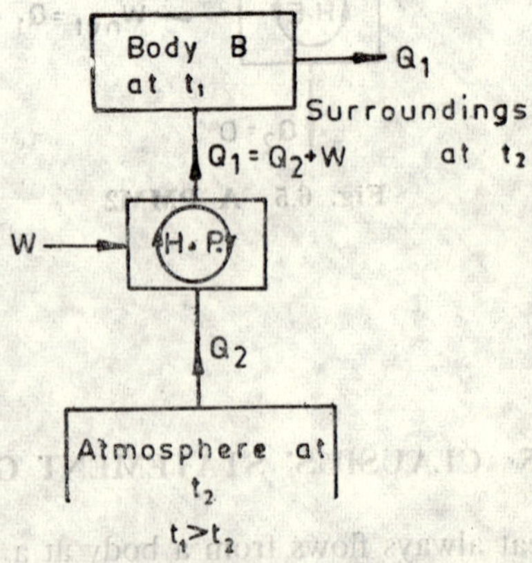

Fig. 6.8 Heat pump

at the constant temperature t_1, if heat is discharged into the body at the same rate at which heat leaks out of the body. The heat is extracted from the low temperature reservoir, which is nothing but the atmosphere, and discharged into the high temperature body B, with the expenditure of work W in a cyclic device called a heat pump. In a heat pump, attention is confined to the high temperature body B. Here Q_1 and W are of primary interest, and the COP is defined as

$$\text{COP} = \frac{Q_1}{W}$$

$$\therefore \qquad [\text{COP}]_{H \cdot P} = \frac{Q_1}{Q_1 - Q_2} \tag{6.7}$$

From equations (6.6) and (6.7), it is found that

$$[\text{COP}]_{H.P.} = [\text{COP}]_{\text{ref.}} + 1 \tag{6.8}$$

The COP of a heat pump is greater than the COP of a refrigerator by unity.

For heat to flow from a cooler to a hotter body, W cannot be zero, and hence, the COP (both for refrigerator and heat pump) cannot be infinity. Therefore, $W > 0$, and $\text{COP} < \infty$.

6.7 EQUIVALENCE OF KELVIN-PLANCK AND CLAUSIUS STATEMENTS

At first sight, Kelvin-Planck's and Clausius' statements may appear to be unconnected, but it can easily be shown that they are virtually two parallel statements of the second law and are equivalent in all respects.

The equivalence of the two statements will be proved if it can be shown that the violation of one statement implies the violation of the second, and vice versa.

(a) Let us first consider a cyclic heat pump P which transfers heat from a low temperature reservoir (t_2) to a high temperature reservoir (t_1) with no other effect, i.e., with no expenditure of work, violating Clausius statement (Fig. 6.9).

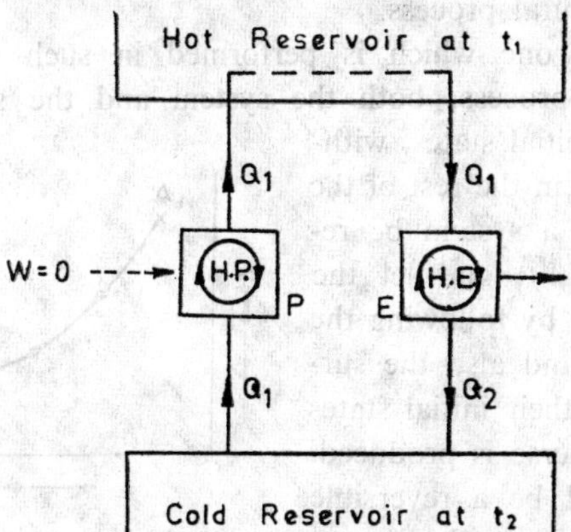

Fig. 6.9 Violation of the Clausius' statement

Let us assume a cyclic heat engine E operating between the same heat reservoirs, producing W_{net} in one cycle. The rate of working of the heat engine is such that it draws an amount of heat Q_1 from the hot reservoir equal to that discharged by the heat pump. Then the hot reservoir may be eliminated, and the heat Q_1 discharged by the heat pump is fed to the heat engine. So we see that the heat pump P and the heat engine E acting together constitute a heat engine operating in cycles and producing net work while exchanging heat only with one body at a single fixed temperature. This violates the Kelvin-Planck statement.

(b) Let us now consider a perpetual motion machine of the second kind (E) which produces net work in a cycle by exchanging heat with only one heat reservoir (at t_1) and thus violates the Kelvin-Planck statement (Fig. 6.10).

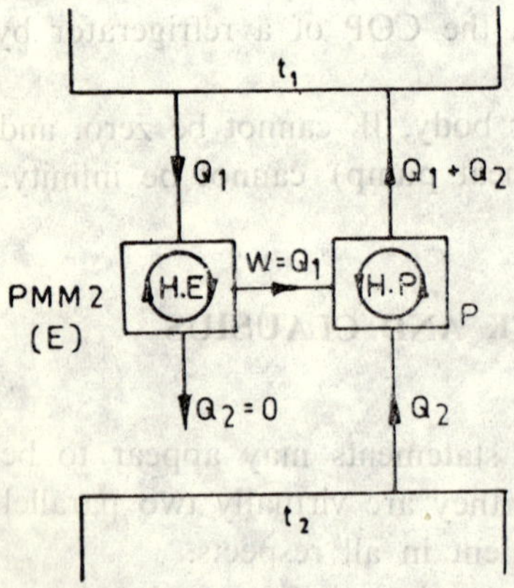

Fig. 6.10 Violation of the Kelvin-Planck statement

Let us assume a cyclic heat pump (P) extracting heat Q_2 from a low temperature reservoir at t_2 and discharging heat to the high temperature reservoir at t_1 with the expenditure of work W equal to what the PMM2 delivers in a complete cycle. So E and P together constitute a heat pump working in cycles and producing the sole effect of transferring heat from a lower to a higher temperature body, thus violating the Clausius statement.

6.8 REVERSIBILITY AND IRREVERSIBILITY

The second law of thermodynamics enables us to divide all processes into two classes:

(a) Reversible or ideal process.

(b) Irreversible or natural process.

A *reversible* process is one which is performed in such a way that at the conclusion of the process, both the system and the surroundings may be restored to their initial states, without producing any changes in the rest of the universe. Let the state of a system be represented by A (Fig. 6.11), and let the system be taken to state B by following the path A-B. If the system and also the surroundings are restored to their initial states and no change in the universe is produced, then the process A-B will be a reversible process. In the reverse process, the system has to be taken from state B to A by fol-

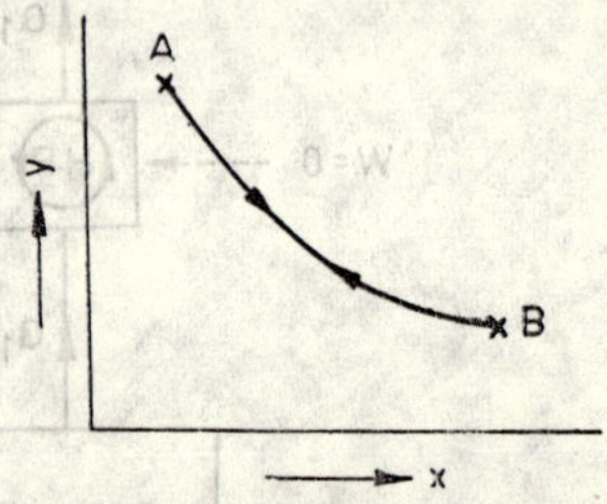

Fig. 6.11 Reversible process

lowing the same path B-A. A reversible process should not leave any trace to show that the process had ever occurred.

A reversible process is carried out infinitely slowly with an infinitesimal gradient, so that every state passed through by the system is an equilibrium state. So a reversible process coincides with a quasi-static process.

Any natural process carried out with a finite gradient is an irreversible process. A reversible process, which consists of *a succession of equilibrium states,* is an idealized hypothetical process, approached only as a limit. It is said to be an *asymptote to reality.* All spontaneous processes are irreversible.

6.9 CAUSES OF IRREVERSIBILITY

The irreversibility of a process may be due to either one or both of the following:

(a) Lack of equilibrium during the process.
(b) Involvement of dissipative effects.

6.9.1 Irreversibility due to lack of equilibrium

The lack of equilibrium (mechanical, thermal or chemical) between the system and its surroundings, or between two systems, or two parts of the same system, causes a spontaneous change which is irreversible. The following are specific examples in this regard:

(a) Heat transfer through a finite temperature difference

A heat transfer process approaches reversibility as the temperature difference between two bodies approaches zero. We define a reversible heat transfer process as one in which heat is transferred through an infinitesimal temperature difference. So to transfer a finite amount of heat through an infinitesimal temperature difference would require an infinite amount of time, or infinite area. All actual heat transfer processes are through a finite temperature difference and are, therefore, irreversible, and the greater the temperature difference, the greater is the irreversibility.

We can demonstrate by the second law that the heat transfer through a finite temperature difference is irreversible. Let us assume that a source at t_A and a sink at t_B ($t_A > t_B$) are available, and let Q_{A-B} be the amount of heat flowing from A to B (Fig. 6.12). Let us assume an engine operating between A and B, taking heat Q_1 from A and discharging heat Q_2 to B. Let the heat transfer process be reversed, and Q_{B-A} be the heat flowing from B to A, and let the rate of working of the engine be such that

$$Q_2 = Q_{B-A}$$

(Fig. 6.13). Then the sink B may be eliminated. The net result is that E produces net work W in a cycle by exchanging heat only with A, thus violating the Kelvin-Planck statement. So the heat transfer process Q_{A-B} is irreversible, and Q_{B-A} is not possible.

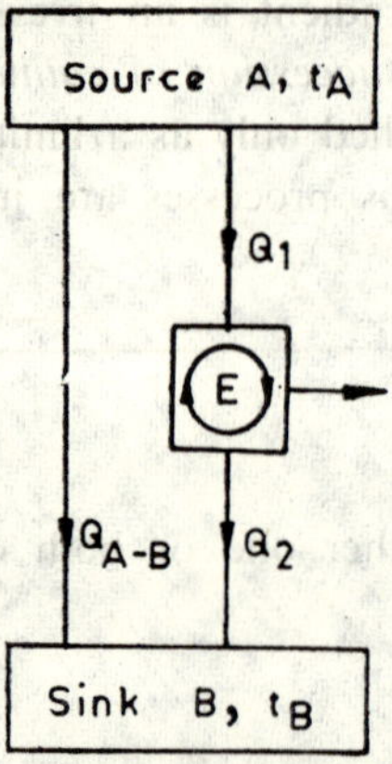

Fig. 6.12 Heat transfer through a finite temperature difference

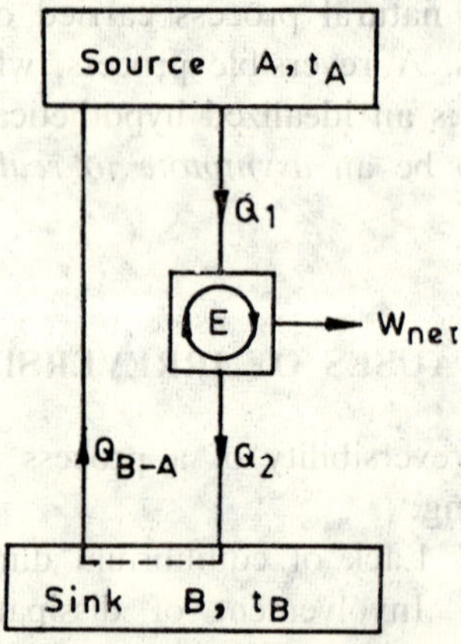

Fig. 6.13 Heat transfer through a finite temperature difference is irreversible

(b) Lack of pressure equilibrium within the interior of the system or between the system and the surroundings

When there exists a difference in pressure between the system and the surroundings, or within the system itself, then both the system and its surroundings or the system alone, will undergo a change of state which will cease only when mechanical equilibrium is established. The reverse of this process is not possible spontaneously without producing any other effect. That the reverse process will violate the second law becomes obvious from the following illustration.

(c) Free expansion

Let us consider an insulated container (Fig. 6.14) which is divided into two compartments A and B by a thin diaphragm. Compartment A contains a mass of gas, while compartment B is completely evacuated. If the diaphragm is punctured, the gas in A will expand into B until the pressures in A and B become equal. This is known as free or unrestrained expansion. We can demonstrate by the second law, that the process of free expansion is irreversible.

To prove this, let us assume that free expansion is reversible, and that the gas in B returns into A with an increase in pressure, and B becomes evacuated as before (Fig. 6.15). There is no other effect. Let us install an engine (a machine, not a cyclic heat engine) between A and B, and permit the gas to expand through the engine from A to B. The engine develops a work output W at the expense of the internal energy of the gas. The internal energy of the gas (system) in B can be restored to its initial value by heat transfer Q $(= W)$ from a source. Now, by the use of the reversed free expansion,

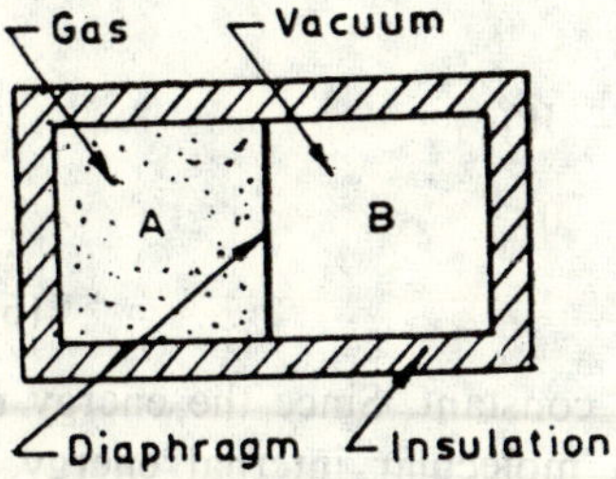

Fig. 6.14 Free expansion

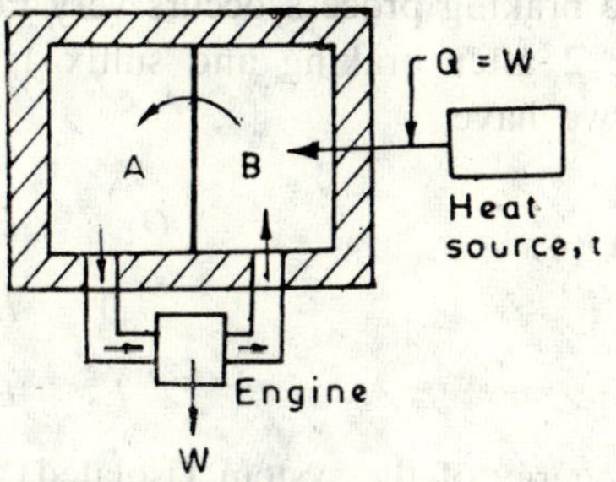

Fig. 6.15 Second law demonstrates that free expansion is irreversible

the system can be restored to the initial state of high pressure in A and vacuum in B. The net result is a cycle, in which we observe that net work output W is accomplished by exchanging heat with a single reservoir. This violates the Kelvin-Planck statement. Hence, free expansion is irreversible.

6.9.2 Irreversibility due to dissipative effects

The irreversibility of a process may be due to the *dissipative effects* in which work is done without producing an equivalent increase in the kinetic or potential energy of any system. The transformation of work into molecular internal energy either of the system or of the reservoir takes place through the agency of such phenomena as friction, viscosity, inelasticity, electrical resistance, and magnetic hysteresis. These effects are known as dissipative effects, and work is said to be dissipated.

(a) Friction

Friction is always present in moving devices. Friction may be reduced by suitable lubrication, but it can never be completely eliminated. If this were possible, a movable device could be kept in continual motion without violating either of the two laws of thermodynamics. The continual motion of a movable device in the complete absence of friction is known as *perpetual motion of the third kind.*

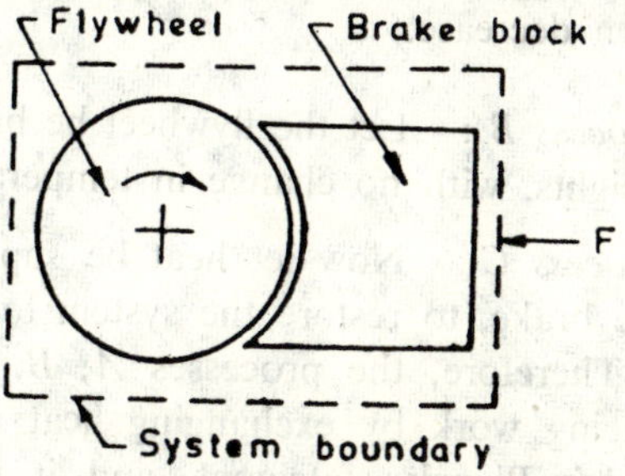

Fig. 6.16 Irreversibility due to dissipative effect like friction

That friction makes a process irreversible can be demonstrated by the second law. Let us consider a system consisting of a flywheel and a brake block (Fig. 6.16). The flywheel was rotating with a certain rpm, and it was brought to rest by applying the friction brake. The distance moved by the brake block is very small, so work transfer is very nearly equal to zero. If the braking process occurs very rapidly, there is little heat transfer. Using suffix 2 after braking and suffix 1 before braking, and applying the first law, we have

$$Q_{1-2} = E_2 - E_1 + W_{1-2}$$

$$0 = E_2 - E_1 + 0$$

$$\therefore \quad E_2 = E_1 \tag{6.9}$$

The energy of the system (isolated) remains constant. Since the energy may exist in the forms of kinetic, potential, and molecular internal energy, we have

$$U_2 + \frac{\mathbf{V}_2^2}{2} + Z_2 g = U_1 + \frac{\mathbf{V}_1^2}{2} + Z_1 g$$

Since the wheel is brought to rest, $\mathbf{V}_2 = 0$, and there is no change in P.E.

$$U_2 = U_1 + \frac{\mathbf{V}_1^2}{2} \tag{6.10}$$

Therefore, the molecular internal energy of the system (i.e., of the brake and the wheel) increases by the absorption of the K.E. of the wheel. The reverse process, i.e., the conversion of this increase in molecular internal energy into K.E. within the system to cause the wheel to rotate is not possible. To prove it by the second law, let us assume that it is possible, and imagine the following cycle with three processes:

Process A : Initially, the wheel and the brake are at high temperature as a result of the absorption of the K.E. of the wheel. and the flywheel is at rest. Let the flywheel now start rotating at a particular rpm at the expense of the internal energy of the wheel and brake, the temperature of which will then decrease.

Process B : Let the flywheel be brought to rest by using its K.E. in raising weights, with no change in temperature.

Process C : Now let heat be supplied from a source to the flywheel and the brake, to restore the system to its initial state.

Therefore, the processes *A*, *B*, and *C* together constitute a cycle producing work by exchanging heat with a single reservoir. This violates the Kelvin-Planck statement, and it will become a PMM2. So the braking process, i.e., the transformation of K.E. into molecular internal energy, is irreversible.

(b) Paddle-wheel work transfer

Work may be transferred into a system in an insulated container by means of a paddle wheel (Fig. 6.17) which is also known as stirring work. Here work transferred is dissipated adiabatically into an increase in the molecular internal energy of the system. To prove the irreversibility of the process, let us assume that the same amount of work is delivered by the system at the expense of its molecular internal energy, and the temperature of the system goes down (Fig. 6.18). The system is brought back to its initial state by heat transfer from a source. These two processes together constitute a cycle in which there is work output and the system exchanges heat with a single

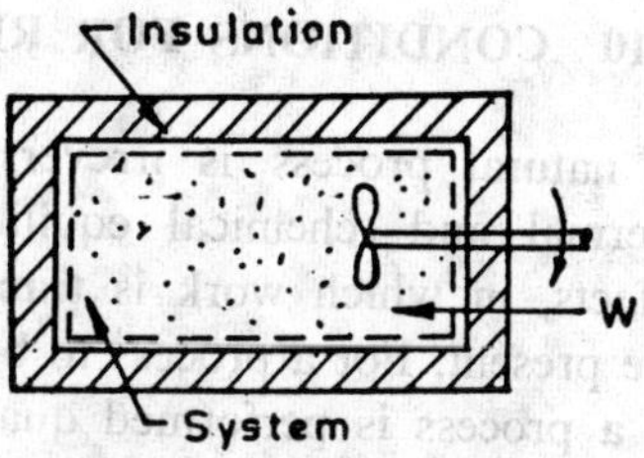

Fig. 6.17 Adiabatic work transfer

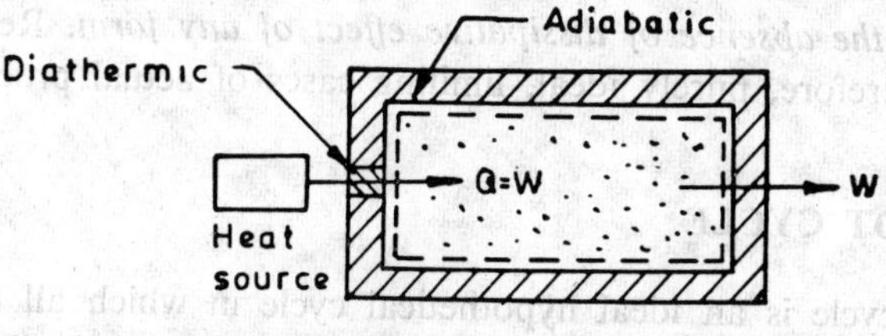

Fig. 6.18 Irreversibility due to dissipation of stirring work into internal energy

reservoir. It becomes a PMM2, and hence the dissipation of stirring work to internal energy is irreversible.

(c) Transfer of electricity through a resistor

The flow of electric current through a wire represents work transfer, because the current can drive a motor which can raise a weight. Taking the wire or the resistor as the system (Fig. 6.19) and writing the first law

$$Q_{1-2} = U_2 - U_1 + W_{1-2}$$

Here both W_{1-2} and Q_{1-2} are negative.

$$\therefore \; W_{1-2} = U_2 - U_1 + Q_{1-2} \quad (6.11)$$

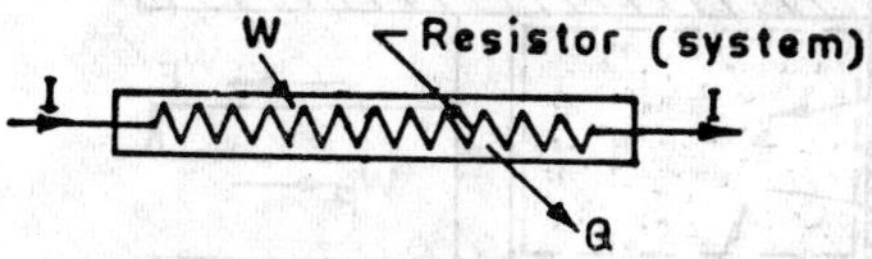

Fig. 6.19 Irreversibility due to dissipation of electrical work into internal energy

A part of the work transfer is stored as an increase in the internal energy of the wire (to give an increase in its temperature), and the remainder leaves the system as heat. At steady state, the internal energy and hence the temperature of the resistor becomes constant with respect to time and

$$W_{1-2} = Q_{1-2} \quad (6.12)$$

The dissipation of electrical work into internal energy or heat is irreversible, which can be proved as in (b).

6.10 CONDITIONS FOR REVERSIBILITY

A natural process is irreversible because the conditions for mechanical, thermal and chemical equilibrium are not satisfied, and the dissipative effects, in which work is transformed into an increase in internal energy, are present. For a process to be reversible, it must not possess these features. If a process is performed quasi-statically, the system passes through states of thermodynamic equilibrium, which may be traversed as well in one direction as in the opposite direction. *If there are no dissipative effects, all the work done by the system during the performance of a process in one direction can be returned to the system during the reverse process.*

A process will be reversible when it is performed in such a way *that the system is at all times infinitesimally near a state of thermodynamic equilibrium and in the absence of dissipative effect of any form.* Reversible processes are, therefore, purely ideal, limiting cases of actual processes.

6.11 CARNOT CYCLE

A reversible cycle is an ideal hypothetical cycle in which all the processes constituting the cycle are reversible. Carnot cycle is a reversible cycle. For a stationary system, as in a piston and cylinder machine, the cycle consists of the following four successive processes (Fig. 6.20):

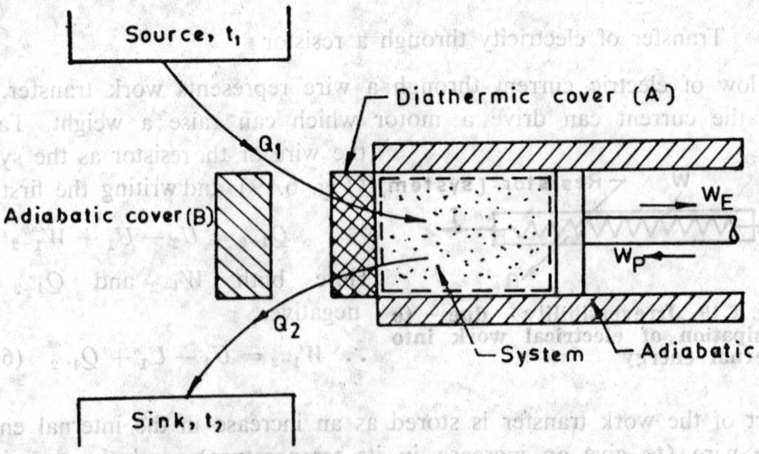

Fig. 6.20 Carnot heat engine—stationary system

(a) *A reversible isothermal process* in which heat Q_1 enters the system at t_1 reversibly from a constant temperature source at t_1 when the cylinder cover is in contact with the diathermic cover A. The internal energy of the system increases.

(b) *A reversible adiabatic process* in which the diathermic cover A is replaced by the adiabatic cover B, and work W_E is done by the system adiabatically and reversibly at the expense of its internal energy, and the temperature of the system decreases from t_1 to t_2.

(c) *A reversible isothermal process* in which B is replaced by A and heat Q_2 leaves the system at t_2 to a constant temperature sink at t_2 reversibly, and the internal energy of the system further decreases.

(d) *A reversible adiabatic process* in which B again replaces A, and work W_p is done upon the system reversibly and adiabatically, and the internal energy of the system increases and the temperature rises from t_2 to t_1

Two reversible isotherms and two reversible adiabatics constitute a Carnot cycle, which is represented in p-v coordinates in Fig. 6.21.

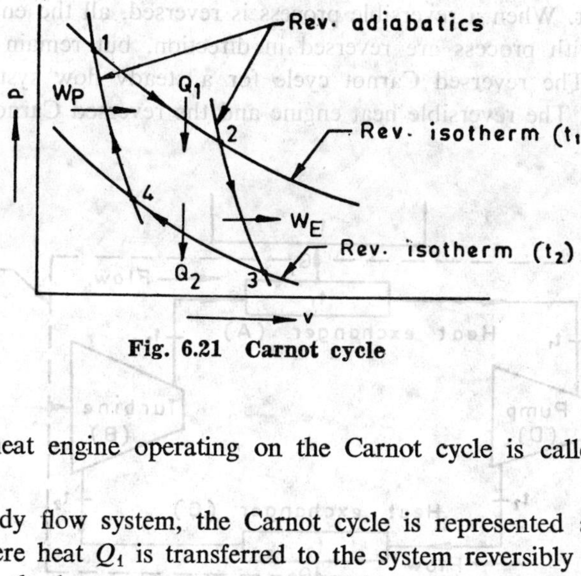

Fig. 6.21 Carnot cycle

A cyclic heat engine operating on the Carnot cycle is called a Carnot heat engine.

For a steady flow system, the Carnot cycle is represented as shown in Fig. 6.22. Here heat Q_1 is transferred to the system reversibly and isothermally at t_1 in the heat exchanger A, work W_T is done by the system reversibly and adiabatically in the turbine (B), then heat Q_2 is transferred from the system reversibly and isothermally at t_2 in the heat exchanger (C), and then work W_p is done upon the system reversibly and adiabatically by the pump (D). To satisfy the conditions for the Carnot cycle, there must not by any friction or heat transfer in the pipelines through which the working fluid flows.

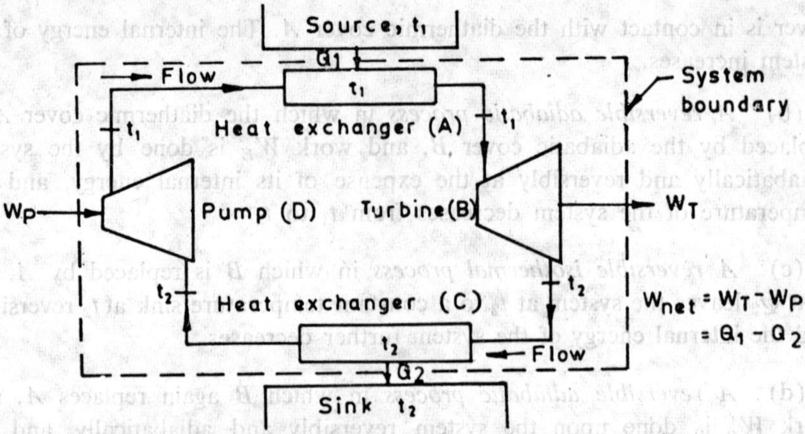

Fig. 6.22 Carnot heat engine—steady flow system

6.12 REVERSED HEAT ENGINE

Since all the processes of the Carnot cycle are reversible, it is possible to imagine that the processes are individually reversed and carried out in reverse order. When a reversible process is reversed, all the energy transfers associated with process are reversed in direction, but remain the same in magnitude. The reversed Carnot cycle for a steady flow system is shown in Fig. 6.23. The reversible heat engine and the reversed Carnot heat engine

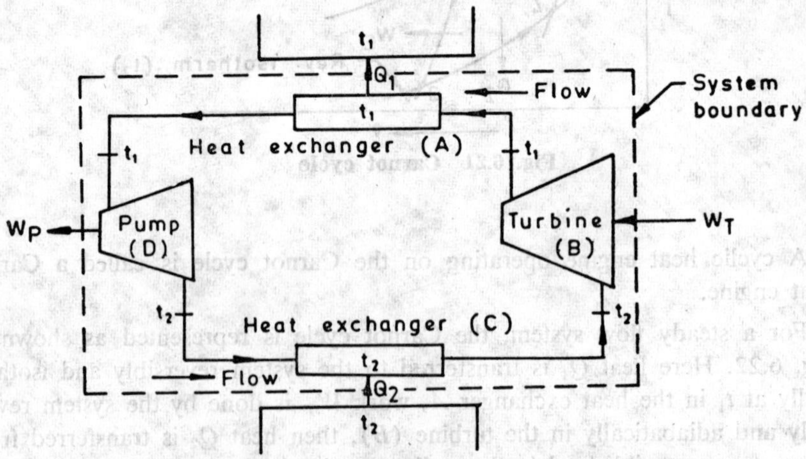

Fig. 6.23 Reversed Carnot heat engine—steady flow process

are represented in block diagrams in Fig. 6.24. If E is a reversible heat engine (Fig. 6.24a), and if it is reversed (Fig. 6.24b), the quantities Q_1, Q_2 and W remain the same in magnitude, and only their directions are reversed. The reversed heat engine **E** takes heat from a low temperature body, discharges heat to a high temperature body, and receives an inward flow of network.

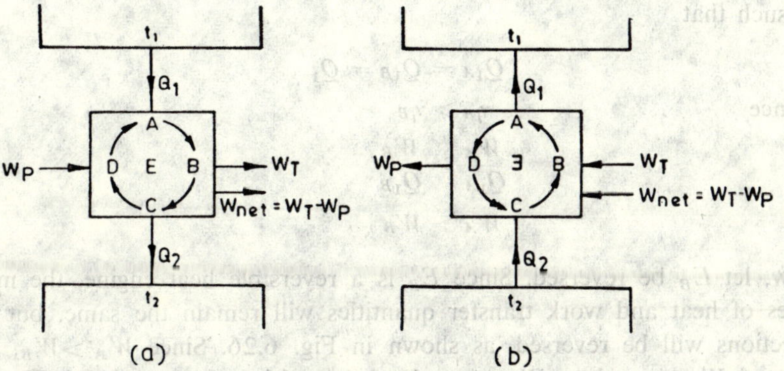

Fig. 6.24 Carnot heat engine and reversed Carnot heat engine shown in block diagrams

The names *heat pump* and *refrigerator* are applied to the reversed heat engine, which have already been discussed earlier.

6.13 CARNOT'S THEOREM

It states that of all heat engines operating between a given constant temperature source and a given constant temperature sink, none has a higher efficiency than a reversible engine.

Let two heat engines E_A and E_B operate between the given source at temperature t_1 and the given sink at temperature t_2 as shown in Fig. 6.25.

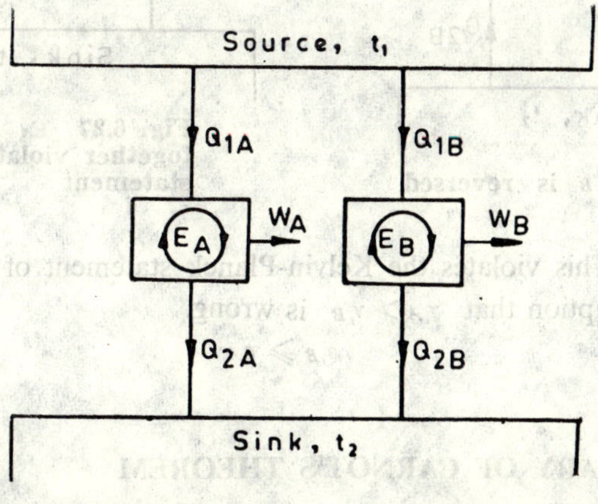

Fig. 6.25 Two cyclic heat engines E_A and E_B operating between the same source and sink, of which E_B is reversible

Let E_A be *any* heat engine, and E_B be *any reversible* heat engine. We have to prove that the efficiency of E_B is more than that of E_A. Let us assume that this is not true and $\eta_A > \eta_B$. Let the rates of working of the engines be such that

$$Q_{1A} = Q_{1B} = Q_1$$

Since
$$\eta_A > \eta_B$$
$$\frac{W_A}{Q_{1A}} > \frac{W_B}{Q_{1B}}$$
$\therefore$
$$W_A > W_B$$

Now, let E_B be reversed. Since E_B is a reversible heat engine, the magnitudes of heat and work transfer quantities will remain the same, but their directions will be reversed, as shown in Fig. 6.26. Since $W_A > W_B$, some part of W_A (equal to W_B) may be fed to drive the reversed heat engine $\exists_B$. Since $Q_{1A} = Q_{1B} = Q_1$ the heat discharged by $\exists_B$ may be supplied to E_A. The source may, therefore, be eliminated (Fig. 6.27). The net result is that E_A and $\exists_B$ together constitute a heat engine which, operating in a cycle, produces net work $W_A - W_B$, while exchanging heat with a single

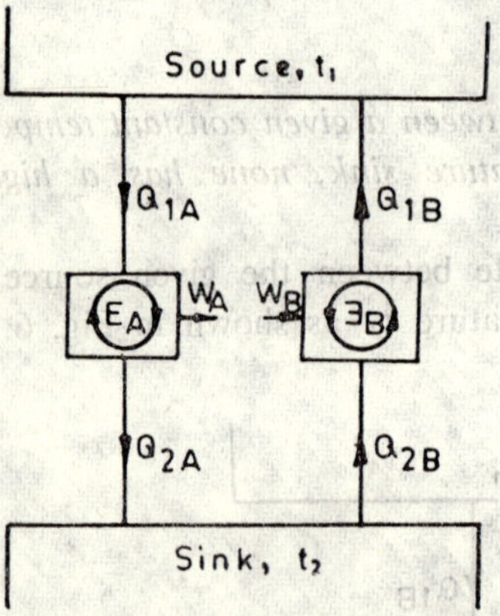

Fig. 6.26 *EB* **is reversed**

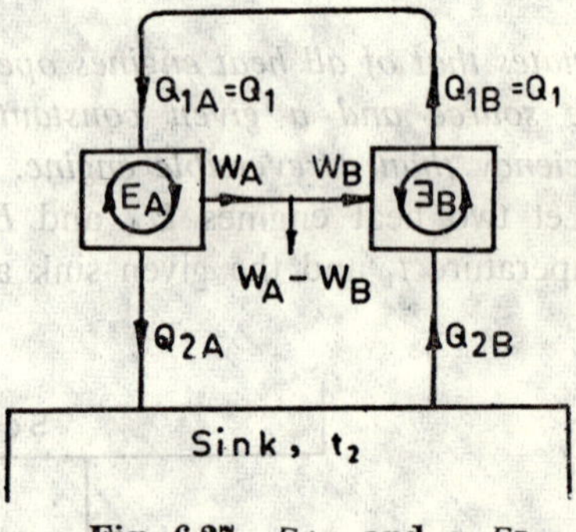

Fig. 6.27 *EA* **and** *EB* **together violate the K-P statement**

reservoir at t_2. This violates the Kelvin-Planck statement of the second law. Hence the assumption that $\eta_A > \eta_B$ is wrong.

Therefore
$$\eta_B \geqslant \eta_A$$

6.14 COROLLARY OF CARNOT'S THEOREM

The efficiency of all reversible heat engines operating between the same temperature levels is the same.

Let both the heat engines E_A and E_B (Fig. 6.25) be reversible. Let us assume $\eta_A > \eta_B$. Similar to the procedure outlined in the preceding article, if E_B is reversed to run, say, as a heat pump using some part of the work output (W_A) of engine E_A, we see that the combined system of heat pump E_B and engine E_A, becomes a PMM2. So η_A cannot be greater than η_B. Similarly, if we assume $\eta_B > \eta_A$ and reverse the engine E_A, we observe that η_B cannot be greater that η_A.

Therefore $$\eta_A = \eta_B$$

Since the efficiencies of all reversible heat engines operating between the same heat reservoirs are the same, *the efficiency of a reversible engine is independent of the nature or amount of the working substance undergoing the cycle.*

6.15 ABSOLUTE THERMODYNAMIC TEMPERATURE SCALE

The efficiency of any heat engine cycle receiving heat Q_1 and rejecting heat Q_2 is given by

$$\eta = \frac{W_{\text{net}}}{Q_1} = \frac{Q_1 - Q_2}{Q_1} = 1 - \frac{Q_2}{Q_1} \tag{6.13}$$

By the second law, it is necessary to have a temperature difference $(t_1 - t_2)$ to obtain work for any cycle. We know that the efficiency of all heat engines operating between the same temperature levels is the same, and it is independent of the working substance. Therefore, for a reversible cycle (Carnot cycle), the efficiency will depend solely upon the temperatures t_1 and t_2, at which heat is transferred, or

$$\eta_{\text{rev.}} = f(t_1, t_2) \tag{6.14}$$

where f signifies some function of the temperatures. From equations (6.13) and (6.14)

$$1 - \frac{Q_2}{Q_1} = f(t_1, t_2)$$

In terms of a new function F

$$\frac{Q_1}{Q_2} = F(t_1, t_2) \tag{6.15}$$

If some functional relationship is assigned between t_1, t_2 and Q_1/Q_2, the equation becomes the definition of a temperature scale.

Let us consider two reversible heat engines, E_1 receiving heat from the source at t_1, and rejecting heat at t_2 to E_2 which, in turn, rejects heat to the sink at t_3 (Fig. 6.28).

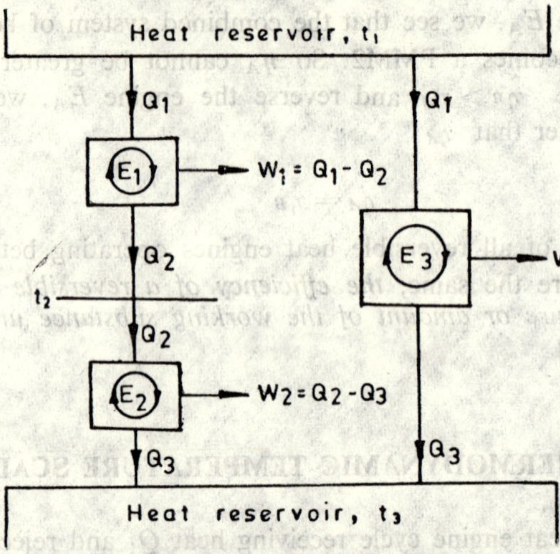

Fig. 6.28 Three Carnot engines

Now
$$\frac{Q_1}{Q_3} = F(t_1, t_2); \quad \frac{Q_2}{Q_3} = F(t_2, t_3)$$

E_1 and E_2 together constitute another heat engine, E_3 operating between t_1 and t_3.

$$\therefore \quad \frac{Q_1}{Q_3} = F(t_1, t_3)$$

Now
$$\frac{Q_1}{Q_2} = \frac{Q_1/Q_3}{Q_2/Q_3}$$

or
$$\frac{Q_1}{Q_2} = F(t_1, t_2) = \frac{F(t_1, t_3)}{F(t_2, t_3)} \tag{6.16}$$

The temperatures t_1, t_2, and t_3 are arbitrarily chosen. The ratio Q_1/Q_2 depends only on t_1 and t_2, and is independent of t_3. So t_3 will drop out from the ratio on the right in equation (6.16). After it has been cancelled, the numerator can be written as $\phi(t_1)$, and the denominator as $\phi(t_2)$, where ϕ is another unknown function. Thus

$$\frac{Q_1}{Q_2} = F(t_1, t_2) = \frac{\phi(t_1)}{\phi(t_2)}$$

Since $\phi(t)$ is an arbitrary function, the simplest possible way to define the *absolute thermodynamic temperature* T is to let $\phi(t) = T$, as proposed by Kelvin. Then, by definition

$$\frac{Q_1}{Q_2} = \frac{T_1}{T_2} \tag{6.17}$$

The absolute thermodynamic temperature scale is also known as the *Kelvin scale.* Two temperatures on the Kelvin scale bear the same relationship to each other as do the heats absorbed and rejected respectively by a Carnot engine operating between two reservoirs at these temperatures. The Kelvin temperature scale is, therefore, independent of the peculiar characteristics of any particular substance.

The heat absorbed Q_1 and the heat rejected Q_2 during the two reversible isothermal processes bounded by two reversible adiabatics in a Carnot engine can be measured. In defining the Kelvin temperature scale also, the triple point of water is taken as the standard reference point. For a Carnot engine operating between reservoirs at temperatures T and T_t, T_t being the triple point of water (Fig. 6.29), arbitrarily assigned the value 273.16°K,

$$\frac{Q}{Q_t} = \frac{T}{T_t}$$

$$\therefore \quad T = 273.16 \frac{Q}{Q_t} \tag{6.18}$$

If this equation is compared with the equations given in Article 2.3, it is seen that *in the Kelvin scale, Q plays the role of thermometric property.*

That the absolute thermodynamic temperature scale has a definite zero point can be shown by imagining a series of reversible engines, extending from a source at T_1 to lower temperatures (Fig. 6.30).

Since

$$\frac{T_1}{T_2} = \frac{Q_1}{Q_2}$$

$$\therefore \quad \frac{T_1 - T_2}{T_2} = \frac{Q_1 - Q_2}{Q_2}$$

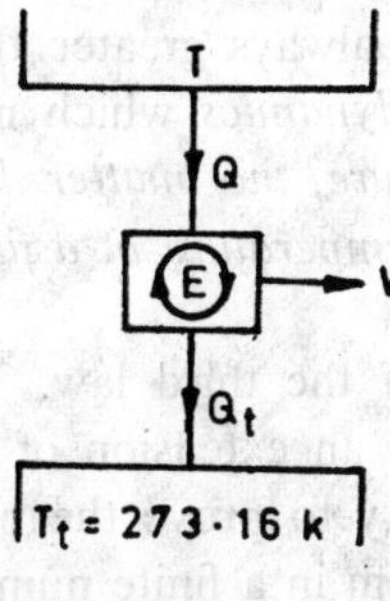

Fig. 6.29 Carnot heat engine with sink at triple point of water

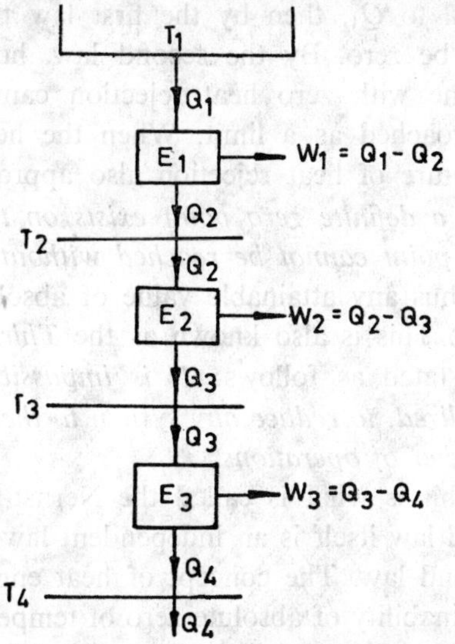

Fig. 6.30 Heat engines operating in series

or
$$T_1 - T_2 = (Q_1 - Q_2)\frac{T_2}{Q_2}$$

Similarly

$$T_2 - T_3 = (Q_2 - Q_3)\frac{T_3}{Q_3}$$

$$= (Q_2 - Q_3)\frac{T_2}{Q_2}$$

$$T_3 - T_4 = (Q_3 - Q_4)\frac{T_2}{Q_2}$$

and so on

If $T_1 - T_2 = T_2 - T_3 = T_3 - T_4 = \ldots\ldots$, assuming equal temperature intervals

$$Q_1 - Q_2 = Q_2 - Q_3 = Q_3 - Q_4 = \ldots$$

or
$$W_1 = W_2 = W_3 = \ldots$$

Conversely, by making the work quantities performed by the engines in series equal ($W_1 = W_2 = W_3 = \ldots$), we will get

$$T_1 - T_2 = T_2 - T_3 = T_3 - T_4 = \ldots$$

at equal temperature intervals. A scale having one hundred equal intervals between the steam point and the ice point could be realized by a series of one hundred Carnot engines operating as in Fig. 6.30. Such a scale would be independent of the working substance.

If enough engines are placed in series to make the total work output equal to Q_1, then by the first law the heat rejected from the last engine will be zero. By the second law, however, the operation of a cyclic heat engine with zero heat rejection cannot be achieved, although it may be approached as a limit. When the heat rejected approaches zero, the temperature of heat rejection also approaches zero as a limit. *Thus it appears that a definite zero point exists on the absolute temperature scale but that this point cannot be reached without a violation of the second law.*

Thus any attainable value of absolute temperature is always greater than zero. This is also known as the *Third Law of Thermodynamics* which may be stated as follows: *It is impossible by any procedure, no matter how idealized, to reduce any system to the absolute zero of temperature in a finite number of operations.*

This is what is called the Nernst-Simon statement of the third law. The third law itself is an independent law of nature, and not an extension of the second law. The concept of heat engine is not necessary to prove the non-attainability of absolute zero of temperature by any system in a finite number of operations.

6.16 EFFICIENCY OF THE REVERSIBLE HEAT ENGINE

The efficiency of a reversible heat engine in which heat is received solely at T_1 is found to be

$$\eta_{rev.} = \eta_{max} = 1 - \left(\frac{Q_2}{Q_1}\right)_{rev.} = 1 - \frac{T_2}{T_1}$$

or

$$\eta_{rev.} = \frac{T_1 - T_2}{T_1}$$

It is observed here that as T_2 decreases, and T_1 increases, the efficiency of the reversible cycle increases.

Since η is always less than unity, T_2 is always greater than zero and positive.

The COP of a refrigerator is given by

$$(COP)_{refr.} = \frac{Q_2}{Q_1 - Q_2} = \frac{1}{\dfrac{Q_1}{Q_2} - 1}$$

For a reversible refrigerator, using

$$\frac{Q_1}{Q_2} = \frac{T_1}{T_2}$$

$$[COP_{refr}]_{rev.} = \frac{T_2}{T_1 - T_2} \tag{6.19}$$

Similarly, for a reversible heat pump

$$[COP_{H.P.}]_{rev.} = \frac{T_1}{T_1 - T_2} \tag{6.20}$$

EXAMPLE 6.1

A cyclic heat engine operates between a source temperature of 800°C and a sink temperature of 30°C. What is the least rate of heat rejection per kW net output of the engine?

Solution

For a reversible engine, the rate of heat rejection will be minimum (Fig. Ex. 6.1).

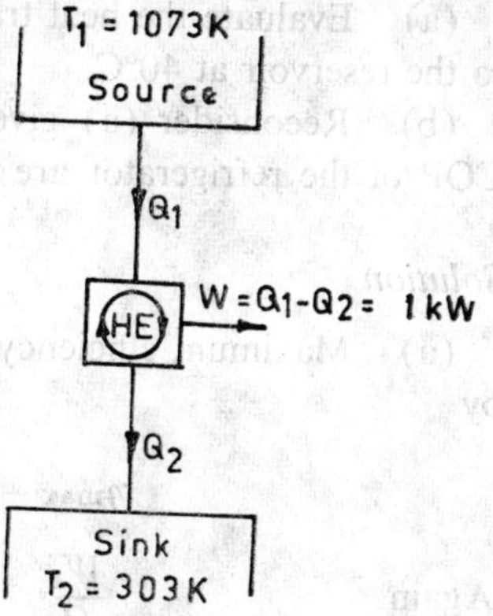

Fig. Ex 6.1

$$\eta_{max.} = \eta_{rev.} = 1 - \frac{T_2}{T_1}$$

$$= 1 - \frac{30 + 273}{800 + 273}$$

$$= 1 - 0.282 = 0.718$$

Now $\dfrac{W_{net}}{Q_1} = \eta_{max} = 0.718$

$\therefore$ $Q_1 = \dfrac{1}{0.718} = 1.392$ kW

Now $Q_2 = Q_1 - W_{net} = 1.392 - 1$

$= 0.392$ kW

This is the least rate of heat rejection. Ans.

EXAMPLE 6.2

A domestic food freezer maintains a temperature of −15°C. The embient air temperature is 30°C. If heat leaks into the freezer at the continuous rate of 1.75 kJ/s what is the least power necessary to pump this heat out continuously?

Solution

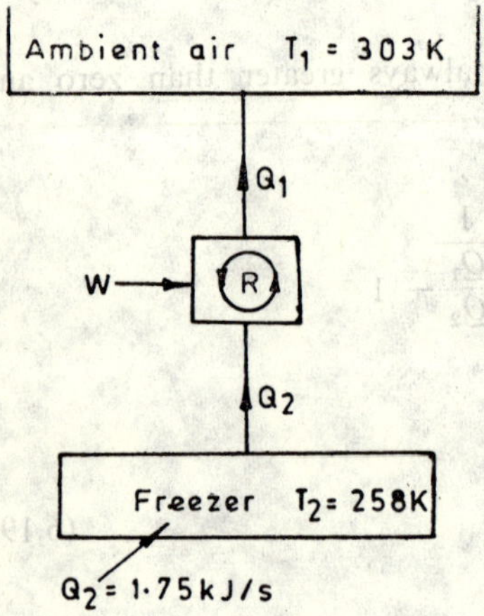

Fig. Ex. 6·2

Freezer temperature,

$$T_2 = -15 + 273 = 258 \text{ K}$$

Ambient air temperature,

$$T_1 = 30 + 273 = 303 \text{ K}$$

The refrigerator cycle removes heat from the freezer at the same rate at which heat leaks into it (Fig. Ex. 6.2.)

For minimum power requirement

$$\frac{Q_2}{T_2} = \frac{Q_1}{T_1}$$

$$\therefore \quad Q_1 = \frac{1.75}{258} \times 303 = 2.06 \text{ kJ/s}$$

$$\therefore \quad W = Q_1 - Q_2$$

$$= 2.06 - 1.75 = 0.31 \text{ kJ/s}$$

$$= 0.31 \text{ kW}$$

EXAMPLE 6.3

A reversible heat engine operates between two reservoirs at temperatures of 600°C and 40°C. The engine drives a reversible refrigerator which operates between reservoirs at temperatures of 40°C and −20°C. The heat transfer to the heat engine is 2000 kJ and the net work output of the combined engine refrigerator plant is 360 kJ.

(a) Evaluate the heat transfer to the refrigerant and the net heat transfer to the reservoir at 40°C.

(b) Reconsider (a) given that the efficiency of the heat engine and the COP of the refrigerator are each 40% of their maximum possible values.

Solution

(a) Maximum efficiency of the heat engine cycle (Fig. Ex. 6.3) is given by

$$\eta_{max} = 1 - \frac{T_2}{T_1} = 1 - \frac{313}{873} = 1 - 0.358 = 0.642$$

Again $\quad \dfrac{W_1}{Q_1} = 0.642$

$\therefore \quad W_1 = 0.642 \times 2000 = 1284 \text{ kJ}$

Maximum COP of the refrigerator cycle

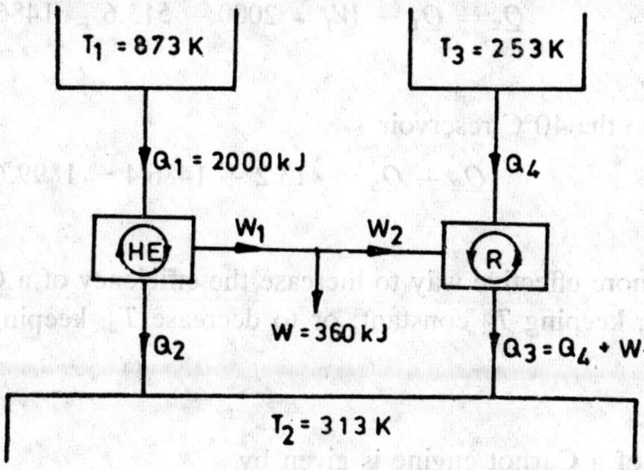

Fig. Ex. 6·3

$$(\text{COP})_{\text{max}} = \frac{T_3}{T_2 - T_3} = \frac{253}{313 - 253} = 4.22$$

Also $\quad \text{COP} = \dfrac{Q_4}{W_2} = 4.22$

Since $\quad W_1 - W_2 = W = 360 \text{ kJ}$

$\therefore \quad W_2 = W_1 - W = 1284 - 360 = 924 \text{ kJ}$

$\therefore \quad Q_4 = 4.22 \times 924 = 3899 \text{ kJ}$

$\therefore \quad Q_3 = Q_4 + W_2 = 924 + 3899 = 4823 \text{ kJ}$

$\quad Q_2 = Q_1 - W_1 = 2000 - 1284 = 716 \text{ kJ}$

Heat rejection to the 40°C reservoir

$= Q_2 + Q_3 = 716 + 4823 = 5539 \text{ kJ}$ Ans. (a)

(b) Efficiency of the actual heat engine cycle

$$\eta = 0\ 4\ \eta_{\text{max}} = 0.4 \times 0.642$$

$\therefore \quad W_1 = 0.4 \times 0.642 \times 2000 = 1284 \times 0.4$

$\quad = 513.6 \text{ kJ}$

$\therefore \quad W_2 = 513.6 - 360 = 153.6 \text{kJ}$

COP of the actual refrigerator cycle

$$\text{COP} = \frac{Q_4}{W_4} = 0.4 \times 4.22 = 1.69$$

Therefore

$Q_4 = 153.6 \times 1.69 = 259.6 \text{ kJ}$ Ans. (b)

$$Q_3 = 259.6 + 153.6 = 413.2 \text{ kJ}$$

$$Q_2 = Q_1 - W_1 = 2000 - 513.6 = 1486.4 \text{ kJ}$$

Therefore

Heat rejected to the 40°C reservoir

$$= Q_2 + Q_3 = 413.2 + 1486.4 = 1899.6 \text{ kJ} \quad \text{Ans. (b)}$$

EXAMPLE 6.4

Which is the more effective way to increase the efficiency of a Carnot engine: to increase T_1, keeping T_2 constant; or to decrease T_2, keeping T_1 constant?

Solution

The efficiency of a Carnot engine is given by

$$\eta = 1 - \frac{T_2}{T_1}$$

If T_2 is constant

$$\left(\frac{\partial \eta}{\partial T_1}\right)_{T_2} = \frac{T_2}{T_1^2}$$

As T_1 increases, η increases, and the slope $\left(\frac{\partial \eta}{\partial T_1}\right)_{T_2}$ decreases (Fig. Ex. 6.4.1).

If T_1 is constant,

$$\left(\frac{\partial \eta}{\partial T_2}\right)_{T_1} = -\frac{1}{T_1}$$

As T_2 decreases, η increases, but the slope $\left(\frac{\partial \eta}{\partial T_2}\right)_{T_1}$ remains constant (Fig. Ex. 6.4.2).

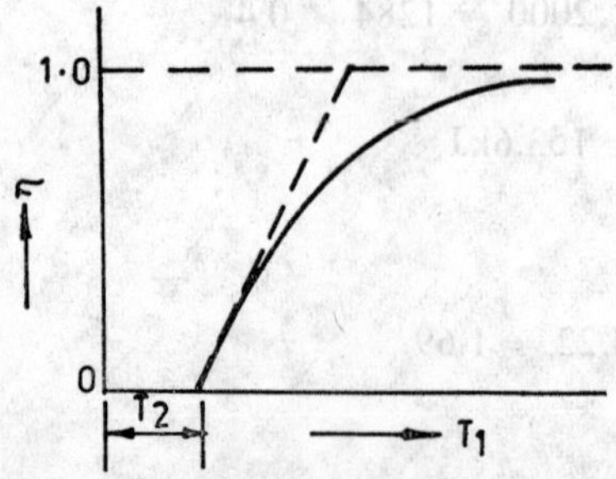

Fig. Ex. 6·4·1

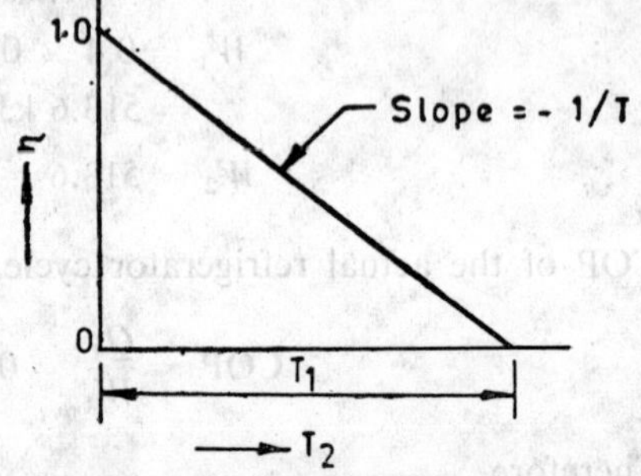

Fig. Ex. 6·4·2

Also
$$\left(\frac{\partial \eta}{\partial T_1}\right)_{T_2} = \frac{T_2}{T_1^2} \text{ and } \left(\frac{\partial \eta}{\partial T_2}\right)_{T_1} = -\frac{T_1}{T_1^2}$$

Since
$$T_1 > T_2, \left(\frac{\partial \eta}{\partial T_2}\right)_{T_1} > \left(\frac{\partial \eta}{\partial T_1}\right)_{T_2}$$

So, the more effective way to increase the efficiency is to decrease T_2. Alternatively, let T_2 be decreased by ΔT with T_1 remaining the same

$$\eta_1 = 1 - \frac{T_2 - \Delta T}{T_1}$$

If T_1 is increased by the same ΔT, T_2 remaining the same

$$\eta_2 = 1 - \frac{T_2}{T_1 + \Delta T}$$

Then

$$\eta_1 - \eta_2 = \frac{T_2}{T_1 + \Delta T} - \frac{T_2 - \Delta T}{T_1}$$

$$= \frac{(T_1 - T_2)\Delta T + (\Delta T)^2}{T_1(T_1 + \Delta T)}$$

Since
$$T_1 > T_2, \ (\eta_1 - \eta_2) > 0$$

The more effective way to increase the cycle efficiency is to decrease T_2.

EXAMPLE 6.5

Kelvin was the first to point out the thermodynamic wastefulness of burning fuel for the direct heating of a house. It is much more economical to use the high temperature heat produced by combustion in a heat engine and then to use the work so developed to pump heat from outdoors up to the temperature desired in the house. In Fig. Ex. 6.5 a boiler furnishes heat Q_1 at the high temperature T_1. This heat is absorbed by a heat engine, which extracts work W and rejects the waste heat Q_2 into the house at T_2. Work W is in turn used to operate a mechanical refrigerator or heat pump, which extracts Q_3 from outdoors at temperature T_3 and rejects Q_2 (where $Q_2' = Q_3 + W$) into the house. As a result of this cycle of operations, a total quantity of heat equal to $Q_2 + Q_2'$ is liberated in the house, against Q_1 which would be provided directly by the ordinary combustion of the fuel. Thus the ratio $Q_2 + Q_2/Q_1$ represents the heat multiplication factor of this method. Determine this multiplication factor if $T_1 = 473$ K, $T_2 = 293$ K, and $T_3 = 273$ K.

Solution

For the reversible heat engine (Fig. Ex. 6.5)

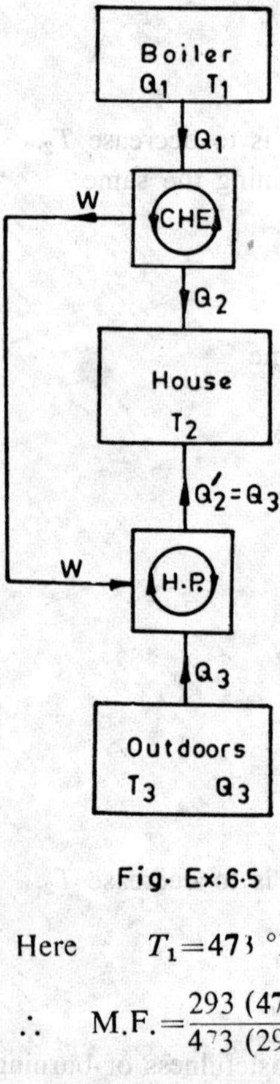

Fig. Ex. 6.5

$$\frac{Q_2}{Q_1} = \frac{T_2}{T_1}$$

$$\therefore \quad Q_2 = Q_1 \left(\frac{T_2}{T_1}\right)$$

Also
$$\eta = \frac{W}{Q_1} = \frac{T_1 - T_2}{T_1}$$

or
$$W = \frac{T_1 - T_2}{T_1} \cdot Q_1$$

For the reversible heat pump

$$\text{COP} = \frac{Q'_2}{W} = \frac{T_2}{T_2 - T_3}$$

$$\therefore \quad Q'_2 = \frac{T_2}{T_2 - T_3} \cdot \frac{T_1 - T_2}{T_1} \cdot Q_1$$

$\therefore$ Multiplication factor (M.F.)

$$= \frac{Q_2 + Q'_2}{Q_1} = \frac{Q_1 \dfrac{T_2}{T_1} + Q_1 \cdot \dfrac{T_2}{T_2 - T_3} \cdot \dfrac{T_1 - T_2}{T_1}}{Q_1}$$

or
$$\text{M.F.} = \frac{T_2^2 - T_2 T_3 + T_2 T_1 - T_2^2}{T_1 (T_2 - T_3)}$$

or
$$\text{M.F.} = \frac{T_2 (T_1 - T_3)}{T_1 (T_2 - T_3)}$$

Here $T_1 = 473$ °K, $T_2 = 293$ °K and $T_3 = 273$ °K

$$\therefore \quad \text{M.F.} = \frac{293\,(473 - 273)}{473\,(293 - 273)} = \frac{2930}{473} = 6.3 \qquad \textbf{Ans.}$$

which means that every kg of coal burned would deliver the heat equivalent to over 6 kg. Of course, in an actual case, the efficiencies would be less than Carnot efficiencies, but even with a reduction of 50%, the possible savings would be quite significant.

Example 6.6

It is proposed that solar energy be used to warm a large collector plate. This energy would, in turn, be transferred as heat to a fluid within a heat engine, and the engine would reject energy as heat to the atmosphere. Experiments indicate that about 1880 kJ/m² hr of energy can be collected when the plate is operating at 90°C. Estimate the minimum collector area that would be required for a plant producing 1 kW of useful shaft power. The atmospheric temperature may be assumed to be 20°C.

Solution

The maximum efficiency for the heat engine operating between the collector plate temperature and the atmospheric temperature (Fig. Ex. 6.6) is

$$\eta_{max} = 1 - \frac{T_2}{T_1} = 1 - \frac{293}{363} = 0.192$$

The efficiency of any actual heat engine operating between these temperatures would be less than this efficiency.

$$\therefore \quad Q_{min} = \frac{W}{\eta_{max}} = \frac{1 \text{ kJ/s}}{0.192} = 5.21 \text{ kJ/s}$$

$$= 18{,}800 \text{ kJ/hr}$$

$\therefore$ Minimum area required for the collector plate

$$= \frac{18{,}800}{1880} = 10 \text{ m}^2 \qquad \text{Ans.}$$

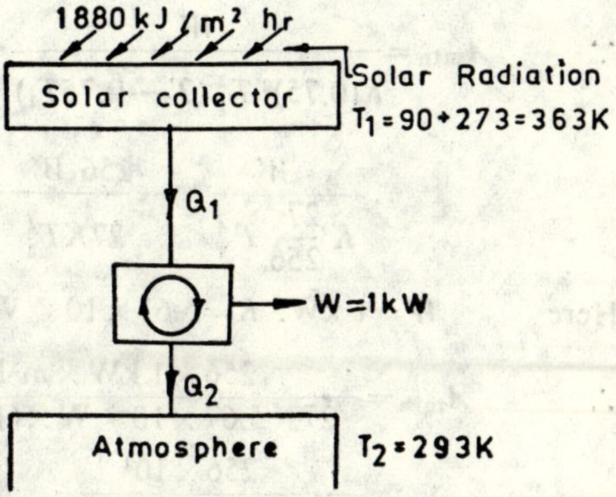

Fig. Ex. 6·6

EXAMPLE 6.7

A reversible heat engine in a satellite operates between a hot reservoir at T_1 and a radiating panel at T_2. Radiation from the panel is proportional to its area and to T_2^4. For a given work output and value of T_1 show that the area of the panel will be minimum when $\frac{T_2}{T_1} = 0.75$.

Determine the minimum area of the panel for an output of 1 kW if the constant of proportionality is 5.67×10^{-8} W/m²K⁴ and T_1 is 1000 K.

Solution

For the heat engine (Fig. Ex. 6.7), the heat rejected Q_2 to the panel (at T_2) is equal to the energy emitted from the panel to the surroundings by radiation. If A is the area of the panel, $Q_2 \propto AT_2^4$, or $Q_2 = KAT_2^4$, where K is a constant.

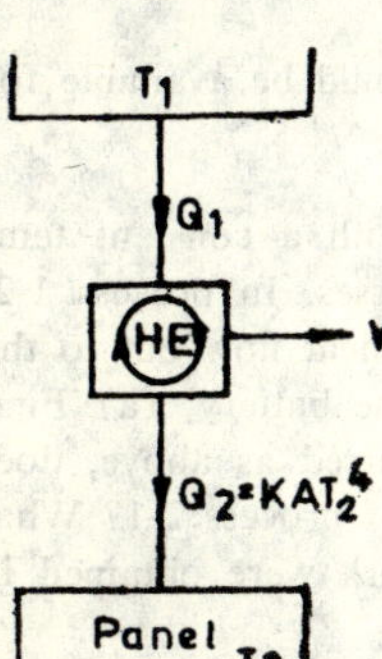

Fig. Ex. 6·7

Now,
$$\eta = \frac{W}{Q_1} = \frac{T_1 - T_2}{T_1}$$

or
$$\frac{W}{T_1 - T_2} = \frac{Q_1}{T_1} = \frac{Q_2}{T_1} = \frac{KAT_2^4}{T_2}$$

$$= KAT_2^3$$

$$\therefore \quad A = \frac{W}{KT_2^3 (T_1 - T_2)} = \frac{W}{K(T_1 T_2^3 - T_2^4)}$$

For a given W and T_1, A will be minimum when

$$\frac{dA}{dT_2} = -\frac{W}{K}(3T_1 T_2^2 - 4T_2^3).\ (T_1 T_2^3 - T_2^4)^{-2} = 0$$

Since $\quad (T_1T_2^3 - T_2^4)^{-2} \neq 0, \; 3T_1T_2^2 = 4T_2^3$

$\therefore \quad \dfrac{T_2}{T_1} = 0.75$ **Proved.**

$$\because \quad A_{\min} = \frac{W}{K(0.75)^3 T_1^3 (T_1 - 0.75T_1)}$$

$$= \frac{W}{K\dfrac{27}{256} T_1^4} = \frac{256\, W}{27 K T_1^4}$$

Here $\quad W = 1$ kW, $K = 5.67 \times 10^{-8}$ W/m²K⁴, and $T_1 = 1000$ K

$$\therefore \quad A_{\min} = \frac{256 \times 1 \text{ kW} \times \text{m}^2\text{K}^4}{27 \times 5.67 \times 10^{-8} \text{ W} \times (1000)^4 K^4}$$

$$= \frac{256 \times 10^3}{27 \times 5.67 \times 10^{-8} \times 10^{12}} \text{ m}^2$$

$$= 0.1672 \text{ m}^2$$ **Ans.**

PROBLEMS

6.1 An inventor claims to have developed an engine that takes in 105 MJ at a temperature of 400 K, rejects 42 MJ at a temperature of 200 K, and delivers 15 kWhr of mechanical work. Would you advise investing money to put this engine in the market?

6.2 If a refrigerator is used for heating purposes in winter so that the atmosphere becomes the cold body and the room to be heated becomes the hot body, how much heat would be available for heating for each kW input to the driving motor? The COP of the refrigerator is 5, and the electromechanical efficiency of the motor is 90%. How does this compare with resistance heating?

6.3 Using an engine of 30% thermal efficiency to drive a refrigerator having a COP of 5, what is the heat input into the engine for each MJ removed from the cold body by the refrigerator?

If this system is used as a heat pump, how many MJ of heat would be available for heating for each MJ of heat input to the engine?

6.4 An electric storage battery which can exchange heat only with a constant temperature atmosphere goes through a complete cycle of two processes. In process 1-2, 2.8 kWhr of electrical work flow into the battery while 732 kJ of heat flow out to the atmosphere. During process 2-1, 2.4 kWhr of work flow out of the battery. (a) Find the heat transfer in process 2-1. (b) If the process 1-2 has occurred as above, does the first law or the second law limit the maximum possible work of process 2-1? What is the maximum possible work? (c) If the maximum possible work were obtained in process 2-1, what will be the heat transfer in the process?

6.5 A household refrigerator is maintained at a temperature of 2°C. Every time the door is opened, warm material is placed inside, introducing an average of 420 kJ, but making only a small change in the temperature of the refrigerator. The door is opened 20 times a day, and the refrigerator operates at 15% of the ideal COP. The cost of work is 32 paise per kWhr. What is the monthly bill for this refrigerator? The atmosphere is at 30°C. Ans. Rs. 15.20

6.6 A heat pump working on the Carnot cycle takes in heat from a reservoir at 5°C and delivers heat to a reservoir at 60°C. The heat pump is driven by a reversible heat engine which takes in heat from a reservoir at 840°C and rejects heat to a reservoir at 60°C. The reversible heat engine also drives a machine that absorbs 30 kW. If the heat pump extracts 17 kJ/s from the 5°C reservoir, determine (a) the rate of heat supply from the 840°C source, and (b) the rate of heat rejection to the 60°C sink.
Ans. (a) 47.61 kW; (b) 34.61 kW.

6.7 A refrigeration plant for a food store operates as a reversed Carnot heat engine cycle. The store is to be maintained at a temperature of −5°C and the heat transfer from the store to the cycle is at the rate of 5 kW. If heat is transferred from the cycle to the atmosphere at a temperature of 25°C, calculate the power required to drive the plant. Ans. 0.56 kW.

6.8 A heat engine is used to drive a heat pump. The heat transfers from the heat engine and from the heat pump are used to heat the water circulating through the radiators of a building. The efficiency of the heat engine is 27% and the COP of the heat pump is 4. Evaluate the ratio of the heat transfer to the circulating water to the heat transfer to the heat engine. Ans. 1.81.

6.9 If 20 kJ are added to a Carnot cycle at a temperature of 100°C and 14.6 kJ are rejected at 0°C, determine the location of absolute zero on the Celsius scale.

6.10 Two reversible heat engines A and B are arranged in series, A rejecting heat directly to B. Engine A receives 200 kJ at a temperature of 421°C from a hot source, while engine B is in communication with a cold sink at a temperatue of 4.4°C. If the work output of A is twice that of B, find (a) the intermediate temperature between A and B, (b) the efficiency of each engine, and (c) the heat rejected to the cold sink. Ans. 143.4°C, 40% & 33.5%, 80 kJ.

6.11 A heat engine operates between the maximum and minimum temperatures of 671°C and 60°C respectively, with an efficiency of 50% of the appropriate Carnot efficiency. It drives a heat pump which uses river water at 4.4°C to heat a block of flats in which the temperature is to be maintained at 21.1°C. Assuming that a temperature difference of 11.1°C exists between the working fluid and the river water, on the one hand, and the required room temperature on the other, and assuming the heat pump to operate on the reversed Carnot cycle, but with a COP of 50% of the ideal COP, find the heat input to the engine per unit heat output from the heat pump. Why is direct heating thermodynamically more wasteful? Ans. 0.82 kJ/kJ heat input.

6.12 An ice-making plant produces ice at atmospheric pressure and at 0°C from water. The mean temperature of the cooling water circulating through the condenser of the refrigerating machine is 18°. Evaluate the minimum electrical work in kWhr required to produce 1 tonne of ice. (The enthalpy of fusion of ice at atmospheric pressure is 333.5 kJ/kg). Ans. 6.11 kWhr.

6.13 A reversible engine works between three thermal reservoirs, A, B, and C. The engine absorbs an equal amount of heat from the thermal reservoirs A and B kept at temperatures T_A and T_B respectively, and rejects heat to the thermal reservoir C kept at temperature T_C. The efficiency of the engine is α times the efficiency of the reversible engine, which works between the two reservoirs A and C. Prove that

$$\frac{T_A}{T_B} = (2\alpha - 1) + 2(1-\alpha)\frac{T_A}{T_C}$$

6.14 A reversible engine operates between temperatures T_1 and T ($T_1 > T$). The energy rejected from this engine is received by a second reversible engine at the same temperature. T. The second engine rejects energy at temperature T_2 ($T_2 < T$). Show that (a) temperature T is the arithmetic mean of temperatures T_1 and T_2 if the engines produce the same amount of work output, and (b) temperature T is the geometric mean of temperatures T_1 and T_2 if the engines have the same cycle efficiencies.

6.15 Two Carnot engines A and B are connected in series between two thermal reservoirs maintained at 1000 °K and 100 °K respectively. Engine A receives 1680 kJ of heat from the high-temperature reservoir and rejects heat to the Carnot engine B. Engine B takes in heat rejected by engine A and rejects heat to the low-temperature reservoir. If engines A and B have equal thermal efficiencies, determine (a) the heat rejected by engine B, (b) the temperature at which heat is rejected by engine A, and (c) the work done during the process by engines A and B respectively. If engines A and B deliver equal work, determine (d) the amount of heat taken in by engine B, and (e) the efficiencies of engines A and B.

6.16 A heat pump is to be used to heat a house in winter and then reversed to cool the house in summer. The interior temperature is to be maintained at 20°C. Heat transfer through the walls and roof is estimated to be 0.525 kJ/s per degree temperature difference between the inside and outside. (a) If the outside temperature in winter is 5°C, what is the minimum power required to drive the heat pump? (b) If the power output is the same as in part (a), what the maximum outer temperature for which the inside can be maintained at 20°C?

6.17 Consider an engine in outer space which operates on the Carnot cycle. The only way in which heat can be transferred from the engine is by radiation. The rate at which heat is radiated is proportional to the fourth power of the absolute temperature. and to the area of the radiating surface. Show that for a given power output and a given T_1, the area of the radiator will be a minimum when

$$\frac{T_2}{T_1} = \frac{3}{4}$$

6.18 It takes 10 KW to keep the interior of a certain house at 20°C when the outside temperature is 0°C. This heat flow is usually obtained directly by burning gas or oil. Calculate the power required if the 10 kW heat flow were supplied by operating a reversible engine with the house as the upper reservoir and the outside surroundings as the lower reservoir, so that the power were used only to perform work needed to operate the engine.

6.19 Prove that the COP of a reversible refrigerator operating between two given temperatures is the maximum.

6.20 A house is to be maintained at a temperature of 20°C by means of a heat pump pumping heat from the atmosphere. Heat losses through the walls of the house are estimated at 0.65 kW per unit of temperature difference between the inside of the house and the atmosphere. (a) If the atmospheric temperature is −10°C, what is the minimum power required to drive the pump? (b) It is proposed to use the same heat pump to cool the house in summer. For the same room temperature, the same heat loss rate, and the same power input to the pump, what is the maximum permissible atmospheric temperature? Ans. 2 kW, 50°C.

7

Entropy

7.1 INTRODUCTION

The first law of thermodynamics was stated in terms of cycles first and it was shown that the cyclic integral of heat is equal to the cyclic integral of work. When the first law was applied for thermodynamic processes, the existence of a property, the internal energy, was found. Similarly, the second law was also first stated in terms of cycles executed by systems. When applied to processes, the second law also leads to the definition of a new property, known as entropy. If the first law is said to be the law of internal energy, then second law may be stated to be the law of entropy. In fact, *thermodynamics is the study of three E's, namely, energy, equilibrium and entropy.*

7.2 TWO REVERSIBLE ADIABATIC PATHS CANNOT INTERSECT EACH OTHER

Let it be assumed that two reversible adiabatics *AC* and *BC* intersect each other at point *C* (Fig. 7.1). Let a reversible isotherm *AB* be drawn in such a way that it intersects the reversible adiabatics at *A* and *B*. The three reversible processes *AB, BC,* and *CA* together constitute a reversible cycle, and the area included represents the net work output in a cycle. But such a cycle is impossible, since net work is being produced in a cycle by a heat engine by exchanging heat with a single reservoir in the process *AB*, which violates the Kelvin-Planck statement of the second law. Therefore, the assumption of the intersection of the reversible adiabatics is wrong. *Through one point, there can pass only one reversible adiabatic.*

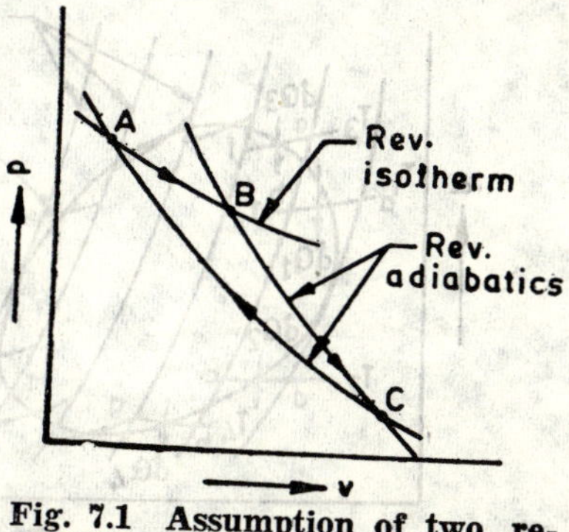

Fig. 7.1 Assumption of two reversible adiabatics intersecting each other

7.3 CLAUSIUS' THEOREM

Let a system be taken from an equilibrium state i to another equilibrium state f by following the reversible path i-f (Fig. 7.2). Let a reversible adiabatic i-a be drawn through i and another reversible adiabatic b-f be drawn through f. Then a reversible isotherm a-b is drawn in such a way that the area under i-a-b-f is equal to the area under i-f. Applying the first law for

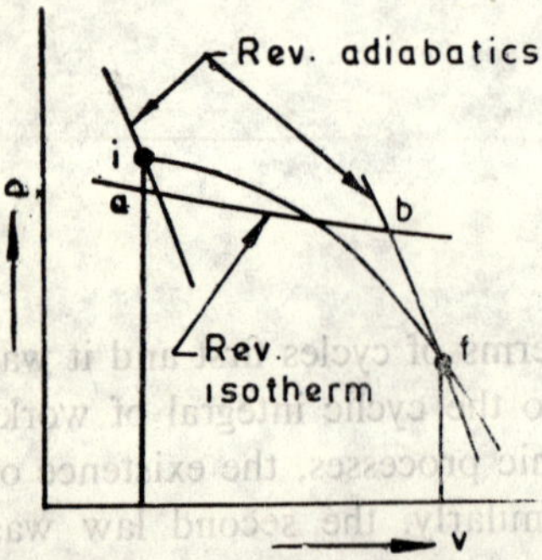

Fig. 7.2 Reversible path substituted by two reversible adiabatics and a reversible isotherm

Process i-f

$$Q_{i-f} = U_f - U_i + W_{if} \tag{7.1}$$

Process i-a-b-f

$$Q_{iabf} = U_f - U_i + W_{iabf} \tag{7.2}$$

Since

$$W_{if} = W_{iabf}$$

$\therefore$ From equations (7.1) and (7.2)

$$Q_{if} = Q_{iabf}$$
$$= Q_{ia} + Q_{ab} + Q_{bf}$$

Since $Q_{ia} = 0$ and $Q_{bf} = 0$

$$Q_{if} = Q_{ab}$$

Heat transferred in the process i-f is equal to the heat transferred in the isothermal process a-b.

Thus any reversible path may be substituted by a reversible zigzag path, between the same end states, consisting of a reversible adiabatic followed by a reversible isotherm and then by a reversible adiabatic, such that the heat transferred during the isothermal process is the same as that transferred during the original process.

Let a smooth closed curve representing a reversible cycle (Fig. 7.3) be considered. Let the closed cycle be divided into a large number of strips by

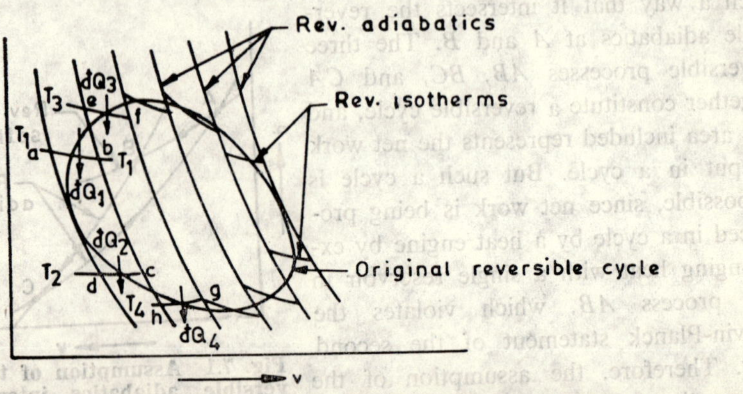

Fig. 7.3 A reversible cycle split into a large number of Carnot cycles

means of reversible adiabatics. Each strip may be closed at the top and bottom by reversible isotherms. The original closed cycle is thus replaced by a zigzag closed path consisting of alternate adiabatic and isothermal processes, such that the heat transferred during all the isothermal processes is equal to the heat transferred in the original cycle. Thus the original cycle is replaced by a large number of Carnot cycles. If the adiabatics are close to one another and the number of Carnot cycles is large, the saw-toothed zigzag line will coincide with the original cycle.

For the elemental cycle *abcd* $đQ_1$ heat is absorbed reversibly at T_1, and $đQ_2$ heat is rejected reversibly at T_2

$$\frac{đQ_1}{T_1}=\frac{đQ_2}{T_2}$$

If heat supplied is taken as positive and heat rejected as negative

$$\frac{đQ_1}{T_1}+\frac{đQ_2}{T_2}=0$$

Similarly, for the elemental cycle *efgh*

$$\frac{đQ_3}{T_3}+\frac{đQ_4}{T_4}=0$$

If similar equations are written for all the elemental Carnot cycles, then for the whole original cycle

$$\frac{đQ_1}{T_1}+\frac{đQ_2}{T_2}+\frac{đQ_3}{T_3}+\frac{đQ_4}{T_4}+\ldots\ldots=0$$

or

$$\oint_R \frac{đQ}{T}=0 \tag{7.3}$$

The cyclic integral of $\frac{đQ}{T}$ for a reversible cycle is equal to zero. This is known as *Clausius' theorem.* The letter R emphasizes the fact that the equation is valid only for a reversible cycle.

7.4 THE PROPERTY OF ENTROPY

Let a system be taken from an initial equilibrium state i to a final equilibrium state f by following the reversible path R_1 (Fig. 7.4). The system is brought back from f to i by following another reversible path R_2. Then the two paths R_1 and R_2 together constitute a reversible cycle. From Clausius' theorem

$$\oint_{R_1R_2} \frac{đQ}{T}=0$$

The above integral may be replaced as the sum of two integrals, one for path R_1 and the other for path R_2

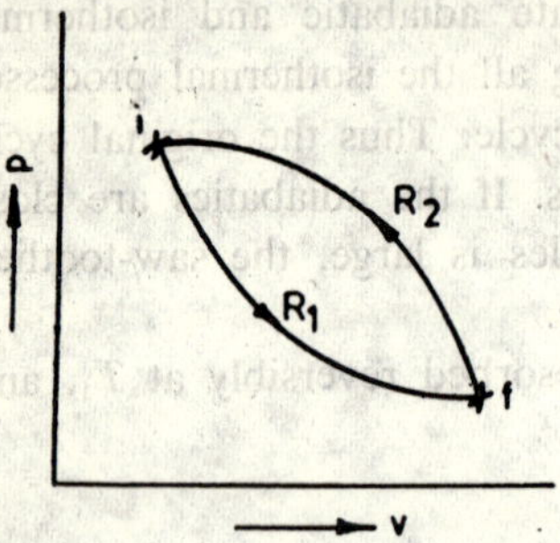

Fig. 7.4 Two reversible paths R_1 and R_2 between two equilibrium states i and f

$$\int_{R_1\, i}^{f} \frac{đQ}{T} + \int_{R_2\, f}^{i} \frac{đQ}{T} = 0$$

or

$$\int_{R_1\, i}^{f} \frac{đQ}{T} = -\int_{R_2\, f}^{i} \frac{đQ}{T}$$

Since R_2 is a reversible path

$$\int_{R_1\, i}^{f} \frac{đQ}{T} = \int_{R_2\, i}^{f} \frac{đQ}{T}$$

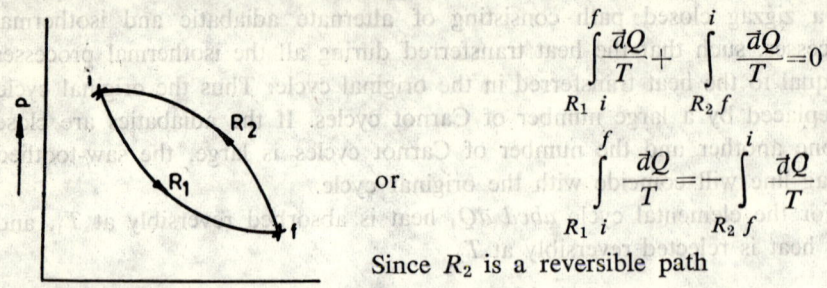

Since R_1 and R_2 represent any two reversible paths, $\int_{R\, i}^{f} \frac{đQ}{T}$ is independent of the reversible path connecting i and f. Therefore, there exists a property of a system whose value at the final state f minus its value at the initial state i is equal to $\int_{R\, i}^{f} \frac{đQ}{T}$. This property is called *entropy*, and is denoted by S. If S_i is the entropy at the initial state i, and S_f is the entropy at the final state f, then

$$\int_{R\, i}^{f} \frac{đQ}{T} = S_f - S_i \tag{7.4}$$

When the two equilibrium states are infinitesimally near

$$\frac{đQ_R}{T} = dS \tag{7.5}$$

where dS is an *exact differential* because S is a point function and a property. The subscript R in $đQ$ indicates that heat $đQ$ is transferred *reversibly*.

The word 'entropy' was first used by Clausius, taken from the Greek word 'tropee' meaning 'transformation'. It is an extensive property, and has the unit J/K. The specific entropy

$$s = \frac{S}{m} \text{ J/kg K}$$

If the system is taken from an initial equilibrium state i to a final equilibrium state f by an *irreversible path,* since entropy is a point or state function, and the entropy change is independent of the path followed, the non-reversible path is to be replaced by a reversible path to integrate for the evaluation of entropy change in the irreversible process (Fig. 7.5).

$$S_f - S_i = \int_i^f \frac{đQ_{\text{rev.}}}{T} = (\Delta S)_{\text{irrev. path}} \tag{7.6}$$

Integration can be performed only on a reversible path.

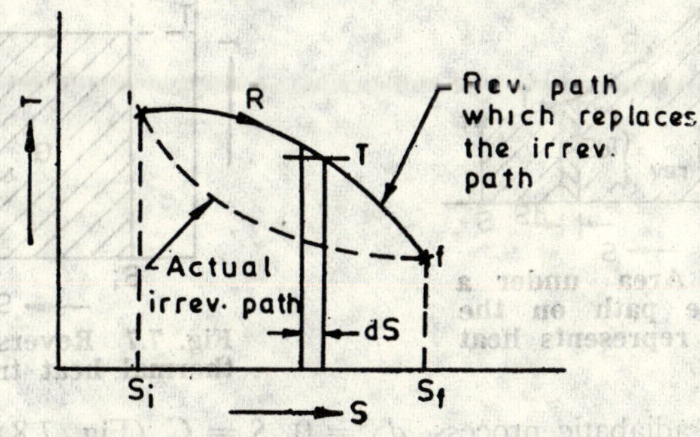

Fig. 7.5 Integration can be done only on a reversible path

7.5 TEMPERATURE-ENTROPY PLOT

The infinitesimal change in entropy dS due to reversible heat transfer $đQ$ at temperature T is

$$dS = \frac{đQ_{\text{rev.}}}{T}$$

If $đQ_{\text{rev}} = 0$, i.e., the process is reversible and adiabatic

$$dS = 0$$

and $$S = \text{constant}$$

A reversible adiabatic process is, therefore, an isentropic process.

Now

$$đQ_{\text{rev.}} = TdS$$

or

$$Q_{\text{rev.}} = \int_i^f TdS$$

The system is taken from i to f reversibly (Fig. 7.6). The area under the curve $\int_i^f TdS$ is equal to the heat transferred in the process.

For reversible isothermal heat transfer (Fig. 7.7), T = constant.

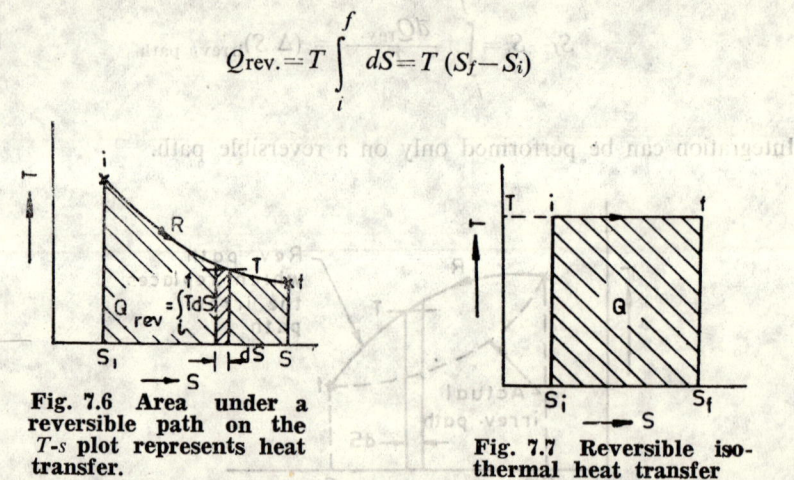

$$\therefore \qquad Q_{\text{rev.}} = T \int_i^f dS = T\,(S_f - S_i)$$

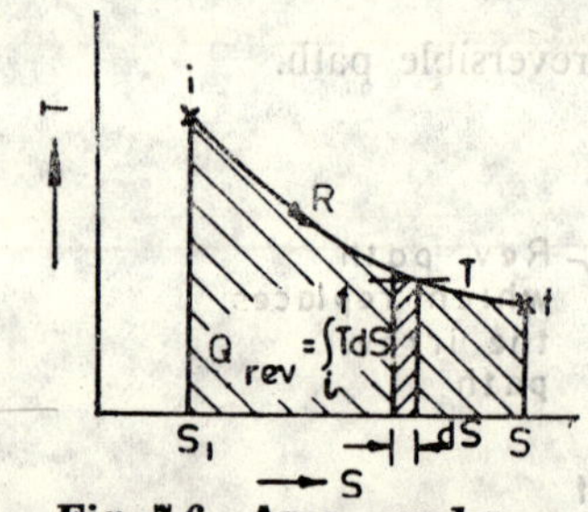

Fig. 7.6 Area under a reversible path on the *T-s* plot represents heat transfer.

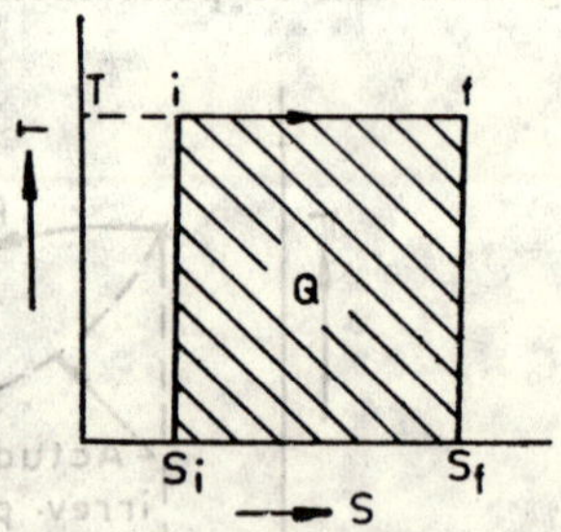

Fig. 7.7 Reversible isothermal heat transfer

For a reversible adiabatic process, $dS = 0$, $S = C$ (Fig. 7.8).

The *Carnot cycle* comprising two reversible isotherms and two reversible adiabatics forms a rectangle in the *T-S* plane (Fig. 7.9). Process 4-1

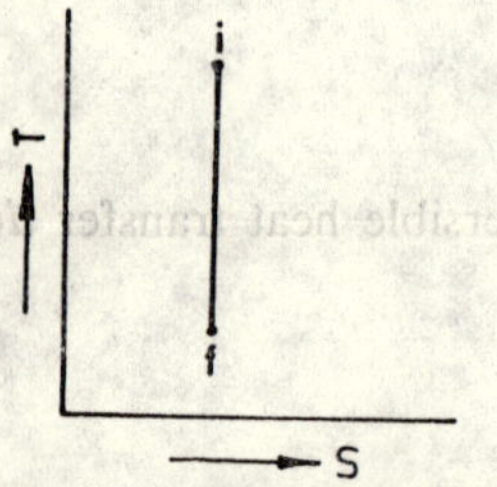

Fig. 7.8 Reversible adiabatic is isentropic.

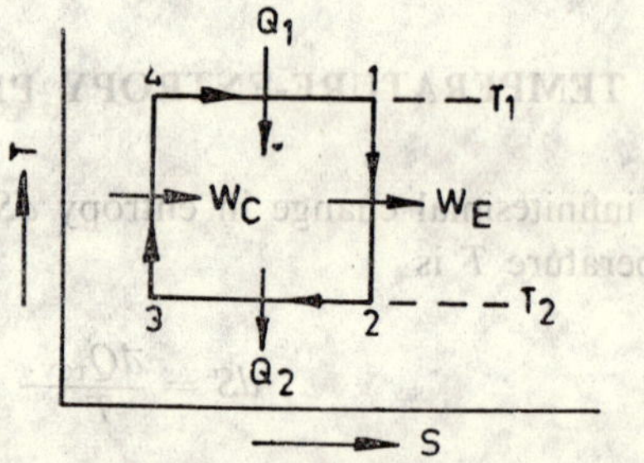

Fig. 7.9 Carnot cycle

represents reversible isothermal heat addition Q_1 to the system at T_1 from an external source, process 1-2 is the reversible adiabatic expansion of the system producing W_E amount of work, process 2-3 is the reversible isothermal heat rejection from the system to an external sink at T_2, and process 3-4 represents reversible adiabatic compression of the system consuming W_c amount of work. Area 1234 represents the net work output per cycle and the area under 4-1 indicates the quantity of heat added to system Q_1.

$$\therefore \qquad \eta_{\text{carnot}} = \frac{Q_1 - Q_2}{Q_1} = \frac{T_1\,(S_1 - S_4) - T_2\,(S_2 - S_3)}{T_1\,(S_1 - S_4)}$$

$$= \frac{T_1 - T_2}{T_1} = 1 - \frac{T_2}{T_1}$$

and
$$W_{\text{net}} = Q_1 - Q_2 = (T_1 - T_2)\,(S_1 - S_4)$$

7.6 THE INEQUALITY OF CLAUSIUS

Let us consider a cycle *ABCD* (Fig. 7.10). Let *AB* be a general process, either reversible or irreversible, while the other processes in the cycle are

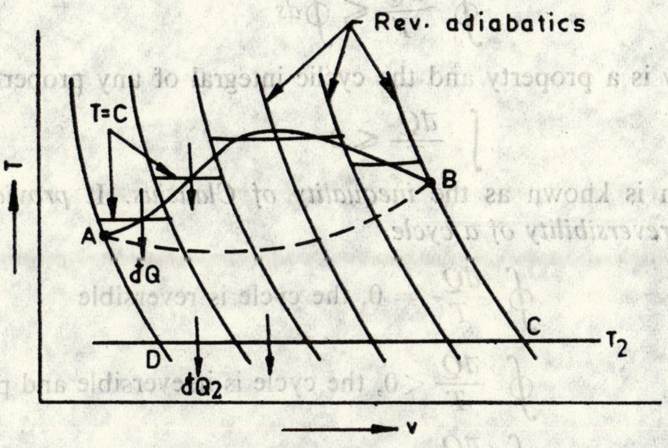

Fig. 7.10 Inequality of Clausius

reversible. Let the cycle be divided into a number of elementary cycles, as shown. For one of these elementary cycles

$$\eta = 1 - \frac{đQ_2}{đQ}$$

where $đQ$ is the heat supplied at T, and $đQ_2$ the heat rejected at T_2.

Now, the efficiency of a general cycle will be equal to or less than the efficiency of a reversible cycle.

$$\therefore \qquad 1 - \frac{đQ_2}{đQ} \leqslant \left(1 - \frac{đQ_2}{đQ}\right)_{\text{rev.}}$$

or
$$\frac{đQ_2}{đQ} \geqslant \left(\frac{đQ_2}{đQ}\right)_{\text{rev.}}$$

or
$$\frac{đQ}{đQ_2} \leqslant \left(\frac{đQ}{đQ_2}\right)_{\text{rev.}}$$

Since
$$\left(\frac{đQ}{đQ_2}\right)_{\text{rev.}} = \frac{T}{T_2}$$

$$\therefore \qquad \frac{đQ}{đQ_2} \leqslant \frac{T}{T_2}$$

or
$$\frac{đQ}{T} \leqslant \frac{đQ_2}{T_2}$$, for any process AB, reversible or irreversible.

For a reversible process

$$ds = \frac{đQ_{\text{rev.}}}{T} = \frac{đQ_2}{T_2} \qquad (7.7)$$

Hence, for any process AB

$$\frac{đQ}{T} \leqslant ds \tag{7.8}$$

Then for any cycle

$$\oint \frac{đQ}{T} \leqslant \oint ds$$

Since entropy is a property and the cyclic integral of any property is zero

$$\int \frac{đQ}{T} \leqslant 0 \tag{7.9}$$

This equation is known as the *inequality of Clausius.* It *provides the criterion of the reversibility of a cycle.*

If $\oint \frac{đQ}{T} = 0$, the cycle is reversible

$\oint \frac{đQ}{T} < 0$, the cycle is irreversible and possible

$\oint \frac{đQ}{T} > 0$, the cycle is impossible, since it violates the second law.

7.7 ENTROPY PRINCIPLE

For any process undergone by a system, we have from equation (7.8)

$$\frac{đQ}{T} \leqslant ds$$

or

$$ds \geqslant \frac{đQ}{T} \tag{7.10}$$

If the process is reversible

$$ds = \frac{đQ}{T}$$

and if it is irreversible

$$ds > \frac{đQ}{T}$$

An isolated system does not undergo any energy interaction with its surroundings, and so its energy is always constant. Also, for an isolated system since $đQ = 0$, from equation (7.10)

$$ds \geqslant 0 \tag{7.11}$$

For a reversible process

$$ds = 0$$

or

$$s = \text{constant}$$

For an irreversible or a real process

$$ds > 0$$

i.e. the entropy of the isolated system can never decrease. This is known as the *principle of increase of entropy*.

An isolated system can always be formed by including any system and its surroundings within a single boundary (Fig. 7.11). Sometimes the original system which is then only a part of the isolated system is called a 'subsystem'. The system and the surroundings together (the universe or the isolated system) include everything which is affected by the process. Entropy may decrease locally at some region within the isolated system, but it must be compensated by a greater increase of entropy somewhere within the system so that the net effect of an irreversible process is an entropy increase of the whole system.

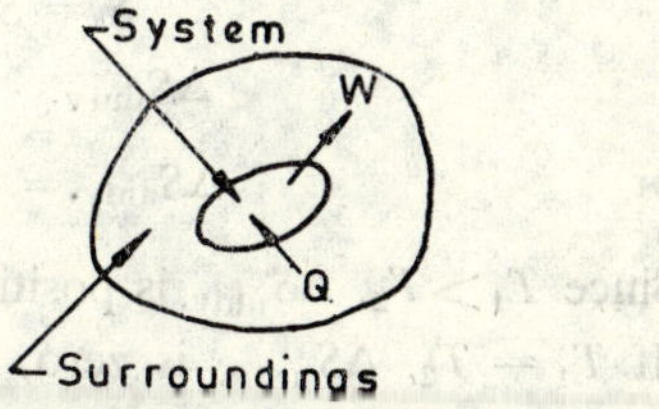

Fig. 7.11 Isolated system

The entropy increase of an isolated system is a measure of the extent of irreversibility of the process undergone by the system.

Rudolf Clausius summarized the first and second laws of thermodynamics in the following words:

(a) Die Energie der Welt ist konstant.

(b) Die Entropie der Welt strebt einem Maximum zu.

[(a) The energy of the world (universe) is constant.
(b) The entropy of the world tends towards a maximum.]

The entropy of an isolated system always increases and becomes a maximum at the state of equilibrium. *When the system is at equilibrium, any conceivable change in entropy would be zero.*

7.7.1 Transfer of heat through a finite temperature difference

Let Q be the rate of heat transfer from reservoir A at T_1 to reservoir B at T_2, $T_1 > T_2$ (Fig. 7.12).

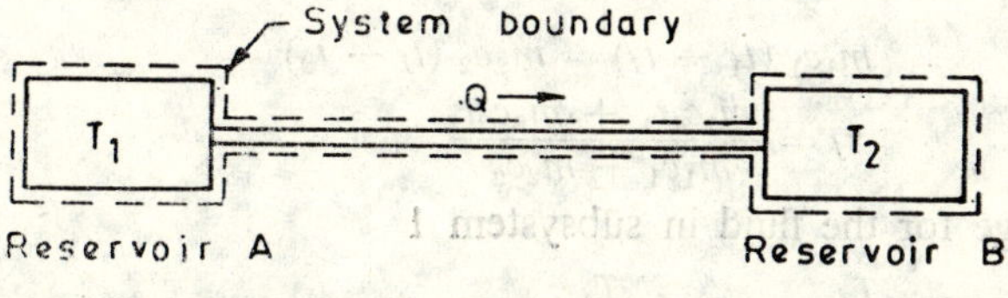

Fig. 7.12 Heat transfer through finite temperature difference

For reservoir A, $\Delta S_A = -Q/T_1$. It is negative because heat Q flows out of the reservoir. For reservoir B, $\Delta S_B = +Q/T_2$. It is positive because heat flows into the reservoir. The rod connecting the reservoirs suffers no entropy change because, once in the steady state, its coordinates do not change.

Therefore, for the isolated system comprising the reservoirs and the rod, and since entropy is an additive property

$$S = S_A + S_B$$

$$\Delta S_{\text{univ.}} = \Delta S_A + \Delta S_B$$

or $$\Delta S_{\text{univ.}} = -\frac{Q}{T_1} + \frac{Q}{T_2} = Q \cdot \frac{T_1 - T_2}{T_1 T_2}$$

Since $T_1 > T_2$, $\Delta S_{\text{univ.}}$ is positive, and the process is irreversible and possible.

If $T_1 = T_2$, $\Delta S_{\text{univ.}}$ is zero, and the process is reversible.

If $T_1 < T_2$, $\Delta S_{\text{univ.}}$ is negative and the process is impossible.

7.7.2 Mixing of two fluids

Subsystem 1 having a fluid of mass m_1, specific heat c_1, and temperature t_1, and subsystem 2 consisting of a fluid of mass m_2, specific heat c_2, and temperature t_2, comprise a composite system in an adiabatic enclosure (Fig. 7.13). When the partition is removed, the two fluids mix together, and

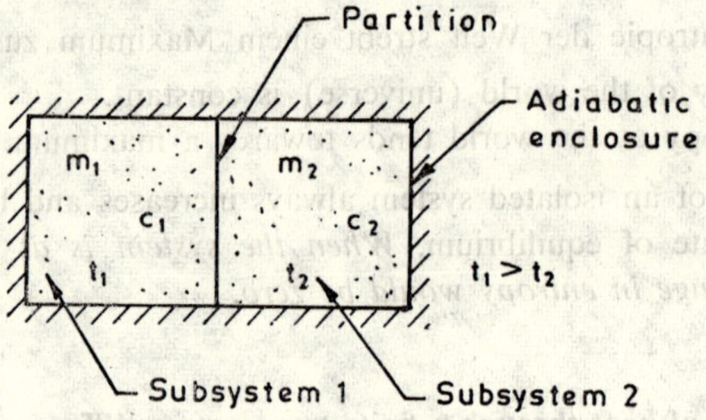

Fig. 7.13 Mixing of two fluids

at equilibrium let t_f be the final temperature, and $t_2 < t_f < t_1$. Since energy interaction is exclusively confined to the two fluids, the system being isolated

$$m_1 c_1 (t_1 - t_f) = m_2 c_2 (t_f - t_2)$$

$$\therefore \quad t_f = \frac{m_1 c_1 t_1 + m_2 c_2 t_2}{m_1 c_1 + m_2 c_2}$$

Entropy change for the fluid in subsystem 1

$$\Delta S_1 = \int_{T_1}^{T_f} \frac{đQ_{\text{rev.}}}{T} = \int_{T_1}^{T_f} \frac{m_1 c_1 \, dT}{T} = m_1 c_1 \ln \frac{T_f}{T_1}$$

$$= m_1 c_1 \ln \frac{t_f + 273}{t_1 + 273}$$

This will be negative, since $T_1 > T_f$.

Entropy change for the fluid in subsystem 2

$$\Delta S_2 = \int_{T_2}^{T_f} \frac{m_2 c_2 \, dT}{T} = m_2 c_2 \ln \frac{T_f}{T_2} = m_2 c_2 \ln \frac{t_f + 273}{t_2 + 273}$$

This will be positive, since $T_2 < T_f$.

$$\therefore \quad \Delta S_{\text{univ.}} = \Delta S_1 + \Delta S_2$$

$$= m_1 c_1 \ln \frac{T_f}{T_1} + m_2 c_2 \ln \frac{T_f}{T_2}$$

$\Delta S_{\text{univ.}}$ will be positive definite, and the mixing process is irreversible.

Although the mixing process is irreversible, to evaluate the entropy change for the subsystems, the irreversible path was replaced by a reversible path on which the integration was performed.

f $m_1 = m_2 = m$ and $c_1 = c_2 = c$.

$$\Delta S_{\text{univ.}} = mc \ln \frac{T_f^2}{T_1 . T_2}$$

and

$$T_f = \frac{m_1 c_1 T_1 + m_2 c_2 T_2}{m_1 c_1 + m_2 c_2} = \frac{T_1 + T_2}{2}$$

$$\therefore \quad \Delta S_{\text{univ.}} = 2\, mc \ln \frac{(T_1 + T_2)/2}{\sqrt{T_1 . T_2}}$$

This is always positive, since the arithmetic mean of any two numbers is always greater than their geometric mean.

7.7.3 Maximum work obtainable from two finite bodies at temperatures T_1 and T_2

Let us consider two identical bodies of constant heat capacity at temperatures T_1 and T_2 respectively, T_1 being higher than T_2. If the two bodies are merely brought together into thermal contact, delivering no work, the final temperature T_f reached would be the maximum

$$T_f = \frac{T_1 + T_2}{2}$$

If a heat engine is operated between the two bodies acting as heat reservoirs (Fig. 7.14), part of the heat withdrawn from body 1 is converted to work W by the heat engine, and the remainder is rejected to body 2. The lowest attainable final temperature T_f corresponds to the delivery of the largest possible amount of work, and is associated with a reversible process.

As work is delivered by the heat engine, the temperature of body 1 will be decreasing and that of body 2 will be increasing. When both the bodies attain the final temperature T_f, the heat engine will stop operating. Let the bodies remain at constant pressure and undergo no change of phase.

Total heat withdrawn from body 1

$$Q_1 = C_p\ (T_1 - T_f)$$

where C_p is the heat capacity of the two bodies at constant pressure.

Total heat rejected to body 2

$$Q_2 = C_p\ \ (T_f - T_2)$$

$\therefore$ Amount of total work delivered by the heat engine

$$W = Q_1 - Q_2$$

$$= C_p\ (T_1 + T_2 - 2T_f) \qquad (7.12)$$

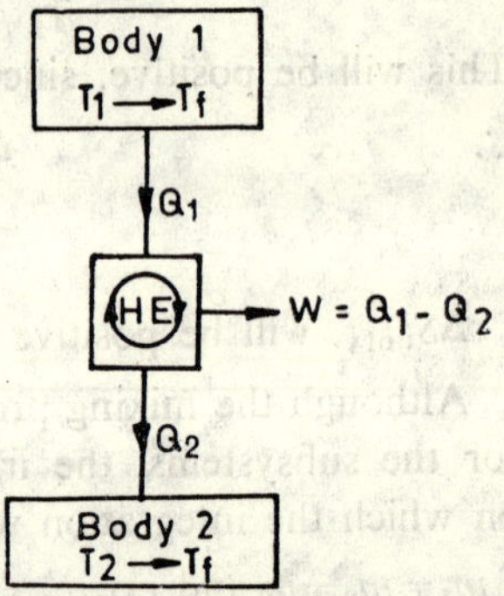

Fig. 7.14 Maximum work obtainable from two finite bodies

Now, for body 1, entropy change ΔS_1 is given by

$$\Delta S_1 = \int_{T_1}^{T_f} C_p \frac{dT}{T} = C_p \ln \frac{T_f}{T_1}$$

For body 2, entropy change ΔS_2 would be

$$\Delta S_2 = \int_{T_2}^{T_f} C_p \frac{dT}{T} = C_p \ln \frac{T_f}{T_2}$$

Since the working fluid operating in the heat engine cycle does not undergo any entropy change, ΔS of the working fluid in heat engine $= \oint dS = 0$. Applying the entropy principle

$$\Delta S_{\text{univ.}} \geqslant 0$$

$$\therefore \qquad C_p \ln \frac{T_f}{T_1} + C_p \ln \frac{T_f}{T_2} \geqslant 0$$

$$C_p \ln \frac{T_f^2}{T_1 T_2} \geqslant 0 \qquad (7.13)$$

From equation (7.12), it is seen that W is a maximum when T_f is a minimum.

From equation (7.13), for T_f to be a minimum

$$C_p \ln \frac{T_f^2}{T_1 \cdot T_2} = 0$$

$$\text{or} \qquad \ln \frac{T_f^2}{T_1 T_2} = 0 = \ln 1$$

$$\therefore \qquad T_f = \sqrt{T_1 \cdot T_2} \qquad (7.14)$$

For W to be a maximum, T_f will be $\sqrt{T_1 T_2}$. From equation (7.12)

$$W_{\max} = C_p\,(T_1 + T_2 - 2\sqrt{T_1 T_2})$$

$$= C_p(\sqrt{T_1} - \sqrt{T_2})^2$$

The final temperatures of the two bodies, initially at T_1 and T_2, can range from $(T_1 + T_2)/2$ to $\sqrt{T_1 T_2}$.

7.8 CAUSES OF ENTROPY INCREASE

The entropy of any closed system can increase in two ways:

(a) By external interaction

(b) By internal irreversibilities (or by dissipative effects, in which work is dissipated into internal energy increase).

Since $dS = \frac{\bar{d}Q_{\text{rev.}}}{T}$, when heat is added to a system, $\bar{d}Q$ is positive, and entropy of the system increases, and when heat is transferred from the system, $\bar{d}Q$ is negative, and entropy of the system decreases. To ca'culate the entropy change of the system, the irreversible path, however, has to be replaced by a reversible one.

If T is the temperature of the system, and $\bar{d}Q$ is the infinitesimal amount of heat transferred to the system during any process (Fig. 7.15), then the entropy increase dS of the system in the process will be

$$dS = d_e S + d_i S$$

$$= \frac{\bar{d}Q}{T} + d_i S \qquad (7.15)$$

where $d_e S$ is the entropy increase due to external interaction, and $d_i S$ is the entropy increase due to internal effects. Hence from equation (7.15)

$$dS > \frac{\bar{d}Q}{T}$$

or $\qquad d_i S > 0$

For a reversible process

$$dS = \frac{\bar{d}Q}{T}$$

and $\qquad d_i S = 0$

Therefore, for any process

$$d_i S > 0 \qquad (7.16)$$

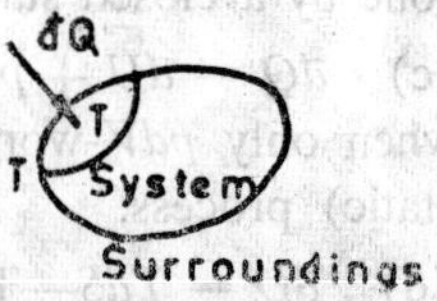

Fig. 7.15 Entropy increase due to heat addition

For an isolated system, $đQ = 0$, and therefore, $dS = d_iS$ from equation (7.15) and for any process $d_iS > 0$.

It may so happen that in a process (e.g., the expansion of a hot fluid in a turbine) the entropy decrease of the system due to heat loss to the surroundings $\left(-\int \frac{đQ}{T}\right)$ is equal to the entropy increase of the system due to internal irreversibilities such as friction, etc. ($\int d_iS$), in which case the entropy of the system before and after the process will remain the same ($\int ds = 0$). *Therefore, an isentropic process need not be adiabatic or reversible. But if the isentropic process is reversible, it must be adiabatic. Also, if the isentropic process is adiabatic, it cannot but be reversible. An adiabatic process need not be isentropic, since entropy can also increase due to friction etc. But if the process is adiabatic and reversible, it must be isentropic.*

7.9 FIRST AND SECOND LAWS COMBINED

By the second law

$$đQ_{\text{rev.}} = TdS$$

and by the first law, for a closed non-flow system,

$$đQ = dU + pdV$$

$$TdS = dU + pdV \tag{7.17}$$

Again, the enthalpy

$$H = U + pV$$

$$dH = dU + pdV + Vdp$$

$$= TdS + Vdp$$

$$TdS = dH - Vdp \tag{7.18}$$

Equations (7.17) and (7.18) are the thermodynamic equations relating the properties of the system.

Let us now examine the following equations as obtained from the first and second laws:

(a) $đQ = dE + đW$—This equation holds good for any process, reversible or irreversible, and for any system.

(b) $đQ = dU + đW$—This equation holds good for any process undergone by a closed stationary system.

(c) $đQ = đU + pdV$—This equation holds good for a closed system when only pdV-work is present. This is true only for a reversible (quasi-static) process.

(d) $đQ = TdS$—This equation is true only for a reversible process.

(e) $TdS = dU + pdV$—This equation holds good for any process reversible or irreversible, undergone by a closed system, since it is a relation among properties which are independent of the path.

(f) $TdS = dH - Vdp$—This equation also relates only the properties of the closed system. There is no path function term in the equation. Hence the equation holds good for any process.

7.10 ENTROPY AND DIRECTION : THE SECOND LAW A DIRECTIONAL LAW OF NATURE

Since the entropy of an isolated system can never decrease, it follows that only those processes are possible in nature which would give an entropy increase for the system and the surroundings together (the universe). All spontaneous processes in nature occur only in one direction from a higher to a lower potential, and these are accompanied by an entropy increase of the universe. When the potential gradient is infinitesimal (or zero in the limit), the entropy change of the universe is zero, and the process is reversible. The second law indicates the direction in which a process takes place. *A process always occurs in such a direction as to cause an increase in the entropy of the universe.* The macroscopic change ceases only when the potential gradient disappears and the equilibrium is reached when the entropy of the universe assumes a maximum value. To determine the equilibrium state of an isolated system it is necessary to express the entropy as a function of certain properties of the system and then render the function a maximum. At equilibrium, the system (isolated) exists at the peak of the entropy-hill, and $dS = 0$.

7.11 ENTROPY AND DISORDER

Work is a macroscopic concept. Work involves order or the orderly motion of molecules, as in the expansion or compression of a gas. The kinetic energy and potential energy of a system represent orderly forms of energy. The kinetic energy of a gas is due to the coordinated motion of all the molecules with the same average velocity in the same direction. The potential energy is due to the vantage position taken by the molecules or displacements of molecules from their normal positions. Heat or thermal energy is due to the random thermal motion of molecules in a completely disorderly fashion and the average velocity is zero. Orderly energy can be readily converted into disorderly energy, e.g., mechanical and electrical energies into internal energy (and then heat) by friction and Joule effect. Orderly energy can also be converted into one another. *But there are natural limitations on the conversion of disorderly energy into orderly energy, as delineated by the second law.* When work is dissipated into internal energy, the disorderly motion of molecules is increased. Two gases, when mixed, represent a higher degree of disorder than when they are separated. An irreversible process always tends to take the system (isolated) to a state of greater disorder. It is a tendency on the part of nature to proceed to a state of greater disorder. An isolated system always tends to a state of greater entropy. So there is a close link between entropy and disorder. It may be stated roughly that *the entropy of a system is a measure of the degree of molecular disorder existing in the system.* When heat is imparted to a system, the disorderly motion of molecules increases, and so the entropy of the system increases. The reverse occurs when heat is removed from the system.

Ludwig Boltzmann (1877) introduced statistical concepts to define disorder by attaching to each state a *thermodynamic probability*, expressed by the quantity W, which is greater the more disordered the state is. The increase of entropy implies that the system proceeds by itself from one state to another with a higher thermodynamic probability (or disorder number). An irreversible process goes on until the *most probable state* (equilibrium state when W is maximum) corresponding to the maximum value of entropy is reached. Boltzmann assumed a functional relation between S and W. while entropy is additive, probability is multiplicative. If the two parts A and B of a system in equilibrium are considered, the entropy is the sum

$$S = S_A + S_B$$

and the thermodynamic probability is the product

$$W = W_A \cdot W_B$$

Again, $S = S(W)$, $S_A = S(W_A)$, and $S_B = S(W_B)$

$$\therefore \quad S(W) = S(W_A W_B) = S(W_A) + S(W_B)$$

which is a well-known functional equation for the logarithm. Thus the famous relation is reached

$$S = K \ln W$$

where K is a constant, known as Boltzmann constant. This is engraved upon Boltzmann's tombstone in Vienna.

When $W = 1$, which represents the greatest order, $S = 0$. This occurs only at $T = O$ K. This state cannot be reached in a finite number of operations. This is called the third law of thermodynamics. In the case of a gas, W increases due to an increase in volume V or temperature T. In the reversible adiabatic expansion of a gas the increase in disorder due to an increase in volume is just compensated by the decrease in disorder due to a decrease in temperature, so that the disorder number or entropy remains constant.

7.12 ABSOLUTE ENTROPY

It is important to note that one is interested only in the amount by which the entropy of the system changes in going from an initial to a final state, and not in the value of absolute entropy. In cases where it is necessary, a zero value of entropy of the system at an arbitrarily chosen standard state is assigned, and the entropy changes are calculated with reference to this standard state.

7.13 POSTULATORY THERMODYNAMICS

The property 'entropy' plays the central role in thermodynamics. In the *classical approach,* as followed in this book, entropy is introduced via the concept of the heat engine. It follows the way in which the subject of thermodynamics developed historically mainly through the contributions of Sadi Carnot, James Prescott Joule, William Thomson (Lord Kelvin), Rudolf Clausius, and Max Planck.

In the *postulatory approach,* as developed by H.B. Callen (see Reference), entropy is introduced at the beginning. The development of the subject has been based on four postulates. Postulate I defines the equilibrium state. Postulate II introduces the property 'entropy' which is rendered maximum at the final equilibrium state. Postulate III refers to the additive nature of entropy which is a monotonically increasing function of energy. Postulate IV mentions that the entropy of any system vanishes at the absolute zero of temperature. With the help of these postulates the conditions of equilibrium under different constraints have been developed.

EXAMPLE 7.1

Water flows through a turbine in which friction causes the water temperature to rise from 35°C to 37°C. If there is no heat transfer, how much does the entropy of the water change in passing through the turbine? (Water is incompressible and the process can be taken to be at constant volume.)

Solution

The presence of friction makes the process irreversible and causes an entropy increase for the system. The flow process is indicated by the dotted line joining the equilibrium states 1 and 2 (Fig. Ex. 7.1). Since entropy is a state property and the entropy change depends only on the two end states and is independent of the path the system follows, to find the entropy change, the irreversible path has to be replaced by a reversible path, as shown in the figure, because no integration can be made on a path other than a reversible path,

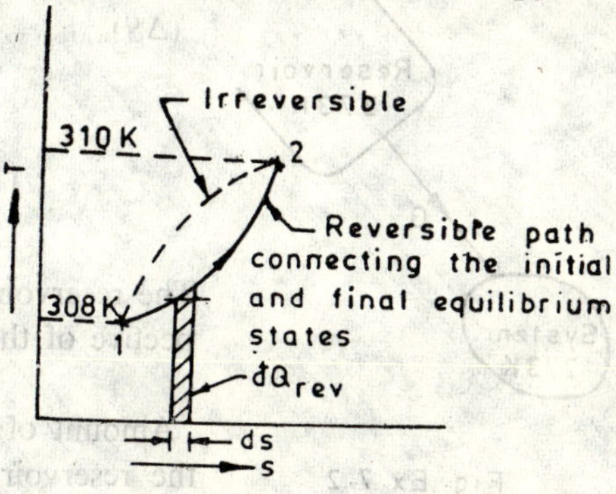

Fig. Ex. 7.1

$$T_2 = 37 + 273 = 310 \text{ K}$$
$$T_1 = 35 + 273 = 308 \text{ K}$$

We have

$$đQ_{\text{rev.}} = TdS$$

$$dS = \frac{m\, c_v\, dT}{T}$$

$$S_2 - S_1 = m\, c_v \ln \frac{T_2}{T_1}$$

$$= 1 \times 4.187 \ \ln \frac{310}{308}$$

$$= 0.0243 \ \text{kJ/K}$$ Ans.

EXAMPLE 7.2

(a) One kg of water at 273 K is brought into contact with a heat reservoir at 373 K. When the water has reached 373 K, find the entropy change of the water, of the heat reservoir, and of the universe.

(b) If water is heated from 273 K to 373 K by first bringing it in contact with a reservoir at 323 K and then with a reservoir at 373 K, what will the entropy change of the universe be?

(c) Explain how water might be heated from 273 K to 373 K with almost no change in the entropy of the universe.

Solution

(a) Water is being heated through a finite temperature difference (Fig. Ex. 7.2). The entropy of water would increase and that of the reservoir would decrease so that the net entropy change of the water (system) and the reservoir together would be positive definite. Water is being heated irreversibly, and to find the entropy change of water, we have to assume a reversible path between the end states which are at equilibrium.

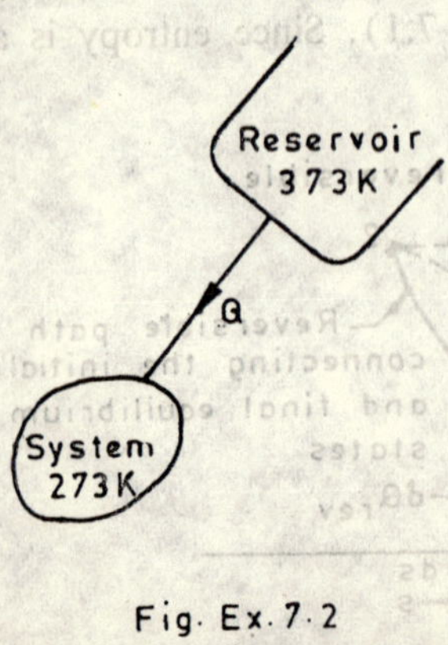

Fig. Ex. 7.2

$$(\Delta S)_{\text{water}} = \int_{T_1}^{T_2} \frac{đQ_r}{T} = \int_{T_1}^{T_2} \frac{mc\,dT}{T} = m\,c \ln \frac{T_2}{T_1}$$

$$= 1 \times 4.187 \ \ln \frac{373}{273} = 1.305 \ \text{kJ/K}$$

The reservoir temperature remains constant irrespective of the amount of heat withdrawn from it.

Amount of heat absorbed by the system from the reservoir,

$$Q = 1 \times 4.187 \times (373 - 273) = 418.7 \ \text{kJ}.$$

$\therefore$ Entropy change of the reservoir

$$(\Delta S)_{res.} = -\frac{Q}{T} = -\frac{418.7}{373} = -1.122 \text{ kJ/K}$$

$\therefore$ Entropy change of the universe

$$(\Delta S)_{univ.} = (\Delta S)_{system} + (\Delta S)_{res.}$$

$$= 1.305 - 1.122 = 0.183 \text{ kJ/K} \qquad \text{Ans.(a).}$$

(b) Water is being heated in two stages, first from 273 K to 323 K by bringing it in contact with a reservoir at 323 K, and then from 323 K to 373 K by bringing it in contact of a second reservoir at 373 K.

$$(\Delta S)_{water} = \int_{273\text{ K}}^{323\text{ K}} m\,c\,\frac{dT}{T} + \int_{323\text{ K}}^{373\text{ K}} m\,c\,\frac{dT}{T}$$

$$= 4.187\left(\ln\frac{323}{273} + \ln\frac{373}{323}\right) = (0.1673 + 0.1441)\ 4.187$$

$$= 1.305 \text{ kJ/K}$$

$$(\Delta S)_{res\cdot I} = -\frac{1 \times 4.187 \times (323 - 273)}{323} = -0.647 \text{ kJ/K}$$

$$(\Delta S)_{res\cdot II} = -\frac{1 \times 4.187 \times (373 - 323)}{373} = -0.56 \text{ kJ/K}$$

$$\therefore \quad (\Delta S)_{univ.} = (\Delta S)_{water} + (\Delta S)_{res\cdot I} + (\Delta S)_{res\cdot II}$$

$$= 1.305 - 0.647 - 0.56$$

$$= 0.098 \text{ kJ/K} \qquad \text{Ans.(b).}$$

(c) The entropy change of the universe would be less and less, if the water is heated in more and more stages, by bringing the water in contact successively with more and more heat reservoirs, each succeeding reservoir being at a higher temperature than the preceding one.

When water is heated in infinite steps, by bringing it in contact with an infinite number of reservoirs in succession, so that at any instant the temperature difference between the water and the reservoir in contact is infinitesimally small, then the net entropy change of the universe would be zero, and the water would be *reversibly* heated.

EXAMPLE 7.3

One kg of ice at −5°C is exposed to the atmosphere which is at 20°C. The ice melts and comes into thermal equilibrium with the atmosphere. (a) Determine the entropy increase of the universe. (b) What is the minimum amount of work necessary to convert the water back into ice at −5°C? c_p of ice is 2.093 kJ/kg°C and the latent heat of fusion of ice is 333.3 kJ/kg.

Solution

Heat absorbed by ice Q from the atmosphere (Fig. Ex. 7.3.1)

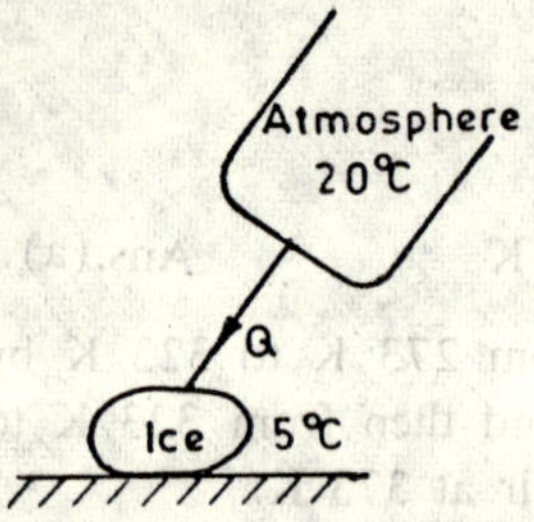

Fig. Ex. 7.3.1

$$= \text{Heat absorbed in solid phase} + \text{Latent heat} + \text{Heat absorbed in liquid phase}$$

$$= 1 \times 2.093 \times [0-(-5)] + 1 \times 333.3 + 1 \times 4.187 \times (20-0)$$

$$= 427.5 \text{ kJ}$$

Entropy change of the atmosphere

$$(\Delta S)_{\text{atm.}} = -\frac{Q}{T} = -\frac{427.5}{293} = -1.46 \text{ kJ/K}$$

Entropy change of the system (ice) as it gets heated from −5°C to 0°C

$$(\Delta S_I)_{\text{system}} = \int_{268}^{273} m\, c_p \frac{dT}{T} = 1 \times 2.093 \ln \frac{273}{268} = 2.093 \times 0.0186$$

$$= 0.0389 \text{ kJ/K}$$

Entropy change of the system as ice melts at 0°C to become water at 0°C

$$(\Delta S_{II})_{\text{system}} = \frac{333.3}{273} = 1.22 \text{ kJ/K}$$

Entropy change of water as it gets heated from 0°C to 20°C

$$(\Delta S_{III})_{\text{system}} = \int_{273}^{293} m\, c_p \frac{dT}{T} = 1 \times 4.187 \ln \frac{293}{273} = 0.296 \text{ kJ/K}$$

Total entropy change of ice as it melts into water

$$(\Delta S)_{\text{total}} = \Delta S_I + \Delta S_{II} + \Delta S_{III}$$

$$= 0.0389 + 1.22 + 0.296$$

$$= 1.5549 \text{ kJ/K}$$

The entropy-temperature diagram for the system as ice at −5°C converts to water at 20°C is shown in Fig. Ex. 7.3.2.

∴ Entropy increase of the universe

$$(\Delta S)_{\text{univ.}} = (\Delta S)_{\text{system}} + (\Delta S)_{\text{atm.}}$$

$$= 1.5549 - 1.46 = 0.0949 \text{ kJ/K} \qquad \text{Ans. (a).}$$

(b) To convert 1 kg of water at 20°C to ice −5°C, 427.5 kJ of heat have to be removed from it, and the system has to be brought from state 4 to state 1 (Fig. Ex. 7.3.2). A refrigerator cycle, as shown in Fig. Ex. 7.3.3, is assumed to accomplish this.

The entropy change of the system would be the same, i.e., $S_4 - S_1$, with the only difference that its sign will be negative, because heat is removed from the system (Fig. Ex. 7.3.2).

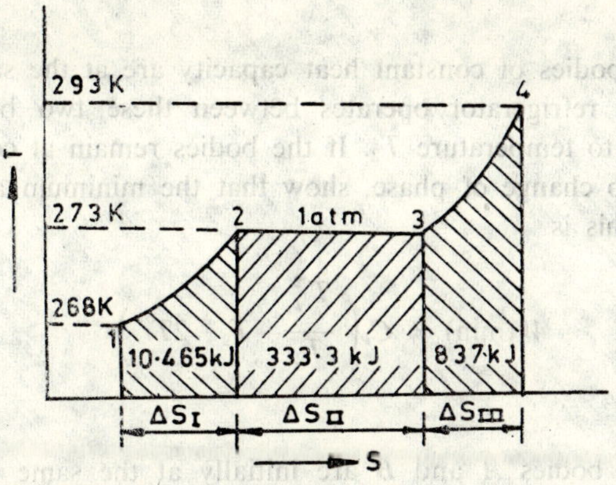

Fig. Ex. 7·3·2

$$(\Delta S)_{\text{system}} = \underset{(\text{negative})}{S_1 - S_4}$$

The entropy change of the working fluid in the refrigerator would be zero, since it is operating in a cycle, i.e.,

$$(\Delta S)_{\text{ref.}} = 0$$

The entropy change of the atmosphere (positive)

$$(\Delta S)_{\text{atm.}} = \frac{Q+W}{T}$$

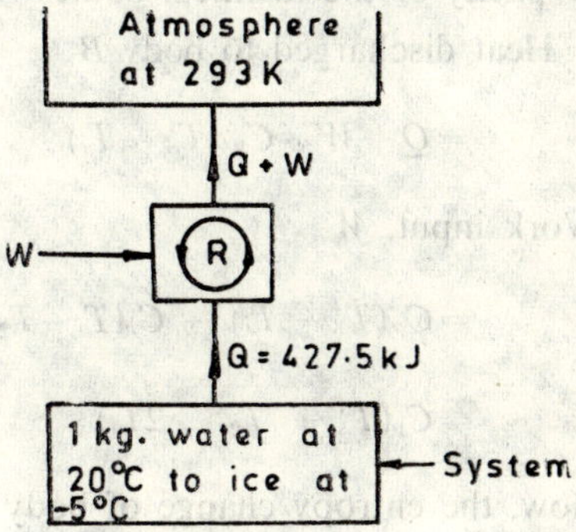

Fig. Ex. 7·3·3

$\therefore$ Entropy change of the universe

$$(\Delta S)_{\text{univ.}} = (\Delta S)_{\text{system}} + (\Delta S)_{\text{ref.}} + (\Delta S)_{\text{atm}}$$

$$= (S_1 - S_4) + \frac{Q+W}{T}$$

By the principle of increase of entropy

$$(\Delta S)_{\text{univ. or isolated system}} \geqslant 0$$

$$\therefore \quad \left[(S_1 - S_4 +) \frac{Q+W}{T}\right] \geqslant 0$$

$$\therefore \quad \frac{Q+W}{T} \geqslant (S_4 - S_1)$$

$$W \geqslant T(S_4 - S_1) - Q$$

$$\therefore \quad W(\text{min}) = T(S_4 - S_1) - Q$$

Here $\quad Q = 427.5$ kJ

$T = 293$ K

$S_4 - S_1 = 1.5549$ kJ/K

$$\therefore \quad W(\text{min}) = 293 \times 1.5549 - 427.5$$

$$= 28.5 \text{ kJ}$$

Ans. (b)

EXAMPLE 7.4

Two identical bodies of constant heat capacity are at the same initial temperature T_i. A refrigerator operates between these two bodies until one body is cooled to temperature T_2. If the bodies remain at constant pressure and undergo no change of phase, show that the minimum amount of work needed to do this is

$$W(\text{min}) = C_p\left(\frac{T_i^2}{T_2}+T_2-2T_i\right)$$

Solution

Both the finite bodies A and B are initially at the same temperature T_i. Body A is to be cooled to temperature T_2 by operating the refrigerator cycle, as shown in Fig. Ex. 7.4. Let T_2' be the final temperature of body B.

Heat removed from body A to cool it from T_i to T_2

$$Q = C_p\ (T_i - T_2)$$

where C_p is the constant pressure heat capacity of the identical bodies A and B.

Heat discharged to body B

$$= Q + W = C_p\ (T_2' - T_i)$$

Work input, W

$$= C_p(T_2' - T_i) - C_p(T_i - T_2)$$

$$= C_p(T_2' + T_2 - 2T_i) \qquad (7.4.1)$$

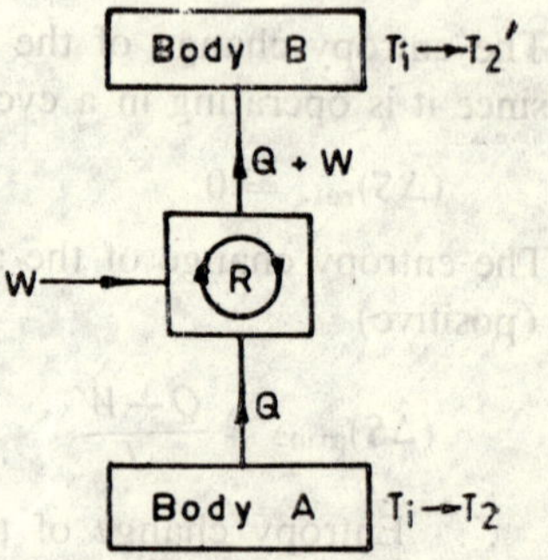

Fig. Ex. 7.4

Now, the entropy change of body A

$$\Delta S_A = \int_{T_i}^{T_2} C_p \frac{dT}{T} = C_p \ln \frac{T_2}{T_i} \text{ (negative)}$$

The entropy change of body B

$$\Delta S_B = \int_{T_i}^{T_2'} C_p \frac{dT}{T} = C_p \ln \frac{T_2'}{T_i} \text{ (positive)}$$

Entropy change (cycle) of refrigerant = 0

$\therefore$ Entropy change of the universe

$$(\Delta S)_{\text{univ.}} = \Delta S_A + \Delta S_B$$

$$= C_p \ln \frac{T_2}{T_i} + C_p \ln \frac{T_2'}{T_i}$$

By the entropy principle

$$(\Delta S)_{\text{univ.}} \geqslant 0$$

$$\left(C_p \ln \frac{T_2}{T_i} + C_p \ln \frac{T'_2}{T_i} \right) \geqslant 0$$

$$C_p \ln \frac{T_2 T'}{T_i^2} \geqslant 0 \tag{7.4.2}$$

In equation (7.4.1) with C_p, T_2, and T_i being given, W will be a minimum when T_2' is a minimum. From equation (7.4.2), the minimum value of T_2' would correspond to

$$C_p \ln \frac{T_2 T'_2}{T_i^2} = 0 = \ln 1$$

$$\therefore \qquad T'_2 = \frac{T_i^2}{T_2}$$

From equation (7.4.1)

$$W\,(\text{min}) = C_p \left(\frac{T_i^2}{T_2} + T_2 - 2\,T_i \right) \qquad \text{Proved.}$$

EXAMPLE 7.5

Three identical finite bodies of constant heat capacity are at temperatures 300, 300, and 100 K. If no work or heat is supplied from outside, what is the highest temperature to which any one of the bodies can be raised by the operation of heat engines or refrigerators?

Solution

Let the three identical bodies A, B, and C having the same heat capacity C be respectively at 300 K, 100 K and 300 K initially, and let us operate a heat engine and a refrigerator, as shown in Fig. Ex. 7.5. Let T_f be the final temperature of bodies A and B, and T_f be the final temperature of body C. Now

$$(\Delta S)_A = C \ln \frac{T_f}{300}$$

$$(\Delta S)_B = C \ln \frac{T_f}{100}$$

$$(\Delta S)_C = C \ln \frac{T_f'}{300}$$

$$(\Delta S)_{H.E.} = 0$$

$$(\Delta S)_{\text{ref.}} = 0$$

where C is the heat capacity of each of the three bodies.

Since $(\Delta S)_{\text{univ.}} \geq 0$

$$\left(C \ln \frac{T_f}{300} + C \ln \frac{T_f}{100} + C \ln \frac{T'_f}{300}\right) \geq 0$$

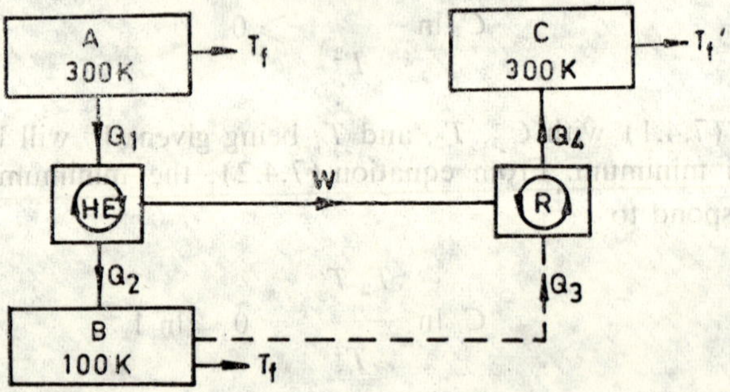

Fig. Ex. 7.5

Since the entropy of an isolated system (or universe) always increases and becomes a maximum at the state of equilibrium, any conceivable change in the entropy of the isolated system will be zero when the system is at equilibrium, T_f and T_f' being the final equilibrium temperatures.

$$\therefore \quad C \ln \frac{T_f}{300} + C \ln \frac{T_f}{100} + C \ln \frac{T'_f}{300} = 0$$

$$\therefore \quad C \ln \frac{T_f^2 \, T'_f}{9{,}000{,}000} = 0$$

$$\therefore \quad T_f^2 \, T'_f = 9{,}000{,}000 \tag{7.5.1}$$

Now

$$Q_1 = C(300 - T_f)$$

$$Q_2 = C(T_f - 100)$$

$$Q_4 = C(T'_f - 300)$$

Again

$$Q_1 = \text{Heat removed from body } A$$
$$= \text{Heat discharged to bodies } B \text{ and } C$$
$$= Q_2 + Q_4$$

$$\therefore \quad C(300 - T_f) = C(T_f - 100) + C(T'_f - 300)$$

$$\therefore \quad T'_f = 700 - 2T_f \tag{7.5.2}$$

From equations (7.5.1) and (7.5.2)

$$T_f^2\,(700-2T_f) = 9{,}000{,}000$$

$$\therefore \quad 2T_f^3 - 700\,T_f^2 + 9{,}000{,}000 = 0$$

or $$T_f = 150 \text{ K}$$

From equation (7.5.2)

$$T_f' = (700-2\times150) \text{ K}$$

$$= 400 \text{ K}$$ Ans.

EXAMPLE 7.6

A system has a heat capacity at constant volume

$$C_V = AT^2$$

where $A = 0.042$ J/K³

The system is originally at 200 K, and a thermal reservoir at 100 K is available. What is the maximum amount of work that can be recovered as the system is cooled down to the temperature of the reservoir?

Solution

Heat removed from the system (Fig. Ex. 7.6)

$$Q_1 = \int_{T_1}^{T_2} C_V dT = \int_{T_1=200\text{ K}}^{T_2=100\text{ K}} 0.042\, T^2 dT$$

$$= 0.042 \left[\frac{T^3}{3}\right]_{200\text{K}}^{100\text{K}}$$

$$= \frac{0.042}{3} \text{ J/K}^3\, (100^3 - 200^3)\text{K}^3 = -98\times10^3 \text{ J}$$

$$(\Delta S)_{\text{system}} = \int_{200\text{ K}}^{100\text{ K}} C_V \frac{dT}{T} = \int_{200\text{ K}}^{100\text{ K}} 0.042\, T^2 \frac{dT}{T}$$

$$= \frac{0.042}{2} \text{ J/K}^3\, [100^2 - 200^2]\text{ K}^2$$

$$= -630 \text{ J/K}$$

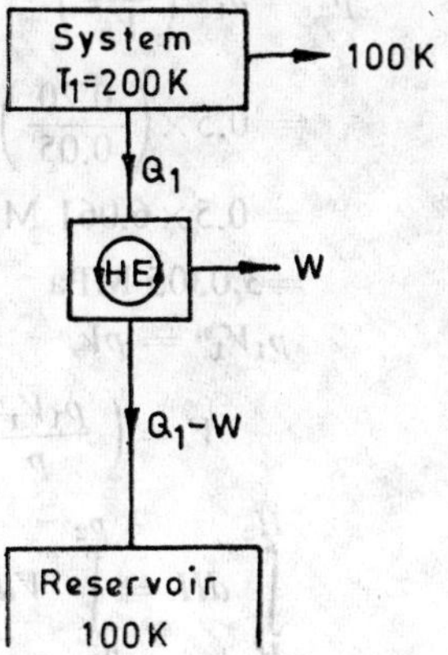

Fig. Ex. 7·6

$$(\Delta S)_{\text{res.}} = \frac{Q_1 - W}{T_{\text{res.}}} = \frac{98\times10^3 - W}{100} \text{ J/K}$$

$$(\Delta S)_{\text{working fluid in H.E.}} = 0$$

$$\therefore \quad (\Delta S)_{\text{univ.}} = (\Delta S)_{\text{system}} + (\Delta S)_{\text{res.}}$$

$$= -630 + \frac{98\times10^3 - W}{100}$$

$$\text{Since} \quad (\Delta S)_{\text{univ.}} \geqslant 0$$

$$\therefore \quad -630 + \frac{98 \times 10^3 - W}{100} \geqslant 0$$

$$980 - \frac{W}{100} - 630 \geqslant 0$$

$$\frac{W}{100} \leqslant 350$$

$$W(\text{max}) = 35{,}000 \text{ J} = 35 \text{ kJ} \qquad \text{Ans.}$$

EXAMPLE 7.7

A fluid undergoes a reversible adiabatic compression from 0.5 MPa, 0.2 m³ to 0.05 m³ according to law, $pv^{1.3}$ = constant. Determine the change in enthalpy, internal energy and entropy, and the heat transfer and work transfer during the process.

Solution

$$TdS = dH - VdP$$

For the reversible adiabatic process, (Fig. Ex. 7.7)

$$dH = VdP$$

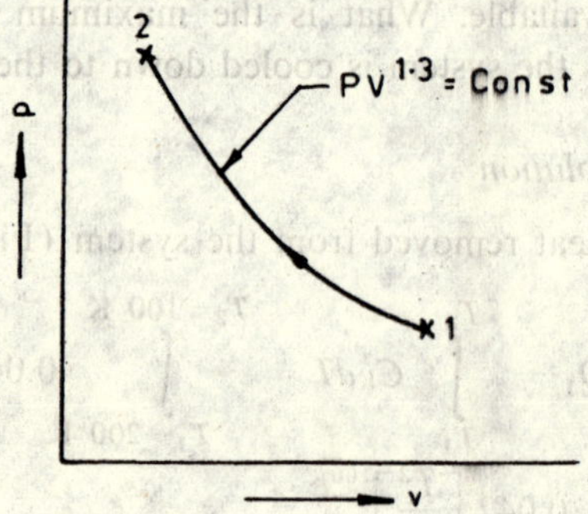

Fig. Ex. 7·7

$$p_1 = 0.5 \text{ MPa}, \quad V_1 = 0.2 \text{ m}^3$$

$$V_2 = 0.05 \text{ m}^3, \quad p_1 V_1^n = p_2 V_2^n$$

$$\therefore \quad p_2 = p_1 \left(\frac{V_1}{V_2}\right)^n$$

$$= 0.5 \times \left(\frac{0.20}{0.05}\right)^{1.3} \text{ MPa}$$

$$= 0.5 \times 6.061 \text{ MPa}$$

$$= 3.0305 \text{ MPa}$$

$$p_1 V_1^n = p V^n$$

$$\therefore \quad V = \left(\frac{p_1 V_1^n}{p}\right)^{1/n}$$

$$\int_{H}^{H_2} dH = \int_{p_1}^{p_2} V dp$$

$$H_2 - H_1 = \int_{p_1}^{p_2} \left[\left(\frac{p_1 V_2^n}{p}\right)\right]^{1/n} dp$$

$$= (p_1 V_1^n)^{1/n} \left(\frac{p_2^{1-1/n} - p_1^{1-1/n}}{1 - 1/n}\right)$$

$$= \frac{n(p_2 V_2 - p_1 V_1)}{n - 1}$$

$$= \frac{1.3\,(3030.5 \times 0.05 - 500 \times 0.2)}{1.3 - 1}$$

$$= 223.3 \text{ kJ}$$

$$H_2 - H_1 = (U_2 + p_2V_2) - (U_1 - p_1V_1)$$

$$= (U_2 - U_1) + (p_2V_2 - p_1V_1)$$

$$\therefore \quad U_2 - U_1 = (H_2 - H_1) - (p_2V_2 - p_1V_1)$$

$$= 223.3 - 51.53$$

$$= 171.77 \text{ kJ}$$ Ans.

$$S_2 - S_1 = 0$$ Ans.

$$Q_{1-2} = 0$$ Ans.

$$Q_{1-2} = U_2 - U_1 + W_{1-2}$$

$$\therefore \quad W_{1-2} = U_1 - U_2 = -171.77 \text{ kJ}$$ Ans.

EXAMPLE 7.8

Air is flowing steadily in an insulated duct. The pressure and temperature measurements of the air at two stations A and B are given below. Establish the direction of the flow of air in the duct. Assume that for air, specific heat c_p is constant at 1.005 kJ/kg-K, $h = c_p\,T$, and $\frac{v}{T} = \frac{0.287}{P}$, where p, v, and T are pressure (in kPa), volume (in m³/kg) and temperature (in K) respectively.

	Station A	*Station B*
Pressure	130 kPa	100 kPa
Temperature	50°C	13°C

Solution

From property relation

$$T ds = dh - v dp$$

$$dS = \frac{dh}{T} - v\,\frac{dp}{T}$$

For two states at A and B the entropy change of the system

$$\int_{S_A}^{S_B} ds = \int_{T_A}^{T_B} \frac{c_p\,dT}{T} - \int_{p_A}^{p_B} 0.287 \;\frac{dP}{p}$$

$$\therefore \quad s_B - s_A = 1.005 \ln \frac{T_B}{T_A} - 0.287 \ln \frac{p_B}{p_A}$$

$$= 1.005 \;\ln \frac{273+13}{273+50} - 0.287 \ln \frac{100}{130}$$

$$= -0.1223 + 0.0753$$

$$= -0.047 \text{ kJ/kg K}$$

$$\therefore \quad (\Delta s)_{\text{system}} = -0.047 \text{ kJ/kg K}$$

Since the duct is insulated, $(\Delta s)_{surr.} = 0$

$\therefore \quad (\Delta s)_{univ.} = -0.047$ kJ/kg K

This is impossible. So the flow must be from B to A.

PROBLEMS

7.1 On the basis of the first law fill in the blank spaces in the following table of imaginary heat engine cycles. On the basis of the second law classify each cycle as reversible, irreversible, or impossible.

Cycle	*Temperature* *Source*	*Sink*	*Rate of Heat Flow* *Supply*	*Rejection*	*Rate of work output*	*Efficiency*
(a)	327°C	27°C	420 kJ/s	230 kJ/s	...kW	—
(b)	1000°C	100°C	... kJ/min	4.2 MJ/min	...kW	65%
(c)	750 K	300 K	... kJ/s	... kJ/s	26 kW	60%
(d)	700 K	300 K	2500 kcal/hr	.., kcal/hr	1 kW	—

7.2 The latent heat of fusion of water at 0°C is 335 kJ/kg. How much does the entropy of 1 kg of ice change as it melts into water in each of the following ways: (a) Heat is supplied reversibly to a mixture of ice and water at 0°C. (b) A mixture of ice and water at 0°C is stirred by a paddle wheel.

7.3 Two kg of water at 80°C are mixed adiabatically with 3 kg of water at 30°C in a constant pressure process of 1 atmosphere. Find the increase in the entropy of the total mass of water due to the mixing processs (c_p of water = 4.187 kJ/kg K).

Ans. 0.0576 kJ/K.

7.4 In a Carnot cycle, heat is supplied at 350°C and rejected at 27°C. The working fluid is water which, while receiving heat, evaporates from liquid at 350°C to steam at 350°C. The associated entropy change is 1.44 kJ/kg K. (a) If the cycle operates on a stationary mass of 1 kg of water, how much is the work done per cycle, and how much is the heat supplied? (b) If the cycle operates in steady flow with a power output of 20 kW, what is the steam flow rate?

Ans. (a) 465.12, 897.12 kJ/kg, (b) 0.043 kg/s.

7.5 A heat engine receives reversibly 420 kJ/cycle of heat from a source at 327°C, and rejects heat reversibly to a sink at 27°C. There are no other heat transfers. For each of the three hypothetical amounts of heat rejected, in (a), (b), and (c) below, compute the cyclic integral of $\frac{đQ}{T}$. From these results show which case is irreversible, which reversible, and which impossible (a) 210 kJ/cycle rejected, (b) 105 kJ/cycle rejected, (c) 315 kJ/cycle rejected.

Ans. (a) Reversible, (b) Impossible, (c) Irreversible.

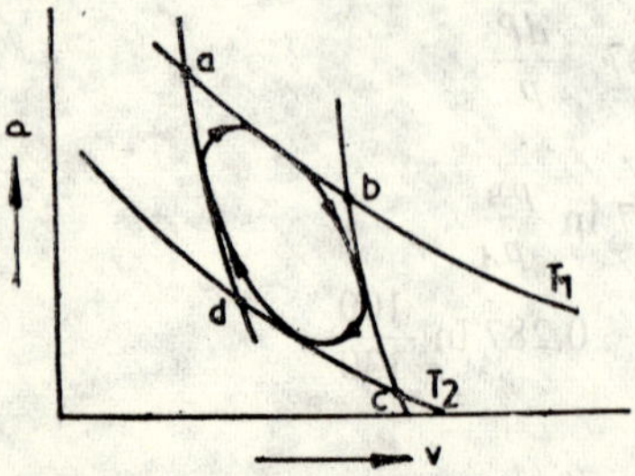

Fig. P. 7.6

7.6 In Fig. P. 7.6, abcd represents a Carnot cycle bounded by two reversible adiabatics and two reversible isotherms at temperatures T_1 and T_2 ($T_1 < T_2$). The oval figure is a reversible cycle, where heat is absorbed at temperatures less than, or equal to, T_1, and rejected at temperatures greater than, or equal to, T_2. Prove that the efficiency of the oval cycle is less than that of the Carnot cycle.

7.7 Water is heated at a constant pressure of 0.7 MPa. The boiling point is 164.97°C. The initial temperature of water is 0°C. The latent heat of evaporation is 2066.3 kJ/kg. Find the increase of entropy of water, if the final state is steam.

Ans. 6.6967 kJ/kg K.

7.8 One kg of air initially at 0.7 MPa, 20°C changes to 0.35 MPa, 60°C by the three reversible non-flow processes, as shown in Fig. P. 7.8. Process 1-*a*-2 consists of a constant pressure expansion followed by a constant volume cooling, process 1-*b*-2 an isothermal expansion followed by a constant pressure expansion, and process 1-*c*-2 an adabatic expansion followed by a constant volume heating. Determine the change of internal energy, enthalpy, and entropy for each process, and find the work transfer and heat transfer for each process. Take $c_p = 1.005$ and $c_v = 0.718$ kJ/kg, K, and assume the specific heats to be constant. Also assume for air $pv = 0{,}287$ T, where p is the pressure in kPa, v the specific volume in m³/kg, and T the temperature in K.

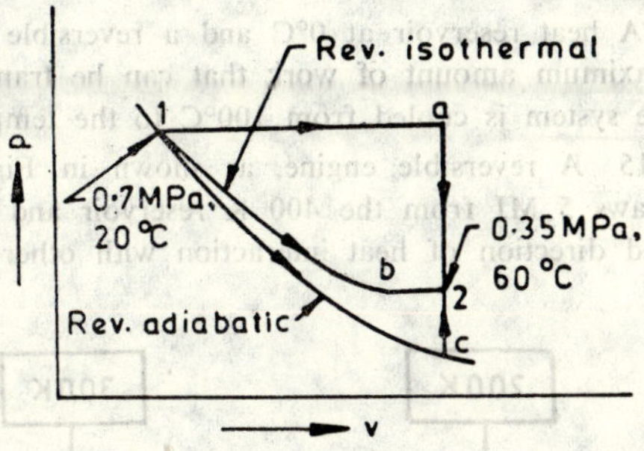

Fig. P. 7·8

7.9 Ten grammes of water at 20°C is converted into ice at −10°C at constant atmospheric pressure. Assuming the specific heat of liquid water to remain constant at 4.2 J/g °C and that of ice to be half of this value, and taking the latent heat of fusion of ice at 0°C to be 335 J/g, calculated the total entropy change of the system.

Ans. −0.1601 kJ/K.

7.10 Calculate the entropy change of the universe as a result of the following processes:

(a) A copper block of 600 g mass and with C_p of 150 J/°C at 100°C is placed in a lake at 8°C.

(b) The same block, at 8°C, is dropped from a height of 100 m into the lake.

(c) Two such blocks, at 100 and 0°C, are joined together.

Ans. (a) 6.69 J/K, (b) 2.095 J/K, (c) 3.64 J/K.

7.11 A system maintained at constant volume is initially at temperature T_1, and a heat reservoir at the lower temperature T_0 is available. Show that the maximum work recoverable as the system is cooled to T_0 is

$$W = C_v \left[(T_1 - T_0) - T_1 \ln \frac{T_1}{T_0} \right]$$

7.12 A body of finite mass is originally at temperature T_1 which is higher than that of a reservoir at temperature T_2. Suppose an engine operates in a cycle between the body and the reservoir until it lowers the temperature of the body from T_1 to T_2, thus extracting heat Q from the body. If the engine does work W, then it will reject heat Q-W to the reservoir at T_2. Applying the entropy principle, prove that the maximum work obtainable from the engine is

$$W(\max) = Q - T_2 \ (S_1 - S_2)$$

where $S_1 - S_2$ is the entropy decrease of the body.

If the body is maintained at constant volume having constant volume heat capacity $C_v = 8.4$ kJ/K which is independent of temperature, and if $T_1 = 373$ K and $T_2 = 303$ K, determine the maximum work obtainable.

7.13 Each of three identical bodies satisfies the equation $U = CT$, where C is the heat capacity of each of the bodies. Their initial temperatures are 200 K, 250 K, and 540 K. If $C = 8.4$ kJ/K, what is the maximum amount of work that can be extracted in a process in which these bodies are brought to a final common temperature?

7.14 In the temperature range between 0°C and 100°C a particular system maintained at constant volume has a heat capacity.

$$C_V = A + 2BT$$

with $A = 0.014$ J/K and $B = 4.2 \times 10^{-4}$ J/K²

A heat reservoir at 0°C and a reversible work source are available. What is the maximum amount of work that can be transferred to the reversible work source as the system is cooled from 100°C to the temperature of the reservoir?

7.15 A reversible engine, as shown in Fig. P. 7.15, during a cycle of operation draws 5 MJ from the 400 K reservoir and does 840 kJ of work. Find the amount and direction of heat interaction with other reservoirs.

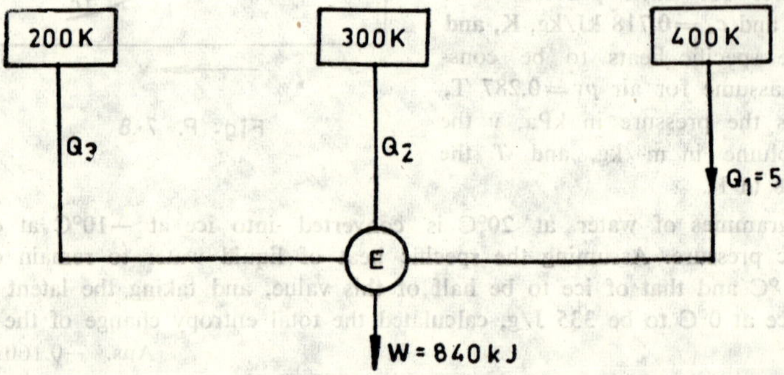

Fig. P. 7.15

7.16 For a fluid for which pv/T is a constant quantity equal to R, show that the change in specific entropy between two states A and B is given by

$$s_B - s_A = \int_{T_A}^{T_B} \left(\frac{C_p}{T}\right) dT - R \ln \frac{p_B}{p_A}$$

A fluid for which R is a constant and equal to 0.287 kJ/kg K, flows steadily through an adiabatic machine, entering and leaving through two adiabatic pipes. In one of these pipes the pressure and temperature are 5 bar and 450 K and in the other pipe the pressure and temperature are 1 bar and 300 K respectively. Determine which pressure and temperature refer to the inlet pipe.

For the given temperature range, c_p is given by

$$c_p = a \ln T + b$$

where T is the numerical value of the absolute temperature and $a = 0.026$ kJ/kg K, $b = 0.86$ kJ/kg K.

Ans. $s_B - s_A = 0.0509$ kJ/kg K,

A is the inlet pipe.

7.17 Two vessels, A and B, each of volume 3 m^3 may be connected by a tube of negligible volume. Vessel A contains air at 0.7 MPa, 95°C, while vessel B contains air at 0.35 MPa, 205°C. Find the change of entropy when A is connected to B by working from the first principles and assuming the mixing to be complete and adiabatic. For air take the relations as given in Example 7.8. Ans. 0.975 kJ/kg K.

7.18 (a) An aluminium block (c_p=400 J/kg K) with a mass of 5 kg is initially at 40°C in room air at 20°C. It is cooled reversibly by transferring heat to a completely reversible cyclic heat engine until the block reaches 20°C. The 20°C room air serves as a constant temperature sink for the engine. Compute (i) the change in entropy for the block, (ii) the change in entropy for the room air, (iii) the work done by the engine.

(b) If the aluminium block is allowed to cool by natural convection to room air, compute (i) the change in entropy for the block, (ii) the change in entropy for the room air (iii) the net change in entropy for the universe.

Ans. (a)—134 J/K, +134 J/K, 740 J,
(b)—134 J/K, +136.5 J/K, 2.5 J/K.

7.19 Two bodies of equal heat capacities C and temperatures T_1 and T_2 form an adiabatically closed system. What will the final temperature be if one lets this system come to equilibrium (a) freely? (b) reversibly? (c) What is the maximum work which can be obtained from this system?

7.20 A solar-powered heat pump receives heat from a solar collector at T_h, rejects heat to the atmosphere at T_a, and pumps heat from a cold space at T_c. The three heat transfer rates are Q_h, Q_a, and Q_c respectively. Derive an expression for the minimum ratio Q_h/Q_c, in terms of the three temperatures.

If T_h=400 K, T_a=300 K, T_c =200 K, Q_c=12 kW, what is the minimum Q_h? If the collector captures 0.2 kW/m^2, what is the minimum collector area required?

7.21 A resistor of 30 ohms is maintained at a constant temperature of 27°C while a current of 10 amperes is allowed to flow for 1 sec. Determine the entropy change of the resistor and the universe.

If the resistor initially at 27°C is now insulated and the same current is passed for the same time, determine the entropy change of the resistor and the universe. The specific heat of the resistor is 0.9 kJ/kgK and the mass of the resistor is 10 g.

8

Available Energy, Availability and Irreversibility

8.1 AVAILABLE ENERGY

The sources of energy can be divided into two groups, viz. high grade energy and low grade energy. The conversion of high grade energy to shaft work is exempt from the limitations of the second law, while conversion of low grade energy is subject to them.

The examples of two kinds of energy are:

High grade energy	*Low grade energy*
(a) Mechanical work	(a) Heat or thermal energy
(b) Electrical energy	(b) Heat derived from nuclear fission or fusion
(c) Water power	(c) Heat derived from combustion of fossil fuels
(d) Wind power	
(e) Kinetic energy of a jet	
(f) Tidal power	

The bulk of the high grade energy in the form of mechanical work or electrical energy is obtained from sources of low grade energy, such as fuels, through the medium of the cyclic heat engine. The complete conversion of low grade energy, heat, into high grade energy, shaft-work, is impossible by virtue of the second law of thermodynamics. That part of the low grade energy which is available for conversion is referred to as *available energy*, while the part which, according to the second law, must be rejected, is known as *unavailable energy*.

Josiah Willard Gibbs is accredited with being the originator of the availability concept. He indicated that environment plays an important part in evaluating the available energy.

8.2 AVAILABLE ENERGY REFERRED TO A CYCLE

The maximum work output obtainable from a certain heat input in a cyclic heat engine (Fig. 8.1) is called the *available energy* (A.E.), or the available part of the energy supplied. The minimum energy that has to be rejected to the sink by the second law is called the *unavailable energy* (U.E.), or the unavailable part of the energy supplied.

Therefore, $Q_1 = \text{A.E.} + \text{U.E.}$ (8.1)

or $W_{\max} = \text{A.E.} = Q_1 - \text{U.E.}$

For the given T_1 and T_2,

$$\eta_{\text{rev.}} = 1 - \frac{T_2}{T_1}$$

For a given T_1, $\eta_{\text{rev.}}$ will increase with the decrease of T_2. The lowest practicable temperature of heat rejection is the *temperature of the surroundings*, T_0.

$$\therefore \quad \eta_{\max} = 1 - \frac{T_0}{T_1}$$

and

$$W_{\max} = \left(1 - \frac{T_0}{T_1}\right) Q_1$$

Let us consider a finite process x-y, in which heat is supplied reversibly to a heat engine (Fig 8.2). Taking an elementary cycle, if $đQ_1$ is the heat received by the engine reversibly at T_1, then

$$đW_{\max} = \frac{T_1 - T_0}{T_1} đQ_1 = đQ_1 - \frac{T_0}{T_1} đQ_1 = \text{A.E.}$$

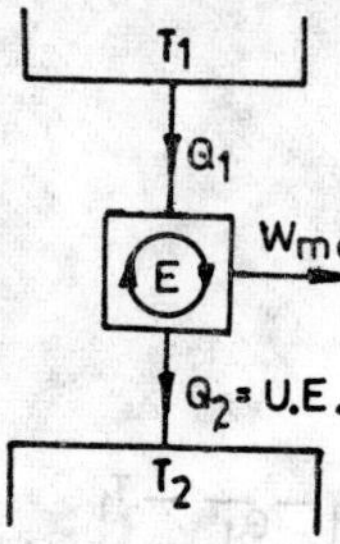

Fig. 8.1 Available and unavailable energy in a cycle

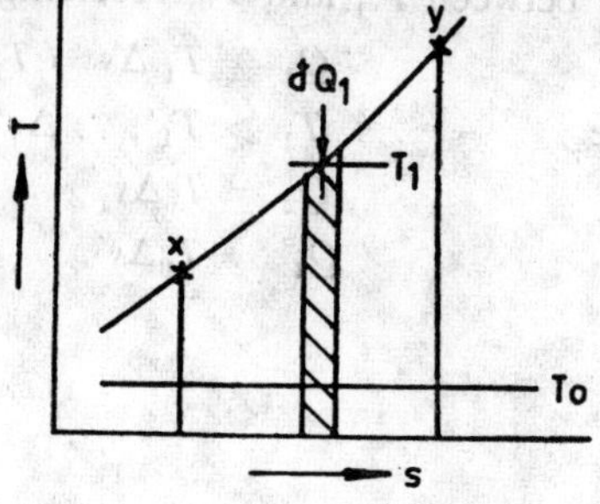

Fig. 8.2 Availability of energy

For the heat engine receiving heat for the whole process x-y, and rejecting heat at T_0

$$\int_x^y đW_{\max} = \int_x^y đQ_1 - \int_x^y \frac{T_0}{T_1} đQ_1$$

$\therefore \quad W_{max} = \text{A.E.}$

$\quad = Q_{xy} - T_0 (s_y - s_x) \quad (8.2)$

or $\quad \text{U.E.} = Q_{xy} - W_{max}$

or $\quad \text{U.E.} = T_0 (s_y - s_x)$

The unavailable energy is thus the product of the lowest temperature of heat rejection, and the change of entropy of the system during the process of supplying heat (Fig. 8.3).

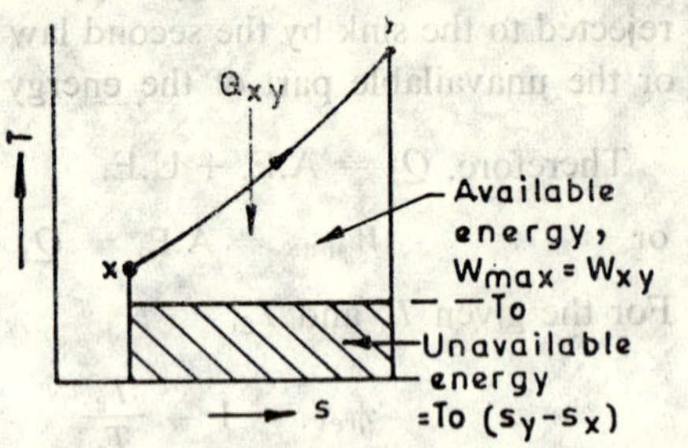

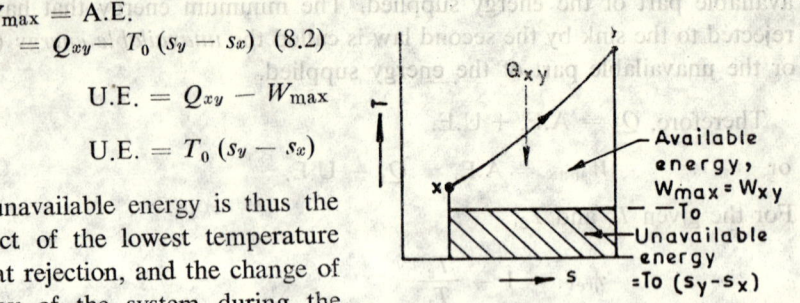

Fig. 8.3 Unavailable energy by the second law

8.2.1 Decrease in available energy when heat is transferred through a finite temperature difference

Whenever heat is transferred through a finite temperature difference, there is a decrease in the availability of energy so transferred.

Let us consider a reversible heat engine operating between T_1 and T_0 (Fig. 8.4). Then

$$Q_1 = T_1 \Delta s, \; Q_2 = T_0 \Delta s, \text{ and } W = \text{A.E.} = (T_1 - T_0)\,\Delta s$$

Let us now assume that heat Q_1 is transferred through a finite temperature difference from the reservoir or source at T_1 to the engine absorbing heat at T'_1, lower than T_1 (Fig. 8.5). The availability of Q_1 as received by the engine at T'_1 can be found by allowing the engine to operate reversibly in a cycle between T'_1 and T_0, receiving Q_1 and rejecting Q'_2.

Now $\quad Q_1 = T_1 \,\Delta s = T_1' \Delta s'$

Since $\quad T_1 > T_1', \; \therefore \; \Delta s' > \Delta s$

$\quad Q_2 = T_0 \Delta s$

$\quad Q_2' = T_0 \Delta s'$

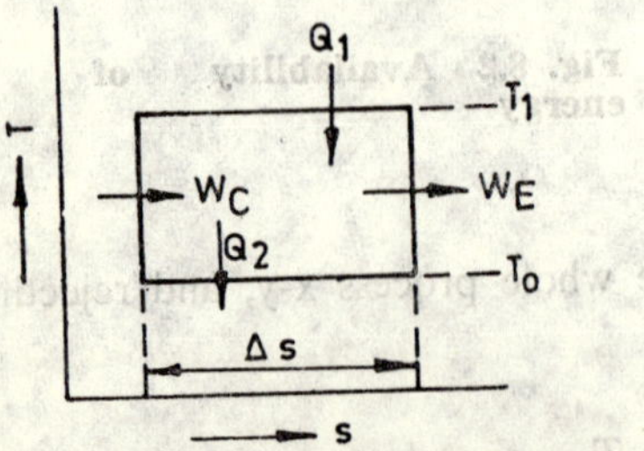

Fig. 8.4 Carnot cycle

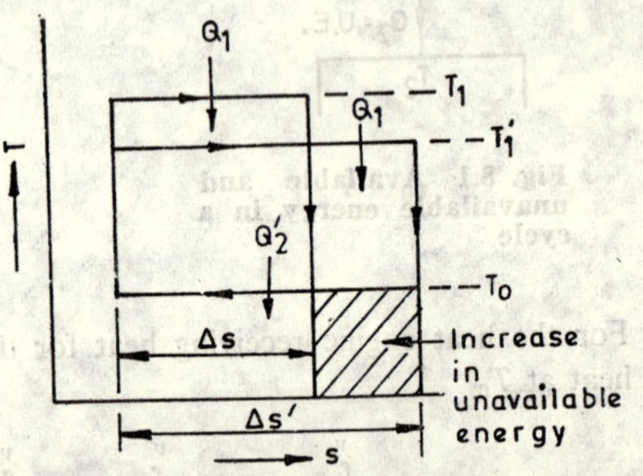

Fig. 8.5 Increase in unavailable energy due to heat transfer through a finite temperature difference

Since $\Delta s' > \Delta s' \therefore Q'_2 > Q_2$

$\therefore \quad W' = Q_1 - Q'_2 = T'_1 \Delta s' - T_0 \Delta s'$

and $\quad W = Q_1 - Q_2 = T_1 \Delta s - T_0 \Delta s$

$\therefore \quad W' < W$, because $Q_2' > Q_2$

Available energy lost due to irreversible heat transfer through finite temperature difference between the source and the working fluid during the heat addition process is given by

$$W - W' = Q_2' - Q_2$$

$$= T_0 (\Delta s' - \Delta s)$$

or, decrease in A.E. $= T_0 (\Delta s' - \Delta s)$

The decrease in available energy is thus the product of the lowest feasible temperature of heat rejection and the additional entropy change in the system while receiving heat irreversibly, compared to the case of reversible heat transfer from the same source.

The greater is the temperature difference $(T_1 - T'_1)$, the greater is the heat rejection Q'_2 and the greater will be the unavailable part of the energy supplied (Fig. 8.5). Energy is said to be degraded each time it flows through a finite temperature difference. That is why the second law is sometimes called the *law of the degradation of energy,* and energy is said to 'run downhill'.

8.3 THE HELMHOLTZ FUNCTION AND THE GIBBS FUNCTION

From the first law, when a system performs an infinitesimal process, reversible or irreversible, between two equilibrium states the work $đW$ is given by (Fig. 8.6).

$$đW = -dU + đQ \qquad (8.5)$$

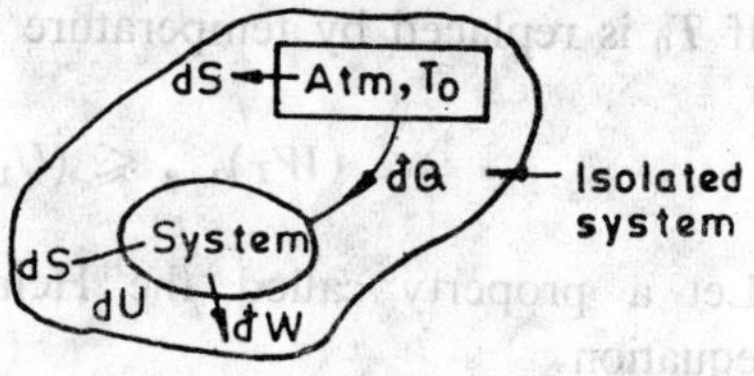

Fig. 8.6 Maximum work equal to decrease in Helmholtz function.

The energy converted to work is thus provided in part by the system, whose internal energy decreases by $-dU$, and in part by the heat reservoirs with which the system is in contact and which give up a quantity of heat $đQ$.

The expressions for the maximum amount of work that can be obtained when a system undergoes a process between two equilibrium states will now be derived. It is assumed that the system exchanges energy only with the environment at temperature T_0.

From the principle of the increase of entropy, the sum of the increase of entropy of the system, dS, and that of the surroundings, dS_0, is equal to, or greater than, zero.

$$dS + dS_0 \geqslant 0$$

Since $đQ$ is flowing out of the reservoir, i.e., the environment

$$dS_0 = -\frac{đQ}{T_0}$$

Hence

$$dS - \frac{đQ}{T_0} \geqslant 0$$

or $$T_0\, dS \geqslant đQ$$

or $$đQ \leqslant T_0\, dS \tag{8.6}$$

∴ From the first law (equation 8.5)

$$đW \leqslant -dU + T_0\, dS$$

$$đW \leqslant -(dU - T_0\, dS)$$

For a finite process between two equilibrium states, this equation can be integrated to give

$$W_{1-2} \leqslant (U_1 - U_2) - T_0 (S_1 - S_2) \tag{8.7}$$

where U_1 and S_1 are the initial values of internal energy and entropy of the system, and U_2 and S_2 are their final values.

Let the special case in which the initial and final temperatures are equal and are the same as that of the environment (Fig. 8.7), be considered. If T_0 is replaced by temperature T of the system

$$(W_T)_{1-2} \leqslant (U_1 - U_2)_T - T (S_1 - S_2)_T \tag{8.8}$$

Let a property called the Helmholtz function F be introduced by the equation

$$F = U - TS$$

Then for two equilibrium states (1) and (2) at the same temperature T

$$(F_1 - F_2)_T = (U_1 - U_2)_T - T (S_1 - S_2)_T$$

From equation (8.8)

$$W_T \leqslant (F_1 - F_2)_T \tag{8.9}$$

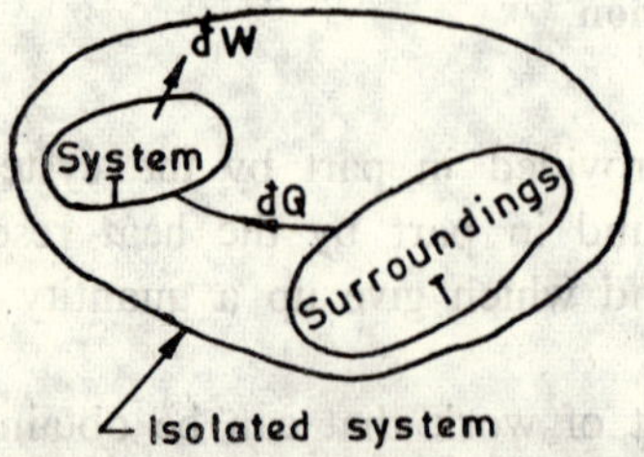

Fig. 8.7 Maximum work equal to decrease in Gibbs function

The work done by a system in any process between two equilibrium states at the same temperature during which the system exchanges heat only with the environment is equal to or less than the decrease in the Helmholtz function of the system during the process. The maximum work is done when the process is reversible and the equality sign holds. If the process is irreversible, the work is less than the maximum.

$$\therefore \qquad (W_T)_{\max} = (W_T)_{\text{rev.}} = (F_1 - F_2)_T \tag{8.10}$$

Let it now be supposed that a system can do other forms of work, in addition to pdV work, such that

$$đW_0 = \int pdV + W$$

where the symbol W represents non-pdV work. In a process in which the pressure at the boundary of the system is equal to the constant external pressure p_0, the work done in the process between two equilibrium states 1 and 2

$$\text{or} \qquad (W_0)_{1-2} = \int_{V_1}^{V_2} pdV + W$$

$$= p_0 (V_2 - V_1) + W$$

It then follows from equation (8.7) that for a system that exchanges heat only with the environment at T_0

$$p_0 (V_2 - V_1) + W \leqslant (U_1 - U_2) - T_0 (S_1 - S_2)$$

$$W_{1-2} \leqslant (U_1 - U_2) - T_0 (S_1 - S_2) + p_0 (V_1 - V_2) \tag{8.11}$$

If the initial and final equilibrium states of a system are at the same temperature and pressure as their surroundings, the temperature and pressure of the surroundings, T_0 and p_0, can be replaced by the temperature and pressure of the system, T and p.

$$(W)_{T,p} \leqslant (U_1 - U_2)_{T,p} - T (S_1 - S_2)_{T,p} + p (V_1 - V_2)_{T,p} \tag{8.12}$$

The Gibbs function G is defined as

$$G = U - TS + pV$$

$$= H - TS$$

Then for two equilibrium states at the same temperature T and pressure p

$$(G_1 - G_2)_{T,p} = (U_1 - U_2)_{T,p} - T (S_1 - S_2)_{T,p} + p (V_1 - V_2)_{T,p}$$

$\therefore$ From equation (8.12)

$$W_{T,p} \leqslant (G_1 - G_2)_{T,p} \tag{8.13}$$

The decrease in the Gibbs function of a system sets an upper limit to the work that can be performed, exclusive of pdV work, in any process between two equilibrium states at the same temperature and pressure, provided the system exchanges heat only with the environment which is at the same temperature and pressure as the end states of the system.

$$(W_{T,p})_{\max} = (W_{T,p})_{\text{rev.}} = (G_1 - G_2)_{T,p} \tag{8.14}$$

If the process is irreversible, the work is less than this maximum.

8.4 AVAILABILITY

Whenever useful work is obtained during a process in which a finite system undergoes a change of state, the process must terminate when the pressure and temperature of the system have become equal to the pressure and temperature of the surroundings, e.g., a cylinder of compressed air operating an engine to do useful work, or a certain quantity of exhaust gases from an internal combustion engine as the high-temperature source of a heat engine.

The availability (A) of a given system is defined as *the maximum useful work (total work minus pdV work) that is obtainable in a process in which the system comes to equilibrium with its surroundings*. Availability is thus a composite property depending on the state of both the system and surroundings.

Let U, S, and V be the initial values of the internal energy, entropy, and volume of a system and U_0, S_0, and V_0 their final values when the system has come to equilibrium with its environment. The system exchanges heat only with the environment, and the process may be either reversible or irreversible. The useful work obtained in the process is from equation (8.12)

$$W \leqslant (U - T_0S + p_0V) - (U_0 - T_0S_0 + p_0V_0) \tag{8.15}$$

Let $\phi = U - T_0S + p_0V$

where ϕ is called the availability function and is a composite property of both the system and its environment, with U, S, and V being properties of the system at some equilibrium state, and T_0 and p_0 the temperature and pressure of the environment. (In the Gibbs function, $G = U - TS + pV$, T, and p refer to the system.)

The decrease in the availability function in a process in which the system comes to equilibrium with its environment is

$$\phi - \phi_0 = (U - T_0S + p_0V) - (U_0 - T_0S_0 + p_0V_0)$$

$$\therefore \quad W \leqslant \phi - \phi_0 \tag{8.16}$$

Thus the useful work is equal to or less than the decrease in the availability function.

The availability A of a given system in a given environment is the maximum useful work obtainable in a reversible process

$$A = W_{max} = \phi - \phi_0 \tag{8.17}$$

This work is obtained in part from a decrease in the internal energy of the system, and in part from the heat withdrawn from the environment.

Let a system be taken from an equilibrium state 1, in which its availability is A_1, to a second equilibrium state 2, in which its availability is A_2. The end state 2 is not in equilibrium with the environment. The maximum useful work that could be obtained in the process

$$W_{max} = A_1 - A_2$$

$$= (\phi_1 - \phi_0) - (\phi_2 - \phi_0)$$

$$\therefore \quad W_{max} = \phi_1 - \phi_2 \tag{8.18}$$

8.5 AVAILABILITY IN STEADY FLOW

Work done by a system comprising mass m of a substance flowing through a device in steady flow is given by

$$W = (H_1 + \frac{1}{2} m \mathbf{V}^2 + mgZ_1) - (H_2 + \frac{1}{2} m \mathbf{V}_2^2 + mgZ_2) + Q \quad (8.19)$$

The subscripts 1 and 2 refer to the entrance and exit, respectively. At the exit let the system be in equilibrium with the environment at pressure p_0 and temperature T_0. Let symbols without subscripts refer to the entrance conditions of the system. Let it be assumed that the changes in K.E. and P.E. are negligible. Then the useful work obtained when a mass m flows through the device is

$$W = (H - H_0) + Q \quad (8.20)$$

The greater the value of Q, the larger will be the useful work W, and W will be maximum when Q is a maximum.

Let S and S_0 be the entropies of the system at the entrance and exit of the device, then

$$(\Delta S)_{\text{system}} = S_0 - S$$

and $$(\Delta S)_{\text{surr.}} = -\frac{Q}{T_0}$$

From the entropy principle

$$(S_0 - S) - \frac{Q}{T_0} \geqslant 0$$

$$\therefore \quad Q \leqslant T_0 (S_0 - S) \quad (8.21)$$

$\therefore$ The useful work

$$W \leqslant (H - H_0) + T_0 (S_0 - S)$$

$$\leqslant (H - T_0 S) - (H_0 - T_0 S_0) \quad (8.22)$$

Let a quantity B, called the *availability function for steady flow*, be defined by the equation

$$B = H - T_0 S$$

The function B, like the function ϕ, is a composite property of a system and its environment. This is also known as the *Keenan function*.

$$\therefore \quad B - B_0 = (H - T_0 S) - (H_0 - T_0 S_0)$$

From equation (8.22)

$$W \leqslant B - B_0 \quad (8.23)$$

For given entrance and exit conditions, the useful work is a maximum when the heat absorbed is a maximum, i.e., when the internal irreversibility is zero and the equality sign holds. *The maximum work obtainable from a system at the entrance of a device, when the pressure and temperature at the exit are those of the environment is called the availability of the system in steady flow, and is given by*

$$A = W_{\max} = B - B_0 \tag{8.24}$$

Let the general case in which the pressure and temperature at the exit of the steady flow device are not those of the environment be now considered. The maximum useful work that could be obtained is the difference between the availabilities at the entrance and exit, or

$$A = A_1 - A_2 = (B_1 - B_0) - (B_2 - B_0)$$

$$\therefore \quad A = W_{\max} = B_1 - B_2 \tag{8.25}$$

or per unit mass

$$a = b_1 - b_2 \tag{8.26}$$

The alternative names for availability and for unavailable quantity $T_0\Delta S$ are *exergy* and *anergy* respectively.

8.6 ENTROPY EQUATION FOR A FLOW PROCESS

Let $đQ_0$ and $đW$ be the heat and work interactions with the environment. Writing the energy equation for the control volume bounded by the control surface (Fig. 8.8)

$$dE_V = đQ_0 + dm_1\left(h_1 + \frac{\mathbf{V}_1^2}{2} + Z_1 g\right) - đW - dm_2\left(h_2 + \frac{\mathbf{V}_2^2}{2} + Z_2 g\right)$$

$$= d\left(U + \frac{m\mathbf{V}^2}{2} + mgZ\right)$$

where dm_1 and dm_2 are the mass entering and leaving the control volume.

$$đW = đQ_0 + dm_1\left(h_1 + \frac{\mathbf{V}_1^2}{2} + Z_1 g\right) - dm_2\left(h_2 + \frac{\mathbf{V}_2^2}{2} + Z_2 g\right) - d\left(U + \frac{m\mathbf{V}^2}{2} + mgZ\right) \tag{8.27}$$

This is the work done by the control volume other than flow work.

Heat is transferred only at those locations where the system and the environment are at the same temperature, so that reversible heat interaction is achieved and

$$đQ_0 = T_0 dS$$

In a system where both heat and mass transfer occur across the control surface the total entropy change is

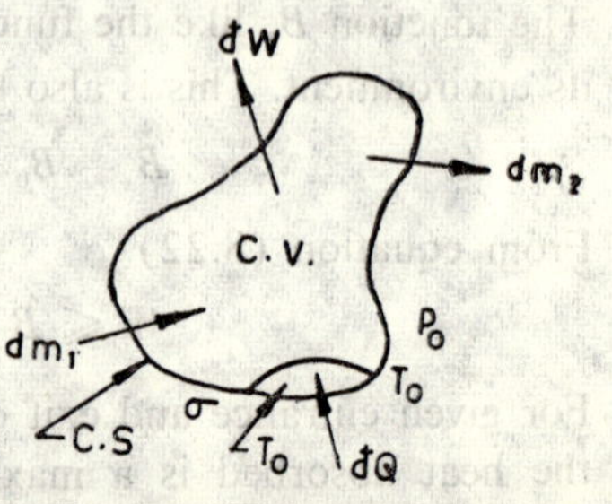

Fig. 8.8 Entropy equation for a flow process

$$dS_\sigma = \frac{đQ_0}{T_0} + dm_1 s_1 - dm_2 . s_2$$

$$= dS + dm_1 s_1 - dm_2 s_2$$

$$\therefore \quad đQ_0 = T_0 (dS_\sigma - dm_1 s_1 + dm_2 s_2) \quad (8.28($$

Substituting for $đQ_0$ in equation (8.27).

$$đW_{\text{rev.}} = T_0 (dS_\sigma - dn_1 s_1 + dm_2 s_2) + dn_1 \left(h_1 + \frac{\mathbf{V}_1^2}{2} + Z_1 g \right)$$

$$- dm_2 \left(h_2 + \frac{\mathbf{V}_2^2}{2} + Z_2 g \right) - d \left(U + \frac{m\mathbf{V}^2}{2} + mgZ \right)_\sigma$$

$$= dm_1 \left(h_1 - T_0 s_1 + \frac{\mathbf{V}_1^2}{2} + Z_1 g \right) - dm_2 \left(h_2 - T_0 s_2 + \frac{\mathbf{V}_2^2}{2} + Z_2 g \right)$$

$$- d \left(U - T_0 S + \frac{m\mathbf{V}^2}{2} + mgZ \right)_\sigma \quad (8.29)$$

This is the reversible work done by a control volume other than flow work.

8.6.1 Entropy equation for a steady flow process

In a steady flow process

$$dm_1 = dm_2 = dm$$

and $$d \left(U - T_0 S + \frac{m\mathbf{V}^2}{2} + mgZ \right)_\sigma = 0$$

$$\therefore \quad đW_{\text{rev}} = dm_1 \left(h_1 - T_0 s_1 + \frac{\mathbf{V}_1^2}{2} + Z_1 g \right)$$

$$- dm_2 \left(h_2 - T_0 s_2 + \frac{\mathbf{V}_2^2}{2} + Z_2 g \right)$$

or $$W_{\text{rev.}} = \left(H_1 - T_0 S_1 + \frac{m\mathbf{V}_1^2}{2} + mgZ_1 \right)$$

$$- \left(H_2 - T_0 S_2 + \frac{m\mathbf{V}_2^2}{2} + mgZ_2 \right) \quad (8.30)$$

The function $(H - T_0S)$ is the Keenan function for steady flow (symbol B).

$$\therefore \quad W_{\text{rev.}} = \left(B_1 + \frac{m\mathbf{V}_1^2}{2} + mgZ_1 \right) - \left(B_2 + \frac{m\mathbf{V}_2^2}{2} + mgZ_2 \right) \quad (8.31)$$

$$= \psi_1 - \psi_2$$

where ψ_1 and ψ_2 are the availability functions for steady flow at the inlet and exit respectively.

When more than two streams entering and leaving the control surface are involved

$$W_{\text{rev.}} = \sum_{\text{net}} \left(B + \frac{m\mathbf{V}^2}{2} + mgZ \right) \quad (8.32)$$

8.6.2 Reversible work in a non-flow process

In a non-flow process

$$dm_1 = dm_2 = 0$$

Equation (8.29) reduces to

$$(đW)_{\text{rev.}} = -d\left(U - T_0S + \frac{m\mathbf{V}^2}{2} + mgZ \right)_\sigma$$

Between two equilibrium end states 1 and 2

$$(W_{\text{rev.}})_{1-2} = U_1 - U_2 - T_0(S_1 - S_2) + m\frac{\mathbf{V}_1^2 - \mathbf{V}_2^2}{2} + mg(Z_1 - Z_2) \quad (8.33)$$

If the system does not possess any K.E. and P.E.

$$(W_{\text{rev.}})_{1-2} = U_1 - U_2 - T_0(S_1 - S_2) \quad (8.34)$$

8.7 IRREVERSIBILITY

The actual work done by a system is always less than the idealized reversible work, and the difference between the two is called the irreversibility of the process.

$$I = W_{\max} - W \quad (8.35)$$

This is also sometimes referred to as *'degradation'* or *'dissipation'*.

For a non-flow process between the equilibrium states, when the system exchanges heat only with the environment

$$I = [(U_1 - U_2) - T_0(S_1 - S_2)] - [(U_1 - U_2) + Q]$$

$$= T_0 (S_2 - S_1) - Q$$
$$= T_0 (\Delta S)_{\text{system}} + T_0 (\Delta S)_{\text{surr.}}$$
$$= T_0 [(\Delta S)_{\text{system}} + (\Delta S)_{\text{surr.}}] \quad (8.36)$$
$$\therefore \quad I \geqslant 0$$

Similarly, for the steady flow process

$$I = W_{\max} - W$$
$$= \left[\left(B_1 + \frac{m\mathbf{V}_1^2}{2} + mgZ_1\right) - \left(B_2 + \frac{m\mathbf{V}^2}{2} + mgZ_2\right)\right]$$
$$- \left[\left(H_1 + \frac{m\mathbf{V}_1^2}{2} + mgZ_1\right) - \left(H_2 + \frac{m\mathbf{V}_2^2}{2} + mgZ_2\right) + Q\right]$$
$$= T_0 (S_1 - S_2) - Q$$
$$= T_0 (\Delta S)_{\text{system}} + T_0 (\Delta S)_{\text{surr.}}$$
$$= T_0 (\Delta S_{\text{system}} + \Delta S_{\text{surr.}}) \quad (8.37)$$

The same expression for irreversibility applies to both flow and non-flow processes. The quantity T_0 $(\Delta S_{\text{system}} + \Delta S_{\text{surr.}})$ represents an increase in unavailable energy (or anergy).

Example 8.1

In a certain process, a vapour, while condensing at 420°C, transfers heat to water evaporating at 250°C. The resulting steam is used in a power cycle which rejects heat at 35°C. What is the fraction of the available energy in the heat transferred from the process vapour at 420°C that is lost due to the irreversible heat transfer at 250°C?

Solution

ABCD (Fig. Ex. 8.1) would have been the power cycle, if there was no temperature difference between the vapour condensing and the water evaporating, and the area under *CD* would have been the unavailable energy.

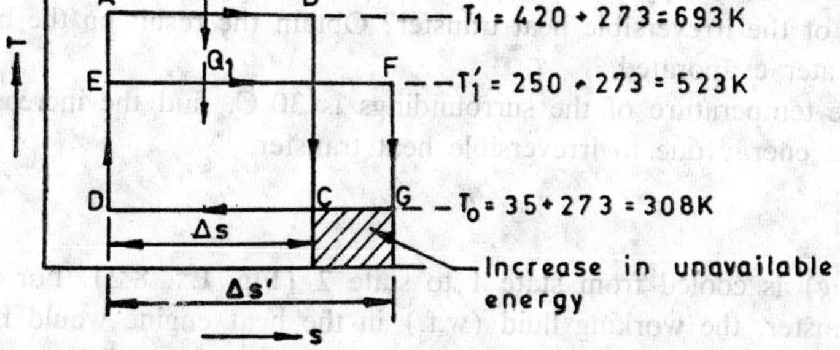

Fig. Ex. 8·1

EFGD is the power cycle when the vapour condenses at 420°C and the water evaporates at 250°C. The unavailable energy becomes the area under *DG*. Therefore, the increase in unavailable energy due to irreversible heat transfer is represented by the area under *CG*.

Now $$Q_1 = T_1\,\Delta S = T_1'\,\Delta S'$$

$$\frac{\Delta S'}{\Delta S} = \frac{T_1}{T_1'}$$

$$\begin{aligned} W' &= \text{work done in cycle } ABCD \\ &= (T_1 - T_0)\,\Delta S \\ W' &= \text{Work done in cycle } EFGD \\ &= (T_1' - T_0)\,\Delta S' \end{aligned}$$

The fraction of energy that becomes unavailable due to irreversible heat transfer

$$= \frac{W - W'}{W} = \frac{T_0(\Delta S' - \Delta S)}{(T_1 - T_0)\,\Delta S} = \frac{T_0\left(\dfrac{\Delta S'}{\Delta S'} - 1\right)}{(T_1 - T_0)}$$

$$= \frac{T_0(T_1 - T_1')}{T_1'(T_1 - T_0)} = \frac{308\,(693 - 523)}{523\,(693 - 308)}$$

$$= 0.26$$ Ans.

EXAMPLE 8.2

In a steam boiler, hot gases form a fire transfer heat to water which vaporizes at constant temperature. In a certain case, the gases are cooled from 1100°C to 550°C while the water evaporates at 220°C. The specific heat of gases is 1.005 kJ/kg-K, and the latent heat of water at 220°C. is 1858.5 kJ/kg. All the heat transferred from the gases goes to the water. How much does the total entropy of the combined system of gas and water increase as a result of the irreversible heat transfer? Obtain the result on the basis of 1 kg of water evaporated.

If the temperature of the surroundings is 30°C, find the increase in unavailable energy due to irreversible heat transfer.

Solution

Gas ($\dot{m}g$) is cooled from state 1 to state 2 (Fig. Ex. 8.2). For reversible heat transfer, the working fluid (w.f.) in the heat engine would have been heated along 2-1, so that at any instant, the temperature difference between gas and the working fluid is zero. Then 1-*b* would have been the expansion of the working fluid down to the lowest possible temperature T_0, and the amount of heat rejection would have been given by the area *abcd*.

When water evaporates at 220°C as the gas gets cooled from 1100°C to 550°C, the resulting power cycle has an unavailable energy represented by the area *aefd*. The increase in unavailable energy due to irreversible heat transfer is thus given by area *befc*.

Entropy increase of 1 kg water

$$(\Delta S)_{\text{water}} = \frac{\text{Latent heat absorbed}}{T} = \frac{1858.5}{(273+220)} = 3.77 \text{ kJ/kg-K}$$

Q_1 = Heat transferred from the gas
= Heat absorbed by water during evaporation

$$= \dot{m}_g c_{p_g} (1100 - 550)$$

$$= 1 \times 1858.5 \text{ kJ}$$

$$\therefore \quad \dot{m}_g c_{pg} = \frac{1858.5}{550} = 3.38 \text{ kJ/°C}$$

$$\Delta S_{\text{gas}} = \int_{T_{g_1}}^{T_{g_2}} \frac{dQ}{T} = \int_{T_{g_1}}^{T_{g_2}} \dot{m}_g c_{p_g} \frac{dT}{T}$$

$$= \dot{m}_g c_{p_g} \ln \frac{T_{g_2}}{T_{g_1}} = 3.38 \ln \frac{823}{1373}$$

$$= -3.38 \times 0.51$$

$$= -1.725 \text{ kJ/K}$$

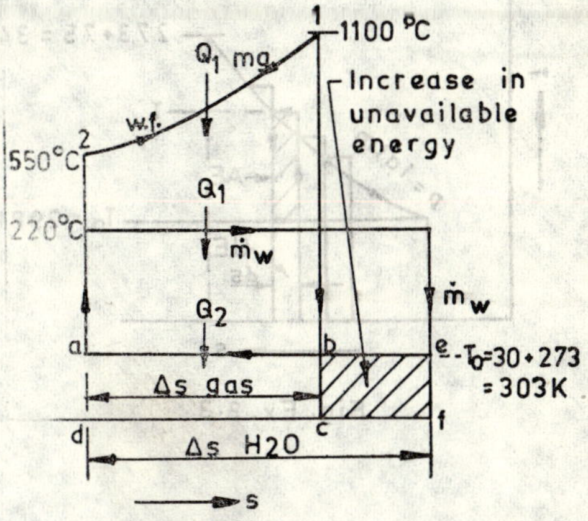

Fig. Ex. 8·2

$$\therefore \Delta S_{\text{total}} = (\Delta S)_{\text{water}} + (\Delta S)_{\text{gas}}$$

$$= 3.77 - 1.725 = 2.045 \text{ kJ/K}$$ **Ans.**

$\therefore$ Increase in unavailable energy

$$= T_0 (\Delta S)_{\text{total}} = 303 \times 2.045$$

$$= 620 \text{ kJ}$$ **Ans.**

Example 8.3

Calculated the available energy in 40 kg of water at 75°C with respect to the surroundings at 5°C, the pressure of water being 1 atm.

Solution

If the water is cooled at a constant pressure of 1 atm from 75°C to 5°C (Fig. Ex. 8.3) the heat given up may be used as a source for a series of Carnot engines each using the surroundings as a sink. It is assumed that the amount of energy received by any engine is small relative to that in the source and the temperature of the source does not change while heat is being exchanged with the engine.

Let us consider that the source has fallen to temperature T, at which level there operates a Carnot engine which takes in heat at this temperature and rejects heat at $T_0 = 278$ K. If δs is the entropy change of water, the work obtainable is $\delta W = -\dot{m}(T - T_0)\,\delta s$

where δs is negative.

$$\therefore \quad \delta W = -40\,(T - T_0)\,\frac{c_p\,\delta T}{T}$$

$$= -40\,c_p\left(1 - \frac{T_0}{T}\right)\delta T$$

With a very great number of engines in the series, the total work (maximum) obtainable when the water is cooled from 348 K to 278 K would be

$$W(\max) = \text{A.E.} = -\lim \sum_{348}^{278} 40\,c_p\left(1 - \frac{T_0}{T}\right)\delta T$$

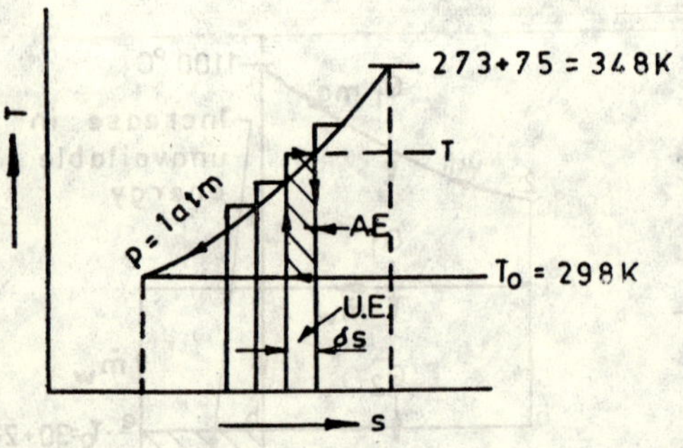

Fig. Ex. 8.3

$$= \int_{278}^{348} 40\,c_p\left(1 - \frac{T_0}{T}\right)dT$$

$$= 40 c_p\left[(348 - 278) - 278 \ln \frac{348}{278}\right]$$

$$= 40 \times 4.2\,(70 - 62)$$

$$= 1340 \text{ kJ} \qquad \text{Ans.}$$

$$Q_1 = 40 \times 4.2\,(348 - 278)$$

$$= 11{,}760 \text{ kJ}$$

$$\text{U.E.} = Q_1 - W(\max)$$

$$= 11{,}760 - 1340 = 10{,}420 \text{ kJ}$$

Example 8.4

Calculate the decrease in available energy when 25 kg of water at 95°C mix with 35 kg of water at 35°C, the pressure being taken as constant and the temperature of the surroundings being 15°C (c_p of water = 4.2 kJ/kg-K).

Solution

The available energy of a system of mass m, specific heat c_p, and at temperature T, is given by

$$\text{A.E.} = m\,c_p \int_{T_0}^{T}\left(1 - \frac{T_0}{T}\right)dT$$

$\therefore$ $(\text{A.E.})_{25}$ = Available energy of 25 kg of water at 95°C

$$= 25 \times 4.2 \int_{273+15}^{273+95}\left(1 - \frac{288}{T}\right)dT$$

$$= 105\left[(368 - 288) - 288 \ln \frac{368}{288}\right]$$

$$= 105\,[80 - 71]$$

$$= 945 \text{ kJ}$$

$$(A.E.)_{35} = \text{Available energy of 35 kg of water at } 35°C$$

$$= 147\left[(308 - 288) - 288 \ln \frac{308}{288}\right]$$

$$= 147\ (20 - 19.5)\ \text{kJ}$$

$$= 73.5\ \text{kJ}$$

Total available energy

$$(A.E.)_{total} = (A.E.)_{25} + (A.E.)_{35}$$

$$= 945 + 73.5$$

$$= 1018.5\ \text{kJ}$$

After mixing, if t is the final temperature

$$25 \times 4.2\ (95\text{-}t) = 35 \times 4.2\ (t - 35)$$

$$\therefore \quad t = \frac{25 \times 95 + 35 \times 35}{25 + 35}$$

$$= 60°C$$

Total mass after mixing $= 25 + 35 = 60$ kg

$$(A.E.)_{60} = \text{Available energy of 60 kg of water at } 60°C$$

$$= 4.2 \times 60\left[(333 - 288) - 288 \ln \frac{333}{288}\right]$$

$$= 252\ (45 - 42)$$

$$= 756\ \text{kJ}$$

$\therefore$ Decrease in available energy due to mixing

= Total available energy before mixing

− Total available energy after mixing

$= 1018.5 - 756$

$= 262.5$ kJ **Ans.**

EXAMPLE 8.5

The moment of inertia of a flywheel is 0.54 kg.m^2 and it rotates at a speed 3000 RPM in a large heat insulated system, the temperature of which is 15°C. If the kinetic energy of the flywheel is dissipated as frictional heat at the shaft bearings which have a water equivalent of 2 kg, find the rise in the temperature of the bearings when the flywheel has come to rest. Calculate the greatest possible amount of this heat which may be returned to the flywheel as high-grade energy, showing how much of the original kinetic energy is now unavailable. What would be the final RPM of the flywheel, if it is set in motion with this available energy?

Solution

Initial angular velocity of the flywheel

$$\omega_1 = \frac{2\pi N_1}{60} = \frac{2\pi \times 3000}{60} = 314.2 \text{ rad/sec}$$

Initial available energy of the flywheel

$$= (\text{K.E.})_{\text{initial}} = \frac{1}{2} I \omega_1^2$$

$$= 0.54 \text{ kg m}^2 \times (314.2)^2 \frac{\text{rad}^5}{\text{sec}^2}$$

$$= 2.66 \times 10^4 \text{ Nm} = 26.6 \text{ kJ}$$

When this K.E. is dissipated as frictional heat, if Δt is the temperature rise of the bearings, we have

water equivalent of the bearings $\times$ rise in temperature = 26.6 kJ

$$\therefore \quad \Delta t = \frac{26.6}{2 \times 4.187} = 3.19°\text{C} \qquad \textbf{Ans.}$$

$\therefore$ Final temperature of the bearings

$$t_f = 15 + 3.19 = 18.19°\text{C}$$

The maximum amount of energy which may be returned to the flywheel as high-grade energy is

$$\text{A.E.} = 2 \times 4.187 \int_{288}^{291.19} \left(1 - \frac{288}{T}\right) dT$$

$$= 2 \times 4.187 \left[(291.19 - 288) - 288 \ln \frac{291.19}{288} \right]$$

$$= 2 \times 4.187\ (3.19 - 3.14) = 2 \times 0.05 \times 4\ 187$$

$$= 0.4187 \text{ kJ}$$

The amount of energy rendered unavailable is

$$\text{U.E.} = (\text{A.E.})_{\text{initial}} - (\text{A.E.})_{\text{returnable as high grade energy}}$$

$$= 26.6 - 0.4187$$

$$= 26.18 \text{ kJ}$$

Since the amount of energy returnable to the flywheel is 0.42 kJ, if ω_2 is the final angular velocity, and the flywheel is set in motion with this energy

$$0.41 \times 10^3 = \tfrac{1}{2} \times 0.54\ \omega_2^2$$

$$\therefore \quad \omega_2^2 = \frac{420}{0.27} = 1550$$

$$\therefore \quad \omega^2 = 39.4 \text{ rad/sec}$$

If N_2 is the final RPM of the flywheel

$$\omega_2 = 39.4 = \frac{2\pi N_2}{60}$$

or $$N_2 = \frac{39.4 \times 60}{2 \times \pi} = 376 \text{ RPM}$$ Ans.

EXAMPLE 8.6

Two kg of air at 500 kPa, 80°C expands adiabatically in a closed system until its volume is doubled and its temperature becomes equal to that of the surroundings which is at 100 kPa, 5°C. For this process, determine (a) the maximum work, (b) the change in availability, and (c) the irreversibility. For air, take $c_v = 0.718$ kJ/kgK, $u = c_v T$ where c_v is constant, and $pV = mRT$ where p is pressure in kPa, V volume in m³, m mass in kg, R a constant equal to 0.287 kJ/kgK, and T temperature in K.

Solution

From the property relation

$$TdS = dU + pdV$$

the entropy change of air between the initial and final states is

$$\int_1^2 dS = \int_1^2 \frac{m\, c_v\, dT}{T} + \int_1^2 \frac{mRdV}{V}$$

or $$S_2 - S_1 = mc_v \ln \frac{T_2}{T_1} + mR \ln \frac{V_2}{V_1}$$

From Eq. (8.7),

$$W_{\text{max}} = (U_1 - U_2) - T_0\,(S_1 - S_2)$$

$$= m\left[c_v\,(T_1 - T_2) + T_0\left(c_v \ln \frac{T_2}{T_1} + R \ln \frac{V_2}{V_1}\right)\right]$$

$$= 2\left[0.718(80-5) + 278\left(0.718 \ln \frac{278}{353} + 0.287 \ln \frac{2}{1}\right)\right]$$

$$= 2\,[53.85 + 278\,(-0.172 + 0.199)]$$

$$= 2\,(53.85 + 7.51)$$

$$= 122.72 \text{ kJ.}$$ Ans. (a).

From Eq. (8.18), the change in availability

$$= \phi_1 - \phi_2$$

$$= (U_1 - U_2) - T_0\,(S_1 - S_2) + p_0\,(V_1 - V_2)$$

$$= W_{\text{max}} + p_0\,(V_1 - V_2)$$

$$= 122.72 + p_0\,(V_1 - 2V_1)$$

$$= 122.72 - 100 \times \frac{2 \times 0.287 \times 353}{500}$$

$$= 82.2 \text{ kJ}$$ Ans. (b).

The irreversibility

$$I = W_{\text{max, useful}} - W_{\text{act}}$$

$$= (\phi_1 - \phi_2) - W_{\text{act}}$$

From the first law,

$$W_{\text{act}} = Q - \Delta U = -\Delta U = U_1 - U_2$$

$$\therefore \quad I = U_1 - U_2 - T_0 (S_1 - S_2) - U_1 + U_2$$

$$= T_0 (S_2 - S_1)$$

$$= T_0 (\Delta S)_{\text{system}}$$

For adiabatic process, $(\Delta S)_{\text{surr}} = 0$

$$I = T_0 \left[mc_v \ln \frac{T_2}{T_1} + m R \ln \frac{V_2}{V_1} \right]$$

$$= 278 \times 2 \left[0.718 \ln \frac{278}{353} + 0.287 \ln 2 \right]$$

$$= 278 \times 2 \; (-0.172 + 0.199)$$

$$= 15.2 \text{ kJ}$$

Ans. (c).

Example 8.7

Air expands through a turbine from 500 kPa, 520°C to 100 kPa, 300°C. During expansion 10 kJ/kg of heat is lost to the surroundings which is at 98 kPa, 20°C. Neglecting the K.E. and P.E. changes, determine per kg of air (a) the decrease in availability, (b) the maximum work, and (c) the irreversibility. For air, take $c_p = 1.005$ kJ/kgK, $h = c_p T$ where c_p is constant, and the p, V and T relation as in Example 8.6.

Solution

From the property relation

$$TdS = dH - Vdp$$

the entropy change of air in the expansion process is

$$\int_1^2 dS = \int_1^2 \frac{m\, c_p\, dT}{T} - \int_1^2 \frac{mRdp}{p}$$

or

$$S_2 - S_1 = mc_p \ln \frac{T_2}{T_1} - mR \ln \frac{p_2}{p_1}$$

For 1 kg of air,

$$s_2 - s_1 = c_p \ln \frac{T_2}{T_1} - R \ln \frac{p_2}{p_1}$$

From Eq. (8.26), the change in availability

$$\psi_1 - \psi_2 = b_1 - b_2$$
$$= (h_1 - T_0 s_1) - (h_2 - T_0 s_2)$$
$$= (h_1 - h_2) - T_0 (s_1 - s_2)$$
$$= c_p (T_1 - T_2) - T_0 \left(R \ln \frac{p_2}{p_1} - c_p \ln \frac{T_2}{T_1} \right)$$
$$= 1.005 (520 - 300) - 293 \left(0.287 \ln \frac{1}{5} - 1.005 \ln \frac{573}{793} \right)$$
$$= 1.005 \times 220 - 293 (0.3267 - 0.4619)$$
$$= 221.1 + 39.6$$
$$= 260.7 \quad \text{kJ/kg}$$

The maximum work is Ans. (a).

$$W_{max} = \text{change in availability} = \psi_1 - \psi_2$$
$$= 260.7 \text{ kJ/kg.}$$ Ans. (b).

From S.F.E.E.,

$$Q + h_1 = W + h_2$$
$$W = (h_1 - h_2) + Q$$
$$= c_p (T_1 - T_2) + Q$$
$$= 1.005 (520 - 300) - 10$$
$$= 211.1 \quad \text{kJ/kg}$$

The irreversibility

$$I = W_{max} - W$$
$$= 260.7 - 211.1$$
$$= 49.6 \text{ kJ/kg}$$ Ans. (c).

Alternatively,

$$I = T_0 (\Delta S_{system} + \Delta S_{surr})$$
$$= 293 \left[1.005 \ln \frac{573}{793} - 0.287 \ln \frac{1}{5} + \frac{10}{293} \right]$$
$$= 293 \times 0.1352 + 10$$
$$= 49.6 \text{ kJ/kg}$$ Ans. (d).

PROBLEMS

8.1 What is the maximum useful work which can be obtained when 100 kJ are abstracted from a heat reservoir at 675 K in an environment at 288 K? What is the loss of useful work if (a) a temperature drop of 50°C is introduced between the heat source and the heat engine, on the one hand, and the heat engine and the heat sink, on the other, (b) the source temperature drops by 50°C and the sink temperature rises by 50°C during the heat transfer process according to the linear law $\frac{dQ}{dT} = \pm$ constant?

Ans. (a) 11.2 kJ, (b) 5.25 kJ.

8.2 In a steam generator, water is evaporated at 260°C, while the combustion gas (c_p=1.08 kJ/kg-K) is cooled from 1300°C to 320°C. The surroundings are at 30°C. Determine the loss in available energy due to the above heat transfer per kg of water evaporated. (Latent heat of vaporization of water at 260°C=1662.5 kJ/kg.)

Ans. 443.6 kJ.

8.3 Exhaust gases leave an internal combustion engine at 800°C and 1 atm., after having done 1050 kJ of work per kg of gas in the engine (c_p of gas=1.1 kJ/kgK). The temperature of the surroundings is 30°C. (a) How much available energy per kg of gas is lost by throwing away the exhaust gases? (b) What is the ratio of the lost available energy to the engine work? Ans. (a) 425.58 kJ, (b) 0.405.

8.4 A hot spring produces water at a temperature of 56°C. The water flows into a large lake, with a mean temperature of 14°C, at a rate of 0.1 m³ of water per min. What is the rate of working of an ideal heat engine which uses all the available energy. Ans. 19.5 kW.

8.5 Eighty kg of water at 100°C are mixed with 50 kg of water at 60°C, while the temperature of the surroundings is 15°C. Determine the decrease in available energy due to mixing.

8.6 A lead storage battery used in an automobile is able to deliver 5.2 MJ of electrical energy. This energy is available for starting the car.

Let compressed air be considered for doing an equivalent amount of work in starting the car. The compressed air is to be stored at 7 MPa, 25°C. What is the volume of the tank that would be required to let the compressed air have an availability of 5.2 MJ? For air, pv=0.287 T, where T is in K, p in kPa, and v in m³/kg.

Ans. 288 m³.

8.7 Ice is to be made from water supplied at 15°C by the process should in Fig. P.8.7. The final temperature of the ice is −10°C, and the final temperature of the water that is used as cooling water in the condenser is 30°C. Determine the minimum work required to produce 1000 kg of ice. Ans. 33.37 MJ.

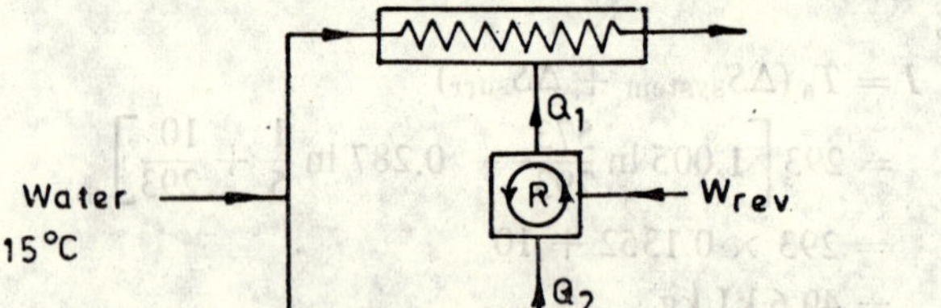

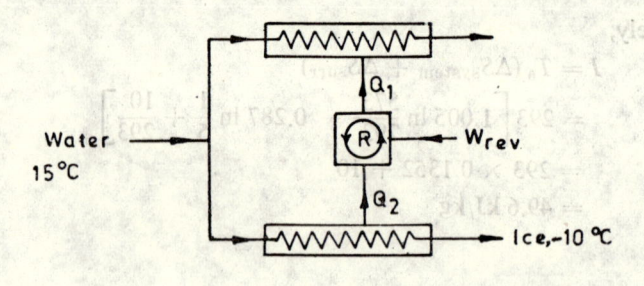

Fig. P. 8.7

8.8 A pressure vessel has a volume of 1 m³ and contains air at 1.4 MPa, 175°C. The air is cooled to 25°C by heat transfer to the surroundings at 25°C. Calculate the availability in the initial and final states and the irreversibility of this process.

Ans. 135 kJ/kg, 114.6 kJ/kg, 222 kJ.

8.9 Air enters an adiabatic compressor at atmospheric conditions of 1 bar, 15°C and leaves at 5.5 bar. The mass flow rate is 0.01 kg/s and the efficiency of the compressor is 75%. After leaving the compressor, the air is cooled to 40°C in an aftercooler. Calculate (a) the power required to drive the compressor, and (b) the rate of irreversibility for the overall process (compressor and cooler). Ans. 2.42 kW, 1 kW.

8.10 In a rotary compressor, air enters at 1.1 bar, 21°C where it is compressed adiabatically to 6.6 bar, 250°C. Calculate the irreversibility and the entropy production for unit mass flow rate. The atmosphere is at 1.03 bar, 20°C. Neglect the K.E. changes. Ans. 19 kJ/kg, 0.064 kJ/kg-K.

8.11 An air preheater is used to heat up the air used for combustion by cooling the outgoing products of combustion from a furnace. The rate of flow of the products is 10 kg/s, and the products are cooled from 300°C to 200°C, and for the products at this temperature $c_p = 1.09$ kJ/kg-K. The rate of air flow is 9 kg/s, the initial air temperature is 40°C, and for the air $c_p = 1.005$ kJ/kg-K.

(a) What is the initial and final availability of the products?

(b) What is the irreversibility for this process?

(c) If the heat transfer from the products were to take place reversibly through heat engines, what would be the final temperature of the air? What power would be developed by the heat engines? Take $T_o = 300$ K.

8.12 A mass of 2 kg of air in a vessel expands from 3 bar, 70°C to 1 bar, 40°C, while receiving 1.2 kJ of heat from a reservoir at 120°C. The environment is at 0.98 bar, 27°C. Calculate the maximum work and the work done on the atmosphere.

Ans. 177 kJ, 112.5 kJ

8.13 Air enters the compressor of a gas turbine at 1 bar, 30°C and leaves the compressor at 4 bar. The compressor has an efficiency of 82%. Calculate per kg of air (a) the work of compression, (b) the reversible work of compression, and (c) the irreversibility. [For air, use

$$\frac{T_{2s}}{T_1} = \left(\frac{p_2}{p_1}\right)^{\gamma - 1/\gamma}$$

where T_{2s} is the temperature of air after isentropic compression and $\gamma = 1.4$. The compressor efficiency is defined as $(T_{2s} - T_1)/(T_2 - T_1)$, where T_2 is the actual temperature of air after compression.]

Ans. (a) 180.5 kJ/kg, (b) 159.5 kJ/kg, (c) 21 kJ/kg,

8.14 A mass of 6.98 kg of air is in a vessel at 200 kPa, 27°C. Heat is transferred to the air from a reservoir at 727°C until the temperature of air rises to 327°C. The environment is at 100 kPa, 17°C. Determine (a) the initial and final availability of air, (b) the maximum useful work associated with the process.

Ans. (a) 103.5, 621.9 kJ (b) 582 kJ.

8.15 Air enters a compressor in steady flow at 140 kPa, 17°C and 70 m/s and leaves it at 350 kPa, 127°C and 110 m/s. The environment is at 100 kPa, 7°C. Calculate per kg of air (a) the actual amount of work required, (b) the minimum work required, and (c) the irreversibility of the process.

Ans. (a) 114.4 kJ, (b) 97.3 kJ, (c) 17.1 kJ.

8.16 Air expands in a turbine adiabatically from 500 kPa, 400 K and 150 m/s to 100 kPa, 300 K and 70 m/s. The environment is at 100 kPa, 17°C. Calculate per kg of air (a) the maximum work output, (b) the actual work output, and (c) the irreversibility. Ans. (a) 159 kJ, (b) 109 kJ, (c) 50 kJ.

9

Properties of Pure Substances

A pure substance is a substance of constant chemical composition throughout its mass. It is a one-component system. It may exist in one or more phases.

9.1 *p-v* DIAGRAM FOR A PURE SUBSTANCE

Assume a unit mass of ice (solid water) at −10°C and 1 atm contained in a cylinder and piston machine (Fig. 9.1). Let the ice be heated slowly so that its temperature is always uniform. The changes which occur in the mass of water would be traced as the temperature is increased while the pressure is held constant. Let the state changes of water be plotted on *p-v* coordinates. The distinct regimes of heating, as shown in Fig. 9.2, are: *1-2*. The temperature of ice increases from −10°C to 0°C. The volume of ice would increase, as would be the case for any solid upon heating. At state 2, i.e. 0°C, the ice would start melting.

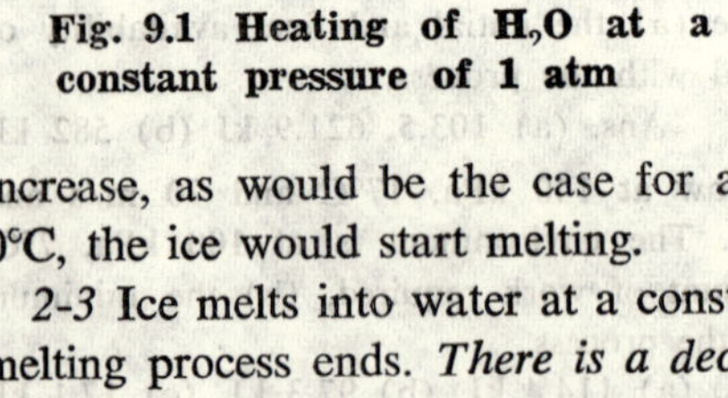

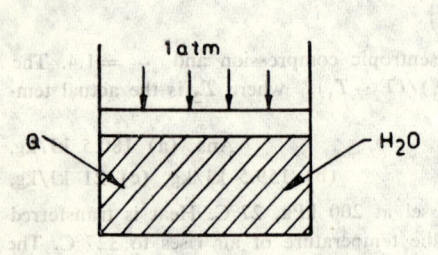

Fig. 9.1 Heating of H_2O at a constant pressure of 1 atm

2-3 Ice melts into water at a constant temperature of 0°C. At state 3, the melting process ends. *There is a decrease in volume, which is a peculiarity of water.*

3-4 The temperature of water increases, upon heating, from 0°C to 100°C. The volume of water increases because of thermal expansion.

4-5 The water starts boiling at state 4 and boiling ends at state 5. This phase change from liquid to vapour occurs at a constant temperature of 100°C (the pressure being constant at 1 atm.). There is a large increase in volume.

5-6 The vapour is heated to, say, 250°C (state 6). The volume of vapour increases from v_5 to v_6.

Water existed in the solid phase between 1 and 2, in the liquid phase between 3 and 4, and in the gas phase beyond 5. Between 2 and 3, the solid changed into the liquid phase by absorbing the latent heat of fusion, and between 4 and 5, the liquid changed into the vapour phase by absorbing the latent heat of vaporization, both at constant temperature and pressure.

The states 2, 3, 4 and 5 are known as *saturation states*. A saturation state is a state from which a change of phase may occur without a change of pressure or temperature. State 2 is a *saturated solid state* because a solid can change into liquid at constant pressure and temperature from state 2. States 3 and 4 are both saturated liquid states. In state 3, the liquid is saturated with respect to solidification, whereas in state 4, the liquid is saturated with respect to vaporization. State 5 is a *saturated vapour state*, because from state 5, the vapour can condense into liquid without a change of pressure or temperature.

If the heating of ice at −10°C to steam at 250°C were done at a constant pressure of 2 atm., similar regimes of heating would have been obtained with similar saturation states 2, 3, 4, and 5, as shown in Fig. 9.2. All the

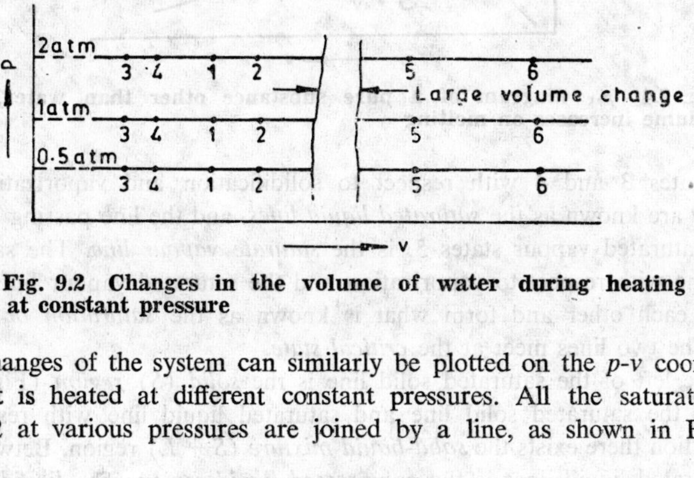

Fig. 9.2 Changes in the volume of water during heating at constant pressure

state changes of the system can similarly be plotted on the *p-v* coordinates, when it is heated at different constant pressures. All the saturated solid states 2 at various pressures are joined by a line, as shown in Fig. 9.3.

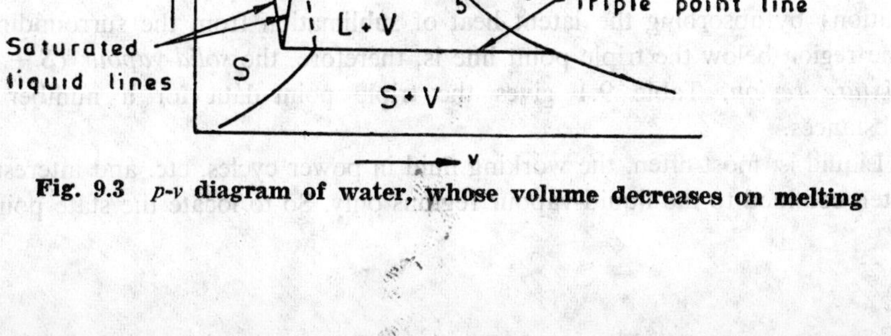

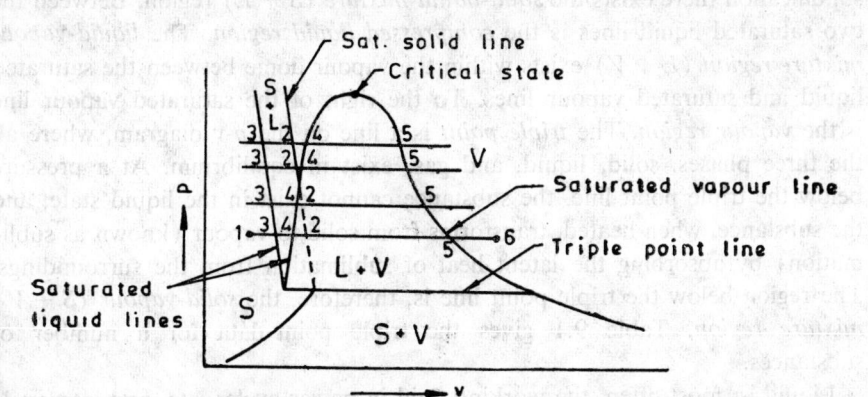

Fig. 9.3 ***p-v*** **diagram of water, whose volume decreases on melting**

Similarly, all the saturated liquid states 3 with respect to solidification, all the saturated liquid states 4 with respect to vaporization, and all the saturated vapour states 5, are joined together.

Figure 9.4 shows state changes of a pure substance other than water whose volume increases on melting.

The line passing through all the saturated solid states 2 (Figs. 9.3 and 9.4) is called the *saturated solid line*. The lines passing through all the saturated

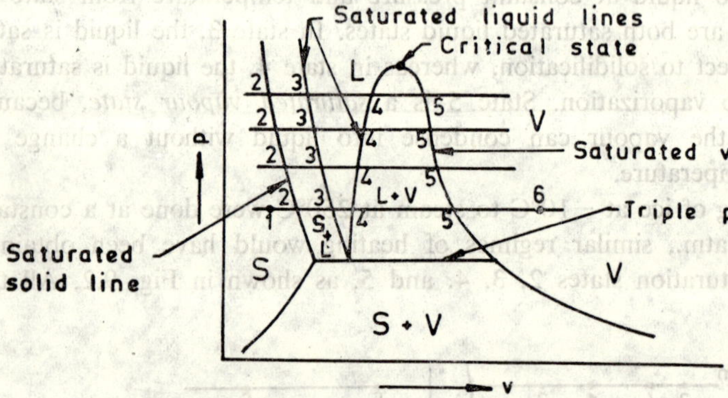

Fig. 9.4 *p-v* **diagram of a pure substance other than water, whose volume increases on melting**

liquid states 3 and 4 with respect to solidification and vaporization respectively are known as the *saturated liquid lines*, and the line passing through all the saturated vapour states 5, is the *saturate vapour line*. The saturated liquid line with respect to vaporization and the saturated vapour line incline towards each other and form what is known as the *saturation or vapour dome*. The two lines meet at the *critical state*.

To the left of the saturated solid line is the *solid (S) region* (Fig. 9.4). Between the saturated solid line and saturated liquid line with respect to solidification there exists the *solid-liquid mixture* $(S + L)$ region. Between the two saturated liquid lines is the *compressed liquid region*. The *liquid-vapour mixture region* $(L + V)$ exists within the vapour dome between the saturated liquid and saturated vapour lines. To the right of the saturated vapour line is the *vapour region*. The *triple point* is a line on the *p-v* diagram, where all the three phases, solid, liquid, and gas, exist in equilibrium. At a pressure below the triple point line, the substance cannot exist in the liquid state, and the substance, when heated, transforms from solid to vapour (known as sublimation) by absorbing the latent heat of sublimation from the surroundings. The region below the triple point line is, therefore, the *solid-vapour* $(S + V)$ *mixture region*. Table 9.1 gives the triple point data for a number of substances.

Liquid is, most often, the working fluid in power cycles, etc. and interest is often confined to the liquid-vapour regions only. So to locate the state points, the solid regions from Figs. 9.3 and 9.4 can be omitted. The *p-v* diagram then becomes as shown in Fig. 9.5. If the vapour at state *A* is compressed

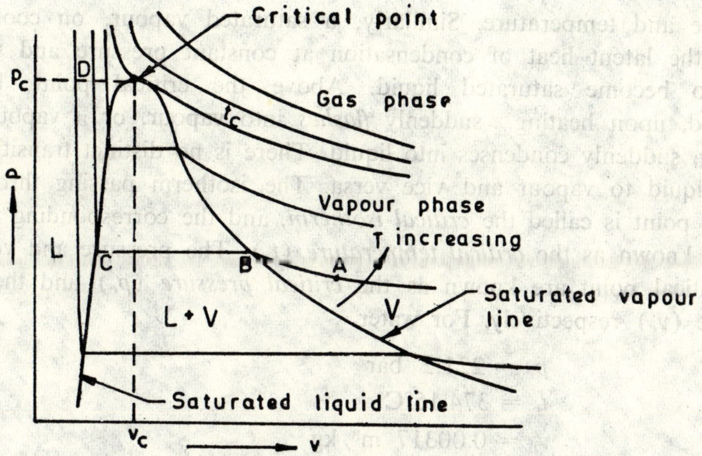

Fig. 9.5 Saturation curve on *p-v* diagram

slowly and isothermally, the pressure will rise until there is saturated vapour at point *B*. If the compression is continued, condensation takes place, the pressure remaining constant so long as the temperature remains constant. At any point between *B* and *C*, the liquid and vapour are in equilibrium. Since a very large increase in pressure is needed to compress the liquid, line *CD* is almost vertical. *ABCD* is a typical *isotherm* of a pure substance on a *p-v* diagram. Some isotherms are shown in Fig. 9.5. As the temperature in-

Table 9.1
Triple-Point Data

Substance	*Temperature, K*	*Pressure, mm Hg*
Acetylene, C_2H_2	192.4	962
Ammonia, NH_3	195.42	45.58
Argon, A	83.78	515.7
Carbon dioxide, CO_2	216.55	3885.1
Carbon monoxide, CO	68.14	115.14
Ethane, C_2H_6	89.88	0.006
Enthylene, C_2H_4	104.00	0.9
Hydrogen, H_2	13.84	52.8
Methane, CH_4	90.67	87.7
Nitrogen, N_2	63.15	94.01
Oxygen, O_2	54.35	1.14
Water, H_2O	273.16	4.587

creases, the liquid-vapour transition, as represented by BC, decreases, and becomes zero at the critical point. Below the critical point only, there is a liquid-vapour transition zone, where a saturated liquid, on heating, absorbs the latent heat of vaporization, and becomes saturated vapour at a constant pressure and temperature. Similarly, a saturated vapour, on cooling, releases the latent heat of condensation at constant pressure and temperature to become saturated liquid. Above the critical point, however, a liquid, upon heating, suddenly *flashes* into vapour, or a vapour, upon cooling, suddenly condenses into liquid. There is no distinct transition zone from liquid to vapour and vice versa. The isotherm passing through the critical point is called the *critical isotherm,* and the corresponding temperature is known as the *critical temperature* (t_c). The pressure and volume at the critical point are known as the *critical pressure* (p_c) and the *critical volume* (v_c) respectively. For water

$$p_c = 221.2 \text{ bar}$$
$$t_c = 374.15°\text{C}$$
$$v_c = 0.00317 \text{ m}^3/\text{kg}$$

The critical point data of certain substances are given in Appendix F. Above the critical point, the isotherms are continuous curves that at large volumes and low pressures approach equilateral hyperbolas.

When a liquid or solid is in equilibrium with its vapour at a given temperature, the vapour exerts a pressure that depends only on the temperature (Fig. 9.6). In general, the greater the temperature, the higher is the *vapour pressure.* The temperature at which the vapour pressure is equal to 760 mm Hg is called the *normal boiling point.*

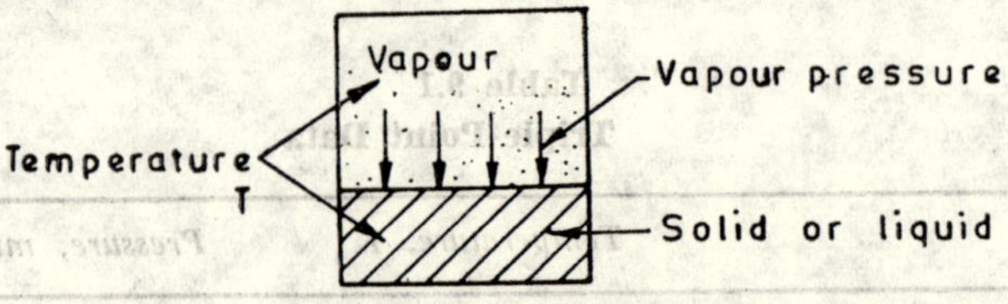

Fig. 9.6 Vapour pressure

Phase change occurs at constant pressure and temperature. A pure liquid at a given pressure will transform into vapour only at a particular temperature, known as *saturation temperature,* which is a function of pressure. Similarly, if the temperature is fixed, the liquid will boil (or condense) only at a particular pressure, called the *saturation pressure,* which is a function of temperature. In Fig. 9.7, if p_1 is the pressure, the corresponding saturation temperature is $(t_{sat.})_1$, or if t_2 is the given temperature, the saturation pressure is $(p_{sat.})_2$. As the pressure increases, the saturation temperature increases. Saturation states exist up to the critical point. At point A, the liquid starts boiling, and at point *B*, the boiling gets completed. At *A*, it is all liquid (saturated) and there is no vapour, while at B, it is all vapour (saturated) and there is no liquid. Vapour content progressively increases as the liquid changes its state from *A* towards *B*.

If v_f is the specific volume of the saturated liquid at a given pressure, and v_g the specific volume of the saturated vapour, then $(v_g - v_f)$ or v_{fg} is the change in specific volume during phase transition (boiling or condensation) at that pressure. As pressure increases, v_{fg} decreases, and at the critical point v_{fg} becomes zero.

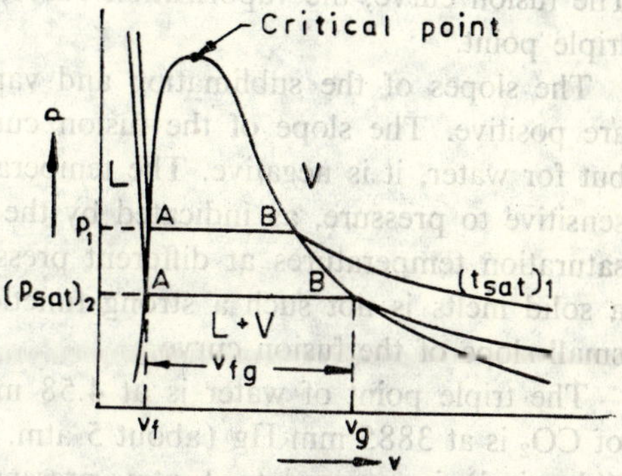

Fig. 9.7 Saturation pressure and temperature

9.2 *p-T* DIAGRAM FOR A PURE SUBSTANCE

The state changes of a pure substance, upon slow heating at different constant pressures, are shown on the *p-v* plane, in Figs. 9.2, 9.3, and 9.4. If these state changes are plotted on *p-T* coordinates, the diagram, as shown in Fig. 9.8, will be obtained. If the heating of ice at −10°C to steam at 250°C at the constant pressure of 1 atm. is considered, 1-2 is the solid (ice) heating, 2-3 is the melting of ice at 0°C, 3-4 is the liquid heating, 4-5 is the

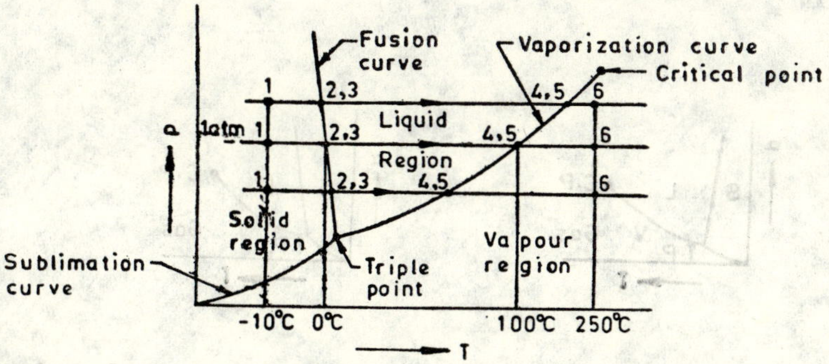

Fig. 9.8 Phase equilibrium diagram on *p-T* coordinates

vaporization of water at 100°C, and 5-6 is the heating in the vapour state. The process will be reversed from state 6 to state 1 upon cooling. The curve passing through the 2, 3 points is called the *fusion curve*, and the curve passing through the 4, 5 points (which indicate the vaporization or condensation at different temperatures and pressures) is called the *vaporization curve*. If the vapour pressure of a solid is measured at different temperatures, and these are plotted, the *sublimation curve* will be obtained. The fusion curve, the vaporization curve, and the sublimation curve meet at triple point.

The slopes of the sublimation and vaporization curves for all substances are positive. The slope of the fusion curve for most substances is positive, but for water, it is negative. The temperature at which a liquid boils is very sensitive to pressure, as indicated by the vaporization curve which gives the saturation temperatures at different pressures, but the temperature at which a solid melts is not such a strong function of pressure, as indicated by the small slope of the fusion curve.

The triple point of water is at 4.58 mm Hg and 273.16 K, whereas that of CO_2 is at 3885 mm Hg (about 5 atm.) and 216.55 K. So when solid CO_2 ('dry ice') is exposed to 1 atm. pressure, it gets transformed into vapour directly, absorbing the latent heat of sublimation from the surroundings, which gets cooled or 'refrigerated'.

9.3 *p-v-T* SURFACE

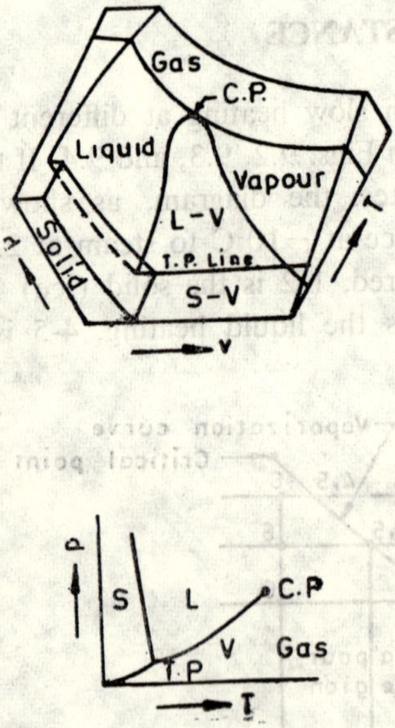

Fig. 9.9 ***p-v-T*** **surface for water which expands on freezing**

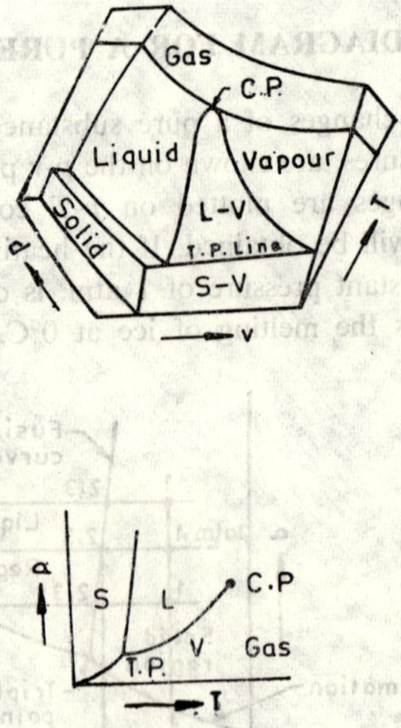

Fig. 9.10 ***p-v-T*** **surface of a substance which contracts on freezing**

The relationships between pressure, specific volume, and temperature can be clearly understood with the aid of a three-dimensional *p-v-T* surface. Figure 9.9 illustrates a substance like water that expands upon freezing and

Fig. 9.10 illustrates substances other than water which contract upon freezing. The projections on the p-T and p-v planes are also shown in these figures. Any point on the p-v-T surface represents an equilibrium state of the substance. The triple point line when projected to the p-T plane becomes a point. The critical isotherm has a point of inflection at the critical point.

9.4 T-s DIAGRAM FOR A PURE SUBSTANCE

The heating of the system of 1 kg of ice at −5°C to steam at 250°C is again considered, the pressure being maintained constant at 1 atm. The entropy increases of the system in different regimes of heating are given below.

1. The entropy increase of ice as it is heated from −5°C to 0°C at 1 atm. ($c_{p\text{ice}}$ = 2.093 kJ/kg-K)

$$\Delta s_1 = s_2 - s_1 = \int \frac{đQ}{T} = \int_{T_1=268}^{T_2=273} \frac{m\,c_p\,dT}{T} = m\,c_p \ln \frac{273}{268}$$

$$= 1 \times 2.093 \ln \frac{273}{268} = 0.0398 \text{ kJ/kg-K}$$

2. The entropy increase of ice as it melts into water at 0°C (latent heat of fusion of ice = 334.96 kJ/kg)

$$\Delta s_2 = s_3 - s_2 = \frac{334.96}{273} = 1.23 \text{ kJ/kg-K}$$

3. The entropy increase of water as it is heated from 0°C to 100°C

($c_{p\text{water}}$ = 4.187 kJ/kg-K)

$$\Delta s_3 = s_4 - s_3 = m\,c_p \ln \frac{T_3}{T_2} = 1 \times 4.187 \ln \frac{373}{273}$$

$$= 1.305 \text{ kJ/kg-K}$$

4. The entropy increase of water as it is vaporized at 100°C, absorbing the latent heat of vaporization (2257 kJ/kg)

$$\Delta s_4 = s_5 - s_4 = \frac{2257}{273} = 6.05 \text{ kJ/kg-K}$$

5. The entropy increase of vapour as it is heated from 100°C to 250°C at 1 atm.

$$\Delta s_5 = s_6 - s_5 = \int_{373}^{523} m\,c_p \frac{dT}{T} = 1 \times 2.093 \ln \frac{523}{373}$$

$$= 0.706 \text{ kJ/kg-K}$$

assuming the average specific heat of steam in the temperature range of 100°C to 250°C as 2.093 kJ/kg-K.

These entropy changes are shown in Fig. 9.11. The curve 1-2-3-4-5-6 is the isobar of 1 atm. If, during the heating process, the pressure had been

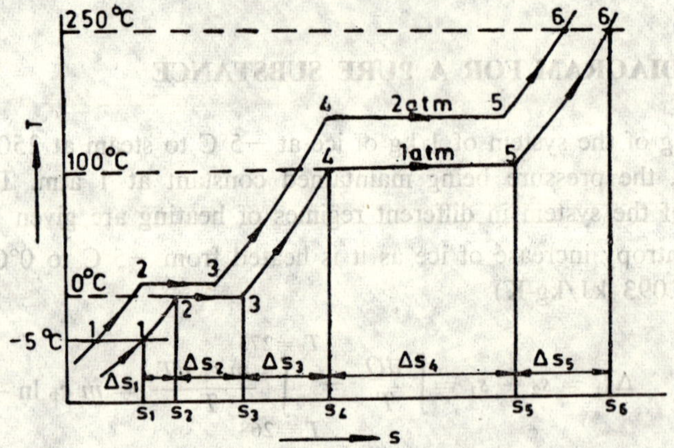

Fig. 9.11 Isobars on *T-s* plot

maintained constant at 2 atm., a similar curve would be obtained. The states 2, 3, 4, and 5 are saturation states. If these states for different pressures are joined, as in Figs. 9.3 and 9.4, the phase equilibrium diagram of a pure substance on the *T-s* coordinates, as shown in Fig. 9.12, would be obtained.

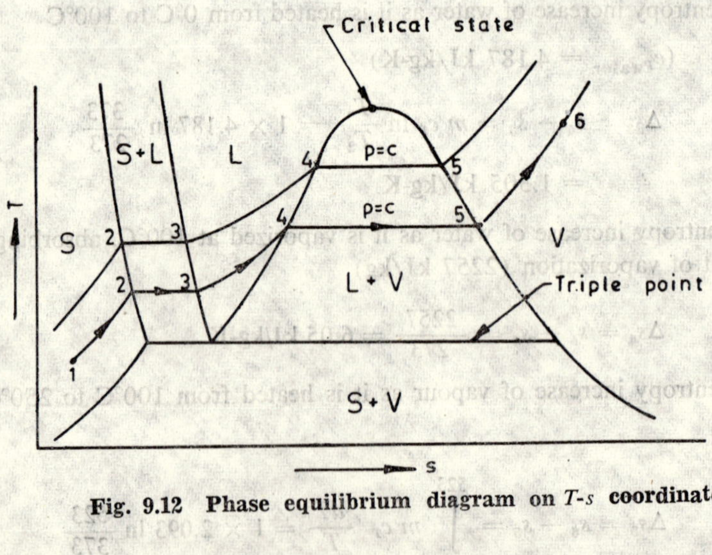

Fig. 9.12 Phase equilibrium diagram on *T-s* coordinates

Most often, liquid-vapour transformations only are of interest, and Fig. 9.13 shows the liquid, the vapour, and the transition zones only. At a particular pressure, s_f is the specific entropy of saturated water, and s_g is that of saturated vapour. The entropy change of the system during the phase change from liquid to vapour at that pressure is s_{fg} $(=s_g - s_f)$. The value of s_{fg} decreases as the pressure increases, and becomes zero at the critical point.

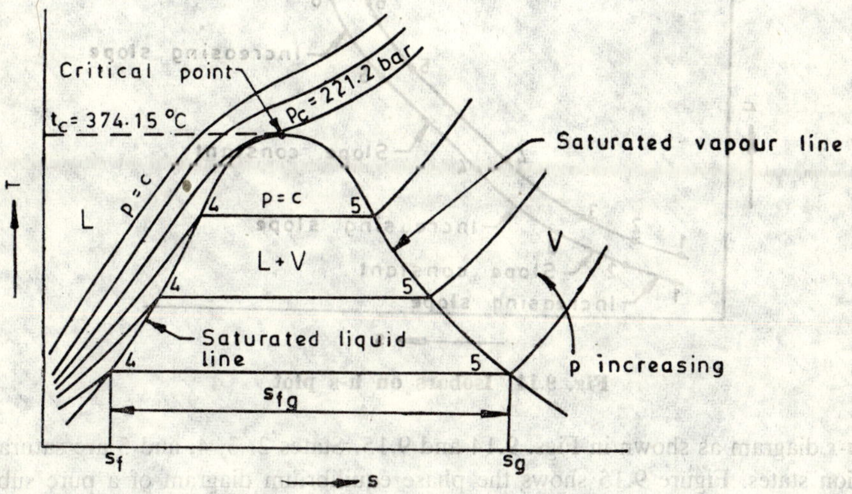

Fig. 9.13 Saturation (or vapour) dome for water.

9.5 *h-s* DIAGRAM OR MOLLIER DIAGRAM FOR A PURE SUBSTANCE

From the first and second laws of thermodynamics, the following property relation was obtained.

$$Tds = dh - vdp$$

or
$$\left(\frac{\partial h}{\partial s}\right)_p = T \tag{9.1}$$

This equation forms the basis of the *h-s* diagram of a pure substance, also called the Mollier diagram. The slope of an isobar on the *h-s* coordinates is equal to the absolute temperature. If the temperature remains constant the slope will remain constant. If the temperature increases, the slope of the isobar will increase.

Consider the heating of a system of ice at −5°C to steam at 250°C, the pressure being maintained constant at 1 atm. The slope of the isobar of 1 atm. on the *h-s* coordinates (Fig. 9.14) first increases as the temperature of the ice increases from −5°C to 0°C (1-2). Its slope then remains constant as ice melts into water at the constant temperature of 0°C (2-3). The slope of the isobar again increases as the temperature of water rises from 0°C to

100°C (3-4). The slope again remains constant as water vaporizes into steam at the constant temperature of 100°C (4-5). Finally, the slope of the isobar continues to increase as the temperature of steam increases to 250°C (5-6) and beyond. Similarly, the isobars of different pressures can be drawn on the

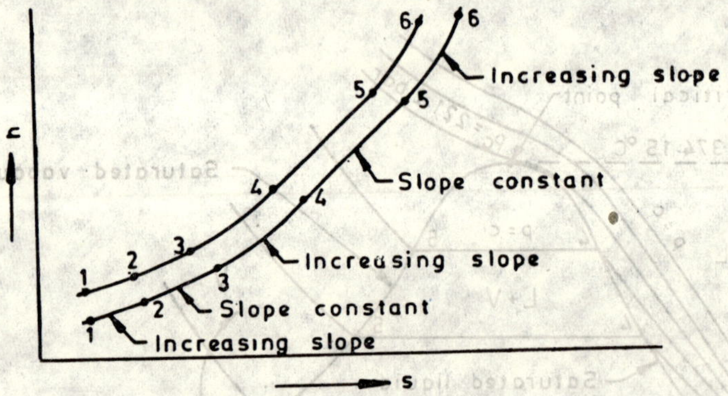

Fig. 9.14 Isobars on h-s plot

h-s diagram as shown in Figs. 9.14 and 9.15. States 2, 3, 4, and 5 are saturation states. Figure 9.15 shows the phase equilibrium diagram of a pure substance on the *h-s* coordinates, indicating the saturated solid line, and saturated liquid lines and saturated vapour line, the various phases, and the transition (mixture) zones.

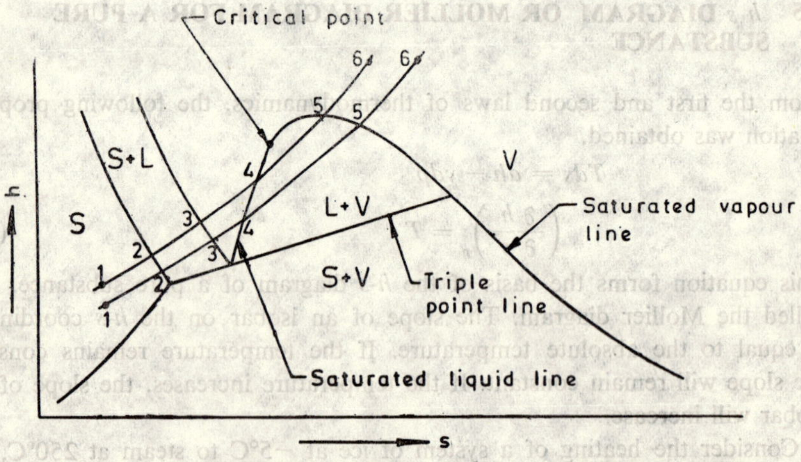

Fig. 9.15 Phase equilibrium diagram on *h-s* coordinates Mollier diagram)

Figure 9.16 is the *h-s* or the Mollier diagram indicating only the liquid and vapour phases. As the pressure increases, the saturation temperature increases, and so the slope of the isobar also increases. Hence, *the constant pressure lines diverge from one another,* and the critical isobar is a tangent at the critical point, as shown. In the vapour region, the states of equal slopes at various pressures are joined by lines, as shown, which are the *constant temperature lines.* At a particular pressure, h_f is the specific enthalpy of saturated water, h_g is that of saturated vapour, and h_{fg} $(= h_g - h_f)$ is the latent heat of vaporization at that pressure. As the pressure increases, h_{fg} decreases, and at the critical pressure, h_{fg} becomes zero.

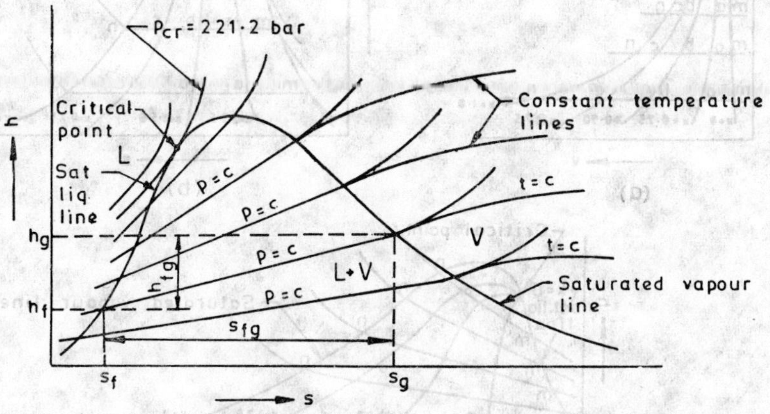

Fig. 9.16 Enthalpyentroy diagram of water

9.6 QUALITY OR DRYNESS FRACTION

If in 1 kg of liquid-vapour mixture, x kg is the mass of vapour and $(1-x)$ kg is the mass of liquid, then x is known as the *quality* or dryness fraction of the liquid-vapour mixture. Therefore, quality indicates the mass fraction of vapour in a liquid-vapour mixture, or

$$x = \frac{m_v}{m_v + m_l}$$

where m_l and m_1 are the masses of vapour and liquid respectively in the mixture. The value of x varies between 0 and 1. For saturated water, when water just starts boiling, $x = 0$, and for saturated vapour, when vaporization is complete, $x = 1$, for which the vapour is said to be *dry saturated.*

Points m in Fig. 9.17 (a), (b), and (c) indicate the saturated liquid states with $x = 0$, and points n indicate the saturated vapour states with $x = 1$, the lines mn indicating the transition from liquid to vapour. Points a, b, and c at various pressures indicate the situations when the masses of vapour reached 25%, 50%, and 75% of the total mass, i.e. at points a, the mass of liquid is 75% and the mass of vapour is 25% of the total mass, at points b, the mixture consists of 50% liquid and 50% vapour by mass, and at points c, the mixture consists of 75% vapour and 25% liquid by mass. The lines passing through points a, b, and c are the constant quality lines of 0.25, 0.50, and 0.75 respectively. Constant quality lines start from the critical point.

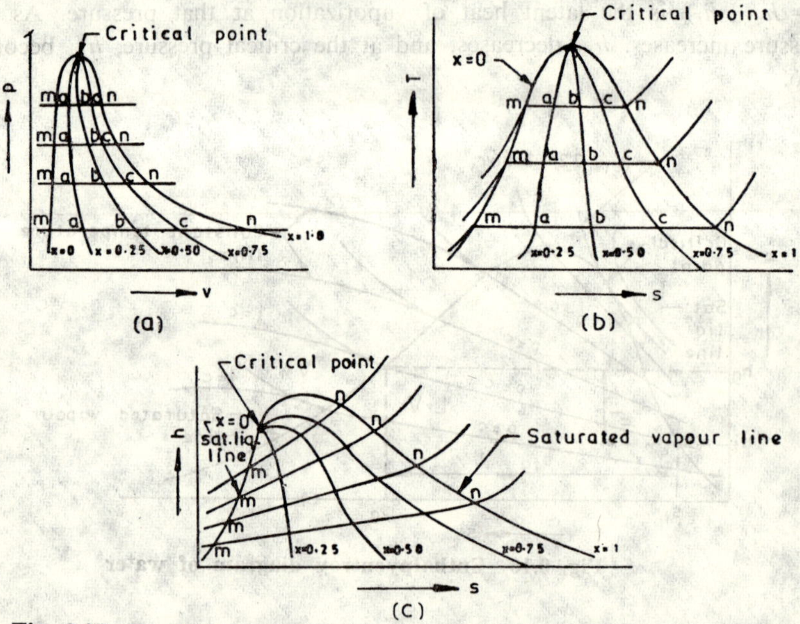

Fig. 9.17 Constant quality lines on p-v, T-s, **and** h-s **diagrams**

Let V be the total volume of a liquid vapour mixture of quality x, V_f the volume of the saturated liquid, and V_g the volume of the saturated vapour, the corresponding masses being m, m_f, and m_g respectively.

Now $$m = m_f + m_g$$

and $$V = V_f + V_g$$

$$mv = m_f v_f + m_g v_g$$

$$= (m - m_g)\, v_f + m_g v_g$$

$$\therefore \quad v = \left(1 - \frac{m_g}{m}\right) v_f + \frac{m_g}{m}\, v_g$$

$$v = (1 - x)\, v_f + x v_g \tag{9.2}$$

where $x = \dfrac{m_g}{m}$, v_f = specific volume of saturated liquid, v_g = specific volume of saturated vapour, and v = specific volume of the mixture of quality x.

Similarly

$$s = (1 - x)\, s_f + x s_g \tag{9.3}$$

$$s = (1 - x)\, h_f + x\, h_g \quad (9.4)$$

$$u = (1 - x)\, u_f + x\, u_g \quad (9.5)$$

where s, h, and u refer to the mixture of quality x, the suffix f and suffix g indicate the conditions of saturated liquid and saturated vapour respectively. From equation (9.2)

$$v = (1 - x)\, v_f + x\, v_g$$

$$= v_f + x\,(v_g - v_f)$$

$$= v_f + x.\, v_{fg}$$

or

$$v = v_f + x\, v_{fg} \quad (9.6)$$

Similarly

$$h = h_f + x\, h_{fg} \quad (9.7)$$

$$s = s_f + x\, s_{fg} \quad (9.8)$$

$$u = u_f + x\, u_{fg} \quad (9.9)$$

9.7 STEAM TABLES

The properties of water are arranged in the steam tables as functions of pressure and temperature. Separate tables are provided to give the properties of water in the saturation states and in the liquid and vapour phases. The *internal energy of saturated water at the triple point (t = 0.01°C) is arbitrarily chosen to be zero.* Since $h = u + pv$, the enthalpy of saturated water at 0.01°C is slightly positive because of the small value of (pv) term. *The entropy of saturated water is also chosen to be zero at the triple point.*

9.7.1 Saturation States

When a liquid and its vapour are in equilibrium at a certain pressure and temperature, only the pressure or the temperature is sufficient to identify the saturation state. If the pressure is given, the temperature of the mixture gets fixed, which is known as the saturation temperature, or if the temperature is given, the saturation pressure gets fixed. Saturated liquid or the saturated vapour has only one independent variable, i.e. only one property is required to be known to fix up the state. Tables A.1(a) and A.1(b) in the appendix give the properties of saturated liquid and saturated vapour. In Table A.1(a), the independent variable is temperature. At a particular temperature, the values of saturation pressure p, and v_f. v_g, h_f, h_{fg}, h_g, s_f, and s_g are given, where v_f, h_f, and s_f refer to the saturated liquid states; v_g, h_g and s_g refer to the saturated vapour state; and v_{fg}, h_{fg}, and s_{fg} refer to the changes in the property values during evaporation (or condensation) at that temperature, where $v_{fg} = v_g - v_f$ and $s_{fg} = s_g - s_f$.

In Table A.1(b), the independent variable is pressure. At a particular pressure, the values of saturation temperature *t*, and v_f, v_g, h_f, h_{fg}, h_g, s_f, and s_g are given. Depending upon whether the pressure or the temperature is given, either Table A.1(a) or Table A.1(b) can be conveniently used for computing the properties of the saturation states.

If data are required for intermediate temperatures or pressures, *linear interpolation is normally accurate*. The reason for the two tables is to reduce the amount of interpolation required.

9.7.2 Liquid-vapour Mixtures

Let us consider a mixture of saturated liquid water and water vapour in equilibrium at pressure *p* and temperature *t*. The composition of the mixture by mass will be given by its quality *x*, and its state will be within the vapour dome (Fig 9.18). The properties of the mixture are as given in Article 9.6, i.e.

$$v = v_f + x\, v_{fg}$$

$$u = u_f + x\, u_{fg}$$

$$h = h_f + x\, h_{fg}$$

$$s = s_f + x\, s_{fg}$$

where v_f, v_{fg}, u_f, u_{fg}, h_f, h_{fg}, s_f and s_{fg} are the saturation properties at the given pressure and temperature.

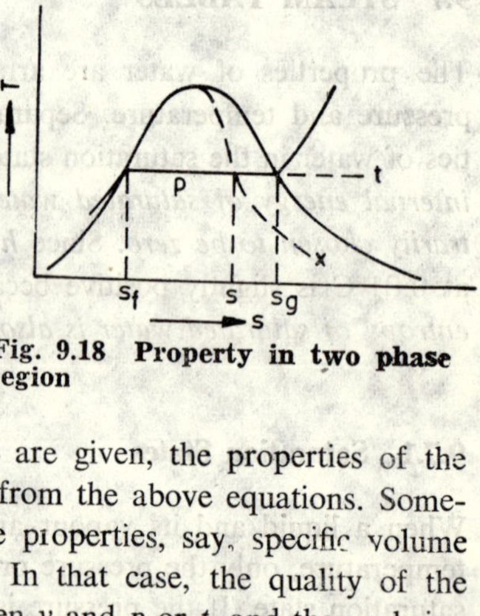

Fig. 9.18 Property in two phase region

If *p* or *t* and the quality of the mixture are given, the properties of the mixture (*v*, *u*, *h*, and *s*) can be evaluated from the above equations. Sometimes, instead of quality, one of the above properties, say, specific volume *v*, and pressure or temperature are given. In that case, the quality of the mixture *x* has to be calculated from the given *v* and *p* or *t*, and then *x* being known, other properties are evaluated.

9.7.3 Superheated Vapour

When the temperature of the vapour is greater than the saturation temperature corresponding to the given pressure, the vapour is said to be *superheated* (state 1 in Fig. 9.19). The difference between the temperature of the superheated vapour and the saturation temperature at that pressure is called the *superheat* or the *degree of superheat*. As shown in Fig. 9.19, the difference $(t_1 - t_{\text{sat.}})$ is the superheat.

In a superheated vapour at a given pressure, the temperature may have different values greater than the saturation temperature. Table A.2 in the appendix gives the values of the properties (volume, enthalpy, and entropy) of superheated vapour for each tabulated pair of values of pressure and temperature, both of which are now independent. Interpolation or extrapolation is to be used for pairs of values of pressure and temperature not given.

9.7.4 Compressed Liquid

When the temperature of a liquid is less than the saturation temperature at the given pressure, the liquid is called *compressed liquid* (state 2 in Fig. 9.19).

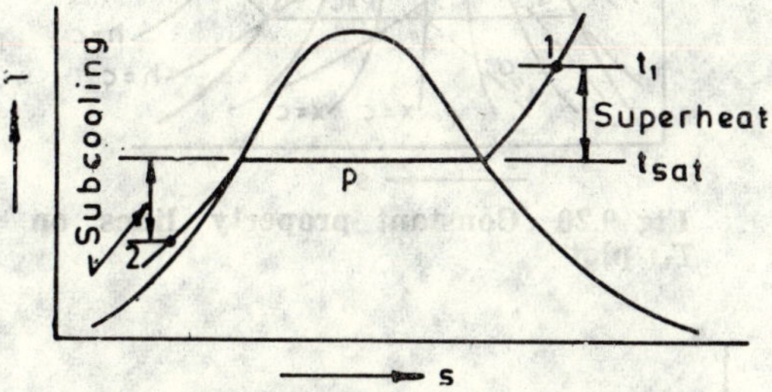

Fig. 9.19 Superheat and subcooling

The pressure and temperature of compressed liquid may vary independently, and a table of properties like the superheated vapour table could be arranged to give the properties at any p and t. However, the *properties of liquids vary little with pressure.* Hence the properties are taken from the saturation tables at the temperature of the compressed liquid. When a liquid is cooled below its saturation temperature at a certain pressure it is said to be *subcooled.* The difference in saturation temperature and the actual liquid temperature is known as the degree of subcooling, or simply, subcooling (Fig. 9.17).

9.8 CHARTS OF THERMODYNAMIC PROPERTIES

The presentation of properties of substances in the form of a chart has certain obvious advantages. The manner of variation of properties is clearly demonstrated in the chart and there is no problem of interpolation. However, the precision is not as much as in steam tables.

The temperature-entropy plot and enthalpy-entropy plot (Figs. 9.20 and 9.21) are commonly used. The temperature-entropy plot shows the vapour dome and the lines of constant pressure, constant volume, constant enthalpy, constant quality, and constant superheat. However, its scale is small and limited in use. The enthalpy-entropy plot or Mollier chart, has a larger scale to provide data suitable for many computations. It contains the same data as does the *T-s* chart. The Mollier chart for water is given in Appendix E.1.

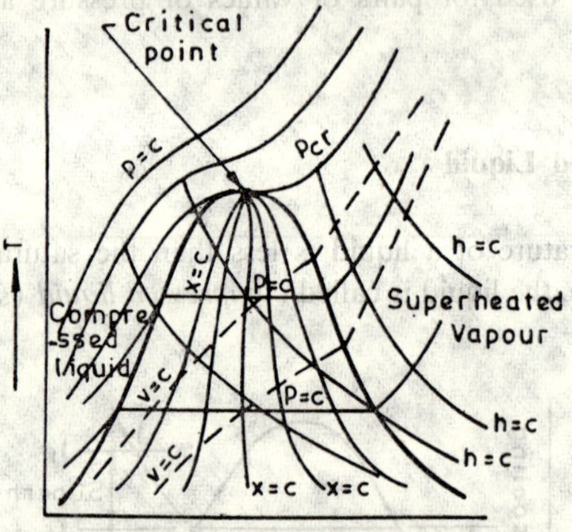

Fig 9.20 Constant property lines on *T-s* plot

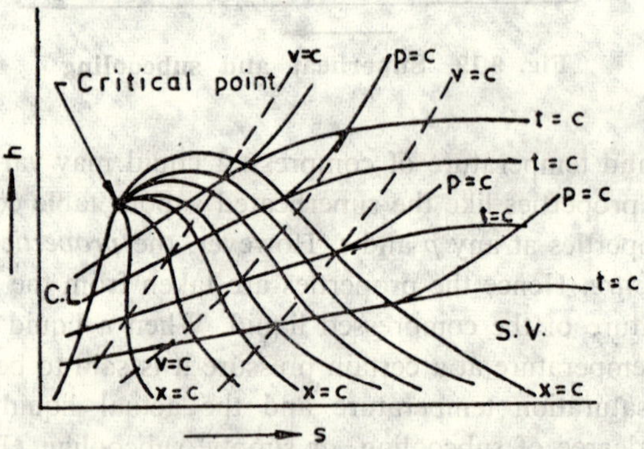

Fig. 9.21 Constant property lines on Mollier diagram

9.9 MEASUREMENT OF STEAM QUALITY

The state of a pure substance gets fixed if two independent properties are given. A pure substance is thus said to have two degrees of freedom. Of all thermodynamic properties, it is easiest to measure the pressure and temperature of a substance. Therefore, whenever pressure and temperature are independent properties, it is the practice to measure them to determine the state of the substance. This is done in the compressed liquid region or the superheated vapour region (Fig. 9.22), where the measured values of pressure and temperature would fix up the state. But when the substance is in the saturation state or two-phase region (Fig. 9.22), the measured values of pressure and temperature could apply equally well to saturated liquid point *f*, saturated vapour point *g*, or to mixtures of any quality, points

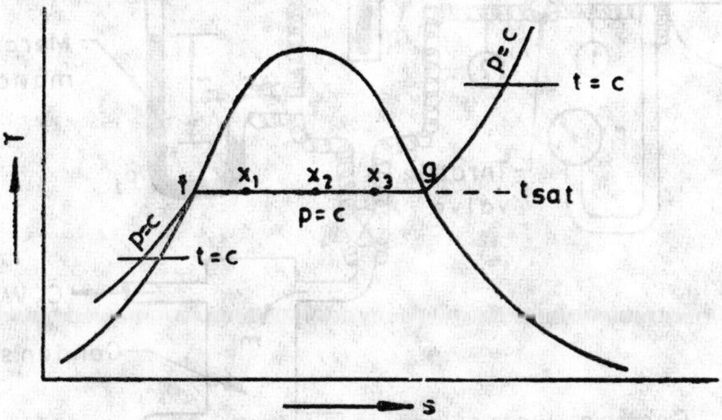

Fig. 9.22 Quality of liquid-vapour mixture

x_1, x_2, or x_3. Of the two properties, *p* and *t*, only one is independent; the other is a dependent property. If pressure is given, the saturation temperature gets automatically fixed for the substance. In order to fix up the state of the mixture, apart from either pressure or temperature, one more property, such as specific volume, enthalpy or composition of the mixture (quality) is required to be known. Since it is relatively difficult to measure the specific volume of a mixture, devices such as calorimeters are used for determining the quality or the enthalpy of the mixture.

In the measurement of quality, the object is always to bring the state of the substance from the two-phase region to the single-phase or superheated region, where both pressure and temperature are independent, and measured to fix the state, either by adiabatic throttling or electric heating.

In the *throttling calorimeter*, a sample of wet steam of mass *m* and at pressure p_1 is taken from the steam main through a perforated sampling tube (Fig. 9.23). Then it is throttled by the partially-opened valve (or orifice) to a pressure p_2, measured by mercury manometer, and temperature t_2, so that after throttling the steam is in the superheated region. The process is shown on the *T-s* and *h-s* diagrams in Fig. 9.24. The steady flow energy equation gives the enthalpy after throttling as equal to enthalpy before throttling. The initial and final equilibrium states 1 and 2 are joined by a dotted line since throttling is irreversible (adiabatic but not isentropic) and the intermediate states are non-equilibrium states not describable by thermodynamic coordinates. The initial state (wet) is given by p_1 and x_1, and the final state by p_2 and t_2 (superheated). Now

since
$$h_1 = h_2$$

$$h_{f_{p1}} + x_1 h_{fg_{p1}} = h_2$$

or
$$x_1 = \frac{h_2 - h_{f_{p1}}}{h_{fg_{p1}}}$$

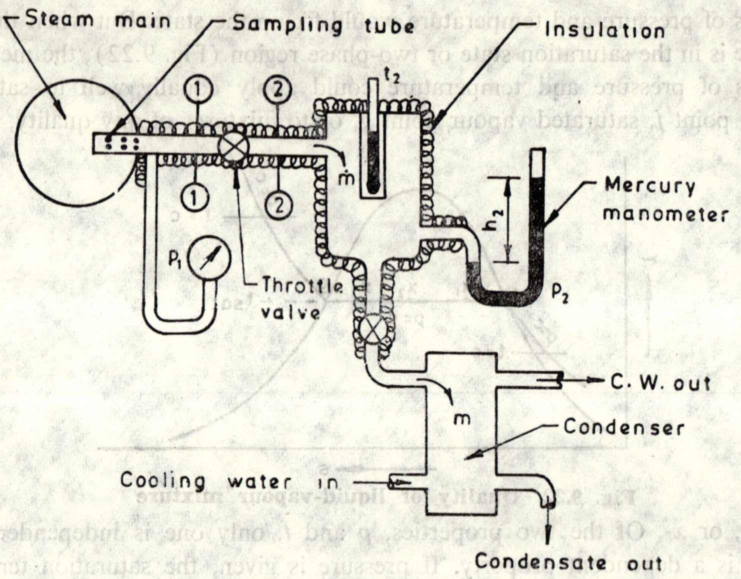

Fig. 9.23 Throttling calorimeter

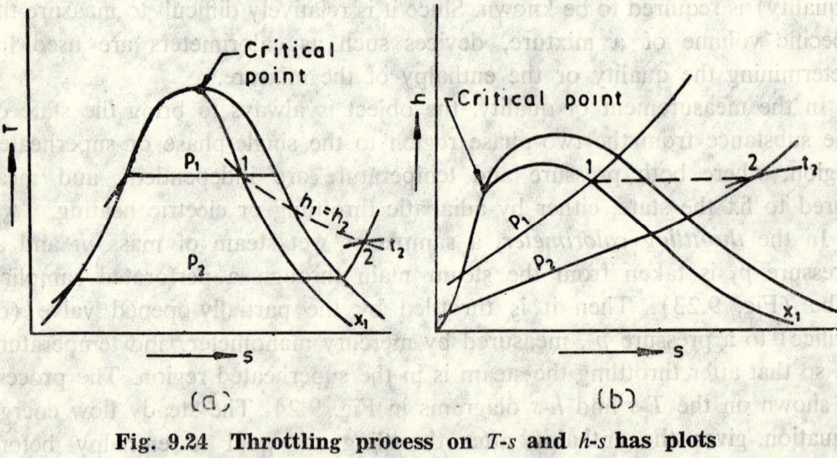

Fig. 9.24 Throttling process on *T-s* and *h-s* has plots

With p_2 and t_2 being known, h_2 can be found out from the superheated steam table. The values of h_f and h_{fg} are taken from the saturated steam table corresponding to pressure p_1. Therefore, the quality of the wet steam x_1 can be calculated.

To be sure that steam after throttling is in the single-phase or superheated region, a minimum of 5°C superheat is desired. So if the pressure after throttling is given and the minimum 5°C superheat is prescribed, then there is the minimum quality of steam (or the maximum moisture content) at the given pressure p_1 which can be measured by the throttling calorimeter.

Fox example, if $p_2 = 1$ atm., then $t_2 = 105°C$ and state 2, after throttling, gets fixed as shown in Fig. 9.25. From state 2, the constant enthalpy line

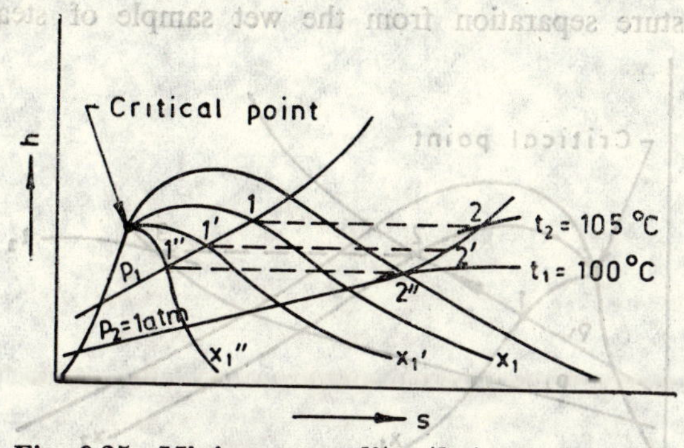

Fig. 9.25 Minimum quality that can be measured only by throttling

intersects the constant pressure p_1 line at 1. Therefore, the quality x_1 is the minimum quality that can be measured simply by throttling. If the quality is, say,, x_1 less than x_1, then after throttling to $p_2 = 1$ atm., the superheat after throttling is less than 5°C. If the quality is x_1'', then throttling to 1 atm. does not give any superheat at all.

When the steam is very wet and the pressure after throttling is not low enough to take the steam to the superheated region, then a *combined separating and throttling calorimeter* is used for the measurement of quality. Steam from the main is first passed through a separator (Fig. 9.26), where some part of the moisture separates out due to the sudden change in

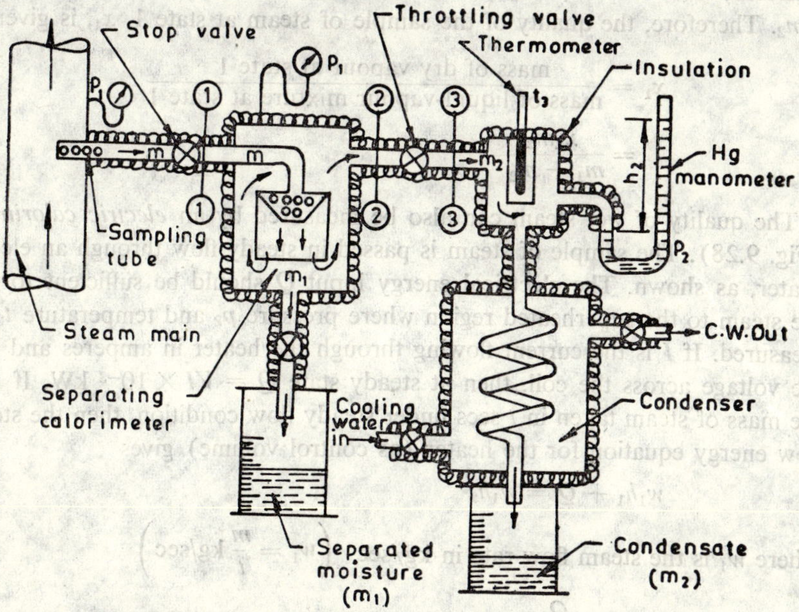

Fig. 9.26 Separating and throttling calorimeter

direction and falls by gravity, and the partially dry vapour is then throttled and taken to the superheated region. In Fig. 9.27, process 1-2 represents the moisture separation from the wet sample of steam at constant

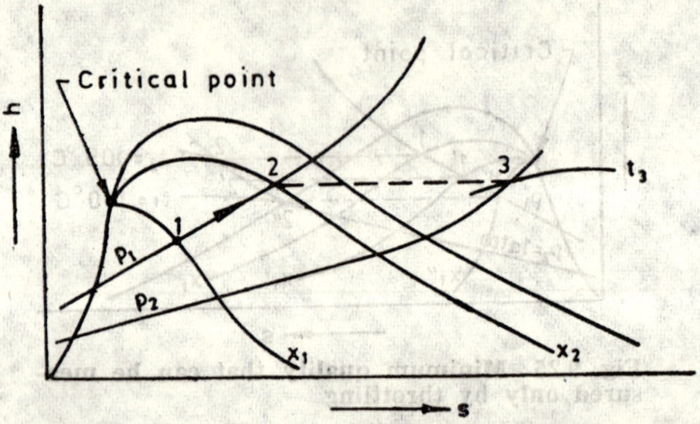

Fig. 9.27 Separating and throttling processes on ***h-s*** **plot.**

pressure p_1 and process 2-3 represents throttling to pressure p_2. With p_2 and t_3 being measured, h_3 can be found out from the superheated steam table. Now,

$$h_3 = h_2 = h_{f_{p1}} + x_2 h_{fg_{p1}}$$

Therefore, x_2, the quality of steam after partial moisture separation, can be evaluated. If m kg of steam is taken through the sampling tube in t secs, m_1 kg of it is separated, and m_2 kg is throttled and then condensed to water and collected, then $m = m_1 + m_2$, and at state 2, the mass of dry vapour will be $x_2 m_2$. Therefore, the quality of the sample of steam at state 1, x_1, is given by

$$x_1 = \frac{\text{mass of dry vapour at state 1}}{\text{mass of liquid-vapour mixture at state 1}}$$

$$= \frac{x_2 m_2}{m_1 + m_2}$$

The quality of wet steam can also be measured by an *electric calorimeter* (Fig. 9.28). The sample of steam is passed in steady flow through an electric heater, as shown. The electrical energy input Q should be sufficient to take the steam to the superheated region where pressure p_2 and temperature t_2 are measured. If I is the current flowing through the heater in amperes and V is the voltage across the coil, then at steady state $Q = VI \times 10^{-3}$ kW. If m is the mass of steam taken in t secs under steady flow condition, then the steady flow energy equation for the heater (as control volume) gives

$$w_1 h_1 + Q = w_1 h_2$$

where w_1 is the steam flow rate in kg/sec $\left(w_1 = \frac{m}{t} \text{ kg/sec}\right)$

$$\therefore \quad h_1 + \frac{Q}{w_1} = h_2$$

With h_2, O, and w_1 being known, h_1 can be computed, Now

$$h_1 = h_{fp_1} + x_1 h_{fgp_1}$$

Hence x_1 can be evaluated.

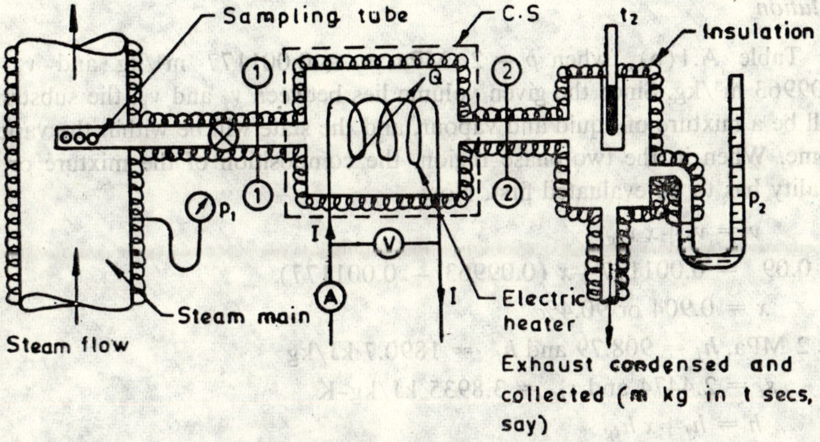

Fig. 9.28 Electrical calorimeter

EXAMPLE 9.1

Find the saturation temperature, the changes in specific volume and entropy during evaporation, and the latent heat of vaporization of steam at 1 MPa.

Solution

At 1 MPa, from Table A.1(b) in the Appendix

$$t_{sat.} = 179.91°C$$

$$v_f = 0.001127 \text{ m}^3/\text{kg} \qquad \text{Ans.}$$

$$v_g = 0.19444 \text{ m}^3/\text{kg}$$

$$\therefore \quad v_{fg} = v_g - v_f = 0.1933 \text{ m}^3/\text{kg} \qquad \text{Ans.}$$

$$s_f = 2.1387 \text{ kJ/kg-K}$$

$$s_g = 6.5865 \text{ kJ/kg-K}$$

$$\therefore \quad s_{fg} = s_g - s_f = 4.4478 \text{ kJ/kg-K} \qquad \text{Ans.}$$

$$h_{fg} = h_g - h_f = 2015.3 \text{ kJ/kg} \qquad \text{Ans.}$$

EXAMPLE 9.2

Saturated steam has an entropy of 6.76 kJ/kg-K. What are its pressure, temperature, specific volume, and enthalpy?

Solution

In Table A.1(b), when $s_g = 6.76$ kJ/kg–K

$$p = 0.6 \text{ MPa}, \quad t = 158.85°C$$

$$v_g = 0.3157 \text{ m}^3/\text{kg, and } h_g = 2756.8 \text{ kJ/kg} \qquad \text{Ans.}$$

EXAMPLE 9.3

Find the enthalpy and entropy of steam when the pressure is 2 MPa and the specific volume is 0.09 m³/kg.

Solution

In Table A.1(b), when $p = 2$ MPa, $v_f = 0.001177$ m³/kg and $v_g = 0.09963$ m³/kg. Since the given volume lies between v_f and v_g, the substance will be a mixture of liquid and vapour, and the state will be within the vapour dome. When in the two-phase region, the composition of the mixture or its quality has to be evaluated first. Now

$$v = v_f + x\, v_{fg}$$

$$0.09 = 0.001177 + x\ (0.09963 - 0.001177)$$

or $x = 0.904$ or 90.4%

At 2 MPa, $h_f = 908.79$ and $h_{fg} = 1890.7$ kJ/kg

$s_f = 2.4474$ and $s_{fg} = 3.8935$ kJ/kg–K

$$h = h_f + x\, h_{fg}$$

$$= 908.79 + 0.904 \times 1890.7 = 2618.79 \text{ kJ/kg}$$ Ans.

$$s = s_f + x\, s_{fg}$$

$$= 2.4474 + 0.904 \times 3.8935$$

$$= 5.9534 \text{ kJ/kg–K}$$ Ans.

EXAMPLE 9.4

Find the enthalpy, entropy, and volume of steam at 1.4 MPa, 380°C.

Solution

At $p = 14$ MPa, in Table A.1(b), $t_{sat.} = 195.07$°C. Therefore, the state of steam must be in the superheated region. In Table A.2, for properties of superheated steam,

at 1.4 MPa, 350°C, $v = 0.2003$ m³/kg

$h = 3149.5$ kJ/kg

$s = 7.1360$ kJ/kg-K

and at 1.4 MPa, 400°C $v = 0.2178$ m³/kg

$h = 3257.5$ kJ/kg

$s = 7.3026$ kJ/kg-K

∴ By interpolation

at 1.4 MPa, 380°C $v = 0.2108$ m³/kg

$h = 3214.3$ kJ/kg

$s = 7.2360$ kJ/kg-K Ans.

EXAMPLE 9.5

A vessel of volume 0.04 m^3 contains a mixture of saturated water and saturated steam at a temperature of 250°C. The mass of the liquid present is 9 kg. Find the pressure, the mass, the specific volume, the enthalpy, the entropy, and the internal energy.

Solution

From Table A.1(a), at 250°C, $p_{sat.} = 3.973$ MPa

$v_f = 0.0012512$ m³/kg, $v_g = 0.05013$ m³/kg

$h_f = 1085.36$ kJ/kg, $h_{fg} = 1716.2$ kJ/kg

$s_f = 2.7927$ kJ/kg-K, $s_{fg} = 3.2802$ kJ/kg-K

Volume of liquid, $V_f = m_f v_f$

$$= 9 \times 0.0012512$$

$$= 0.01126 \text{ m}^3$$

Volume of vapour, $V_g = 0.04 - 0.01126$

$$= 0.02874 \text{ m}^3$$

$\therefore$ Mass of vapour,

$$m_g = \frac{V_g}{v_g} = \frac{0.02874}{0.05013} = 0.575 \text{ kg}$$

$\therefore$ Total mass of mixture,

$$m = m_f + m_g = 9 + 0.575 = 9.575 \text{ kg}$$ Ans.

Quality of mixture,

$$x = \frac{m_g}{m_f + m_g} = \frac{0.575}{9.575} = 0.06$$

$\therefore$ $v = v_f + x v_{fg}$

$$= 0.0012512 + 0.06\,(0.05013 - 0.0012512)$$

$$= 0.00418 \text{ m}^3/\text{kg}$$ Ans.

$$h = h_f + x h_{fg}$$

$$= 1085.36 + 0.06 \times 1716.2$$

$$= 1188.32 \text{ kJ/kg}$$ Ans.

$$s = s_f + x s_{fg}$$

$$= 2.7927 + 0.06 \times 3.2802$$

$$= 2.9895 \text{ kJ/kg-K}$$ Ans.

$$u = h - pv$$

$$= 1188.32 - 3.973 \times 10^3 \times 0.00418$$

$$= 1171.72 \text{ kJ/kg}$$ Ans.

Also, at 250°C, $u_f = 1080.39$, and $u_{fg} = 1522.0$ kJ/kg

$$\therefore \quad u = u_f + xu_{fg}$$
$$= 1080.39 + 0.06 \times 1522$$
$$= 1071.71 \text{ kJ/kg}$$ **Ans.**

EXAMPLE 9.6

Steam initially at 0.3 MPa, 250°C is cooled at constant volume. (a) At what temperature will the steam become saturated vapour? (b) What is the quality at 80°C ? What is the heat transferred per kg of steam in cooling from 250°C to 80°C ?

Solution

At 0.3 MPa, $t_{sat.} = 133.55$°C

Since $t > t_{sat.}$, the state would be in the superheated region (Fig. Ex. 9.6). From Table A.2, for properties of superheated steam, at 0.3 MPa, 250°C

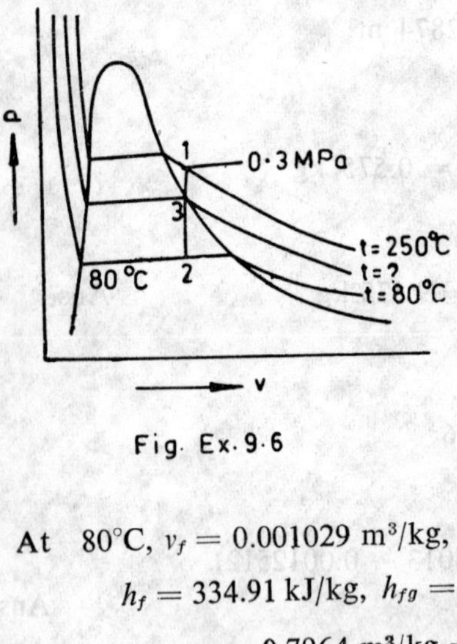

Fig. Ex. 9.6

$$v = 0.7964 \text{ m}^3/\text{kg}$$
$$h = 2967.6 \text{ kJ/kg}$$
$$\therefore \quad v_1 = v_3 = v_2 = 0.7964 \text{ m}^3/\text{kg}$$

In Table A.1

when $v_g = 0.8919$, $t_{sat.} = 120$°C

when $v_g = 0.7706$, $t_{sat.} = 125$°C

Therefore, when $v_g = 0.7964$, $t_{sat.}$, by linear interpolation, would be 123.9°. Steam would become saturated vapour at $t = 123.9$°C **Ans. (a).**

At 80°C, $v_f = 0.001029$ m³/kg, $v_g = 3.407$ m³/kg,

$h_f = 334.91$ kJ/kg, $h_{fg} = 2308.8$ kJ/kg, $p_{sat.} = 47.39$ kPa

$$v_1 = v_2 = 0.7964 \text{ m}^3/\text{kg} = v_{f80°C} + x_2 v_{fg80°C}$$
$$= 0.001029 + x_2 (3.407 - 0.001029)$$

$$\therefore \quad x_2 = \frac{0.79539}{3.40597} = 0.234$$ **Ans. (b)**

$$h_2 = 334.91 + 0.234 \times 2308.8 = 875.9 \text{ kJ/kg}$$
$$h_2 = 2967.6 \text{ kJ/kg}$$

From the first law of thermodynamics

$$dQ = du + pdv$$
$$\therefore \quad (dQ)_v = du$$

or $Q_{1-2} = u_2 - u_1 = (h_2 - p_2v_2) - (h_1 - p_1v_1)$

$$= (h_2 - h_1) + v\,(p_1 - p_2)$$

$$= (875.9 - 2967.6) + 0.7964\,(300 - 47.39)$$

$$= -\,2091.7 + 201.5$$

$$= -\,1890.2 \text{ kJ/kg}$$

Ans. (c)

EXAMPLE 9.7

Steam initially at 1.5 MPa, 300°C expands reversibly and adiabatically in a steam turbine to 40°C. Determine the ideal work output of the turbine per kg of steam.

Solution

The steady flow energy equation for the control volume, as shown in Fig. Ex. 9.7.1, gives (other energy terms being neglected)

$$h_1 = h_2 + W$$

$$\therefore \quad W = h_1 - h_2$$

Work is done by steam at the expense of a fall in its enthalpy value. The process is reversible and adiabatic, so it is isentropic. The process is shown on the *T-s* and *h-s* diagrams in Fig. Ex. 9.7.2.

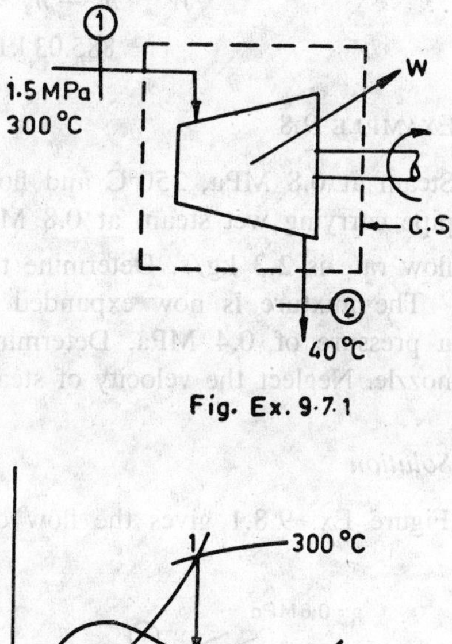

Fig. Ex. 9.7.1

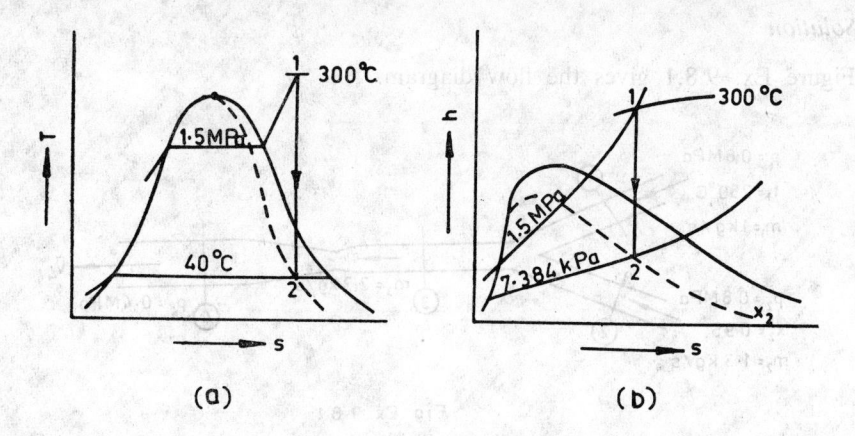

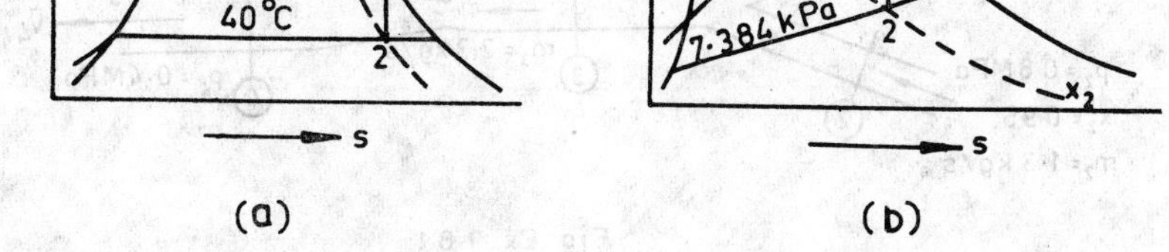

Fig. Ex. 9.7.2

From Table A.1(a), at 40°C

$p_{sat.} = 7.384$ kPa, $s_f = 0.5725$, and $s_{fg} = 7.6845$ kJ/kg -K

$h_f = 167.57$, and $h_{fg} = 2406.7$ kJ/kg

At $p = 1.5$ MPa, $t = 300°C$, from the tabulated properties of superheated steam (Table A.2.)

$$s_1 = 6.9189 \text{ kJ/kg-K}$$
$$h_1 = 3037.6 \text{ kJ/kg}$$

Since
$$s_1 = s_2$$
$$6.9189 = s_f + x_2\, s_{fg_{40°C}}$$
$$= 0.5725 + x_2 \times 7.6845$$
$$x_2 = \frac{6.3464}{7.6845} = 0.826 \text{ or } 82.6\%$$

$\therefore$
$$h_2 = hf_{40°C} + x_2\, h_{fg_{40°C}}$$
$$= 167.57 + 0.826 \times 2406.7$$
$$= 2152.57 \text{ kJ/kg}$$

$\therefore$
$$W = h_1 - h_2 = 3037.6 - 2152.57$$
$$= 885.03 \text{ kJ/kg} \qquad \text{Ans.}$$

EXAMPLE 9.8

Steam at 0.8 MPa, 250°C and flowing at the rate of 1 kg/s passes into a pipe carrying wet steam at 0.8 MPa, 0.95 dry. After adiabatic mixing the flow rate is 2.3 kg/s. Determine the condition of steam after mixing.

The mixture is now expanded in a frictionless nozzle isentropically to a pressure of 0.4 MPa. Determine the velocity of the steam leaving the nozzle. Neglect the velocity of steam in the pipeline.

Solution

Figure Ex. 9.8.1 gives the flow diagram.

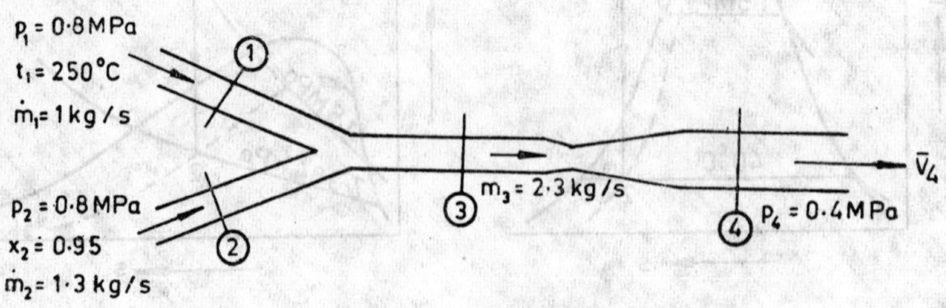

Fig. Ex. 9·8·1

$$w_2 = w_3 - w_1 = 2.3 - 1.0 = 1.3 \text{ kg/s}$$

The energy equation for the adiabatic mixing of the two streams gives

$$w_1 h_1 + w_2 h_2 = w_3 h_3 \qquad (9.8.1)$$

At 0.8 MPa, 250°C, $h_1 = 2950.0$ kJ/kg

At 0.8 MPa, 0.95 dry

$$h_2 = h_{f7\ \text{ata}} + 0.95\, h_{fg7\ \text{ata}}$$

$$= 721.11 + 0.95 \times 2048.0$$

$$= 2666.71 \text{ kJ/kg}$$

∴ From equation (9.8.1)

$$1 \times 2950 + 1.3 \times 2666.71 = 2.3 \times h_3$$

∴ $$h_3 = 2790 \text{ kJ/kg}$$

Since $(h_g)_{0\cdot 8\text{MPa}} = 2769.1$ kJ/kg

and $h_3 > h_g$, the state must be in the superheated region. From the steam tables, when $p = 0.8$ MPa, $t = 200$°C

$$h = 2839.3 \text{ kJ/kg}$$

When $p = 0.8$ MPa, $t_{\text{sat.}} = 170.43$°C

$$h_g = 2769.1 \text{ kJ/kg}$$

By linear interpolation

$$t_3 = 179\text{°C}$$

∴ Degree of superheat = 179 − 170.43 = 8.57°C

∴ Condition of steam after mixing = 0.8 MPa, 179°C Ans.

The energy equation for the nozzle gives

$$h_3 = h_4 + \frac{\mathbf{V}_4^2}{2}$$

since $\mathbf{V}_3$ = velocity of steam in the pipeline = 0

Steam expands isentropically in the nozzle to 0.4 MPa. By interpolation,

$s_3 = 6.7087$ kJ/kg-K $= s_4$

∴ $$6.7087 = 1.7766 + x_4 \times 5.1193$$

$$x_4 = 0.964$$

∴ $$h_4 = 604.74 + 0.964 \times 2133.8 = 2660 \text{ kJ/kg}$$

$$\mathbf{V}_4^2 \times 10^{-3} = 2\ (h_3 - h_4) = 2 \times 130 = 260$$

$$\mathbf{V}_4 = \sqrt{26} \times 100 = 509.9 \text{ m/s}$$ Ans.

The processes are shown on the *h-s* and *T-s* diagrams in Fig. Ex. 9.8.2.

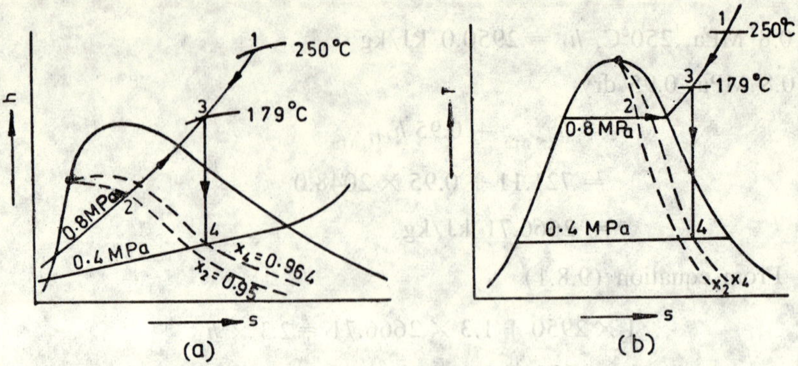

Fig Ex. 9·8·2

EXAMPLE 9.9

Steam flows in a pipeline at 1.5 MPa. After expanding to 0.1 MPa in a throttling calorimeter, the temperature is found to be 120°C. Find the quality of steam in the pipeline. What is the maximum moisture at 1.5 MPa that can be determined with this set-up if at least 5°C of superheat is required after throttling for accurate readings?

Solution

At state 2 (Fig. Ex. 9.9), when $p = 0.1$ MPa, $t = 120$°C by interpolation

$$h_2 = 2716.2 \text{ kJ/kg, and at } p = 1.5 \text{ MPa}$$

$$h_f = 844.89 \text{ and } h_{fg} = 1947.3 \text{ kJ/kg}$$

$$\therefore \quad h_1 = h_2$$

or $$h_{f\,1.5\,\text{MPa}} + x_1\, h_{fg\,1.5\,\text{MPa}} = h_2$$

$$844.89 + x_1 \times 1947.3 = 2716.2$$

$$x_1 = \frac{1871.3}{1947.3} = 0.963 \qquad \text{Ans.}$$

When $$p = 0.1 \text{ MPa and } t = 99.63 + 5 = 104.63°\text{C}$$

$$h_3 = 2685.5 \text{ kJ/kg}$$

since $$h_3 = h_4$$

$$2685.5 = 844.89 + x_4 \times 1947.3$$

$$\therefore \quad x_4 = \frac{1840.6}{1947.3} = 0.948$$

The maximum moisture that can be determined with this set-up is only 5.2%. Ans.

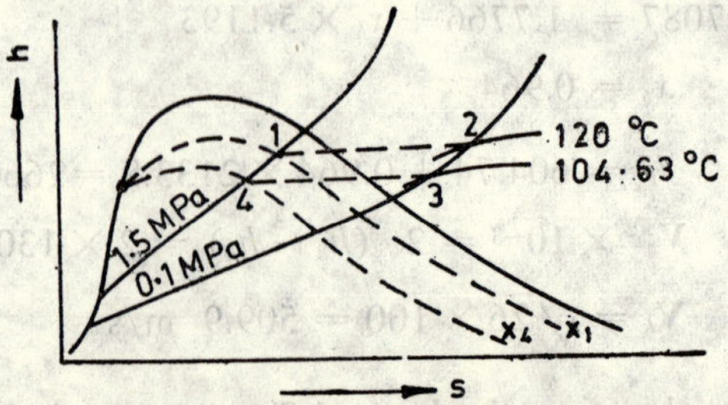

Fig. Ex. 9.9

EXAMPLE 9.10

The following data were obtained with a separating and throttling calorimeter:

Pressure in pipeline	— 1.5 MPa
Condition after throttling	— 0.1 MPa, 110°C
During 5 min moisture collected in the separator	— 0.150 litre at 70°C
Steam condensed after throttling during 5 min	— 3.24 kg

Find the quality of steam in the pipeline.

Solution

As shown in Fig. Ex. 9.10

at 0.1 MPa, 110°C

$$h_3 = 2696.2 \text{ kJ/kg}$$

Now $$h_3 = h_2 = h_{f\,1.5\text{ MPa}} + x_2\, h_{fg\,1.5\text{ MPa}}$$

or $$2696.2 = 844.89 + x_2 \times 1947.3$$

$\therefore$ $$x_2 = \frac{1851.31}{1947.3} = 0.955$$

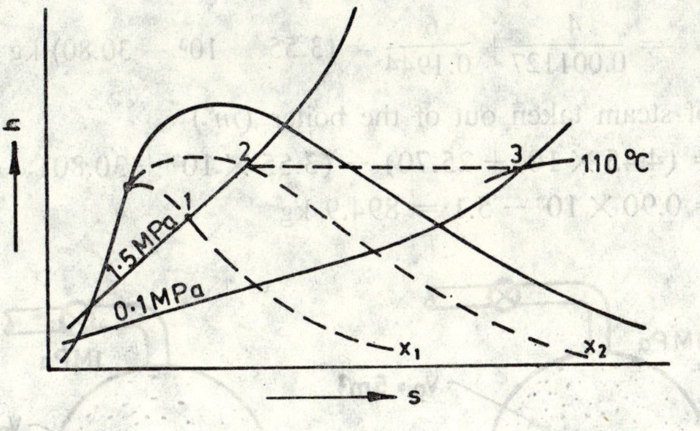

Fig. Ex. 9.10

If m_1 = mass of moisture collected in separator in 5 min and m_2 = mass of steam condensed after throttling in 5 min.

then $$x_1 = \frac{x_2\, m_2}{m_1 + m_2}$$

At 70°C, $v_f = 0.001023 \text{ m}^3/\text{kg}$

$$m_1 = \frac{150 \times 10^{-6} \text{ m}^3}{1023 \times 10^{-6} \text{ m}^3/\text{kg}}$$

$$= 0.1462 \text{ kg}$$

$$m_2 = 3.24 \text{ kg}$$

$$\therefore \quad x_1 = \frac{0.955 \times 3.24}{0.1462 + 3.24} = \frac{3.1}{3.3862} = 0.915 \quad \text{Ans.}$$

EXAMPLE 9.11

A steam boiler initially contains 5 m³ of steam and 5 m³ of water at 1 MPa. Steam is taken out at constant pressure until 4 m³ of water is left. What is the heat transferred during the process ?

Solution

At 1 MPa, $v_f = 0.001127$, and $v_g = 0.1944 \text{ m}^3/\text{kg}$

$$h_g = 2778.1 \text{ kJ/kg}$$

$$h_f = 761.68, \quad u_g = 2583.6 \text{ kJ/kg}$$

$$u_{fg} = 1822 \text{ kJ/kg}$$

The initial mass of saturated water and steam in the boiler (Fig. Ex. 9.11)

$$\frac{V_f}{v_f} + \frac{V_g}{v_g} = \frac{5}{0.001127} + \frac{5}{0.1944} = (4.45 \times 10^3 + 25.70) \text{ kg}$$

where suffix f refers to saturated water and suffix g refers to saturated vapour.
Final mass of saturated water and steam

$$= \frac{4}{0.001127} + \frac{6}{0.1944} = (3.55 \times 10^3 + 30.80) \text{ kg}$$

$\therefore$ Mass of steam taken out of the boiler (m_s)

$$= (4.45 \times 10^3 + 25.70) - (3.55 \times 10^3 + 30.80)$$

$$= 0.90 \times 10^3 - 5.1 = 894.9 \text{ kg}$$

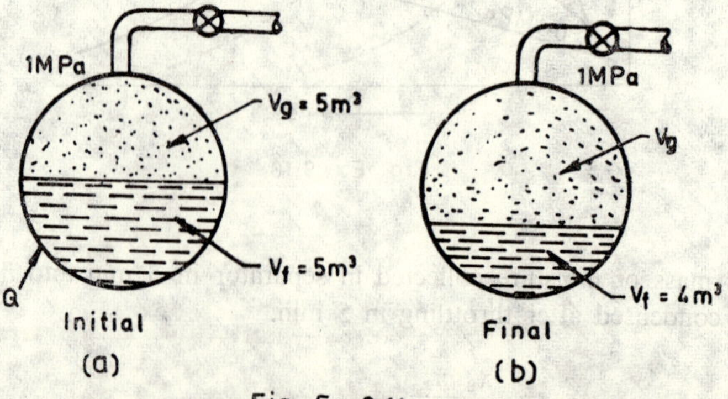

Fig. Ex. 9.11

Making an energy balance, we have: Initial energy stored in saturated water and steam + Heat transferred from the external source = Final energy stored in saturated water and steam + Energy leaving with the steam.

or $$U_1 + Q = U_f + \dot{m}_s h_g$$

assuming that the steam taken out is dry $(x = 1)$

or $$4.45 \times 10^3 \times 761.68 + 25.70 \times 2583.6 + Q$$

$$= 3.55 \times 10^3 \times 761.68 + 30.8 \times 2583.6 + 8749 \times 2778.1$$

or $$Q = 8749 \times 2778.1 - (0.90 \times 10^3)$$

$$\times 761.68 + 5.1 \times 2583.6$$

$$= 2425000 - 685500 + 13176$$

$$= 1752{,}676 \text{ kJ} = 1752.676 \text{ MJ}$$ Ans.

EXAMPLE 9.12

A 280 mm diameter cylinder fitted with a frictionless leakproof piston contains 0.02 kg of steam at a pressure of 0.6 MPa and a temperature of 200°C. As the piston moves slowly outwards through a distance of 305 mm, the steam undergoes a fully-resisted expansion during which the steam pressure p and the steam volume V are related by pV^n = constant, where n is a constant. The final pressure of the steam is 0.12 MPa. Determine (a) the value of n, (b) the work done by the steam, and (c) the magnitude and sign of heat transfer.

Solution

Since the path of expansion (Fig. Ex. 9.12) follows the equation

$$pV^n = C$$

$$p_1 V_1^n = p_2 V_2^n$$

Taking logarithms and arranging the terms

$$n = \frac{\log \frac{p_1}{p_2}}{\log \frac{V_2}{V_1}}$$

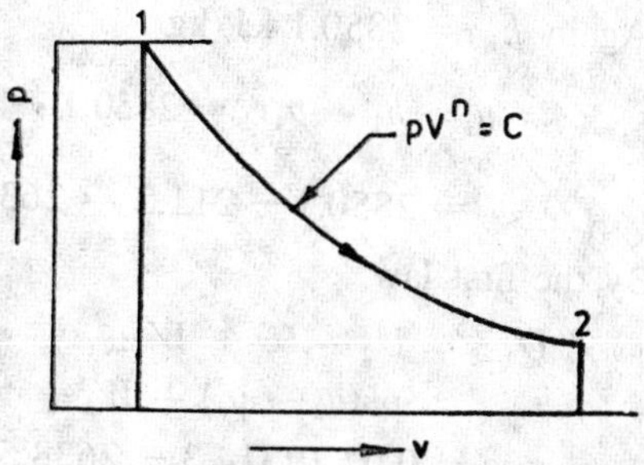

Fig. Ex. 9·12

Now, at 0.6 MPa, 200°C, from Tables A.3

$$v_1 = 0.352 \text{ m}^3/\text{kg}$$

$$h_1 = 2850.1 \text{ kJ/kg}$$

$\therefore$ Total volume, V_1, at state 1 $= 0.352 \times 0.02 = 0.00704\ \text{m}^3$

Displaced volume $= \frac{\pi}{4} d^2 l$

$$= \frac{\pi}{4} \times (0.28)^2 \times 0.305$$

$$= 0.0188\ \text{m}^3$$

$\therefore$ Total volume V_2 after expansion $= 0.0188 + 0.00704 = 0.02584\ \text{m}^3$

$$n = \frac{\log \frac{0.6}{0.12}}{\log \frac{0.02584}{0.00704}} = \frac{\log 5}{\log 3.68} = 1.24 \qquad \text{Ans. (a)}$$

Work done by steam in the expansion process

$$W_{1-2} = \int_{V_1}^{V_2} p dV = \frac{p_1 V_1 - p_2 V_2}{n - 1}$$

$$= \frac{6 \times 10^5\ /\text{Nm}^2 \times 0.00704\ \text{m}^3 - 1.2 \times 10^5\ \text{N/m}^2 \times 0.02584\ \text{m}^3}{1.24 - 1}$$

$$= \frac{4224 - 3100.8}{0.24}\ \text{N m}$$

$$= 4680\ \text{Nm} = 4.68\ \text{kJ} \qquad \text{Ans. (b)}.$$

Now $V_2 = 0.02584\ \text{m}^3$

$$\therefore \quad v_2 = \frac{0.02584}{0.02} = 1.292\ \text{m}^3/\text{kg}$$

Again

$$v_2 = v_{f\,0.12\ \text{MPa}} + x_2\, v_{fg\,0.12\ \text{MPa}}$$

or $\qquad 1.292 = 0.0010476 + x_2 \times 1.4271$

$$\therefore \quad x_2 = \frac{1.291}{1.4271} = 0.906$$

At 0.12 MPa, $u_f = 439.3$ kJ/kg, $u_g = 2512.0$ kJ/kg

$$\therefore \quad u_2 = 439.3 + 0.906\ (2512 - 439.3)$$

$$= 2314.3\ \text{kJ/kg}$$

Again $h_1 = 2850.1$ kJ/kg

$$\therefore \quad u_1 = h_1 - p_1 v_1 = 2850.1 - \frac{0.6 \times 10^6 \times 0.00704 \times 10^{-3}}{0.02}$$

$$= 2850.1 - 211.2 = 2638.9\ \text{kJ/kg}$$

By the first law

$$Q_{1-2} = U_2 - U_1 + W_{1-2}$$

$$= m\,(u_2 - u_1) + W_{1-2}$$

$$= 0.02\ (2314.3 - 2638.5) + 4.68$$

$$= -6.484 + 4.68 = -1.804\ \text{kJ} \qquad \text{Ans. (c)}.$$

EXAMPLE 9.13

A large insulated vessel is divided into two chambers, one containing 5 kg of dry saturated steam at 0.2 MPa and the other 10 kg of steam, 0.8 quality at 0.5 MPa. If the partition between the chambers is removed and the steam is mixed thoroughly and allowed to settle, find the final pressure, steam quality, and entropy change in the process.

Solution

The vessel is divided into chambers, as shown in Fig. Ex. 9.13.1.

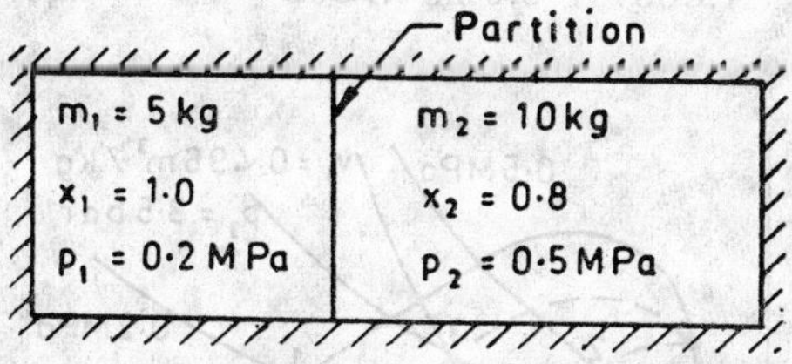

Fig. Ex. 9.13.1

At 0.2 MPa, $v_g = v_1 = 0.8857 \text{ m}^3/\text{kg}$

$$\therefore \quad V_1 = m_1 v_1 = 5 \times 0.8857 = 4.4285 \text{ m}^3$$

At 0.5 MPa, $v_2 = v_f + x_2 v_{fg}$

$$= 0.001093 + 0.8 \times 0.3749 = 0.30101 \text{ m}^3/\text{kg}$$

$$\therefore \quad V_2 = m_2 v_2 = 10 \times 0.30101 = 3.0101 \text{ m}^3$$

$\therefore$ Total volume, $V_m = V_1 + V_2 = 7.4386 \text{ m}^3$ (of mixture)

Total mass of mixture, $m_m = m_1 + m_2 = 5 + 10 = 15$ kg

$\therefore$ Specific volume of mixture

$$v_m = \frac{V_m}{m_m} = \frac{7.4386}{15} = 0.496 \text{ m}^3/\text{kg}$$

By energy balance

$$m_1 h_1 + m_2 h_2 = m_3 h_3$$

At 0.2 MPa, $h_g = h_1 = 2706.7$ kJ/kg

At 0.5 MPa, $h_2 = h_f + x_2 h_{fg}$

$$= 640.23 + 0.8 \times 2108.5 = 2327.03 \text{ kJ/kg}$$

$$\therefore \quad h_3 = h_m = \frac{5 \times 2706.7 + 10 \times 2327.03}{15} = 2453.6 \text{ kJ/kg}$$

Now, for the mixture

$$h_3 = 2453.6 \text{ kJ/kg}$$
$$v_3 = 0.496 \text{ m}^3\text{/kg}$$

From the Mollier diagram, with the given values of h and v, point 3 after mixing is fixed (Fig. Ex. 9.13.2).

$x_3 = 0.870$ Ans. $\qquad s_3 = 6.29$ kJ/kg-K

$p_3 = 3.5$ bar Ans. $\qquad s_1 = s_{g\,0.2\text{ MPa}} = 7.1271$ kJ/kg-K

$$s_2 = s_{f\,0.5\text{ MPa}} + 0.8\, s_{fg\,0.5\text{ MPa}}$$
$$= 1.8607 + 0.8 \times 4.9606 = 5.8292 \text{ kJ/kg-K}$$

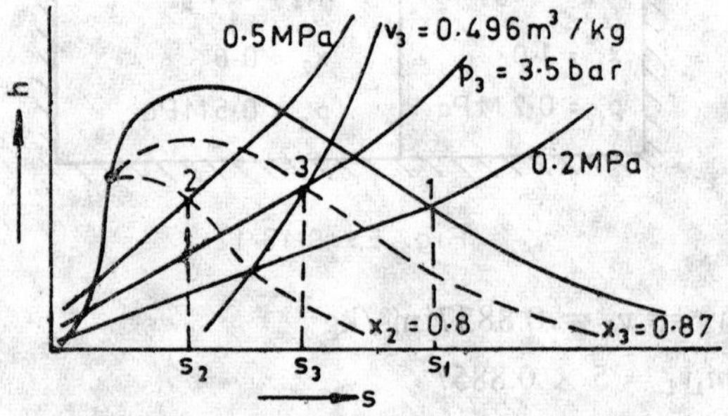

Fig. Ex. 9.13.2

$\therefore$ Entropy change during the process

$$= m_3 s_3 - (m_1 s_1 + m_2 s_2)$$
$$= 15 \times 6.298 - (5 \times 7.1271 + 10 \times 5.8292)$$
$$= 0.43 \text{ kJ/kg} \qquad \textbf{Ans.}$$

EXAMPLE 9.14

Steam generated at a pressure of 6 MPa and a temperature of 400°C is supplied to a turbine via a throttle valve which reduces the pressure to 5 MPa. Expansion in the turbine is adiabatic to a pressure of 0.2 MPa, the isentropic efficiency (actual enthalpy drop/isentropic enthalpy drop) being 82%. The surroundings are at 0.1 MPa, 20°C. Determine the availability of steam before and after the throttle valve and at the turbine exhaust, and calculate the specific work output from the turbine. The K.E. and P.E. changes are negligible.

Solution

Steady flow availability ψ is given by

$$\Psi = (h - h_0) - T_0 (s - s_0) + \frac{1}{2} V_1^2 + g (Z - Z_0)$$

where subscript 0 refers to the surroundings. Since the K.E. and P.E. changes are negligible

$$\Psi_1 = \text{Availability of steam before throttling}$$
$$= (h_1 - h_0) - T_0 (s_1 - s_0)$$

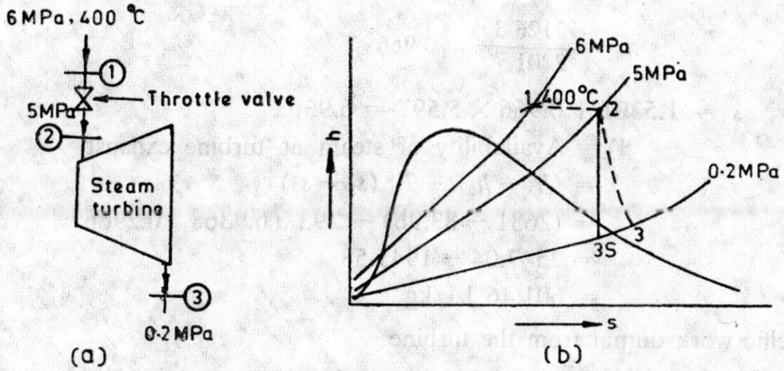

Fig. Ex. 9·14

At 6 MPa, 400°C (Fig. Ex. 9.14)

$$h_1 = 3177.2 \text{ kJ/kg}$$
$$s_1 = 6.5408 \text{ kJ/kg-K}$$

At 20°C

$$h_0 = 83.96 \text{ kJ/kg}$$
$$s_0 = 0.2966 \text{ kJ/kg-K}$$

$$\therefore \quad \Psi_1 = (3177.2 - 83.96) - 293 (6.5408 - 0.2966)$$
$$= 3093.24 - 1829.54 = 1263.7 \text{ kJ/kg}$$ Ans.

Now $h_1 = h_2$, for throttling

At $h = 3177.2$ kJ/kg and $p = 5$ MPa, from the superheated steam table

$$\left.\begin{array}{l} t_2 = 390°\text{C} \\ s_2 = 6.63 \text{ kJ/kg-K} \end{array}\right\} \text{by linear interpolation}$$

$$\therefore \quad \psi_2 = \text{Availability of steam after throttling}$$
$$= (h_2 - h_0) - T_0 (s_2 - s_0)$$
$$= (3177.2 - 83.96) - 293 (6.63 - 0.2966)$$
$$= 3093.24 - 1855.69$$
$$= 1237.55 \text{ kJ/kg}$$

Decrease in availability due to throttling

$$= \psi_1 - \psi_2 = 1263.7 - 1237.55 = 26.15 \text{ kJ/kg}$$

Now

$$s_2 = s_{3S} = 6.63 = 1.5301 + x_{3S} (7.1271 - 1.5301)$$

$$\therefore \quad x_{3s} = \frac{5.10}{5.5970} = 0.9112$$

$h_{3s} = 504.7 + 0.9112 \times 2201.9 = 2511.07$ kJ/kg

$h_2 - h_{3s} = 3177.2 - 2511.07 = 666.13$ kJ/kg

$\therefore \quad h_2 - h_3 = \eta_{is}\,(h_1 - h_{3s}) = 0.82 \times 666.13 = 546.2$ kJ/kg

$\therefore \quad h_3 = 2631$ kJ/kg $= 504.7 + x_3 \times 2201.7$

$$\therefore \quad x_3 = \frac{2126.3}{2201.7} = 0.966$$

$s_3 = 1.5301 + 0.966 \times 5.597 = 6.9368$

$\therefore \quad \Psi_3 =$ Availability of steam at turbine exhaust

$= (h_3 - h_0) - T_0\,(s_3 - s_0)$

$= (2631 - 83.96) - 293\,(6.9368 - 0.2966)$

$= 2547.04 - 1945.58$

$= 601.46$ kJ/kg

Specific work output from the turbine

$= h_2 - h_3 = 3177.2 - 2631 = 546.2$ kJ/kg Ans.

The work done is less than the loss of availability of steam between states 2 and 3, because of the irreversibility accounted for by the isentropic efficiency.

Example 9.15

A steam turbine receives 600 kg/hr of steam at 25 bar, 350°C. At a certain stage of the turbine, steam at the rate of 150 kg/hr is extracted at 3 bar, 200°C. The remaining steam leaves the turbine at 0.2 bar, 0.92 dry. During the expansion process, there is heat transfer from the turbine to the surroundings at the rate of 10 kJ/s. Evaluate per kg of steam entering the turbine (a) the availability of steam entering and leaving the turbine, (b) the maximum work, and (c) the irreversibility. The atmosphere is at 30°C.

Solution

At 25 bar, 350°C

$h_1 = 3125.87$ kJ/kg

$s_1 = 6.8481$ kJ/kg K

At 30°C, $h_0 = 125.79$ kJ/kg

$S_o = S_{f30°C} = 0.4369$ kJ/kg-K

At 3 bar, 200°C

$h_2 = 2865.5$ kJ/kg

$s_2 = 7.3115$ kJ/kg-K

At 0.2 bar (0.92 dry)

$h_f = 251.4$ kJ/kg

$h_{fg} = 2358.3$ kJ/kg

$$s_f = 0.8320 \text{ kJ/kg-K}$$
$$s_g = 7.9085 \text{ kJ/kg-K}$$
$$\therefore \quad h_3 = 251.4 + 0.92 \times 2358.3 = 2421.04 \text{ kJ/kg}$$
$$s_3 = 0.8320 + 0.92 \times 7.0765 = 7.3424 \text{ kJ/kg-K}$$

The states of steam are shown in Fig. Ex. 9.15.

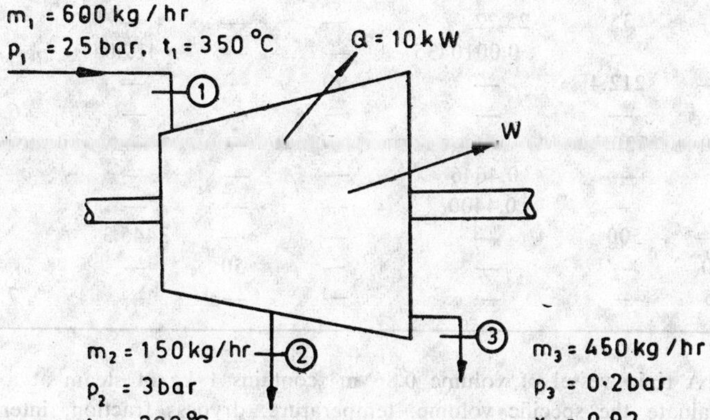

Fig. Ex. 9·15

(a) Availability of steam entering the turbine

$$\Psi_1 = (h_1 - h_0) - T_0\,(s_1 - s_0)$$
$$= (3125.87 - 125.79) - 303\,(6.8481 - 0.4369)$$
$$= 3000.08 - 1942.60 = 1057.48 \text{ kJ/kg} \qquad \text{Ans.}$$

Availability of steam leaving the turbine

$$\Psi_2 = (h_2 - h_0) - T_0\,(s_2 - s_0)$$
$$= (2865.5 - 125.79) - 303\,(7.3115 - 0.4369)$$
$$= 2739.71 - 2083.00 = 656.71 \text{ kJ/kg}$$

(b) Maximum work $= \psi_1 - \psi_2 = 1057.48 - 656.71 = 400.77$ kJ/kg

(c) Irreversibility

$$I = T_0\,(w_2 s_2 + w_3 s_3 - w_1 s_1) - Q$$
$$= 303\,(150 \times 7.3115 + 450 \times 7.3424$$
$$- 600 \times 6.8481) - (-10 \times 3600)$$
$$= 303\,(1096.73 + 3304.08 - 4108.86) + 36000$$
$$= 124{,}460.85 \text{ kJ/hr}$$
$$= 124.461 \text{ MJ/hr} \qquad \text{Ans.}$$

PROBLEMS

9.1 Complete the following table of properties for 1 kg of water (liquid, vapour or mixture)

	p (bar)	t (°C)	v (m³/kg)	x (%)	Super-heat (°C)	h (kJ/kg)	s (kJ/kg-K)
(a)	—	35	25.22	—	—	—	—
(b)	—	—	0.0010435	—	—	419.04	—
(c)	—	212.4	—	90	—	—	—
(d)	1	—	—	—	—	—	6.104
(e)	10	320	—	—	—	—	—
(f)	5	—	0.4646	—	—	—	—
(g)	4	—	0.4400	—	—	—	—
(h)	—	500	—	—	—	3445.3	—
(i)	20	—	—	—	50	—	—
(j)	15	—	—	—	—	—	7.2690

9.2 (a) A rigid vessel of volume 0.86 m³ contains 1 kg of steam at a pressure of 2 bar. Evaluate the specific volume, temperature, dryness fraction, internal energy, enthalpy, and entropy of steam.

(b) The steam is heated to raise its temperature to 150°C. Show the process on a sketch of the p-v diagram, and evaluate the pressure, increase in enthalpy, increase in internal energy, increase in entropy of the steam, and the heat transfer. Evaluate also the pressure at which the steam becomes dry saturated. Ans. (a) 0.86 m³|kg, 120.23°C, 0.97, 2468.54 kJ|kg, 2640.54 kJ|kg, 6.9592 kJ|kgK

(b) 2.3 bar, 126 kJ|kg, 106.6 kJ|kg, 0.2598 kJ|kgK, 106.6 kJ|kg

9.3 Ten kg of water at 45°C is heated at a constant pressure of 10 bar until it becomes superheated vapour at 300°C. Find the changes in volume, enthalpy, internal energy and entropy.

Ans. 2.569 m³, 28627.5 kJ, 26047.6 kJ, 64.842 kJ|K.

9.4 Water at 40°C is continuously sprayed into a pipeline carrying 5 tonnes of steam at 5 bar, 300°C per hour. At a section downstream where the pressure is 3 bar, the quality is to be 95%. Find the rate of water spray in kg/hr. Ans. 912.67 kg|hr.

9.5 A rigid vessel contains 1 kg of a mixture of saturated water and saturated steam at a pressure of 0.15 MPa. When the mixture is heated, the state passes through the critical point. Determine (a) the volume of the vessel, (b) the mass of liquid and of vapour in the vessel initially, (c) the temperature of the mixture when the pressure has risen to 3 MPa, and (d) the heat transfer required to produce the final state (c). Ans. (a) 0.003155 m³, (b) 0.9982 kg, 0.0018 kg, (c) 233.9°C, (d) 581.46 kJ/kg.

9.6 A sample of steam from a boiler drum at 3 MPa is put through a throttling calorimeter in which the pressure and temperature are found to be 0.1 MPa, 120°C. Find the quality of the sample taken from the boiler. Ans. 0.951.

9.7 It is desired to measure the quality of wet steam at 0.5 MPa. The quality of steam is expected to be not more than 0.9.

(a) Explain why a throttling calorimeter to atmospheric pressure will not serve the purpose.

(b) Will the use of a separating calorimeter, ahead of the throttling calorimeter, serve the purpose, if at best 5 C degree of superheat is desirable at the end of throttling? What is the minimum dryness fraction required at the exit of the separating calorimeter to satisfy this condition?

9.8 The following observations were recorded in an experiment with a combined separating and throttling calorimeter:

Pressure in the steam main—15 bar

Mass of water drained from the separator—0.55 kg

Mass of steam condensed after passing through the throttle valve—4.20 kg

Pressure and temperature after throttling—1 bar, 120°C

Evaluate the dryness fraction of the steam in the main, and state with reasons, whether the throttling calorimeter alone could have been used for this test.

Ans. 0.85.

9.9 Steam from an engine exhaust at 1.25 bar flows steadily through an electric calorimeter and comes out at 1 bar, 130°C. The calorimeter has two 1 kW heaters and the flow is measured to be 3.4 kg in 5 min. Find the quality in the engine exhaust. For the same mass flow and pressures, what is the maximum moisture that can be determined if the outlet temperature is at least 105°C? Ans. 0.944, 0.921.

9.10 Steam expands isentropically in a nozzle from 1 MPa, 250°C to 10 kPa. The steam flow rate is 1 kg/s. Find the velocity of steam at the exit from the nozzle, and the exit area of the nozzle. Neglect the velocity of steam at the inlet to the nozzle.

The exhaust steam from the nozzle flows into a condenser and flows out as saturated water. The cooling water enters the condenser at 25°C and leaves at 35°C Determine the mass flow rate of cooling water. Ans. 1224 m|s, 0.0101 m^2, 47.81 kg|s.

9.11 A reversible polytropic process beings with steam at $p_1=10$ bar, $t_1=200$°C, and ends with $p_2=1$ bar. The exponent n has the value 1.15. Find the final specific volume, the final temperature, and the heat transferred per kg of fluid.

9.12 Two streams of steam, one at 2 MPa, 300°C and the other at 2 MPa, 400°C, mix in a steady flow adiabatic process. The rates of flow of the two streams are 3 kg/min and 2 kg/min respectively. Evaluate the final temperature of the emerging stream, if there is no pressure drop due to the mixing process. What would be the rate of increase in the entropy of the universe?

9.13 Boiler steam at 8 bar, 250°C, reaches the engine control valve through a pipeline at 7 bar, 200°C. It is throttled to 5 bar before expanding in the engine to 0.1 bar, 0.9 dry. Determine per kg of steam (a) the heat loss in the pipe line, (b) the temperature drop in passing through the throttle valve, (c) the work output of the engine, (d) the entropy change due to throttling, and (e) the entropy change in passing through the engine. Ans. (a) 105.3 kJ|kg, (b) 5°C, (c) 499.35 kJ|kg, (d) 0.1433 kJ|kg-K, (e) 0.3657 kJ|kg-K.

9.14 Tank A (Fig. P.9.14) has a volume of 0.1 m^3 and contains steam at 200°C, 10% liquid and 90% vapour by volume, while tank B is evacuated. The valve is then opened, and the tanks eventually come to the same pressure, which is found to be 4 bar. During this process, heat is transferred such that the steam remains at 200°C. What is the volume of tank B?

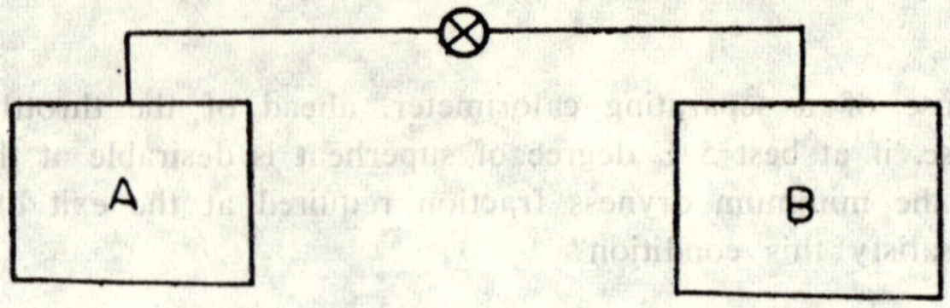

Fig. P. 9.14

9.15 A sample of wet steam from a steam main flows steadily through a partially open valve into a pipeline in which is fitted an electric coil. The valve and the pipeline are well insulated. The steam mass flow rate is 0.008 kg|s while the coil takes 3.91 amperes at 230 volts. The main pressure is 4 bar, and the pressure and temperature of the steam downstream of the coil are 2 bar and 160°C respectively. Steam velocities may be assumed to be negligible.

(a) Evaluate the quality of steam in the main.

(b) State, with reasons, whether an insulated throttling calorimeter could be used for this test.

9.16 Two insulated tanks, *A* and *B*, are connected by a valve. Tank *A* has a volume of 0.70 m^3 and contains steam at 1.5 bar, 200°C. Tank *B* has a volume of 0.35 m^3 and contains steam at 6 bar with a quality of 90%. The valve is then opened, and the two tanks come to a uniform state. If there is no heat transfer during the process, what is the final pressure? Compute the entropy change of the universe.

9.17 A spherical aluminium vessel has an inside diameter of 0.3 m and a 0.62 cm thick wall. The vessel contains water at 25°C with a quality of 1%. The vessel is then heated until the water inside is saturated vapour. Considering the vessel and water together as a system, calculate the heat transfer during this process. The density of aluminium is 2.7 g/cm^3 and its specific heat is 0.896 kJ/kg-K.

9.18 Steam at 10 bar, 250°C flowing with negligible velocity at the rate of 3 kg/min mixes adiabatically with steam at 10 bar, 0.75 quality, flowing also with negligible velocity at the rate of 5 kg/min. The combined stream of steam is throttled to 5 bar and then expanded isentropically in a nozzle to 2 bar. Determine (a) the state of steam after mixing, (b) the state of steam after throttling, (c) the increase in entropy due to throttling, (d) the velocity of steam at the exit from the nozzle, and (e) the exit area of the nozzle. Neglect the K.E. of steam at the inlet to the nozzle.

9.19 Steam at 65 bar, 400°C leaves the boiler to enter a steam turbine fitted with a throttle governor. At a reduced load, as the governor takes action, the pressure of steam is reduced to 59 bar by throttling before it is admitted to the turbine. Evaluate the availabilities of steam before and after the throttling process and the irreversibility due to it.

10

Properties of Gases and Gas Mixtures

10.1 AVOGADRO'S LAW

A mole of a substance has a mass numerically equal to the molecular weight of the substance.

One g mol of oxygen has a mass of 32 gm, 1 kg mol of oxygen has a mass of 32 kg, 1 kg mol of nitrogen has a mass of 28 kg, and so on.

Avogadro's law states that the volume of a g mol of all gases at the pressure of 760 mm Hg and temperature of 0°C is the same, and is equal to 22.4 litres. Therefore, 1 g mol of a gas has a volume of 22.4×10^3 cm³ and 1 kg mol of a gas has a volume of 22.4 m³ at normal temperature and pressure (N.T.P.).

For a certain gas, if m is its mass in kg, and μ its molecular weight, then the number of kg moles of the gas, n, would be given by

$$n = \frac{m \text{ kg}}{\mu \dfrac{\text{kg}}{\text{kgmol}}} = \frac{m}{\mu} \text{kgmoles}$$

The molar volume, v, is given by

$$v = \frac{V}{n} \text{m}^3/\text{kg mol}$$

where V is the total volume of the gas in m³.

10.2 EQUATION OF STATE OF A GAS

The functional relationship among the properties, pressure p, molar or specific volume v, and temperature T, is known as *an equation of state*, which may be expressed in the form,

$$f(p, v, T) = 0$$

If two of these properties of a gas are known, the third can be evaluated from the equation of state.

It was discussed in Chapter 2 that gas is the best-behaved thermometric substance because of the fact that the ratio of pressure p of a gas at any temperature to pressure p_t of the same gas at the triple point, as both p and p_t approach zero, approaches a value independent of the nature of the gas. The ideal gas temperature T of the system at whose temperature the gas exerts pressure p (Article 2.5) was defined as

$$T = 273.16 \lim_{p_t \to 0} \frac{p}{p_t} \qquad \text{(Const. } V\text{)}$$

$$T = 273.16 \lim_{p_t \to 0} \frac{V}{V_t} \qquad \text{(Const. } p\text{)}$$

The relation between pv and p of a gas may be expressed by means of a power series of the form

$$pv = A\,(1 + B'p + C'p^2 + \ldots) \tag{10.1}$$

where A, B', C', etc., depend on the temperature and nature of the gas.

A fundamental property of gases is that $\lim_{p \to 0} (pv)$ is independent of the nature of the gas and depends only on T. This is shown in Fig. 10.1, where the product pv is blotted against p for four different gases in the bulb (nitrogen, air, hydrogen, and oxygen) at the boiling point of sulphur, at steam point and at the triple point of water. In each case, it is seen that as $p \to 0$, pv approaches the same value for all gases at the same temperature. From equation (10.1)

$$\lim_{p \to 0} pv = A$$

Therefore, the constant A is a function of temperature only and independent of the nature of the gas.

$$\lim \frac{p}{p_t}(\text{Const. } V) = \lim \frac{pV}{p_t V} = \frac{\lim pv}{\lim (pv)_t} = \frac{A}{A_t}$$

$$\lim \frac{V}{V_t}(\text{Const. } p) = \lim \frac{pV}{pV_t} = \frac{\lim pv}{\lim (pv)_t} = \frac{A}{A_t}$$

The ideal gas temperature, T, is thus

$$T = 273.16 \frac{\lim (pv)}{\lim (pv)_t}$$

$$\therefore \qquad \lim (pv) = \left[\frac{\lim (pv)_t}{273.16}\right] T$$

The term within bracket is called the *universal gas constant* and is denoted by $\bar{R}$. Thus

$$\bar{R} = \frac{\lim (pv)_t}{273.16} \tag{10.2}$$

The value obtained for $\lim_{p \to 0} (pv)_t$ is 22.4 $\dfrac{\text{litre-atm}}{\text{g mol}}$.

$$\therefore \qquad \bar{R} = \frac{22.4}{273.16} = 0.083 \frac{\text{litre-atm.}}{\text{g mol K}}$$

The equation of state of a gas is thus

$$\lim_{p \to 0} pv = R\,T \tag{10.3}$$

where v is the molar volume.

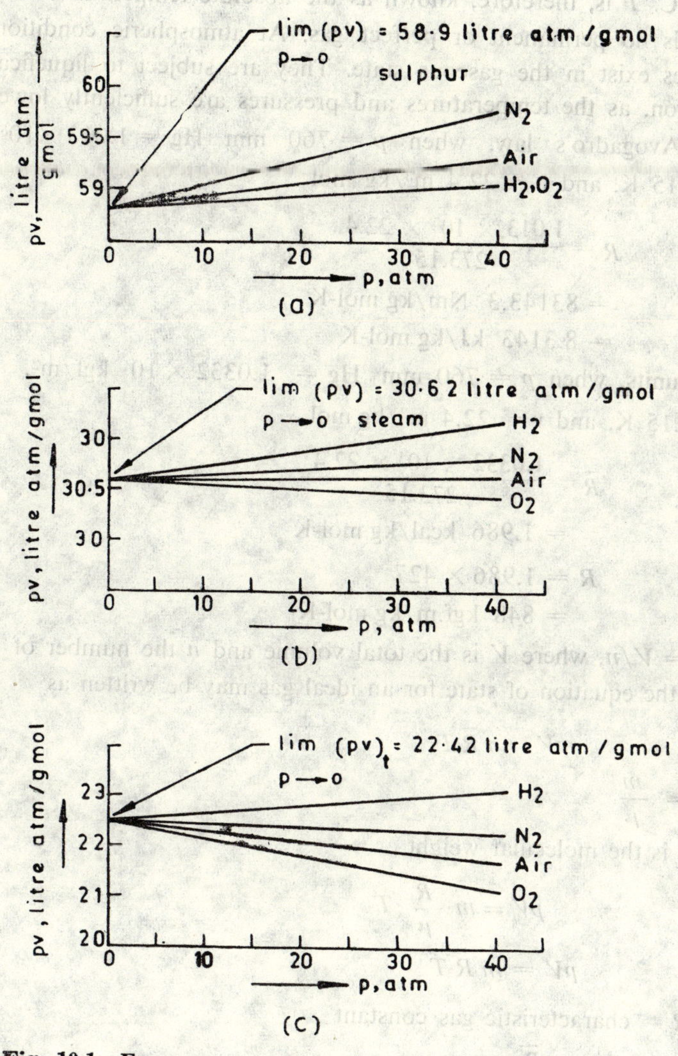

Fig. 10.1 For any gas, $\lim_{p \to 0} (Pv)_T$ is independent of the nature of the gas and depends only on T'.

10.3 IDEAL GAS

A hypothetical gas which obeys the law $pv = \overline{R}\,T$ at all pressures and temperatures is called an *ideal gas*.

Real gases do not conform to this equation of state with complete accuracy. As $p \to 0$, or $T \to \infty$, the real gas approaches the ideal gas behaviour. In the equation $pv = \bar{R}T$, as $T \to 0$, i.e., $t \to -273.15°C$, if v remains constant, $p \to 0$, or if p remains constant, $v \to 0$. Since negative volume or negative pressure is inconceivable, the lowest possible temperature is 0 K or −273.15°C. T is, therefore, known as the absolute temperature.

There is no permanent or perfect gas. At atmospheric conditions only, these gases exist in the gaseous state. They are subject to liquefication or solidification, as the temperatures and pressures are sufficiently lowered.

From Avogadro's law, when $p = 760$ mm Hg $= 1.013 \times 10^5$ N/m², $T - 273.15$ K, and $v = 22.4$ m³/kg mol

$$\bar{R} = \frac{1.013 \times 10^5 \times 22.4}{273.15}$$

$$= 83143.3 \text{ Nm/kg mol-K}$$

$$= 8.3143 \text{ kJ/kg mol-K}$$

In MKS units, when $p = 760$ mm Hg $= 1.0332 \times 10^4$ kgf/m², $T = 273.15$ K, and $v = 22.4$ m³/kg mol

$$\bar{R} = \frac{1.0332 \times 10^4 \times 22.4}{273.15}$$

$$= 1.986 \text{ kcal/kg mol-K}$$

Also
$$\bar{R} = 1.986 \times 427$$

$$= 848 \text{ kgf.m/kg mol-K}$$

Since $v = V/n$, where V is the total volume and n the number of moles of the gas, the equation of state for an ideal gas may be written as

$$pV = n\bar{R}T \tag{10.4}$$

Also $n = \dfrac{m}{\mu}$

where μ is the molecular weight

$$\therefore \quad pV = m \cdot \frac{\bar{R}}{\mu} \cdot T$$

or

$$pV = mRT \tag{10.5}$$

where R = characteristic gas constant

$$= \frac{\bar{R}}{\mu} \tag{10.6}$$

For oxygen, e.g.,

$$R_{O_2} = \frac{8.3143}{32} = 0.262 \text{ kJ/kg-K}$$

For air

$$R_{air} = \frac{848}{28.96} = 29.27 \text{ kgf.m/kg-K}$$

Also

$$R_{air} = \frac{8\,3143}{28.96} = 0.287 \text{ kJ/kg-K}$$

There are 6.023×10^{23} molecules in a g mol of a substance.

This is known as Avogadro's number (A).

$$\therefore \quad A = 6.023 \times 10^{26} \text{ molecules/kg mol}$$

In n kg moles of gas, the total number or molecules, N, are

or $\quad N = n.A$

or $\quad n = N/A$

$$pV = N.\frac{\bar{R}}{A}T$$

$$= NKT \tag{10.7}$$

where K = Boltzmann constant

$$= \frac{\bar{R}}{A} = \frac{8314.3}{6.023 \times 10^{26}} = 1.38 \times 10^{-23} \text{ J/molecule-K}$$

Also $\quad K = \dfrac{1.986}{6.023 \times 10^{26}} = 3.2964 \times 10^{-27}$ kcal/molecule-K

Therefore, the equation of state of an ideal gas is given by

$$pV = m\,R\,T$$

$$= n\,\bar{R}\,T$$

$$= N\,K\,T$$

10.3.1 Specific Heats, Internal Energy, and Enthalpy of an Ideal Gas

An ideal gas not only satisfies the equation of state $pv = RT$, but its *specific heats are constant* also. For real gases, these vary appreciably with temperature, and little with pressure.

The properties of a substance are related by

$$Tds = du + pdv$$

or
$$ds = \frac{du}{T} + \frac{p}{T}dv \tag{10.8}$$

The internal energy u is assumed to be a function of T and v, i.e.

$$u = f\,(T, v)$$

or
$$du = \left(\frac{\partial u}{\partial T}\right)_v dT + \left(\frac{\partial u}{\partial v}\right)_T dv \tag{10.9}$$

From equations (10.8) and (10.9)

$$ds = \frac{1}{T}\left(\frac{\partial u}{\partial T}\right)_v dT + \frac{1}{T}\left[\left(\frac{\partial u}{\partial v}\right)_T + p\right]dv \tag{10.10}$$

Again, let

$$s = f(T, v)$$

$$ds = \left(\frac{\partial s}{\partial T}\right)_v dT + \left(\frac{\partial s}{\partial v}\right)_T dv \tag{10.11}$$

Comparing equations (10.10) and (10.11)

$$\left(\frac{\partial s}{\partial T}\right)_v = \frac{1}{T}\left(\frac{\partial u}{\partial T}\right)_v \tag{10.12}$$

$$\left(\frac{\partial s}{\partial v}\right)_T = \frac{1}{T}\left[\left(\frac{\partial u}{\partial v}\right)_T + p\right] \tag{10.13}$$

Differentiating equation (10.12) with respect to v when T is constant

$$\frac{\partial^2 s}{\partial T \partial v} = \frac{1}{T}\frac{\partial^2 u}{\partial T \partial v} \tag{10.14}$$

Differentiating equation (10.13) with respect to T when v is constant

$$\frac{\partial^2 s}{\partial v \partial T} = \frac{1}{T}\frac{\partial^2 u}{\partial v \partial T} + \frac{1}{T}\left(\frac{\partial p}{\partial T}\right)_v - \frac{1}{T^2}\left(\frac{\partial u}{\partial v}\right)_T - \frac{p}{T^2} \tag{10.15}$$

From equations (10.14) and (10.15)

$$\frac{1}{T}\frac{\partial^2 u}{\partial T . \partial v} = \frac{1}{T}\frac{\partial^2 u}{\partial v . \partial T} + \frac{1}{T}\left(\frac{\partial p}{\partial T}\right)_v - \frac{1}{T^2}\left(\frac{\partial u}{\partial v}\right)_T - \frac{p}{T^2}$$

or

$$\left(\frac{\partial u}{\partial v}\right)_T + p = T\left(\frac{\partial p}{\partial T}\right)_v \tag{10.16}$$

For an ideal gas

$$pv = RT$$

$$v\left(\frac{\partial p}{\partial T}\right)_v = R$$

$$\left(\frac{\partial p}{\partial T}\right)_v = \frac{R}{v} = \frac{p}{T} \tag{10.17}$$

From equations (10.16) and (10.17)

$$\left(\frac{\partial u}{\partial v}\right)_T = 0 \tag{10.18}$$

Therefore, u does not change when v changes at constant temperature.

Similarly, if $u = f(T, p)$, it can be shown that $\left(\frac{\partial u}{\partial p}\right)_T = 0$. Therefore, u does not change with p either, when T remains constant.

u does not change unless T changes.

Then $$u = f(T) \tag{10.19}$$

only for an ideal gas. This is known as *Joule's law*.

If $u = f(T, v)$

$$du = \left(\frac{\partial u}{\partial T}\right)_v dT + \left(\frac{\partial u}{\partial v}\right)_T dv$$

Since the last term is zero by equation (10.18), and by definition

$$c_v = \left(\frac{\partial u}{\partial T}\right)_v$$

$$du = c_v \, dT \tag{10.20}$$

The equation $du = c_v \, dT$ holds good for an ideal gas for any process, whereas for any other substance it is true for a constant volume process only.

Since c_v is constant for an ideal gas,

$$\Delta u = c_v \, \Delta T$$

The enthalpy of any substance is given by

$$h = u + pv$$

For an ideal gas

$$h = u + RT$$

Therefore

$$h = f(T) \tag{10.21}$$

only for an ideal gas.

Now $\qquad dh = du + RdT$

Since R is a constant

$$\Delta h = \Delta u + R\Delta T$$
$$= c_v \Delta T + R\Delta T$$
$$= (c_v + R)\,\Delta T \tag{10.22}$$

Since h is a function of T only, and by definition

$$c_p = \left(\frac{\partial h}{\partial T}\right)_p$$

$$dh = c_p \, dT \tag{10.23}$$

or $\qquad \Delta h = c_p \Delta T \qquad (10.24)$

From equations (10.22) and (10.23)

$$c_p = c_v + R$$

or $\qquad c_p - c_v = R \qquad (10.25)$

The equation $dh = c_p \, dT$ holds good for an ideal gas, even when pressure changes, but for any other substance, this is true only for a constant pressure change.

The ratio of c_p/c_v is of importance in ideal gas computations, and is designated by the symbol γ, i.e.

$$\frac{c_p}{c_v} = \gamma$$

or $\qquad c_p = \gamma c_v$

From equation (10.25)

$$(\gamma - 1)\, c_v = R$$

$$\left.\begin{aligned} c_v &= \frac{R}{\gamma - 1} \\ \text{and} \quad c_p &= \frac{\gamma R}{\gamma - 1} \end{aligned}\right\} \text{kJ/kg-K} \qquad (10.26)$$

If $R = \dfrac{\bar{R}}{\mu}$ is substituted in equation (10.26)

$$\left.\begin{aligned} \mu c_v &= (c_v)_{\text{mol}} = \frac{\bar{R}}{\gamma - 1} \\ \text{and} \quad \mu c_p &= (c_p)_{\text{mol}} = \frac{\gamma \bar{R}}{\gamma - 1} \end{aligned}\right\} \text{kJ/(kgmol) (K)} \qquad (10.27)$$

$\mu\, c_v$ and c_p are the *molar or molal specific heats* at constant volume and at constant pressure respectively.

It can be shown by the classical kinetic theory of gases that the values of γ are 5/3 for monoatomic gases and 7/5 for diatomic gases. When the gas molecule contains more than two atoms (i.e. for polyatomic gages) the value of γ may be taken approximately as 4/3. The minimum value of γ is thus 1 and the maximum is 1.67.

10.3.2 Entropy Change of an Ideal Gas

From the general property relations

$$Tds = du + pdv$$

$$Tds = dh - vdp$$

and for an ideal gas, $du = c_v\, dT$, $dh = c_p\, dT$, and $pv = RT$, the entropy change between any two states 1 and 2 may be computed as given below

$$ds = \frac{du}{T} + \frac{p}{T} dv$$

$$= c_v \frac{dT}{T} + R \frac{dv}{v}$$

$$\therefore \quad s_2 - s_1 = c_v \ln \frac{T_2}{T_1} + R \ln \frac{v_2}{v_1} \qquad (10.28)$$

Also

$$ds = \frac{dh}{T} - \frac{v}{T} dp$$

$$= c_p \frac{dT}{T} - R \frac{dp}{p}$$

or

$$s_2 - s_1 = c_p \ln \frac{T_2}{T_1} - R \ln \frac{p_2}{p_1} \qquad (10.29)$$

Since $R = c_p - c_v$, equation (10.29) may be written as

$$s_2 - s_1 = c_p \ln \frac{T_2}{T_1} - c_p \ln \frac{p_2}{p_1} + c_v \ln \frac{p_2}{p_1}$$

or
$$s_2 - s_1 = c_p \ln \frac{v_2}{v_1} + c_v \ln \frac{p_2}{p_1} \qquad (10.30)$$

Any one of the three equations (10.28), (10.29), and (10.30), may be used for computing the entropy change between any two states of an ideal gas.

10.3.3 Reversible Adiabatic Process

The general property relations for an ideal gas may be written as

$$Tds = du + pdv = c_v dT + pdv$$

and
$$Tds = dh - vdp = c_p dT - vdp$$

For a reversible adiabatic change, $ds = 0$

$\therefore$
$$c_v dT = -pdv \qquad (10.31)$$

and
$$c_p dT = vdp \qquad (10.32)$$

By division

$$\frac{c_p}{c_v} = \gamma = -\frac{vdp}{pdv}$$

or
$$\frac{dp}{p} + \gamma \frac{dv}{v} = 0$$

or
$$d(\ln p) + \gamma\, d(\ln v) = d(\ln c)$$

where c is a constant

$\therefore$
$$\ln p + \gamma \ln v = \ln c$$

$$pv^{\gamma} = c \qquad (10.33)$$

Between any two states 1 and 2

$$p_1 v_1^{\gamma} = p_2 v_2^{\gamma}$$

or
$$\frac{p_2}{p_1} = \left(\frac{v_1}{v_2}\right)^{\gamma}$$

For an ideal gas

$$pv = RT$$

From equation (10.33)

$$p = c.v^{-\gamma}$$

$\therefore$
$$c.v^{-\gamma} \cdot v = RT$$

$$c \cdot v^{1-\gamma} = RT$$

$$Tv^{\gamma-1} = \text{constant} \qquad (10.34)$$

Between any two states 1 and 2, for a reversible adiabatic process in the case of an ideal gas

$$T_1 v_1^{\gamma-1} = T_2 v_2^{\gamma-1}$$

or
$$\frac{T_2}{T_1} = \left(\frac{v_1}{v_2}\right)^{\gamma-1} \qquad (10.35)$$

Similarly, substituting from equation (10.33)

$$v = \left(\frac{c}{p}\right)^{1/\gamma} \quad \text{in the equation } pv = RT$$

$$p \cdot \frac{c'}{p^{1/\gamma}} = RT$$

$$p^{1-(1/\gamma)} \cdot c' = RT$$

$$\therefore \quad T p^{(1-\gamma)/\gamma} = \text{constant} \tag{10.36}$$

Between any two states 1 and 2

$$T_1 p_1^{(1-\gamma)/\gamma} = T_2 p_2^{(1-\gamma)/\gamma}$$

$$\therefore \quad \frac{T_2}{T_1} = \left(\frac{p_2}{p_1}\right)^{(\gamma-1)/\gamma} \tag{10.37}$$

Equations (10.33), (10.34), and (10.36) give the relations among p, v, and T in a reversible adiabatic process for an ideal gas.

The *internal energy change* of an ideal gas for a reversible adiabatic process is given by

$$Tds = du + pdv = 0$$

$$\text{or} \quad \int_1^2 du = -\int_1^2 pdv = -\int_1^2 \frac{c}{v^\gamma} dv$$

$$\text{where} \quad pv^\gamma = p_1 v_1^\gamma = p_2 v_2^\gamma = c$$

$$\therefore \quad u_2 - u_1 = c\frac{v_2^{1-\gamma} - v_1^{1-\gamma}}{\gamma - 1} = \frac{p_2 v_2^\gamma \cdot v_2^{1-\gamma} - p_1 v_1^\gamma \cdot v_1^{1-\gamma}}{\gamma - 1}$$

$$= \frac{p_2 v_2 - p_1 v_1}{\gamma - 1}$$

$$= \frac{R(T_2 - T_1)}{\gamma - 1} = \frac{RT_1}{\gamma - 1}\left(\frac{T_2}{T_1} - 1\right)$$

$$= \frac{RT_1}{\gamma - 1}\left[\left(\frac{p_2}{p_1}\right)^{\gamma-1/\gamma} - 1\right] \tag{10.38}$$

The enthalpy change of an ideal gas for a reversible adiabatic process may be similarly derived.

$$Tds = dh - vdp = 0$$

$$\text{or} \quad \int_1^2 dh = \int_1^2 vdp = \int_1^2 \frac{(c)^{1/\gamma}}{p^{1/\gamma}} dp$$

$$\text{where} \quad p_1 v_1^\gamma = p_2 v_2^\gamma = c$$

$$\therefore \quad h_2 - h_1 = \frac{\gamma}{\gamma - 1} c^{1/\gamma}[p_2^{(\gamma-1)/\gamma} - p_1^{(\gamma-1)/\gamma}]$$

$$= \frac{\gamma}{\gamma - 1}\left(p_1 v_1^{\gamma}\right)^{1/\gamma} \cdot (p_1)^{(\gamma-1)/\gamma}\left[\left(\frac{p_2}{p_1}\right)^{(\gamma-1)/\gamma} - 1\right]$$

$$= \frac{\gamma p_1 v_1}{\gamma - 1}\left[\left(\frac{p_2}{p_1}\right)^{(\gamma-1)/\gamma} - 1\right]$$

$$= \frac{\gamma R T_1}{\gamma - 1}\left[\left(\frac{p_2}{p_1}\right)^{(\gamma-1)/\gamma} - 1\right] \tag{10.39}$$

The work done by an ideal gas in a reversible adiabatic process is given by

$$đQ = dU + đW = 0$$

or $$đW = -dU$$

i.e. work is done at the expense of the internal energy.

$$\therefore \quad W_{1-2} = U_1 - U_2 = m(u_1 - u_2)$$

$$= \frac{m(p_1 v_1 - p_2 v_2)}{\gamma - 1} = \frac{mR(T_1 - T_2)}{\gamma - 1} = \frac{m R T_1}{\gamma - 1}\left[1 - \left(\frac{p_2}{p_1}\right)^{(\gamma-1)/\gamma}\right] \tag{10.40}$$

where m is the mass of gas.

10.3.4 Reversible Isothermal Process

When an ideal gas of mass m undergoes a reversible isothermal process from state 1 to state 2, the work done is given by

$$\int_1^2 đW = \int_{V_1}^{V_2} p dV$$

or $$W_{1-2} = \int_{V_1}^{V_2} \frac{mRT}{V} dV = mRT \ln \frac{V_2}{V_1}$$

$$= m R T \ln \frac{p_1}{p_2} \tag{10.41}$$

The heat transfer involved in the process

$$Q_{1-2} = U_2 - U_1 + W_{1-2}$$

$$= W_{1-2} = mRT \ln V_2/V_1 = T(S_2 - S_1) \tag{10.42}$$

10.3.5 Polytropic Process

An equation of the form pv^n = constant, where n is a constant can be used approximately to describe many processes which occur in practice. Such a process is called a *polytropic process*. For an ideal gas undergoing a polytropic process, the following relations may be derived as it was done in the case of a reversivle adiabatic process (Article 10.3.3).

$$T\,v^{n-1} = \text{constant}$$
$$Tp^{(1-n)/n} = \text{constant}$$

with n replacing γ

The work done in a reversible polytropic process would be

$$W_{1-2} = \int_1^2 pdV = \frac{(p_2V_2 - p_1V_1)}{1-n}$$

$$= \frac{m\,p_1v_1}{1-n}\left[\left(\frac{p_2}{p_1}\right)^{(n-1)/n} - 1\right] \quad (10.43)$$

It is to be noted that.

$\Delta U = -\int pdV$ only for a reversible adiabatic process. Equation (10.43) does not give $-\Delta U$ unless $n = \gamma$. Also for the polytropic

$$\int_{p_1}^{p_2} vdp = \frac{n(p_2v_2 - p_1v_1)}{n-1}$$

$$= \frac{np_1v_1}{n-1}\left[\left(\frac{p_2}{p_1}\right)^{(n-1)/n} - 1\right] \quad (10.44)$$

It is also to be noted that $\int vdp$ is not Δh unless $n = \gamma$.

For a reversible polytropic process with a unit mass of an ideal gas, the first law gives

$$Q = \Delta u + \int pdv$$

$$= c_v\Delta T + \frac{R\,\Delta T}{1-n}$$

$$= c_v\Delta T + \frac{c_v(\gamma - 1)}{1-n}\,\Delta T$$

$$= c_v\left[1 + \frac{\gamma-1}{1-n}\right]\Delta T$$

$$= c_v\,\frac{\gamma - n}{1-n}\cdot \Delta T$$

$$= c_n\,\Delta T \quad (10.45)$$

where $c_n \left(= c_v \dfrac{\gamma - n}{1 - n}\right)$ is called the *polytropic specific heat.*

The polytropic processes for various values of n are shown in Fig. 10.2 on the p-v and T-s diagrams. The values of n for some familiar processes are given below

Isobaric process ($p = c$), $n = 0$

Isothermal process ($T = c$), $n = 1$

Isentropic process ($s = c$), $n = \gamma$

Isometric or isochoric process ($v = c$), $n = \infty$

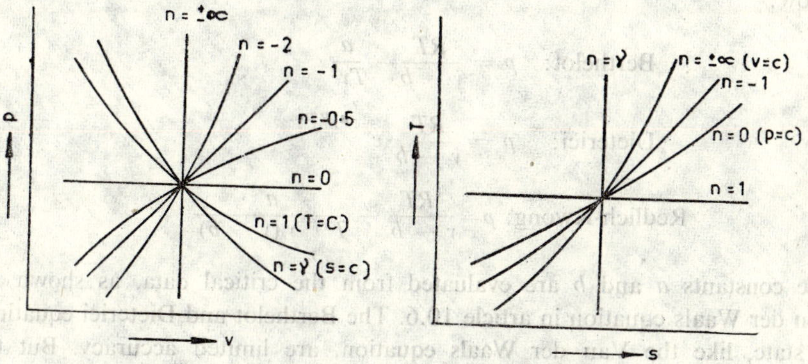

Fig. 10.2 Process in which pv^n = constant.

10.4 EQUATIONS OF STATE

The ideal gas equation of state $pv = \bar{R}T$ can be established from the postulates of the kinetic theory of gases developed by Clerk Maxwell, with two important assumptions that there is little or no attraction between the moleculess of the gas and that the volume occupied by the molecules themselves is negligibly small compared to the volume of the gas. When pressure is very small or temperature very large, the intermolecular attraction and the volume of the molecules compared to the total volume of the gas are not of much significance, and the real gas obeys very closely the ideal gas equation. But as pressure increases, the intermolecular forces of attraction and repulsion increase, and also the volume of the molecules becomes appreciable compared to the total gas volume. So then the real gases deviate considerably from the ideal gas equation. Van der Waals, by applying the laws of mechanics to individual molecules, introduced two correction terms in the equation of ideal gas, and his equation is given below.

$$\left(p + \frac{a}{v^2}\right)(v - b) = RT \qquad (10.46)$$

The coefficient a was introduced to account for the existence of mutual attraction between the molecules. The term a/v_2 is called the *force of cohesion*. The coefficient b was introduced to account for the volumes of the molecules, and is known as *co-volume*.

Real gases conform more closely with the Van der Waals equation of state than the ideal gas equation of state, particularly at higher pressures. But it is not obeyed by a real gas in all ranges of pressures and temperatures. Many more equations of state were later introduced, and notable among these are the equation developed by Berthelot, Dieterici, Beattie-Bridgeman, Kammerlingh Onnes, Hirshfelder-Bird-Spotz-McGee-Sutton, Wohl, Redlich-Kwong, and Martin-Hou.

Apart from the Van der Waals equation, three two-constant equations of state are those of Berthelot, Dieterici, and Redlich-Kwong, as given below:

$$\text{Berthelot:} \quad p = \frac{RT}{v - b} - \frac{a}{Tv^2}$$

$$\text{Dieterici:} \quad p = \frac{RT}{v - b} \cdot e^{-a/RTv}$$

$$\text{Redlich-Kwong:} \quad p = \frac{RT}{v - b} - \frac{a}{T^{1/2}\, v\,(v + b)}$$

The constants a and b are evaluated from the critical data, as shown for Van der Waals equation in article 10.6. The Berthelot and Dieterici equations of state, like the Van der Waals equation, are limited accuracy. But the Redlich-Kwong equation gives good results at high pressures and is fairly accurate for temperatures above the critical value.

Another widely used equation of state with good accuracy is the Beattie-Bridgeman equation:

$$p = \frac{RT(1 - e)}{v^2}(v + B) - \frac{A}{v^2}$$

where

$$A = A_0\left(1 - \frac{a}{v}\right), \; B = B_0\left((1 - \frac{b}{v}\right), \; e = \frac{c}{vT^3}$$

There are five constants, A_0, B_0, a, b, and c, which have to be determined experimentally for each gas. The Beattie-Bridgeman equation does not give satisfactory results in the critical point region.

10.5 VIRIAL EXPANSIONS

The relation between pv and p in the form of power series, as given in equation (10.1), may be expressed as

$$pv = A(1 + B'P + C'p^2 + D'p^3 + \ldots\ldots)$$

For any gas, from equation (10.3)

$$\lim_{p \to 0} pv = A = \bar{R}T$$

$$\therefore \quad \frac{pv}{\bar{R}T} = 1 + B'p + C'p^2 + D'p^3 + \ldots\ldots \tag{10.47}$$

An alternative expression is

$$\frac{pv}{\bar{R}T} = 1 + \frac{B}{v} + \frac{C}{v^2} + \frac{D}{v^3} + \ldots \tag{10.48}$$

Both expressions in equations (10.47) and (10.48) are known as *virial expansions* or virial equations of state, first introduced by the Dutch physicist, Kammerlingh Onnes. B', C', B, C, etc. are called *virial coefficients.* B', and B are called second virial coefficients, C' and C are called third virial coefficients, and so on. For a given gas, these coefficients are functions of temperature only.

The ratio $pv/\bar{R}T$ is called the compressibility factor, Z. For an ideal gas $Z = 1$. The magnitude of Z for a certain gas at a particular pressure and temperature gives an indication of the extent of deviation of the gas from the ideal gas behaviour. The virial expansions become

$$Z = 1 + B'p + C'p^2 + D'p^3 + \ldots\ldots$$

and

$$Z = 1 + \frac{B}{v} + \frac{C}{v^2} + \frac{D}{v^3} + \ldots$$

The relations between B', C', and B, C,... can be derived as given below

$$\frac{pv}{\bar{R}T} = 1 + B'p + C'p^2 + D'p^3 + \ldots$$

$$= 1 + B'\left[\frac{\bar{R}T}{v}\left(1 + \frac{B}{v} + \frac{C}{v^2} + \ldots\right)\right]$$

$$+ C'\left[\left(\frac{\bar{R}T}{v}\right)^2\left(1 + \frac{B}{v} + \frac{C}{v^2} \ldots + \right)^2\right] + \ldots$$

$$= 1 + \frac{B'\bar{R}T}{v} + \frac{B'B\bar{R}T + C'(\bar{R}T)^2}{v^2}$$

$$+ \frac{B'\bar{R}TC + C'(\bar{R}T)^2 + D'(\bar{R}T)^3}{v^3} + \ldots$$

Comparing this equation with equation (10.48) and rearranging

$$B' = \frac{B}{\bar{R}T}, \quad C' = \frac{C - B^2}{(\bar{R}T)^2},$$

$$D' = \frac{D - 3BC + 2B^3}{(\bar{R}T)^3}, \text{ and so on.}$$

Therefore

$$Z = \frac{pv}{\bar{R}T} = 1 + B'p + C'p^2 + \ldots$$

$$= 1 + \frac{B}{\bar{R}T}p + \frac{C - B^2}{(\bar{R}T)^2}p^2 + \ldots \quad (10.49)$$

The terms B/v, C/v^2 etc. of the virial expansion arise on account of molecular interactions. If no such interactions exist (at very low pressures) $B = 0$, $C = 0$, etc., $Z = 1$ and $pv = \bar{R}T$.

10.6 LAW OF CORRESPONDING STATES

For a certain gas, the compressibility factor Z is a function of p and T [equation (10.49)], and so a plot can be made of lines of constant temperature on coordinates of p and Z (Fig. 10.3). From this plot Z can be obtained for any value of p and T, and the volume can then be obtained from the equation $pv = ZRT$. The advantage of using Z instead of a direct plot of v is a smaller range of values in plotting.

For each substance, there is a compressibility factor chart. It would be very convenient if one chart could be used for all substances. The general shapes of the vapour dome and of the constant temperature lines on the p-v plane are similar for all substances, although the scales may be different. This similarity can be exploited by using dimensionless properties called *reduced properties.* The reduced pressure is the ratio of the existing pressure to the critical pressure of the substance, and similarly for reduced temperature and reduced volume. Then

$$p_r = \frac{p}{p_c}, \quad T_r = \frac{T}{T_c}, \quad v_r = \frac{v}{v_c}$$

where subscript r denotes the reduced property, and subscript c denotes the property at the critical state. When T_r is plotted as a function of reduced pressure and Z, a single plot, known as *the generalized compressibility chart,* is found to be satisfactory for a great variety of substances. Although necessarily approximate, the plots are extremely useful in situations where detailed data on a particular gas are lacking but its critical properties are available.

The relation among the reduced properties p_r, T_r, and v_r, is known as the *law of corresponding states.* It can be derived from the various equations of state, such as those of Van der Waals, Berthelot, and Dieterici. For a Van der Waals gas,

$$\left(p + \frac{a}{v^2}\right)(v - b) = RT$$

where a, b, and R are the characteristic constants of the particular gas.

$$\therefore \qquad p = \frac{RT}{v-b} - \frac{a}{v^2}$$

or
$$pv^3 - (pb + RT)\,v^2 + av - ab = 0$$

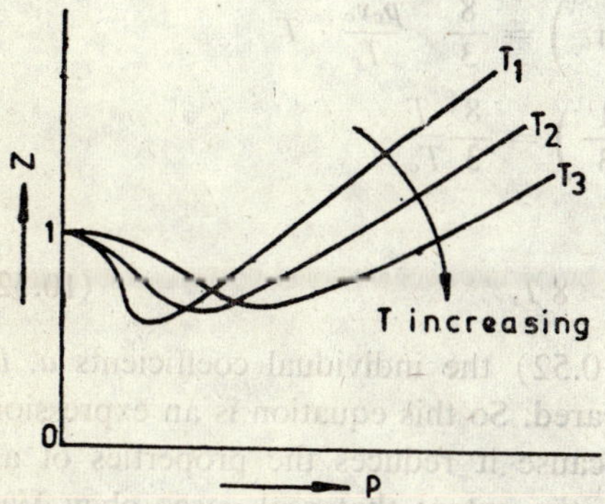

Fig. 10.3 Compressibility factor chart

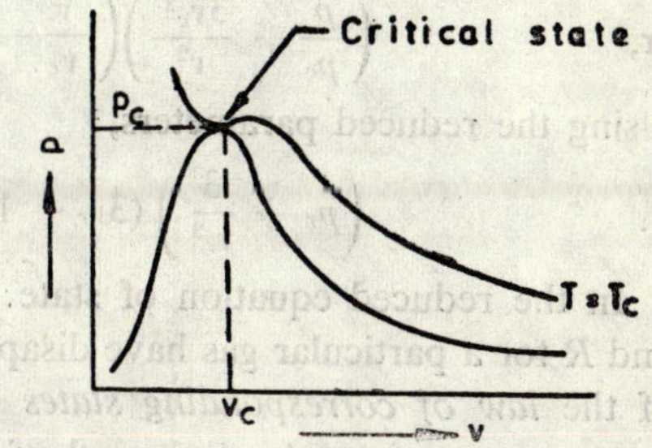

Fig. 10.4 Critical properties on *p-v* diagram

It is therefore a cubic in v and for given values of p and T has three roots of which only one need be real. For low temperatures, three positive real roots exist for a certain range of pressure. As the temperature increases the three real roots approach one another and at the critical temperature they become equal. Above this temperature only one real root exists for all values of p. The critical isotherm T_c at the critical state on the p-v plane (Fig. 10.4), where the three real roots of the Van der Waal equation coincide, not only has a zero slope, but also its slope changes at the critical state (point of inflection), so that the first and second derivatives of p with respect to v at $T = T_c$ are each equal to zero. Therefore

$$\left(\frac{\partial p}{\partial v}\right)_{T = T_c} = -\frac{RT_c}{(v_c - b)} + \frac{2a}{v_c^3} = 0 \tag{10.50}$$

$$\left(\frac{\partial^2 p}{\partial v^2}\right)_{T = T_c} = \frac{2.RT_c}{(v_c - b)^3} - \frac{6a}{v_c^4} = 0 \tag{10.51}$$

From these two equations, by rearranging and dividing, $b = 1/3\ v_c$.
Substituting the value of b in equation (10.50),

$$R = \frac{8a}{9T_c v_c}$$

Substituting the values of b and R in equation (10.46)

$$\left(p_c + \frac{a}{v_c^2}\right)\left(\frac{2}{3}\,v_c\right) = \frac{8a}{9T_c\,v_c}\cdot T_c$$

$$\therefore \qquad a = 3\,p_c v_c^2$$

Therefore, the value of R becomes

$$R = \frac{8}{3} \frac{p_c v_c}{T_c}$$

The values of a, b, and R have thus been expressed in terms of critical properties. Substituting these in the Van der Waals equation of state

$$\left(p + \frac{3\, p_c v_c^2}{v^2} \right)\left(v - \frac{1}{3} v_c \right) = \frac{8}{3} \frac{p_c v_c}{T_c} \cdot T$$

or,
$$\left(\frac{p}{p_c} + \frac{3 v_c^2}{v^2} \right)\left(\frac{v}{v_c} - \frac{1}{3} \right) = \frac{8}{3} \frac{T}{T_c}$$

Using the reduced parameters,

$$\therefore \quad \left(p_r + \frac{3}{v_r^2} \right)(3v_r - 1) = 8\, T_r \qquad (10.52)$$

In the reduced equation of state (10.52) the individual coefficients a, b and R for a particular gas have disappeared. So this equation is an expression of the *law of corresponding states* because it reduces the properties of all gases to one formula. It is a 'law' to the extent that real gases obey Van der Waals equation. Two different substances are considered to be in 'corresponding states, if their pressures, volumes and temperatures are of the same fractions (or multiples) of the critical pressure, volume and temperature of the two substances. The *generalized compressibility* chart in terms of reduced properties is shown in Fig. 10.5. It is very useful in predicting the properties of substances for which more precise data are not available. The value of Z at the critical state for a Van der Waals gas is 0.375 (since $R = \frac{8}{3} \frac{p_c\, v_c}{T_c}$). At very low pressures Z approaches unity, as a real gas approaches the ideal gas behaviour. Equation (10.52) can also be written in the following form

$$\left(p_r\, v_r + \frac{3}{v_r} \right)(3v_r - 1) = 8 T_r v_r$$

$$\therefore \quad p_r v_r = \frac{8}{3} \frac{T_r v_r}{3v_r - 1} - \frac{3}{v_r}$$

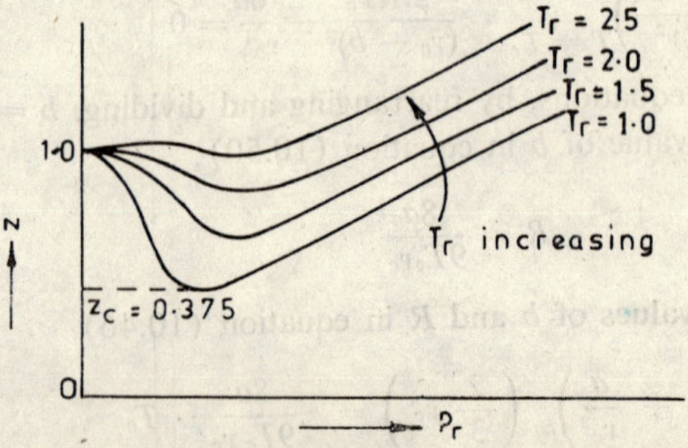

Fig. 10.5 Generalized chart compressibility

Figure 10.6 shows the law of corresponding states in reduced coordinates, $p_r v_r$ vs. p_r. Differentiating equation (10.53) with respect to p_r and making it equal to zero, it is possible to determine the minima of the isotherms as shown below.

$$\frac{\partial}{\partial p_r}\left[\frac{8T_r v_r}{3v_r-1}-\frac{3}{v_r}\right]_{T_r}=0$$

or
$$\frac{\partial}{\partial v_r}\left[\frac{8T_r v_r}{3v_r-1}-\frac{3}{v_r}\right]_{T_r}\left[\frac{\partial v_r}{\partial p_r}\right]_{T_r}=0$$

Since
$$\left[\frac{\partial v_r}{\partial p_r}\right]_{T_r}\neq 0$$

$$\frac{\partial}{\partial v_r}\left[\frac{8T_r v_r}{3v_r-1}-\frac{3}{v_r}\right]_{T_r}=0$$

$$\therefore \quad \frac{8T_r}{(3v_r-1)^2}=\frac{3}{v_r^2}$$

or
$$\frac{3(3v_r-1)^2}{v_r^2}=8\,T_r=\left[p_r+\frac{3}{v_r^2}\right](3v_r-1)$$

Simplifying

$$(p_r v_r)^2-9(p_r v_r)+6p_r=0$$

This is the equation of a parabola passing through the minima of the isotherms (Fig. 10.6).

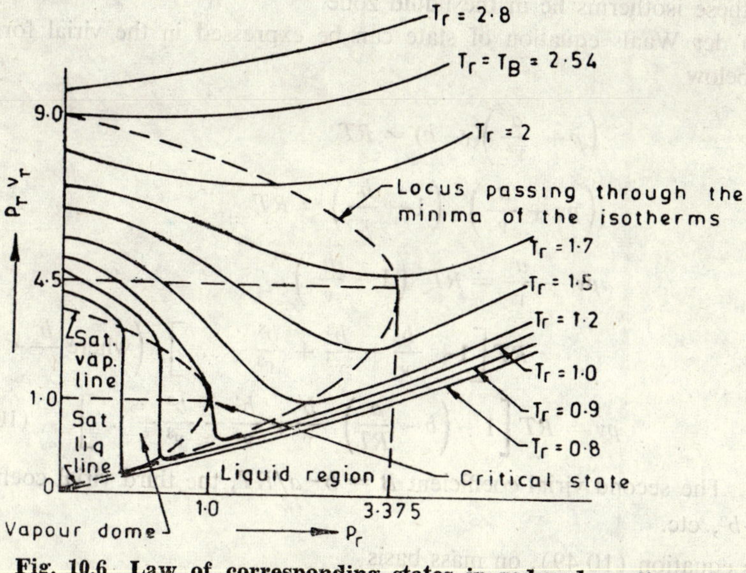

Fig. 10.6 Law of corresponding states in reduced coordinates

When

$$p_r=0, \quad p_r v_r=0,9$$

Again

$$p_r = \frac{9(p_r v_r) - (p_r v_r)^2}{6}$$

$$\frac{dp_r}{d(p_r v_r)} = 9 - 2(p_r v_r) = 0$$

$$\therefore \quad p_r v_r = 4.5$$

$$p_r = \frac{9 \times 4.5 - (4.5)^2}{6} = 3.375$$

The parabola has the vertex at $p_r v_r = 4.5$ and $p_r = 3.375$, and it intersects the ordinate at 0 and 9.

Each isotherm up to the marked T_B has a minimum (Fig. 10.6). The T_B isotherm has an initial horizontal portion ($p_r v_r$ = constant) so that Boyle's law is obeyed fairly accurately up to moderate pressures. Hence the corresponding temperature is called the *Boyle temperature* for that gas. The Boyle temperature T_B can be determined by making

$$\left[\frac{\partial (p_r v_r)}{\partial p_r}\right]_{T_r = T_B} = 0 \qquad \text{when } p_r = 0$$

Above the Boyle temperature, the isotherms slope upward and show no minima.

As T_r is reduced below the critical (i.e. for $T_r < 1$), the gas becomes liquefied, and during phase transition isotherms are vertical. The minima of all these isotherms lie in the liquid zone.

Van der Waals equation of state can be expressed in the virial form as given below

$$\left(p + \frac{a}{v^2}\right)(v - b) = RT$$

$$\left(pv + \frac{a}{v}\right)\left(1 - \frac{b}{v}\right) = RT$$

$$\therefore \quad pv + \frac{a}{v} = RT\left(1 - \frac{b}{v}\right)^{-1}$$

$$= RT\left[1 + \frac{b}{v} + \frac{b^2}{v^2} + \frac{b^3}{v^3} + \ldots\right] \left(\text{where } \frac{b}{v} < 1\right)$$

$$\therefore \quad pv = RT\left[1 + \left(b - \frac{a}{RT}\right)\frac{b}{v} + \frac{b^2}{v^2} + \frac{b^3}{v^3} + \ldots\right] \qquad (10.54)$$

$\therefore$ The second virial coefficient $B = b - a/RT$, the third virial coefficient $C = b^2$, etc.

From equation (10.49), on mass basis

$$pv = RT\left(1 + \frac{B}{RT}p + \frac{C - B^2}{R\,T^2}p^2 + \ldots\right)$$

To determine Boyle temperature, T_B

$$\left[\frac{\partial (pv)}{\partial p}\right]_{\substack{T=C\\p=0}} = 0 = \frac{B}{\bar{R}T}$$

$$\therefore \qquad B = 0$$

or $$T_B = \frac{a}{b\,\bar{R}}, \text{ because } \quad B = b - \frac{a}{\bar{R}T}$$

The point at which B is equal to zero gives the Boyle temperature. The second virial coefficient is the most important. Since $\left[\frac{\partial(pv)}{\partial p}\right]_{p=0} = B$, when B is known, the behaviour of the gas at moderate pressures is completely determined. The terms which contain higher power (C/v^2, D/v^3, etc.) become significant only at very high pressures.

10.7 PROPERTIES OF MIXTURES OF GASES—DALTON'S LAW OF PARTIAL PRESSURES

Let us imagine a homogeneous mixture of inert ideal gases at a temperature T, a pressure p, and a volume V. Let us suppose there are n_1 moles of gas A_1, n_2 moles of gas A_2, ...and upto n_c moles of gas A_c (Fig. 10.7). Since there is no chemical reaction, the mixture is in a state of equilibrium with the equation of state

$$pV = (n_1 + n_2 + \ldots + n_c)\,\bar{R}T$$

where $\bar{R} = 8.3143$ kJ/kgmol-K

$$\therefore \qquad p = \frac{n_1\,\bar{R}T}{V} + \frac{n_2\,\bar{R}T}{V} + \ldots + \frac{n_c\,\bar{R}T}{V}$$

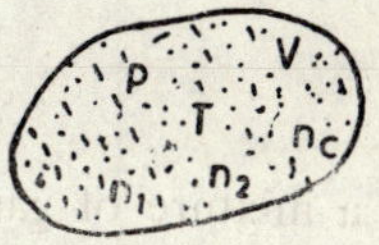

Fig. 10.7 Mixture of gases

The expression $\frac{n_K\,\bar{R}T}{V}$ represents the pressure that the Kth gas would exert if it occupied the volume V alone at temperature T. This is called the *partial pressure* of the Kth gas and is denoted by p_K. Thus

$$p_1 = \frac{n_1\,\bar{R}\,T}{V}, \quad p = \frac{n_2\,\bar{R}\,T}{V}, \quad \ldots, p_c = \frac{n_c\,\bar{R}\,T}{V}$$

and

$$p = p_1 + p_2 + \ldots + p_c \tag{10.55}$$

This is known as *Dalton's law of partial pressures* which states that the total pressure of a mixture of ideal gases is equal to the sum of the partial pressures.

Now

$$V = (n_1 + n_2 + \ldots + n_c) . \frac{\bar{R}T}{p}$$

$$= \sum n_K . \frac{\bar{R}T}{p}$$

and the partial pressure of the *Kth* gas is

$$p_K = \frac{n_K \bar{R} T}{V}$$

Substituting the value of V

$$p_K = \frac{n_K \bar{R} T \cdot p}{\Sigma n_K \cdot \bar{R} T} = \frac{n_K}{\Sigma n_K} \cdot p$$

Now

$$\Sigma n_K = n_1 + n_2 + \ldots + n_c$$

= Total number of moles of gas

The ratio $\frac{n_K}{\Sigma n_K}$ is called the *mole fraction* of the *Kth* gas, and is denoted by x_K. Thus

$$x_1 = \frac{n_1}{\Sigma n_K}, \quad x_2 = \frac{n_2}{\Sigma n_K}, \quad \ldots, x_c = \frac{n_c}{\Sigma n_K}$$

and $\quad p_1 = x_1 p, \; p_2 = x_2 p, \; \ldots, \; p_c = x_c p$

or $\quad p_K = x_K \cdot p \qquad (10.56)$

Also

$$x_1 + x_2 + \ldots + x_c = 1 \qquad (10.57)$$

In a mixture of gases, if all but one mole fraction is determined, the last can be calculated from the above equation. Again, in terms of masses

$$p_1 V = m_1 R_1 T$$
$$p_2 V = m_2 R_2 T$$
$$\cdots\cdots\cdots\cdots\cdots$$
$$p_c V = m_c R_c T$$

Adding, and using Dalton's law

$$pV = (m_1 R_1 + m_2 R_2 + \ldots + m_c R_c) T \qquad (10.58)$$

where $\quad p = p_1 + p_2 + \ldots + p_c$

For the gas mixture

$$pV = (m_1 + m_2 + \ldots + m_c) \; R_m \, T \qquad (10.59)$$

where R_m is the gas constant for the mixture. From equations (10.58) and (10.59)

$$R_m = \frac{m_1 R_1 + m_2 R_2 + \ldots + m_c R_c}{m_1 + m_2 + \ldots + m_c} \qquad (10.60)$$

The gas constant of the mixture is thus the weighted mean, on a mass basis, of the gas constants of the components.

The total mass of gas mixture m is

$$m = m_1 + \ldots + m_c$$

If μ denotes the *equivalent molecular weight* of the mixture having n total number of moles.

$$n\mu = n_1\mu_1 + n_2\mu_2 + \ldots + n_c\mu_c$$

$$\therefore \quad \mu = x_1\mu_1 + x_2\mu_2 + \ldots + x_c\mu_c$$

or
$$\mu = \Sigma x_K \, \mu_K \tag{10.61}$$

A quantity called the *partial volume* of a component of a mixture is the volume that the component alone would occupy at the pressure and temperature of the mixture. Designating the partial volumes by V_1, V_2, etc.

$$pV_1 = m_1R_1T, \; pV_2 = m_2R_2T, \; \ldots, pV_c = m_cR_cT$$

or
$$p(V_1 + V_2 + \ldots + V_c) = (m_1R_1 + m_2R_2 + m_cR_c)T \tag{10.62}$$

From equations (10.58), (10.59), and (10.62)

$$V = V_1 + V_2 + \ldots + V_c \tag{10.63}$$

The total volume is thus equal to the sum of the partial volumes.

The *specific volume of the mixture, v, is* given by

$$v = \frac{V}{m} = \frac{V}{m_1 + m_2 + \ldots + m_c}$$

or
$$\frac{1}{v} = \frac{m_1 + m_2 + \ldots + m_c}{V}$$

$$= \frac{m_1}{V} + \frac{m_2}{V} + \ldots + \frac{m_c}{V}$$

or
$$\frac{1}{v} = \frac{1}{v_1} + \frac{1}{v_2} + \ldots + \frac{1}{v_c} \tag{10.64}$$

where v_1, v_2... denote specific volumes of the components, each component occupying the total volume.

Therefore, the density of the mixture

$$\rho = \rho_1 + \rho_2 + \ldots + \rho_c \tag{10.64a}$$

10.8 INTERNAL ENERGY, ENTHALPY AND SPECIFIC HEATS OF GAS MIXTURES

When gases at equal pressures and temperatures are mixed adiabatically without work, as by inter-diffusion in a constant volume container, the first law requires that the internal energy of the gaseous system remains constant, and experiments show that the temperature remains constant. Hence the internal energy of a mixture of gases is equal to the sum of the internal energies of the individual components, each taken at the temperature and volume of the mixture (i.e. sum of the 'partial' internal energies). This is also true for any of the thermodynamic properties like H. C_v, C_p, S, F, and G, and is known as *Gibbs theorem*. Therefore, on a mass basis

$$mu_m = m_1u_1 + m_2u_2 + \ldots + m_cu_c$$

$$u_m = \frac{m_1u_1 + m_2u_2 + \ldots + m_cu_c}{m_1 + m_2 + \ldots + m_c} \tag{10.65}$$

which is the average specific internal energy of the mixture.

Similarly, the total enthalpy of a gas mixture is the sum of the 'partial' enthalpies

$$mh_m = m_1h_1 + m_2h_2 + \ldots + m_ch_c$$

and $$h_m = \frac{m_1h_1 + m_2h_2 + \ldots + m_ch_c}{m_1 + m_2 + \ldots + m_c} \tag{10.66}$$

From the definitions of specific heats, it follows that

$$c_{vm} = \frac{m_1c_{v1} + m_2\,c_{v2} + \ldots\, m_cc_{pc}}{m_1 + m_2 + \ldots + m_c} \tag{10.67}$$

and $$c_{pm} = -\frac{m_1c_{p1} + m_2c_{p2} + \ldots + m_cc_{pc}}{m_1 + m_2 + \ldots + m_c} \tag{10.68}$$

10.9 ENTROPY OF GAS MIXTURES

Gibbs theorem states that the total entropy of a mixture of gases is the sum of the partial entropies. The partial entropy of one of the gases of a mixture is the entropy that the gas would have if it occupied the whole volume alone at the same temperature. Let us imagine a number of inert ideal gases separated from one another by suitable partitions, all the being at the same temperature T and pressure p. The total entropy (initial)

$$S_i = n_1s_1 + n_2s_2 \quad \ldots + n_cs_c$$
$$= \Sigma\, n_K s_K$$

From property relation

$$Tds = dh - vdp = c_pdT - vdp$$

$$\therefore \quad ds = c_p\frac{dT}{T} - \bar{R}\,\frac{dp}{p}$$

The entropy of 1 mole of the Kth gas at T and p

$$s_K = \int c_{pK}\frac{dT}{T}\bar{R}\ln p + p + s_{0K}$$

where S_{OK} is the constant of integration

$$\therefore \qquad S_i = \bar{R}\,\Sigma\, n_K \left(\frac{1}{\bar{R}} \int c_{pK} \frac{dT}{T} + \frac{s_{0K}}{\bar{R}} - \ln p \right)$$

Let

$$\sigma_K = \frac{1}{\bar{R}} \int c_{pK} \frac{dT}{T} + \frac{s_{0K}}{\bar{R}}$$

then

$$S_i = \bar{R}\,\Sigma\, n_K\,(\sigma_K - \ln p) \tag{10.69}$$

After the partitions are removed, the gases diffuse into one another at the same temperature and pressure, and by Gibbs theorem, the entropy of the mixture, S_f, is the sum of the partial entropies, with each gas exerting its respective partial pressure. Thus

$$S_f = \bar{R}\,\Sigma\, n_K\,(\sigma_K - \ln p_K)$$

Since

$$p_K = x_K\, p$$

$$S_f = \bar{R}\,\Sigma\, n_K\,(\sigma_K - \ln x_K - \ln p) \tag{10.70}$$

A change in entropy due to the diffusion of any number of inert ideal gases is

$$S_f - S_i = -\bar{R}\,\Sigma\, n_K \ln x_K \tag{10.71}$$

or

$$S_f - S_i = -\bar{R}\,(n_1 \ln x_1 + n_2 \ln x_2 + \ldots + n_c \ln x_c)$$

Since the mole fractions are less than unity, $(S_f - S_i)$ is always positive, conforming to the Second Law.

Again

$$S_f - S_i = -\bar{R} \left(n_1 \ln \frac{p_1}{p} + n_2 \ln \frac{p_2}{p} + \ldots + n_c \ln \frac{p_c}{p} \right) \tag{10.72}$$

which indicates that each gas undergoes in the diffusion process a *free expansion* from total pressure p to the respective partial pressure at constant temperature.

Similarly, on a mass basis, the entropy change due to diffusion

$$S_f - S_i = -\Sigma\, m_K R_K \ln \frac{p_K}{p}$$

$$= -\left(m_1 R_1 \ln \frac{p_1}{p} + m_2 R_2 \ln \frac{p_2}{p} + \ldots \quad \ldots + m_c R_c \ln \frac{p_c}{p} \right)$$

10.10 GIBBS FUNCTION OF A MIXTURE OF INERT IDEAL GASES

From the equations

$$dh = c_p dT$$

$$ds = c_p \frac{dT}{T} - \bar{R}\, \frac{dp}{p}$$

the enthalpy and entropy of 1 mole of an ideal gas at temperature T and pressure p are

$$h = h_0 + \int c_p\, dT$$

$$s = \int c_p \frac{dT}{T} + s_0 - \bar{R} \ln p$$

Therefore, the molar Gibbs function

$$g = h - Ts$$

$$= h_0 + \int c_p dT - T \int c_p \frac{dT}{T} - Ts_0 + \bar{R}T \ln p$$

Now $\int d(uv) = \int u dv + \int v du = uv$

Let $u = \frac{1}{T},\ v = \int c_p dT$

Then $$\frac{1}{T}\int c_p dT = \int \frac{1}{T} c_p dT + \int \int c_p dT \left(-\frac{1}{T^2}\right) dT$$

$$= \int \frac{1}{T} c_p dT - \int \frac{\int c_p dT}{T^2} dT$$

$$\int c_p dT - T \int c_p \frac{dT}{T} = -T \int \frac{\int c_p dT}{T^2} dT$$

Therefore

$$g = h_0 - T \int \frac{\int c_p dT}{T^2} dT - Ts_0 + \bar{R}T \ln p$$

$$= \bar{R}T \left(\frac{h_0}{\bar{R}T} - \frac{1}{\bar{R}} \int \frac{\int c_p dT}{T^2} dT - \frac{s_0}{\bar{R}} + \ln p \right)$$

Let

$$\phi = \frac{h_0}{\bar{R}T} - \frac{1}{\bar{R}} \int \frac{\int c_p dT}{T^9} dT - \frac{s_0}{\bar{R}} \qquad (10.73)$$

Thus

$$g = \bar{R}T (\phi + \ln p) \qquad (10.73a)$$

where ϕ is a function of temperature only.

Let us consider a number of inert ideal gases separated from one another at the same T and p

$$G_i = \Sigma\, n_K g_K$$

$$= \bar{R}T\, \Sigma\, n_K (\phi_K + \ln p)$$

After the partitions are removed, the gases will diffuse, and the partial Gibbs function of a particular gas is the value of G, if that gas occupies the same volume at the same temperature exerting a partial pressure p_K. Thus

$$G_f = \bar{R}T\, \Sigma\, n_K (\phi_K + \ln p_K)$$

$$= \bar{R}T\, \Sigma\, n_K (\phi_K + \ln p + \ln x_K)$$

Therefore

$$G_f - G_i = \bar{R}T \, \Sigma \, n_K \ln x_K \tag{10.74}$$

Since $x_K < 1$, $(G_f - G_i)$ is negative because G decreases due to diffusion.

Gibbs function of a mixture of ideal gases at T and p is thus

$$G = \bar{R}T \, \Sigma \, n_K \, (\phi_K + \ln p + \ln x_K) \tag{10.75}$$

EXAMPLE 10.1

Two vessels, A and B, both containing nitrogen, are connected by a valve which is opened to allow the contents to mix and achieve an equilibrium temperature of 27°C. Before mixing the following information is known about the gases in the two vessels.

Vessel A	*Vessel B*
$p = 1.5$ MPa	$p = 0.6$ MPa
$t = 50°$C	$t = 20°$C
Contents = 0.5 kgmol	Contents = 2.5 kg

Calculate the final equilibrium pressure, and the amount of heat transferred to the surroundings. If the vessel had been perfectly insulated, calculate the final temperature and pressure which would have been reached. Take $\gamma = 1.4$.

Solution

For the gas in vessel A (Fig. Ex. 10.1)

$$p_A V_A = n_A \bar{R} T_A$$

where V_A is the volume of vessel A.

$$1.5 \times 10^3 \times V_A$$

$$= 0.5 \times 8.31343 \times 323$$

$$V_A = 0.895 \text{ m}^3$$

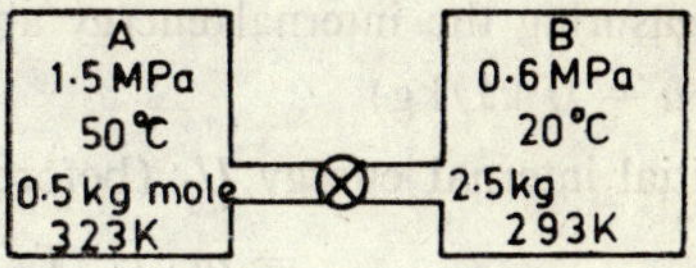

Fig. Ex. 10·1

The mass of gas in vessel A.

$$m_A = n_A \, \mu_A$$

$$= 0.5 \text{ kgmol} \times 28 \text{ kg/kgmol}$$

$$= 14 \text{ kg}$$

Characteristic gas constant R of nitrogen

$$R = \frac{8.3143}{28} = 0.297 \text{ kJ/kg-K}$$

For the vessel B

$$p_B V_B = m_B \cdot RT_B$$

$$0.6 \times 10^3 \times V_B = 2.5 \times 0.297 \times 293$$

$$\therefore \quad V_B = 0.363 \ \text{m}^3$$

Total volume of A and B

$$V = V_A + V_B = 0.895 + 0.363$$
$$= 1.258 \ \text{m}^3$$

Total mass of gas

$$m = m_A + m_B = 14 + 2.5 = 16.5 \ \text{kg}$$

Final temperature after mixing

$$T = 27 + 273 = 300 \ \text{K}$$

For the final condition after mixing

$$pV = mRT$$

where p is the final equilibrium pressure

$$\therefore \quad p \times 1.258 = 16.5 \times 0.297 \times 300$$

$$\therefore \quad p = \frac{16.5 \times 0.297 \times 300}{1.258}$$

$$= 1168.6 \ \text{kPa}$$

$$= 1.168 \ \text{MPa}$$

$$c_v = \frac{R}{\gamma - 1} = \frac{0.297}{0.4}$$

$$= 0.743 \ \text{kJ/kg-K}$$

Since there is no work transfer, the amount of heat transfer

$$Q = \text{change of internal energy}$$
$$= U_2 - U_1$$

Measuring the internal energy above the datum of absolute zero (at $T = 0$ K $u = 0$ kJ/kg)

Initial internal energy U_1 (before mixing)

$$= m_A c_v T_A + m_B c_v T_B$$

$$= (14 \times 323 + 2.5 \times 293) \times 0.743$$

$$= 3904.1 \ \text{kJ}$$

Final internal energy U_2 (after mixing)

$$= m c_v T$$

$$= 16.5 \times 0.743 \times 300$$

$$= 3677.9 \ \text{kJ}$$

$$\therefore \quad Q = 3677.9 - 3904.1 = -\ 6.2 \ \text{kJ}$$ Ans.

If the vessels were insulated

$$U_1 = U_2$$

$$m_A c_v T_A + m_B c_v T_B = m c_v T$$

where T would have been the final temperature.

$$\therefore \quad T = \frac{m_B T_B + m_B T_B}{m}$$

$$= \frac{14 \times 323 + 2.5 \times 293}{16.5} = 318.5 \text{ K}$$

or $\quad t = 45.5°\text{C}.$ Ans.

The final pressure

$$p = \frac{m R T}{V} = \frac{16.5 \times 0.297 \times 318.5}{1.258}$$

$$= 1240.7 \text{ kPa}$$

$$= 1.24 \text{ MPa}$$

EXAMPLE 10.2

A certain gas has $c_p = 1.968$ and $c_v = 1.507$ kJ/kg-K. Find its molecular weight and the gas constant.

A constant volume chamber of 0.3 m³ capacity contains 2 kg of this gas at 5°C. Heat is transferred to the gas until the temperature is 100°C. Find the work done, the heat transferred, and the changes in internal energy, enthalpy and entropy.

Solution

Gas constant, $R = c_p - c_v = 1.968 - 1.507$

$$= 0.461 \text{ kJ/kg-K}$$ Ans.

Molecular weight,

$$\mu = \frac{\bar{R}}{R} = \frac{8.3143}{0.461} = 18.04 \text{ kg/kgmol}$$ Ans.

At constant volume

$$Q_{1-2} = m c_v \,(t_2 - t_1)$$

$$= 2 \times 1.507 \,(100 - 5)$$

$$= 286.33 \text{ kJ}$$ Ans.

$$W_{1-2} = \int_1^2 p dv = 0$$ Ans.

Change in internal energy

$$U_2 - U_1 = Q_{1-2} = 286.33 \text{ kJ}$$ Ans.

Change in enthalpy

$$H_2 - H_1 = m\, c_p\ (t_2 - t_1)$$
$$= 2 \times 1.968\ (100 - 5) = 373.92 \text{ kJ} \qquad \text{Ans.}$$

Change in entropy

$$S_2 - S_1 = m\, c_v \ln \frac{T_2}{T_1} = 2 \times 1.507 \ln \frac{373}{268}$$
$$= 0.921 \text{ kJ/K}$$

EXAMPLE 10.3

(a) The specific heats of a gas are given by $c_p = a + kT$ and $c_v = b + kT$, where a, b, and k are constants and T is in K. Show that for an isentropic expansion of this gas

$$T^b\ v^{a-b}\ e^{kT} = \text{constant}$$

(b) 1.5 kg of this gas occupying a volume of 0.06 m³ at 5.6 MPa expands isentropically until the temperature is 240°C. If $a = 0.946$, $b = 0.662$, and $k = 10^{-4}$, calculate the work done in the expansion.

Solution

(a) $c_p - c_v = a + kT - b - kT$

$$= a - b = R$$

Now $$ds = c_v \frac{dT}{T} + R \frac{dv}{v}$$

$$= (b + kT) \frac{dT}{T} + (a - b) \frac{dv}{v} = b \frac{dT}{T} + k\,dT + (a - b) \frac{dv}{v}$$

For an isentropic process

$$b \ln T + kT + (a - b) \ln v = \text{constant}$$

$\therefore$ $$T^b \cdot v^{a-b} \cdot e^{kT} = \text{cvonstant} \qquad \text{(Q.E.D.)}$$

(b) $R = a - b = 0.946 - 0.662 = 0.284$ kJ/kg-K

$$T_2 = 240 + 273 = 513 \text{ K}$$

$$T_1 = \frac{p_1 V_1}{mR} = \frac{5.6 \times 10^3 \times 0.06}{1.5 \times 0.284} = 788.73 \text{ K} = 789 \text{ K}$$

$$TdS = dU + đW = 0$$

$$\therefore \quad W_{1-2} = -\int_{T_1}^{T_2} m\, c_v dT$$

$$= 1.5 \int_{513}^{789} (0.662 + 0.0001\, T)\, dT$$

$$= 1.5\, [0.662\, (789 - 513) + 10^{-4} \times 0.5\, \{(789)^2 - (513)^3\}]$$

$$= 1.5\ (182.71 + 19.97)$$

$$= 304 \text{ kJ} \qquad \text{Ans.}$$

EXAMPLE 10.4

Show that for an ideal gas, the slope of the constant volume line on the *T-s* diagram is more than that of the constant pressure line.

Solution

We have, for 1 kg of ideal gas

$$Tds = du + pdv$$
$$= c_v dT + pdv$$
$$\therefore \quad \left(\frac{\partial T}{\partial s}\right)_v = \frac{T}{c_v}$$

Also, $Tds = dh - vdp$

$$= c_p dT - vdp$$
$$\therefore \quad \left(\frac{\partial T}{\partial s}\right)_p = \frac{T}{c_p}$$

Since $\quad c_p > c_v, \quad \frac{T}{c_v} > \frac{T}{c_p}$

$$\therefore \quad \left(\frac{\partial T}{\partial s}\right)_v > \left(\frac{\partial T}{\partial s}\right)_p$$

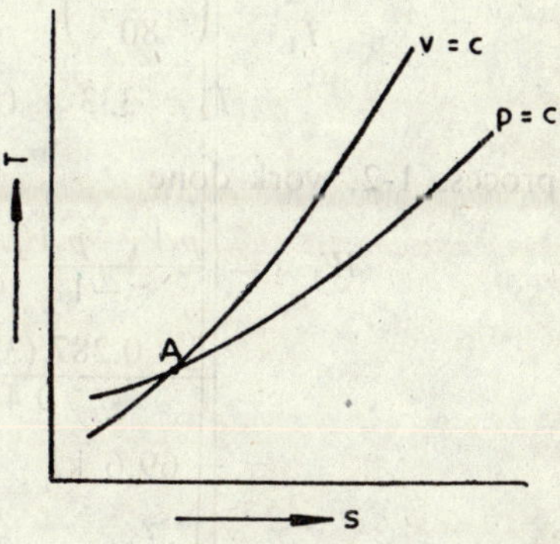

Fig. Ex. 10.4

This is shown in Fig. Ex. 10.4. The slope of the constant volume line passing through point *A* is steeper than that of the constant pressure line passing through the same point. (Q.E.D.)

EXAMPLE 10.5

0.5 kg of air is compressed reversibly and adiabatically from 80 kPa, 60°C to 0.4 MPa, and is then expanded at constant pressure to the original volume. Sketch these processes on the *p-v* and *T-s* planes. Compute the heat transfer and work transfer for the whole path.

Solution

The processes have been shown on the *p-v* and *T-s* planes in Fig. Ex. 10.5.

At state 1

$$p_1 V_1 = mRT_1$$

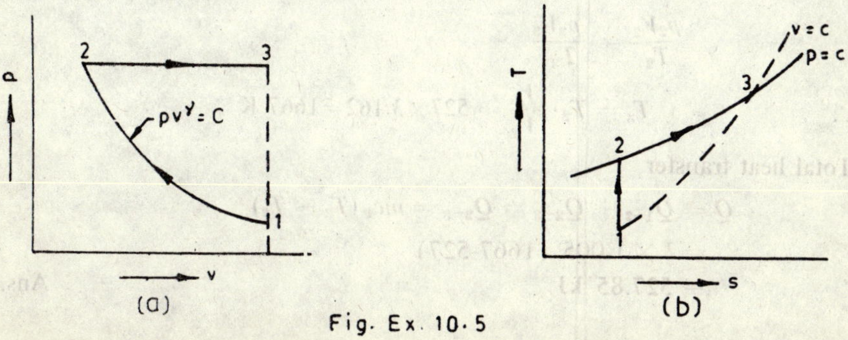

Fig. Ex. 10.5

$\therefore \quad V_1$ = volume of air at state 1

$$= \frac{m\,R\,T_1}{p_1} = \frac{1 \times 0.287 \times 333}{2 \times 80} = 0.597 \text{ m}^3$$

Since the process 1-2 is reversible and adiabatic

$$\frac{T_2}{T_1} = \left(\frac{p_2}{p_1}\right)^{(\gamma-1)/\gamma}$$

$$\therefore \quad \frac{T_2}{T_1} = \left(\frac{400}{80}\right)^{(1.4-1)/1.4} = (5)^{2/7}$$

$$\therefore \quad T_2 = 333 \times (5)^{2/7} = 527 \text{ K}$$

For process 1-2, work done

$$W_{1-2} = \frac{p_1V_1 - p_2V_2}{\gamma - 1} = \frac{m\,R\,(T_1 - T_2)}{\gamma - 1}$$

$$= \frac{\frac{1}{2} \times 0.287\,(333 - 527)}{0.4}$$

$$= -\,69.6 \text{ kJ}$$

Again $\quad p_1v_1^{\gamma} = p_2v_2^{\gamma}$

$$\therefore \quad \left(\frac{v_2}{v_1}\right)^{\gamma} = \frac{p_1}{p_2} = \frac{80}{400} = \frac{1}{5}$$

$$\therefore \quad \frac{v_2}{v_1} = \left(\frac{1}{5}\right)^{1/1.4} = \frac{1}{3.162} = \frac{V_2}{V_1}$$

$$\therefore \quad V_2 = \frac{0.597}{3.162} = 0.189 \text{ m}^3$$

For process 2-3, work done

$$W_{2-3} = p_2\,(V_1 - V_2) = 400\,(0.597 - 0.189)$$

$$= 163.2 \text{ kJ}$$

$\therefore$ Total work transfer

$$W = W_{1-2} + W_{2-3}$$

$$= -\,69.6 + 163.2 = 93.6 \text{ kJ}$$ Ans.

For states 2 and 3

$$\frac{p_2V_2}{T_2} = \frac{p_3V_3}{T_3}$$

$$\therefore \quad T_3 = T_2 \cdot \frac{V_3}{V_2} = 527 \times 3.162 = 1667 \text{ K}$$

Total heat transfer

$$Q = Q_{1-2} + Q_{2-3} = Q_{2-3} = mc_p\,(T_3 - T_2)$$

$$= \tfrac{1}{2} \times 1.005\,\,(1667\text{-}527)$$

$$= 527.85 \text{ kJ}$$ Ans.

EXAMPLE 10.6

A mass of air is initially at 260°C and 700 kPa, and occupies 0.028 m³. The air is expanded at constant pressure to 0.084 m³. A polytropic process with $n = 1.50$ is then carried out, followed by a constant temperature process which completes a cycle. All the processes are reversible (a) Sketch the cycle in the *p-v* and planes. (b) Find the heat received and heat rejected in the cycle. (c) Find the efficiency of the cycle.

Solution

(a) The cycle is sketched on the *p-v* and *T-s* planes in Fig. Ex. 10.6.

Given $p_1 = 700$ kPa, $T_1 = 260 + 273 = 533$ K $= T_3$

$V_1 = 0.028$ m³

$V_2 = 0.084$ m³

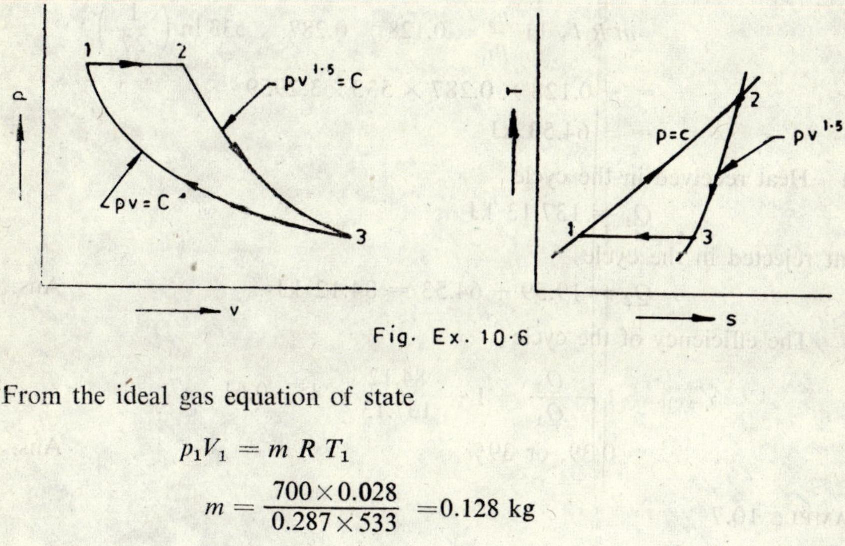

Fig. Ex. 10·6

From the ideal gas equation of state

$$p_1V_1 = m\,R\,T_1$$

$$m = \frac{700 \times 0.028}{0.287 \times 533} = 0.128 \text{ kg}$$

Now $$\frac{T_2}{T_1} = \frac{p_2\,V_2}{p_1\,V_1} = \frac{0.084}{0.028} = 3$$

$\therefore\quad T_2 = 3 \times 533 = 1599$ K

Again $$\frac{p_2}{p_3} = \left(\frac{T_2}{T_3}\right)^{n/(n-1)} = \left(\frac{1599}{533}\right)^{1.5/0.5} = (3)^3 = 27$$

Heat transfer in process 1-2

$$Q_{1-2} = m\,c_p\,(T_2 - T_1)$$
$$= 0.128 \times 1.005\,(1599 - 533)$$
$$= 137.13 \text{ kJ}$$

Heat transfer in process 2-3

$$Q_{2-3} = \Delta U + \int p dv$$

$$= m\, c_v\, (T_3 - T_2) + \frac{m\, R\, (T_2 - T_3)}{n-1}$$

$$= m\, c_v\, \frac{n-\gamma}{n-1}(T_3 - T_2) = 0.128 \times 0.718 \times \frac{1.5-1.4}{1.5-1}(533-1599)$$

$$= 0.128 \times 0.718 \times \frac{0.1}{0.5}(-1066)$$

$$= -19.59 \text{ kJ}$$

For process 3-1

$$đQ = dU + đW = đW$$

$$\therefore \quad Q_{3-1} = W_{3-1} = \int_3^1 pdV = m\, R\, T_1 \ln \frac{V_1}{V_3}$$

$$= m\, R\, T_1 \ln \frac{p_3}{p_1} = 0.128 \times 0.287 \times 533 \ln \left(\frac{1}{27}\right)$$

$$= -0.128 \times 0.287 \times 533 \times 3.2959$$

$$= -64.53 \text{ kJ}$$

(b) Heat received in the cycle

$$Q_1 = 137.13 \text{ kJ}$$

Heat rejected in the cycle

$$Q_2 = 19.59 + 64.53 = 84.12 \text{ kJ}$$ Ans.

(c) The efficiency of the cycle

$$\eta_{\text{cycle}} = 1 - \frac{Q_2}{Q_1} = 1 - \frac{84.12}{137.13} = 1 - 0.61$$

$$= 0.39, \text{ or } 39\%$$ Ans.

EXAMPLE 10.7

A mass of 0.25 kg of an ideal gas has a pressure of 300 kPa, a temperature of 80°C, and a volume of 0.07 m³. The gas undergoes an irreversible adiabatic process to a final pressure of 300 kPa and final volume of 0.10 m³, during which the work done on the gas is 25 kJ. Evaluate the c_p and c_v of the gas and the increase in entropy of the gas.

Solution

From $\quad p_1 V_1 = m\, R\, T_1$

$$R = \frac{300 \times 0.07}{0.25 \times (273 + 80)} = 0.238 \text{ kJ/kg-K}$$

Final temperature

$$T_2 = \frac{p_2 V_2}{m\ R} = \frac{300 \times 0.1}{0.25 \times 0.238} = 505 \text{ K}$$

Now

$$Q = (U_2 - U_1) + W = m\, c_v\, (T_2 - T_1) + W$$

$$0 = 0.25\, c_v\, (505 - 353) - 25$$

$$\therefore \qquad c_v = \frac{25}{0.25 \times 152} = 0.658 \text{ kJ/kg-K}$$

Now $\qquad c_p - c_v = R$

$$c_p = 0.658 + 0.238 = 0.896 \text{ kJ/kg-K}$$

Entropy change

$$S_2 - S_1 = mc_v \ \ln \frac{p_2}{p_1} + mc_p \ \ln \frac{v_2}{v_1}$$

$$= m\, c_p \ln \frac{V_2}{V_1} = 0.25 \times 0.896 \ \ln \frac{0.10}{0.07}$$

$$= 0.224 \times 0.3569 = 0.08 \text{ kJ/kg-K}$$ Ans.

EXAMPLE 10.8

A mixture of ideal gases consists of 3 kg of nitrogen and 5 kg of carbon-dioxide at a pressure of 300 kPa and a temperature of 20°C. Find (a) the mole fraction of each constituent, (b) the equivalent molecular weight of the mixture, (c) the equivalent gas constant of the mixture, (d) the partial pressures and the partial volumes, (e) the volume and density of the mixture, and (f) the c_p and c_v of the mixture.

If the mixture is heated at constant volume to 40°C, find the changes in internal energy, enthalpy and entropy of the mixture. Find the changes in internal energy, enthalpy and entropy of the mixture if the heating is done at constant pressure. Take γ for CO_2 and N_2 to be 1.286 and 1.4 respecively.

Solution

(a) Since mole fraction

$$x_i = \frac{n_i}{\Sigma n_i}$$

$$x_{N_2} = \frac{\frac{3}{28}}{\frac{3}{28} + \frac{5}{44}} = 0.485$$

$$x_{CO_2} = \frac{\frac{5}{44}}{\frac{3}{28} + \frac{5}{44}} = 0.515$$ Ans.

(b) Equivalent molecular weight of the mixture

$$M = x_1\mu_1 + x_2\mu_2$$
$$= 0.485 \times 28 + 0.515 \times 44$$
$$= 36.25 \text{ kg/kgmol}$$ **Ans.**

(c) Total mass,

$$m = m_{N_2} + m_{CO_2} = 3 + 5 = 8 \text{ kg.}$$

Equivalent gas constant of the mixture

$$R = \frac{m_{N_2} R_{N_2} + m_{CO_2} R_{CO_2}}{m}$$

$$= \frac{3 \times \frac{8.3143}{28} + 5 \times \frac{8.3143}{44}}{8} = \frac{0.89 + 0.94}{8}$$

$$= 0.229 \text{ kJ/kg-K}$$ Ans.

(d) $p_{N_2} = x_{N_2} \cdot p = 0.485 \times 300 = 145.5$ kPa

$p_{CO_2} = x_{CO_2} \cdot p = 0.515 \times 300 = 154.5$ kPa

$$V_{N_2} = \frac{m_{N_2} R_{N_2} T}{p} = \frac{3 \times \frac{8.3143}{28} \times 293}{300} = 0.87 \text{ m}^3$$

$$V_{CO_2} = \frac{m_{CO_2} \cdot R_{CO_2} T}{p} = \frac{5 \times \frac{8.3143}{44} \times 293}{300} = 0.923 \text{ m}^3$$ **Ans.**

(e) Total volume of the mixture

$$V = \frac{mRT}{p} = \frac{m_{N_2} R_{N_2} T}{p_{N_2}} = \frac{m_{CO_2} R_{CO_2} T}{p_{CO_2}}$$

$$\therefore \quad V = \frac{8 \times 0.229 \times 293}{300} = 1.79 \text{ m}^3$$

Density of the mixture

$$\rho = \rho_{N_2} + \rho_{CO_2} = \frac{m}{V} = \frac{8}{1.79}$$

$$= 4.46 \text{ kg/m}^3$$ **Ans.**

(f)
$$c_{p_{N_2}} - c_{v_{N_2}} = R_{N_2}$$

$$\therefore \quad c_{v_{N_2}} = \frac{R_{N_2}}{\gamma - 1} = \frac{8.3143}{28 \times (1.4 - 1)}$$

$$= 0.742 \text{ kJ/kg K}$$

$$\therefore \quad c_{p_{N_2}} = 1.4 \times 0.742$$

$$= 1.039 \text{ kJ/kg-K}$$

For CO_2, $\gamma = 1.286$

$$\therefore \quad c_{v\,CO_2} = \frac{R_{CO_2}}{\gamma - 1} = \frac{8.3143}{44 \times 0.286} = 0.661 \text{ kJ/kg-K}$$

$$c_{p\,CO_2} = 1.286 \times 0.661 = 0.85 \text{ kJ/kg-K}$$

For the mixture

$$c_p = \frac{m_{N_2}\, c_{p\,N_2} + m_{CO_2}\, c_{p\,CO_2}}{m_{N_2} + m_{CO_2}}$$

$$= \tfrac{3}{8} \times 1.039 + \tfrac{5}{8} \times 0.85$$

$$= 0.92 \text{ kJ/kg=K}$$

$$c_v = \frac{m_{N_2}\, c_{v\,N_2} + m_{CO_2}\, c_{v\,CO_2}}{m}$$

$$= \tfrac{3}{8} \times 0.742 + \tfrac{5}{8} \times 0.661 = 0.69 \text{ kJ/kg-K} \qquad \text{Ans.}$$

If the mixture is heated at constant volume

$$U_2 - U_1 = m\, c_v\, (T_2 - T_1)$$

$$= 8 \times 0.69 \times (40 - 20)$$

$$= 110.4 \text{ kJ}$$

$$H_2 - H_1 = m\, c_p\, (T_2 - T_1)$$

$$= 8 \times 0.92 \times 20 = 147.2 \text{ kJ}$$

$$S_2 - S_1 = m\, c_v \ln \frac{T_2}{T_1} + m\, R \ln \frac{V_2}{V_1}$$

$$= m\, c_v \ln \frac{T_2}{T_1} = 8 \times 0.69 \times \ln \frac{313}{293}$$

$$= 0.368 \text{ kJ/kg-K} \qquad \text{Ans.}$$

If the mixture is heated at constant pressure, ΔU and ΔH will remain the same. The change in entropy will be

$$S_2 - S_1 = m\, c_p \ln \frac{T_2}{T_1} - mR \ln \frac{p_2}{p_1}$$

$$= m\, c_p \ln \frac{T_2}{T_1} = 8 \times 0.92 \ln \frac{313}{293}$$

$$= 0.49 \text{ kJ/kg-K} \qquad \text{Ans.}$$

Example 10.9

Find the increase in entropy when 2 kg of oxygen at 60°C are mixed with 6 kg of nitrogen at the same temperature. The initial pressure of each constituent is 103 kPa and is the same as that of the mixture.

$$x_{O_2} = \frac{p_{O_2}}{p} = \frac{\frac{2}{32}}{\frac{2}{32}+\frac{6}{28}} = 0.225$$

$$x_{N_2} = \frac{p^{N^2}}{p} = 0.775$$

Entropy increase due to diffusion

$$\Delta S = -m_{O_2} R_{O_2} \ln \frac{p_{O_2}}{p} - m_{N_2} R_{N_2} \ln \frac{p_{N_2}}{p}$$

$$= -2\left(\frac{8.3143}{32}\right) \ln 0.225 - 6\left(\frac{8.3143}{28}\right) \ln 0.775$$

$$= 1.2314 \text{ kJ/kg-K}$$ Ans.

EXAMPLE 10.10

The gas neon has a molecular weight of 20.183 and its critical temperature, pressure and volume are 44.5 K, 2.73 MPa and 0.0416 m³/kgmol. Reading from a compressibility chart for a reduced pressure of 2 and a reduced temperature of 1.3, the compressibility factor Z is 0.7. What are the corresponding specific volume, pressure, temperature, and reduced volume?

Solution

At $p_r = 2$ and $T_r = 1.3$ from chart (Fig. Ex. 10.10)

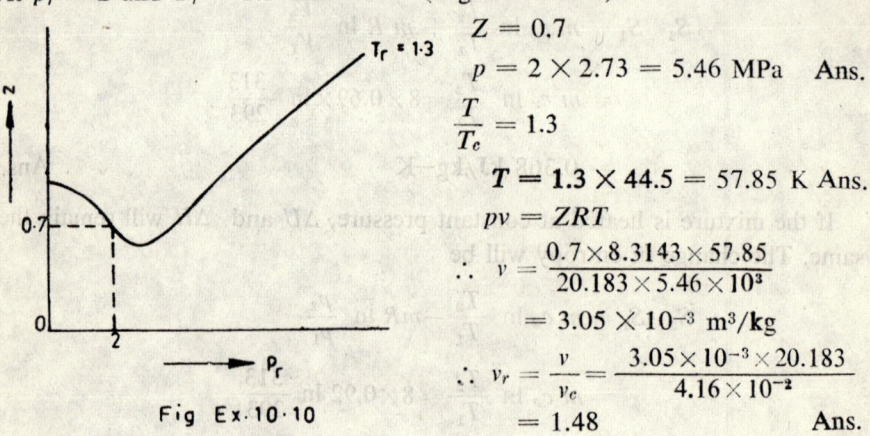

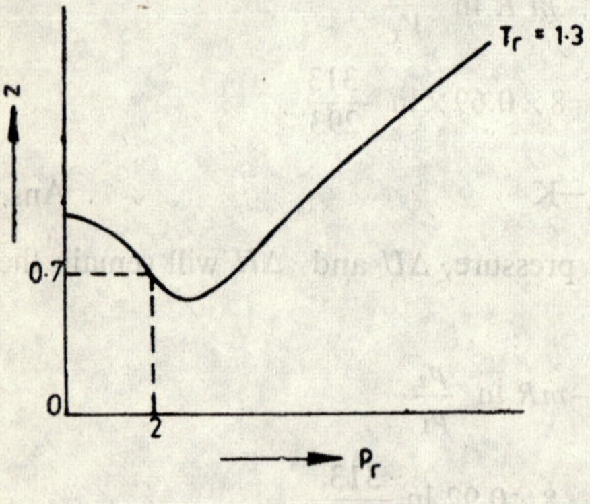

Fig Ex. 10·10

$$Z = 0.7$$

$$p = 2 \times 2.73 = 5.46 \text{ MPa}$$ Ans.

$$\frac{T}{T_c} = 1.3$$

$$T = 1.3 \times 44.5 = 57.85 \text{ K}$$ Ans.

$$pv = ZRT$$

$$\therefore \quad v = \frac{0.7 \times 8.3143 \times 57.85}{20.183 \times 5.46 \times 10^3}$$

$$= 3.05 \times 10^{-3} \text{ m}^3/\text{kg}$$

$$\therefore \quad v_r = \frac{v}{v_c} = \frac{3.05 \times 10^{-3} \times 20.183}{4.16 \times 10^{-2}}$$

$$= 1.48$$ Ans.

EXAMPLE 10.11

For the Berthelot equation of state

$$p = \frac{RT}{v-b} - \frac{a}{T v^2}$$

show that (a) $\lim_{\substack{p \to 0 \\ T \to \infty}} (RT - pv) = 0$

(b) $\lim_{T \to \infty} \frac{v}{T} = \frac{R}{p}$, (c) Boyle temperature, $T_B = \sqrt{\frac{a}{bR}}$,

(d) Critical properties $p_c = \frac{1}{12b}\sqrt{\frac{2\,aR}{3\,b}}$, $v_c = 3b$, $T_c = \sqrt{\frac{8\,a}{27\,bR}}$,

(e) Law of corresponding states

$$\left(p_r + \frac{3}{T_r \cdot v_r^2}\right)(3v_r - 1) = 8T_r$$

Solution

(a) $$p = \frac{RT}{v-b} - \frac{a}{Tv^2}$$

$$\therefore \quad RT = \left(p + \frac{a}{Tv^2}\right)\left(v - b\right)$$

or $$\frac{RT}{p} = v + \frac{a}{p\,v\,T} - b - \frac{ab}{pv^2T}$$

$$\therefore \quad RT - pv = \frac{a}{vT} - bp - \frac{ab}{v^2T}$$

$$\therefore \quad \lim_{\substack{p \to 0 \\ T \to \infty}} (RT - pv) = 0$$

Proved (a)

(b) Now $$v = \frac{RT}{p} - \frac{a}{pvT} + b + \frac{ab}{pv^2T}$$

$$\therefore \quad \frac{v}{T} = \frac{R}{p} - \frac{a}{pvT^2} + \frac{b}{T} + \frac{ab}{pv^2T^2}$$

$$\therefore \quad \lim_{T \to \infty} \frac{v}{T} = \frac{R}{p}$$

Proved (b)

(c) $$pv = RT - \frac{a}{vT} + bp + \frac{ab}{v^2T}$$

The last three terms of the equation are very small, except at very high pressures and small volume. Hence substituting $v\,T\,/\,p$

$$pv = RT - \frac{ab}{RT^2} + bp + \frac{ab\,p^2}{R^2\,T^3}$$

$$\therefore \quad \left[\frac{\partial\,(pv)}{\partial\,p}\right]_T = -\frac{a}{RT^2} + b + \frac{2\,abp}{R^2T^2} = 0$$

when $p = 0$, $T = T_B$, the Boyle temperature

$$\therefore \quad \frac{a}{RT_B^2} = b$$

$$\text{or} \quad T_B = \sqrt{\frac{a}{bR}} \qquad \text{Proved (c)}$$

(d)
$$p = \frac{RT}{v-b} - \frac{a}{Tv^2}$$

$$\left(\frac{\partial p}{\partial v}\right)_{T=T_c} = -\frac{2RT_c}{(v_c-b)^2} + \frac{2a}{T_c \cdot v_c^3} = 0$$

$$\left(\frac{\partial^2 p}{\partial v^2}\right)_{T=T_c} = \frac{2RT_c}{(v_c-b)^3} - \frac{6a}{T_c \cdot v_c^4} = 0$$

$$\left(p_c + \frac{a}{T_c v_c^2}\right)\left(v_c - b\right) = RT_c$$

By solving the three equations, as was done in the case of Van der Waals equation of state in Article 10.6

$$p_c = \frac{1}{12b}\sqrt{\frac{2aR}{3b}}, \quad v_c = 3b, \text{ and } T_c = \sqrt{\frac{8a}{27bR}} \qquad \text{Proved (d)}$$

(e) Solving the above three equations

$$a = \frac{8 v_c^3 p_c^2}{R} = 3p_c \cdot v_c^2 \cdot T_c$$

$$b = \frac{v_c}{3},\ R = \frac{8}{3}\frac{p_c v_c}{T_c} \quad \left(\text{so that } Z_c = \tfrac{3}{8}\right)$$

Substituting in the equation

$$\left(p + \frac{a}{Tv^2}\right)(v-b) = RT$$

$$\left(p + \frac{3p_c v_c^2 T_c}{Tv^2}\right)\left(v - \frac{v_c}{3}\right) = \frac{8p_c v_c}{3T_c}\cdot T$$

$$\left(p_r + \frac{3}{T_r v_r^2}\right)(3v_r - 1) = 8T_r$$

This is the law of corresponding states.

Proved (e)

PROBLEMS

10.1 What is the mass of air contained in a room 6 m×9 m×4 m if the pressure is 101.325 kPa and the temperature is 25°C?

Ans. 256 kg.

10.2 The usual cooking gas (mostly methane) cylinder is about 25 cm in diameter and 80 cm in height. It is charged to 12 MPa at room temperature (27°C). (a) Assuming the ideal gas law, find the mass of gas filled in the cylinder. (b) Explain how the actual cylinder contains nearly 15 kg of gas. (c) If the cylinder is to be protected against excessive pressure by means of a fusible plug, at what temperature should the plug melt to limit the maximum pressure to 15 MPa?

10.3 A certain gas has $c_p = 0.913$ and $c_v = 0.653$ kJ/kg-K. Find the molecular weight and the gas constant R of the gas.

10.4 From an experimental determination the specific heat ratio for acetylene (C_2H_2) is found to 1.26 Find the too specific heats.

10.5 Find the molal specific heats for monatonic, diatomic, and polyatomic gases, if their specific heat ratios are respectively 5/3, 7/5, and 4/3.

10.6 A supply of natural gas is required on a site 800 m above storage level. The gas at −150°C, 1.1 bar from storage is pumped steadily to a point on the site where its pressure is 1.2 bar, its temperature 15°C, and its flow rate 1000 m³/hr. If the work transfer to the gas at the pump is 15 kW, find the heat transfer to the gas between the two points. Neglect the change in K.E. and assume that the gas has the properties of methane (CH_4) which may be treated as an ideal gas having $\gamma = 1.33$ ($g = 9.75$ m/s²). **Ans. 63.9 kW.**

10.7 A constant volume chamber of 0.3 m³ capacity contains 1 kg of air at 5°C. Heat is transferred to the air until the temperature is 100°C. Find the work done, the heat transferred, and the changes in internal energy, enthalpy and entropy.

10.8 One kg of air in a closed system, initially at 5°C and occupying 0.3 m³ volume, undergoes a constant pressure heating process to 100°C. There is no work other than *pdv* work. Find (a) the work done during the process, (b) the heat transferred, and (c) the entropy change of the gas.

10.9 0.1 m³ of hydrogen initially at 1.2 MPa, 200°C undergoes a reversible isothermal expansion to 0.1 MPa. Find (a) the work done during the process, (b) the heat transferred, and (c) the entropy change of the gas.

10.10 Air in a closed stationary system expands in a reversible adiabatic process from 0.5 MPa, 15°C to 0.2 MPa. Find the final temperature, and per kg of air, the change in enthalpy, the heat transferred, and the work done.

10.11 If the above process occurs in an open steady flow system, find the final temperature, and per kg of air, the change in internal energy, the heat transferred, and the shaft work. Neglect velocity and elevation changes.

10.12 The indicator diagram for a certain water-cooled cylinder and piston air compressor shows that during compression $pv^{1.3}$=constant. The compression starts at 100 kPa, 25°C and ends at 600 kPa. If the process is reversible, how much heat is transferred per kg of air ?

10.13 An ideal gas of molecular weight 30 and $\gamma = 1.3$ occupies a volume of 1.5 m³ at 100 kPa and 77°C. The gas is compressed according to the law $pv^{1.25}$=constant to and heating, work done, heat transferred, and the total change of entropy. a pressure of 3 MPa. Calculate the volume and temperature at the end of compression

10.14 Calculate the change of entropy when 1 kg of air changes from a temperature of 330 K and a volume of 0.15 m³ to a temperature of 550 K and a volume of 0.6 m³. If the air expands according to the law, pv^n=constant, between the same end states, calculate the heat given to, or extracted from, the air during the expansion, and show that it is approximately equal to the change of entropy multiplied by the mean absolute temperature.

10.15 0.5 kg of air, initially at 25°C, is heated reversibly at constant pressure until the volume is doubled, and is then heated reversibly at constant volume until the pressure is doubled. For the total path, find the work transfer, the heat transfer, and the change of entropy.

10.16 An ideal gas cycle of three processes uses Argon (Mol. wt. 40) as a working substance. Process 1–2 is a reversible adiabatic expansion from 0.014 m³, 700 kPa,

280°C to 0.056 m³. Process 2–3 is a reversible isothermal process. Process 3–1 is a constant pressure process in which heat transfer is zero. Sketch the cycle in the *p-v* and *T-s* planes, and find (a) the work transfer in process 1-2, (b) the work transfer in process 2-3, and (c) the net work of the cycle. Take $\gamma = 1.67$.

10.17 A gas occupies 0.024 m³ at 700 kPa and 95°C. It is expanded in the non-flow process according to the law $pv^{1.2}$=constant to a pressure of 70 kPa after which it is heated at constant pressure back to its original temperature. Sketch the process on the *p-v* and *T-s* diagrams, and calculate for the whole process the work done, the heat transferred, and the change of entropy. Take $c_p = 1.047$ and $c_v = 0.775$ kJ/kg-K for the gas.

10.18 0.5 kg of air at 600 kPa receives an addition of heat at constant volume so that its temperature rises from 110°C to 650°C. It then expands in a cylinder polytropically to its original temperature and the index of expansion is 1.32. Finally, it is compressed isothermally to its original volume. Calculate (a) the change of entropy during each of the three stages, (b) the pressures at the end of constant volume heat addition and at the end of expansion. Sketch the processes on the *p-v* and *T-s* diagrams.

10.19 0.5 kg of helium and 0.5 kg of nitrogen are mixed at 20°C and at a total pressure of 100 kPa. Find (a) the volume of the mixture, (b) the partial volumes of the components, (c) the partial pressures of the components, (d) the mole fractions of the components, (e) the specific heats c_p and c_v of the mixture, and (f) the gas constant of the mixture.

10.20 A gaseous mixture consists of 1 kg of oxygen and 2 kg of nitrogen at a pressure of 150 kPa and a temperature of 20°C. Determine the changes in internal energy, enthalpy and entropy of the mixture when the mixture is heated to a temperature of 100°C (a) at constant volume, and (b) at constant pressure.

10.21 A closed rigid cylinder is divided by a diaphragm into two equal compartments, each of volume 0.1 m³. Each compartment contains air at a temperature of 20°C. The pressure in one compartment is 2.5 MPa and in the other compartment is 1 MPa. The diaphragm is ruptured so that the air in both the compartments mixes to bring the pressure to a uniform value throughout the cylinder which is insulated. Find the net change of entropy for the mixing process.

10.22 A vessel is divided into three compartments (a), (b), and (c) by two partitions. Part (a) contains oxygen and has a volume of 0.1 m³, (b) has a volume of 0.2 m³ and contains nitrogen, while (c) is 0.05 m³ and holds CO_2. All three parts are at a pressure of 2 bar and a temperature of 13°C. When the partitions are removed and the gases mix, determine the change of entropy of each constituent, the final pressure in the vessel and the partial pressure of each gas. The vessel may be taken as being completely isolated from its surroundings.

Ans. 0.0875, 0.0783, 0.0680 kJ/K; 2 bar; 0.5714, 1.1429, 0.2857 bar.

10.23. A Carnot cycle uses 1 kg of air as the working fluid. The maximum and minimum temperatures of the cycle are 600 K and 300 K. The maximum pressure of the cycle is 1 MPa and the volume of the gas doubles during the isothermal heating process. Show by calculation of net work and heat supplied that the efficiency is the maximum possible for the given maximum and minimum temperatures.

10.24 An ideal gas cycle consists of three reversible processes in the following sequence: (a) constant volume pressure rise, (b) isentropic expansion to *r* times the initial volume, and (c) constant pressure decrease in volume. Sketch the cycle on the *p-v* and *T-s* diagrams. Show that the efficiency of the cycle is

$$\eta_{\text{cycle}} = \frac{r^{\gamma} - 1 - \gamma(r-1)}{r^{\gamma} - 1}$$

Evaluate the cycle efficiency when $\gamma = \frac{4}{3}$ and $r = 8$. Ans. ($\eta = 0.378$)

10.25 Using the Dieterici equation of state

$$p = \frac{RT}{v-b} \cdot \exp\left(-\frac{a}{RTv}\right)$$

(a) show that

$$p_c = \frac{n}{4e^2 b^2},\ v_c = 2b,\ T_c = \frac{a}{4\,Rb}$$

(b) expand in the form

$$pv = RT\left(1 + \frac{B}{v} + \frac{C}{v^2} + \ldots\right)$$

(c) show that

$$T_B = \frac{a}{bR}$$

10.26 The number of moles, the pressures, and the temperatures of gases *a*, *b*, and *c* are given below

Gas	*m (kgmol)*	*p (kPa)*	*t (°C)*
N_2	1	350	100
CO	3	420	200
O_2	2	700	300

If the containers are connected, allowing the gases to mix freely, find (a) the pressure and temperature of the resulting mixture at equilibrium, and (b) the change of entropy of each constituent and that of the mixture.

10.27 Calculate the volume of 2.5 kgmoles of steam at 236.4 atm. and 776.76 K with the help of compressibility factor versus reduced pressure graph. At this volume and the given pressure, what would the temperature be in K, if steam behaved like a Van der Waals gas?

The critical pressure, volume, and temperature of steam are 218.2 atm., 57 cm³/gm-mole, and 647.3 K respectively.

10.28 Two vessels, *A* and *B*, each of volume 3m³ may be connected together by a tube of negligible volume. Vessel *A* contains air at 7 bar, 95°C while *B* contains air at 3.5 bar, 205°C. Find the change of entropy when *A* is connected to *B*. Assume the mixing to be complete and adiabatic. Ans. (0.975 kJ/kgK)

10.29 An ideal gas at temperature T_1 is heated at constant pressure to T_2 and then expanded reversibly, according to the law pv^n = constant, until the temperature is once again T_1. What is the required value of *n*, if the changes of entropy during the separate processes are equal? Ans. ($= n2\gamma/\gamma+1$)

10.30 A certain mass of sulphur dioxide (SO_2) is contained in a vessel of 0.142 m³ capacity, at a pressure and temperature of 23.1 bar and 18°C respectively. A valve is opened momentarily and the pressure falls immediately to 6.9 bar. Sometime later the temperature is again 18°C and the pressure is observed to be 9.1 bar. Estimate the value of specific heat ratio. Ans. 1.29

10.31 A gaseous mixture contains 21% by volume of nitrogen, 50% by volume of hydrogen, and 29% by volume of carbon-dioxide. Calculate the molecular weight of the mixture, the characteristic gas constant R for the mixture and the value of the reversible adiabatic index γ. (At 10°C, the c_p values of nitrogen, hydrogen, and carbon dioxide are 1.039, 14.235, and 0.828/kJ/kg-K respectively.)

A cylinder contains 0.085 m³ of the mixture at 1 bar and 10°C. The gas undergoes a reversible non-flow process during which its volume is reduced to one-fifth of its original value. If the law of compression is $pv^{1.2}$=constant, determine the work and heat transfers in magnitude and sense and the change in entropy.

Ans. 19.64 kg/kg. mol, 0.423 kJ/kg-K, 1.365, —16 kJ, —7.24 kJ, —0.31 kJ/kg-K

10.32 Two moles of an ideal gas at temperature T and pressure p are contained in a compartment. In an adjacent compartment is one mole of an ideal gas at temperature $2T$ pressure p. The gases mix adiabatically but do not react chemically when a partition separating the compartments is withdrawn. Show that the entropy increase due to the mixing process is given by

$$R\left(\ln\frac{27}{4}+\frac{\gamma}{\gamma-1}\ln\frac{32}{27}\right)$$

provided that the gases are different and that the ratio of specific heats γ is the same for both gases and remains constant.

What would the entropy change be if the mixing gases were of the same species?

10.33 n_1 moles of an ideal gas at pressure p_1 and temperature T are in one compartment of an insulated container. In an adjoining compartment, separated by a partition, are n_2 moles of an ideal gas at pressure p_2 and temperature T. When the partition is removed, calculate (a) the final pressure of the mixture, (b) the entropy change when the gases are identical, and (c) the entropy change when the gases are different. Prove that the entropy change in (c) is the same as that produced by two independent free expansions.

11

Combined First and Second Laws

11.1 SOME MATHEMATICAL THEOREMS

Theorem 1. If a relation exists among the variables x, y, and z, then z may be expressed as a function of x and y, or

$$dz = \left(\frac{\partial z}{\partial x}\right)_y dx + \left(\frac{\partial z}{\partial y}\right)_x dy$$

If $$\left(\frac{\partial z}{\partial x}\right)_y = M, \text{ and } \left(\frac{\partial z}{\partial y}\right)_x = N$$

then $dz = M\,dx + N\,dy$,
where z, and N are functions of x and y. Differentiating M partially with respect to y, and N with respect to x

$$\left(\frac{\partial M}{\partial y}\right)_x = \frac{\partial^2 z}{\partial x \cdot \partial y}$$

$$\left(\frac{\partial N}{\partial x}\right)_y = \frac{\partial^2 z}{\partial y \cdot \partial x}$$

$$\therefore \quad \left(\frac{\partial M}{\partial y}\right)_x = \left(\frac{\partial N}{\partial x}\right)_y \tag{11.1}$$

This is the *condition of exact (or perfect) differential.*

Theorem 2. If a quantity f is a function of x, y, and z, and a relation exists among x, y, and z, then f is a function of any two of x, y, and z. Similarly any one of x, y, and z may be regarded to be a function of f and any one of x, y, and z. Thus, *if*

$$x = x(f, y)$$

$$dx = \left(\frac{\partial x}{\partial f}\right)_y df + \left(\frac{\partial x}{\partial y}\right)_f dy$$

Similarly, if

$$y = y\,(f, z)$$

$$dy = \left(\frac{\partial y}{\partial f}\right)_z df + \left(\frac{\partial y}{\partial z}\right)_f dz$$

Substituting the expression of dy in the preceding equation

$$dx = \left(\frac{\partial x}{\partial f}\right)_y df + \left(\frac{\partial x}{\partial y}\right)_f \left[\left(\frac{\partial y}{\partial f}\right)_z df + \left(\frac{\partial y}{\partial z}\right)_f dz\right]$$

$$= \left[\left(\frac{\partial x}{\partial f}\right)_y + \left(\frac{\partial x}{\partial y}\right)_f \left(\frac{\partial y}{\partial f}\right)_z\right] df + \left(\frac{\partial x}{\partial y}\right)_f \left(\frac{\partial y}{\partial z}\right)_f dz$$

Again

$$dx = \left(\frac{\partial x}{\partial f}\right)_z df + \left(\frac{\partial x}{\partial z}\right)_f dz$$

$$\therefore \quad \left(\frac{\partial x}{\partial z}\right)_f = \left(\frac{\partial x}{\partial y}\right)_f \left(\frac{\partial y}{\partial z}\right)_f$$

$$\therefore \quad \left(\frac{\partial x}{\partial y}\right)_f \left(\frac{\partial y}{\partial z}\right)_f \left(\frac{\partial z}{\partial x}\right)_f = 1 \qquad (11.2)$$

Theorem 3. Among the variables x, y, and z, any one variable may be considered as a function of the other two. Thus

$$x = x(y, z)$$

$$dx = \left(\frac{\partial x}{\partial y}\right)_z dy + \left(\frac{\partial x}{\partial z}\right)_y dz$$

Similarly,

$$dz = \left(\frac{\partial z}{\partial x}\right)_y dx + \left(\frac{\partial z}{\partial y}\right)_x dy$$

$$\therefore \quad dx = \left(\frac{\partial x}{\partial y}\right)_z dy + \left(\frac{\partial x}{\partial z}\right)_y \left[\left(\frac{\partial z}{\partial x}\right)_y dx + \left(\frac{\partial z}{\partial y}\right)_x dy\right]$$

$$= \left[\left(\frac{\partial x}{\partial y}\right)_z + \left(\frac{\partial x}{\partial z}\right)_y \left(\frac{\partial z}{\partial y}\right)_x\right] dy + \left(\frac{\partial x}{\partial z}\right)_y \left(\frac{\partial z}{\partial x}\right)_y dx$$

$$= \left[\left(\frac{\partial x}{\partial y}\right)_z + \left(\frac{\partial x}{\partial z}\right)_y \left(\frac{\partial z}{\partial y}\right)_x\right] dy + dx$$

$$\therefore \quad \left(\frac{\partial x}{\partial y}\right)_z + \left(\frac{\partial x}{\partial z}\right)_y \left(\frac{\partial z}{\partial y}\right)_x = 0$$

or

$$\left(\frac{\partial x}{\partial y}\right)_z \left(\frac{\partial z}{\partial x}\right)_y \left(\frac{\partial y}{\partial z}\right)_x = -1 \qquad (11.3)$$

Among the thermodynamic variable p, V, and T, the following relation holds good

$$\left(\frac{\partial p}{\partial v}\right)_T \left(\frac{\partial V}{\partial T}\right)_p \left(\frac{\partial T}{\partial p}\right)_V = -1$$

11.2 MAXWELL'S EQUATIONS

A pure substance existing in a single phase has only two independent variables. Of the eight quantities p, V, T, S, U, H, F (Helmholtz function), and G (Gibbs function) *any one may be expressed as a function of any two others.*

For a pure substance undergoing an infinitesimal reversible process

(a) $dU = TdS - pdV$

(b) $dH = dU + pdV + Vdp = TdS + Vdp$

(c) $dF = dU - TdS - SdT = -pdV - SdT$

(d) $dG = dH - TdS - SdT = Vdp - SdT$

Since $U, H, F,$ and G are thermodynamic properties and exact differentials of the type

$$dz = Mdx + Ndy, \text{ then}$$

$$\left(\frac{\partial M}{\partial y}\right)_x = \left(\frac{\partial N}{\partial x}\right)_y$$

Applying this to the four equations

$$\left(\frac{\partial T}{\partial V}\right)_S = -\left(\frac{\partial p}{\partial S}\right)_V \tag{11.4}$$

$$\left(\frac{\partial T}{\partial p}\right)_S = \left(\frac{\partial V}{\partial S}\right)_p \tag{11.5}$$

$$\left(\frac{\partial p}{\partial T}\right)_V = \left(\frac{\partial S}{\partial V}\right)_T \tag{11.6}$$

$$\left(\frac{\partial V}{\partial T}\right)_p = -\left(\frac{\partial S}{\partial p}\right)_T \tag{11.7}$$

These four equations are known as *Maxwell's equations.*

11.3 TdS EQUATIONS

Let entropy S be imagined as a function of T and V. Then

$$dS = \left(\frac{\partial S}{\partial T}\right)_V dT + \left(\frac{\partial S}{\partial V}\right)_T dV$$

$$\therefore \quad TdS = T\left(\frac{\partial S}{\partial T}\right)_V dT + T\left(\frac{\partial S}{\partial V}\right)_T dV$$

Since $T\left(\frac{\partial S}{\partial T}\right)_V = C_v$, heat capacity at constant volume, and

$\left(\frac{\partial S}{\partial V}\right)_T = \left(\frac{\partial p}{\partial T}\right)_V$, Maxwell's third equation,

$$TdS = C_v dT + T\left(\frac{\partial p}{\partial T}\right)_V dV \tag{11.8}$$

This is known as the *first TdS equation.*

If $S = S(T, p)$

$$dS = \left(\frac{\partial S}{\partial T}\right)_p dT + \left(\frac{\partial S}{\partial p}\right)_T dp$$

$$\therefore \quad TdS = T\left(\frac{\partial S}{\partial T}\right)_p dT + T\left(\frac{\partial S}{\partial p}\right)_T dp$$

Since $T\left(\frac{\partial S}{\partial T}\right)_p = C_p$, and $\left(\frac{\partial S}{\partial p}\right)_T = -\left(\frac{\partial V}{\partial T}\right)_p$

then

$$TdS = C_p dT - T\left(\frac{\partial V}{\partial T}\right)_p dp \tag{11.9}$$

This is known as the *second TdS equation*.

11.4 DIFFERENCE IN HEAT CAPACITIES

Equating the first and second TdS equations

$$TdS = C_p dT - T\left(\frac{\partial V}{\partial T}\right)_p dp = C_v dT + T\left(\frac{\partial p}{\partial T}\right)_V dV$$

$$(C_p - C_v)\, dT = T\left(\frac{\partial p}{\partial T}\right)_V dV + T\left(\frac{\partial V}{\partial T}\right)_p dp$$

$$\therefore \quad dT = \frac{T\left(\frac{\partial p}{\partial T}\right)_V}{C_p - C_v} dV + \frac{T\left(\frac{\partial V}{\partial T}\right)_p}{C_p - C_v} dp$$

Again $\quad dT = \left(\frac{\partial T}{\partial V}\right)_p dV + \left(\frac{\partial T}{\partial p}\right)_V dp$

$$\therefore \quad \frac{T\left(\frac{\partial p}{\partial T}\right)_V}{C_p - C_v} = \left(\frac{\partial T}{\partial V}\right)_p \quad \text{and} \quad \frac{T\left(\frac{\partial V}{\partial T}\right)_p}{C_p - C_v} = \left(\frac{\partial T}{\partial p}\right)_V$$

Both these equations give

$$C_p - C_v = T\left(\frac{\partial p}{\partial T}\right)_V \left(\frac{\partial V}{\partial T}\right)_p$$

But $\quad \left(\frac{\partial p}{\partial T}\right)_V \left(\frac{\partial T}{\partial V}\right)_p \left(\frac{\partial V}{\partial p}\right)_T = -1$

$$\therefore \quad C_p - C_v = -T\left(\frac{\partial V}{\partial T}\right)_p^2 \left(\frac{\partial p}{\partial V}\right)_T \tag{11.10}$$

This is a very important equation in thermodynamics. It indicates the following important facts.

(a) Since $\left(\frac{\partial V}{\partial T}\right)_p^2$ is always positive, and $\left(\frac{\partial p}{\partial V}\right)_T$ for any substance is negative, $(C_p - C_v)$ is always positive. Therefore, C_p is always greater than C_v.

(b) As $T \to 0$ K, $C_p \to C_v$, or at absolute zero, $C_p = C_v$.

(c) When $\left(\frac{\partial V}{\partial T}\right)_p = 0$ (e.g., for water at 4°C, when density is maximum, or specific volume minimum), $C_p = C_v$.

(d) For an ideal gas, $pV = mRT$

$$\left(\frac{\partial V}{\partial T}\right)_p = \frac{mR}{p} = \frac{V}{T}$$

and
$$\left(\frac{\partial p}{\partial V}\right)_T = -\frac{mRT}{V^2}$$

$\therefore$
$$C_p - C_v = mR$$

or
$$c_p - c_v = R$$

Equation (11.10) may also be expressed in terms of volume expansivity (β), defined as

$$\beta = \frac{1}{V}\left(\frac{\partial V}{\partial T}\right)_p$$

and isothermal compressibility (k), defined as

$$k = -\frac{1}{V}\left(\frac{\partial V}{\partial p}\right)_T$$

$$C_p - C_v = \frac{TV\left[\frac{1}{V}\left(\frac{\partial V}{\partial T}\right)_p\right]^2}{-\frac{1}{V}\left(\frac{\partial V}{\partial p}\right)_T}$$

$\therefore$
$$C_p - C_v = \frac{TV\beta^2}{k} \qquad (11.11)$$

11.5 RATIO OF HEAT CAPACITIES

At constant S, the two TdS equations become

$$C_p dT_S = T\left(\frac{\partial V}{\partial T}\right)_p dp_S$$

$$C_v dT_S = -T\left(\frac{\partial p}{\partial T}\right)_V dV_S$$

$\therefore$
$$\frac{C_p}{C_v} = -\left(\frac{\partial V}{\partial T}\right)_p\left(\frac{\partial T}{\partial p}\right)_V\left(\frac{\partial p}{\partial V}\right)_S = \frac{\left(\frac{\partial p}{\partial V}\right)_S}{\left(\frac{\partial p}{\partial V}\right)_T}$$

The adiabatic compressibility (k_S) is defined as

$$k_S = -\frac{1}{V}\left(\frac{\partial V}{\partial p}\right)_S$$

$$\therefore \quad \frac{C_p}{C_v} = \frac{-\frac{1}{V}\left(\frac{\partial V}{\partial p}\right)_T}{-\frac{1}{V}\left(\frac{\partial V}{\partial p}\right)_S} = \gamma$$

or

$$\gamma = \frac{k}{k_S} \tag{11.12}$$

11.6 ENERGY EQUATION

For a system undergoing an infinitesimal reversible process between two equilibrium states, the change of internal energy is

$$dU = TdS - pdV$$

Substituting the first TdS equation

$$dU = C_v dT + T\left(\frac{\partial p}{\partial T}\right)_V dV - pdV$$

$$= C_v dT + \left[T\left(\frac{\partial p}{\partial T}\right)_V - p\right] dV$$

If $\quad U = U(T, V)$

$$dU = \left(\frac{\partial U}{\partial T}\right)_V dT + \left(\frac{\partial U}{\partial V}\right)_T dV$$

$$\left(\frac{\partial U}{\partial V}\right)_T = T\left(\frac{\partial p}{\partial T}\right)_V - p \tag{11.13}$$

This is known as the *energy equation*. Two applications of the equation are given below:

(a) For an ideal gas, $\quad p = \frac{n\bar{R}T}{V}$

$$\therefore \quad \left(\frac{\partial p}{\partial T}\right)_V = \frac{n\bar{R}}{V} = \frac{p}{T}$$

$$\therefore \quad \left(\frac{\partial U}{\partial V}\right)_T = T \cdot \frac{p}{T} - p = 0$$

U does not change when V changes at $T = C$.

$$\left(\frac{\partial U}{\partial p}\right)_T \left(\frac{\partial p}{\partial V}\right)_T \left(\frac{\partial V}{\partial U}\right)_T = 1$$

$$\therefore \quad \left(\frac{\partial U}{\partial p}\right)_T \left(\frac{\partial p}{\partial V}\right)_T = \left(\frac{\partial U}{\partial V}\right)_T = 0$$

Since $\quad \left(\frac{\partial p}{\partial V}\right)_T \neq 0, \; \left(\frac{\partial U}{\partial p}\right)_T = 0$

U does not change either when p changes at $T = C$. So the internal energy of an ideal gas is a function of temperature only, as shown earlier in Chapter 10 .

Similarly

$$dH = TdS + Vdp$$

and
$$TdS = C_p dT - T \left(\frac{\partial V}{\partial T}\right)_p dp$$

$$\therefore \quad dH = C_p dT + \left[V - T\left(\frac{\partial V}{\partial T}\right)_p\right] dp$$

$$\therefore \quad \left(\frac{\partial H}{\partial p}\right)_T = V - T\left(\frac{\partial V}{\partial T}\right)_p \tag{11.14}$$

As shown for internal energy, it can be similarly proved from equation (11.14) that the enthalpy of an ideal gas is not a function of either volume or pressure

$$\left[\text{i.e.} \left(\frac{\partial H}{\partial p}\right)_T = 0 \text{ and } \left(\frac{\partial H}{\partial V}\right)_T = 0\right]$$

but a function of temperature alone.

(b) Thermal radiation in equilibrium with the enclosing walls possesses an energy that depends only on the volume and temperature. The energy density (u), defined as the ratio of energy to volume, is a function of temperature only, or

$$u = \frac{U}{V} = f(T) \quad \text{only}$$

The electromagnetic theory of radiation states that radiation exerts a pressure, and that the pressure exerted by the black-body radiation in an enclosure is given by

$$p = \frac{u}{3}$$

Black-body radiation is thus specified by the pressure, volume, and temperature of the radiation.

Since
$$U = uV \text{ and } p = \frac{u}{3}$$

$$\left(\frac{\partial U}{\partial V}\right)_T = u \text{ and } \left(\frac{\partial p}{\partial T}\right)_V = \frac{1}{3}\frac{du}{dT}$$

By substituting in the energy equation (11.13)

$$u = \frac{T}{3} \cdot \frac{du}{dT} - \frac{u}{3}$$

$$\therefore \quad \frac{du}{u} = 4\frac{dT}{T}$$

or $\qquad \ln u = \ln T^4 + \ln b$

or $\qquad u = b\, T^4$

where b is a constant. This is known as the *Stefan-Bolzmann Law*

Since $\qquad U = u\ V = V\ b\ T^4$

$$\left(\frac{\partial U}{\partial T}\right)_V = C_v = 4\ VbT^3$$

and $$\left(\frac{\partial p}{\partial T}\right)_V = \frac{1}{3}\frac{du}{dT} = \frac{4}{3} bT^3$$

from the first TdS equation

$$TdS = C_v dT + T\left(\frac{\partial p}{\partial T}\right)_V dV$$

$$= 4VbT^3 dT + \frac{4}{3}\ bT^4 \cdot dV$$

For a reversible isothermal change of volume, the heat to be supplied reversibly to keep temperature constant

$$Q = \frac{4}{3} bT^4 \Delta V$$

For a reversible adiabatic change of volume

$$\frac{4}{3} bT^4 dV = -\ 4VbT^3\ dT$$

$$\therefore \qquad \frac{dV}{V} = -\ 3\ \frac{dT}{T}$$

or $\qquad VT^3 = \text{const.}$

If the temperature is one-half the original temperature, the volume of black-body radiation is to be increased adiabatically eight times its original volume so that the radiation remains in equilibrium with matter at that temperature.

11.7 JOULE-KELVIN EFFECT

A gas is made to undergo continuous throttling process by a valve, as shown in Fig. 11.1. The pressures and temperatures of the gas in the insulated pipe upstream and downstream of the valve are measured with suitable manometers and thermometers.

Let p_i and T_i be the arbitrarily chosen pressure and temperature before throttling and let them be kept constant. By operating the valve manually, the gas is throttled successively to different pressures and temperatures pf, T_{f_1}; p_{f_2}, T_{f_2}; p_{f_3}, T_{f_3} and so on. These are then plotted on the T-p coordinates as shown in Fig. 11.2. All the points represent equilibrium states of some constant mass of gas, say, 1 kg, at which the gas has the *same enthalpy*. The curve passing through all these points is an isenthalpic curve

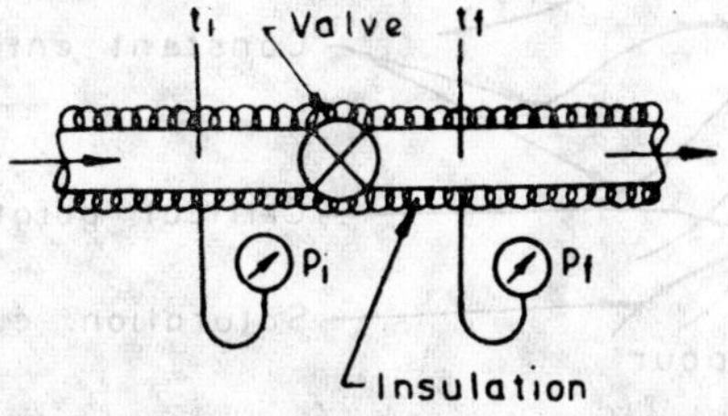

Fig. 11.1 Joule-Thomson expansion

or an *isenthalpe*. It is not the graph of a throttling process, but the graph through points of equal enthalpy.

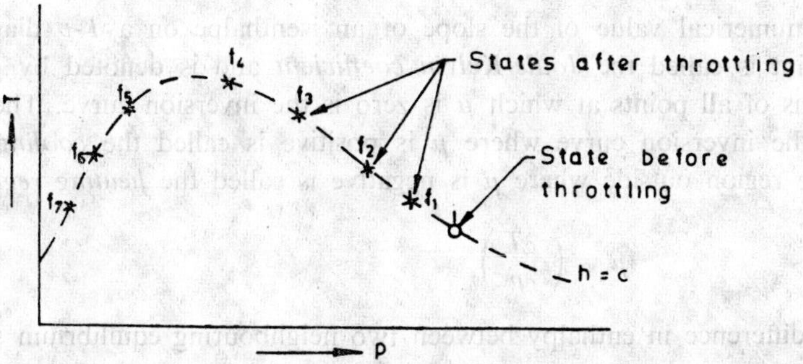

Fig. 11.2 Isenthalpic states of a gas

The initial temperature and pressure of the gas (before throttling) are then set to new values, and by throttling to different states, a *family of isenthalpes* is obtained for the gas, as shown in Figs. 11.3 and 11.4. The curve

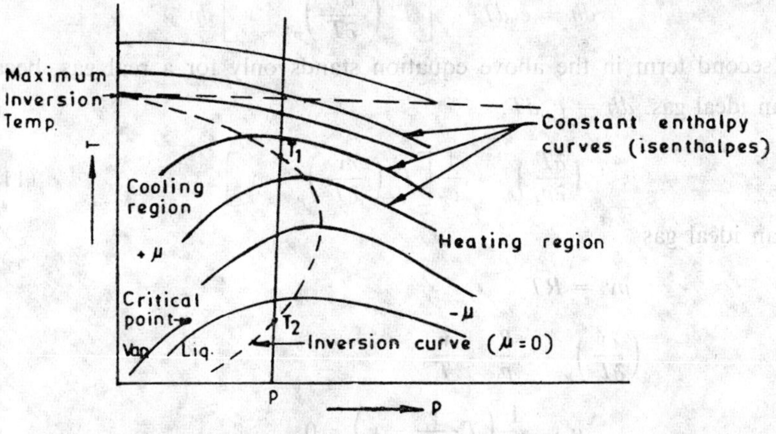

Fig. 11.3 Isenthalpic curves and the inversion curve

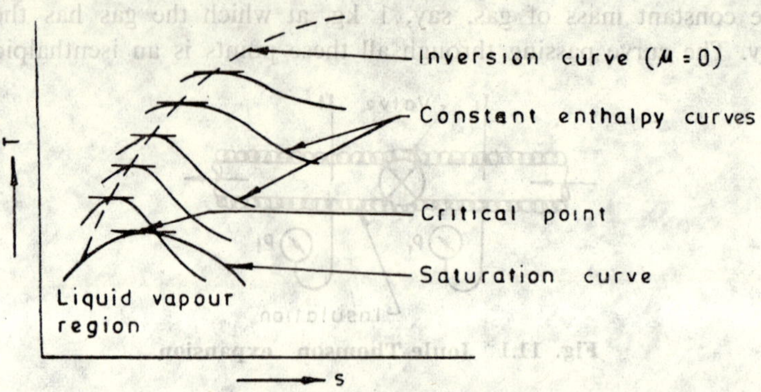

Fig. 11.4 Inversion and saturation curves on T-s plot

passing through the maxima of these isenthalpes is called the *inversion curve*.

The numerical value of the slope of an isenthalpe on a T-p diagram at any point is called the *Joule-Kelvin coefficient* and is denoted by μ. Thus the locus of all points at which μ is zero is the inversion curve. The region inside the inversion curve where μ is positive is called the *cooling region* and the region outside where μ is negative is called the *heating region*. So,

$$\mu = \left(\frac{\partial T}{\partial p}\right)_h$$

The difference in enthalpy between two neighbouring equilibrium states is

$$dh = Tds + vdp$$

and the second TdS equation (per unit mass)

$$Tds = c_p dT - T\left(\frac{\partial v}{\partial T}\right)_p dp$$

$$\therefore \quad dh = c_p dT - \left[T\left(\frac{\partial v}{\partial T}\right)_p - v\right] dp$$

The second term in the above equation stands only for a real gas, because for an ideal gas, $dh = c_p dT$.

$$\therefore \quad \mu = \left(\frac{\partial T}{\partial p}\right)_h = \frac{1}{c_p}\left[T\left(\frac{\partial v}{\partial T}\right)_p - v\right] \tag{11.15}$$

For an ideal gas

$$pv = RT$$

$$\therefore \quad \left(\frac{\partial v}{\partial T}\right)_p = \frac{R}{p} = \frac{v}{T}$$

$$\therefore \quad \mu = \frac{1}{c_p}\left(T \cdot \frac{v}{T} - v\right) = 0$$

There is no change in temperature when an ideal gas is made to undergo a Joule-Kelvin expansion (i.e. throttling).

For achieving the effect of cooling by Joule-Kelvin expansion, the initial temperature of the gas must be below the point where the inversion curve intersects the temperature axis i.e. below the *maximum inversion temperature*. For nearly all substances, the maximum inversion temperature is above the normal ambient temperature, and hence cooling can be obtained by the Joule-Kelvin effect. In the case of hydrogen and helium, however, the gas is to be precooled in heat exchangers below the maximum inversion temperature before it is throttled. For liquefaction the gas has to be cooled below the critical temperature.

Let the iniial state of gas before throttling be at A (Fig. 11.5). The change in temperature may be positive, zero, or negative, depending upon

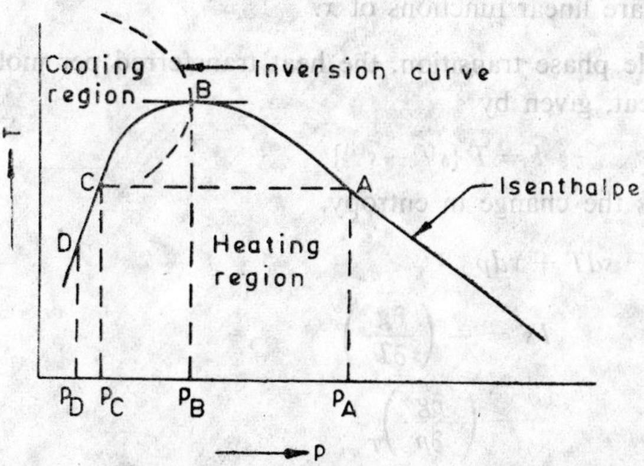

Fig. 11.5 Maximum cooling by Joule-Kelvin expansion

the final pressure after throttling. If the final pressure lies between A and B, there will be a rise in temperature or heating effect. If it is at C, there will be no change in temperature. If the final pressure is below p_C, there will be a cooling effect, and if the final pressure is p_D, the temperature drop will be $(T_A - T_D)$.

Maximum temperature drop will occur if the initial state lies on the inversion curve.

The volume expansivity is

$$\beta = \frac{1}{v}\left(\frac{\partial v}{\partial T}\right)_p$$

So the Joule-Kelvin coefficient μ is given by, from equation (11.15)

$$\mu = \frac{1}{c_p}\left[T\left(\frac{\partial v}{\partial T}\right)_p - v\right]$$

or

$$\mu = \frac{v}{c_p}[\beta T - 1]$$

For an ideal gas, $\beta = \frac{1}{T}$ and $\mu = 0$

There are two inversion temperatures for each pressure, e.g. T_1 and T_2 at pressure p (Fig. 11.3).

11.8 CLAUSIUS-CLAPEYRON EQUATION

During phase transitions like melting, vaporization and sublimation, the temperature and pressure remain constant, while the entropy and volume change. If x is the fraction of initial phase i which has been transformed into final phase f, then

$$s = (1-x)\, s^{(i)} + x s^{(f)}$$
$$v = (1-x)\, v^{(i)} + x v^{(f)}$$

where s and v are linear functions of x.

For reversible phase transition, the heat transferred per mole (or per kg) is the latent heat, given by

$$l = T\,\{s^{(f)} - s^{(i)}\}$$

which indicates the change in entropy.

Now $\quad dg = -\,sdT + vdp$

or
$$ls = -\left(\frac{\partial g}{\partial T}\right)^{p}$$

and
$$v = \left(\frac{\partial g}{\partial p}\right)_T$$

A *phase change of the first order* is known as any phase change that satisfies the following requirements:

(a) There are changes of entropy and volume.

(b) The first-order derivatives of Gibbs function change discontinuously.

Let us consider the first-order phase transtition of one mole of a substance from phase i to phase f. Using the first TdS equation

$$Tds = c_v dT + T\left(\frac{\partial p}{\partial T}\right)_V dv$$

for the phase transition which is reversible, isothermal and isobaric, and integrating over the whole change of phase, and since $\left(\frac{\partial p}{\partial T}\right)_V$ is independent of v

$$T\,\{s^{(f)} - s^{(i)}\} = T\,\frac{dp}{dT}\cdot\{v^{(f)} - v^{(i)}\}$$

$$\therefore \quad \frac{dp}{dT} = \frac{s^{(f)} - s^{(i)}}{v^{(f)} - v^{(i)}} = \frac{l}{T\{v^{(f)} - v^{(i)}\}} \tag{11.16}$$

The above equation is known as the *Clausius-Clapeyron equation*.

The Clausius-Clapeyron equation can also be derived in another way.

For a reversible process at constant T and p, the Gibbs function remains constant. Therefore, for the first-order phase change at T and p

$$g^{(i)} = g^{(f)}$$

and for a phase change at $T + dT$ and $p + dp$ (Fig. 11.6)

$$g^{(i)} + dg^{(i)} = g^{(f)} + dg^{(f)}$$

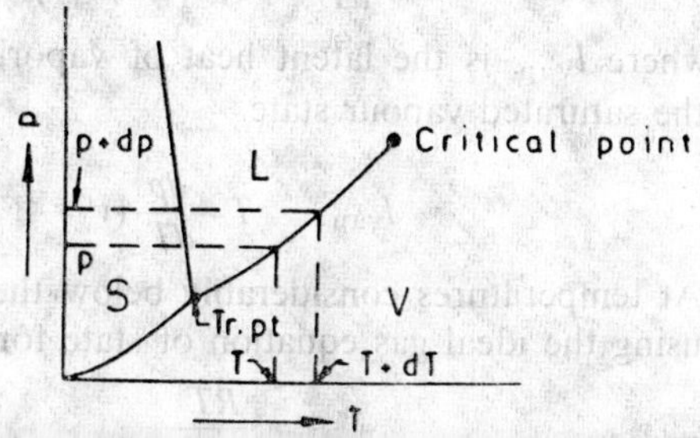

Fig. 11.6 First order phase transition.

Subtracting

$$dg^{(i)} = dg^{(f)}$$

or

$$-s^{(i)}\, dT + v^{(i)}\, dp = -s^{(f)}\, dT + v^{(f)}\, dp$$

$$\therefore \quad \frac{dp}{dT} = \frac{s^{(f)} - s^{(i)}}{v^{(f)} - v^{(i)}} = \frac{l}{T\,(v^{(f)} - v^{(i)})}$$

For *fusion*

$$\frac{dp}{dT} = \frac{l_{fu}}{T\,(v'' - v')}$$

where l_{fu} is the latent heat of fusion, the first prime indicates the saturated solid state, and the second prime the saturated liquid state. The slope of the fusion curve is determined by $(v'' = v')$, since l_{fu} and T are positive. If the substance expands on melting, $v'' > v'$, the slope is positive. This is the usual case. Water, however, contracts on melting and has the fusion curve with a negative slope (Fig. 11.7).

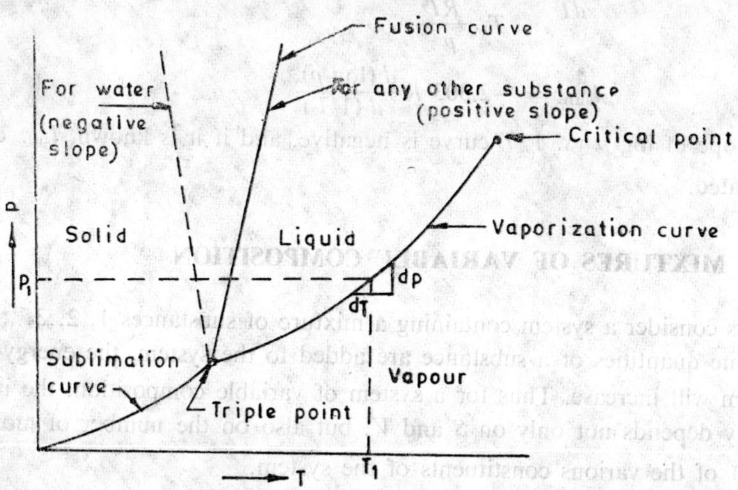

Fig. 11.7 Phase diagram for water and any other substance on p-T coordinates

For *vaporization*

$$\frac{dp}{dT} = \frac{l_{vap}}{T\,(v''' - v'')}$$

where $l_{vap.}$ is the latent heat of vaporization, and the third prime indicates the saturated vapour state.

$$\therefore \qquad l_{vap.} = T\,\frac{dp}{dT}\,(v''' - v'')$$

At temperatures considerably below the critical temperature, $v''' \gg v''$ and using the ideal gas equation of state for vapour

$$v''' = \frac{RT}{p}$$

$$l_{vap.} \simeq T \cdot \frac{dp}{dT} \cdot \frac{RT}{p}$$

or

$$l_{vap.} = \frac{RT^2}{p}\,\frac{dp}{dT} \tag{11.17}$$

If the slope dp/dT at any state (e.g. point p_1, T_1 in Fig. 11.7) is known, the latent heat of vaporization can be computed from the above equation.

For *sublimation*

$$\frac{dp}{dT} = \frac{l_{sub.}}{T\,(v''' - v')}$$

where $l_{sub.}$ is the latent heat of sublimation.

Since $v''' \gg v'$, and vapour pressure is low, $v''' = \dfrac{RT}{p}$

$$\frac{dp}{dT} = \frac{l_{sub.}}{T.\,\dfrac{RT}{p}}$$

or

$$l_{sub.} = -2.303\,R\,\frac{d\,(\log p)}{d\,(1/T)}$$

the slope of log p *vs.* $1/T$ curve is negative, and if it is known, $l_{sub.}$ can be estimated.

11.9 MIXTURES OF VARIABLE COMPOSITION

Let us consider a system containing a mixture of substances 1, 2, 3, . . . , K. If some quantities of a substance are added to the system, the energy of the system will increase. Thus for a system of variable composition, the internal energy depends not only on S and V, but also on the number of moles (or mass) of the various constituents of the system.

Thus

$$U = U\,(S, V, n_1, n_2, \ldots, n_K)$$

where $n_1, n_2, \ldots, n_K$ are the number of moles of substances 1, 2,..., K. The composition may change not only due to addition or subtraction, but also due to chemical reaction and inter-phase mass transfer. For a small change in U, assuming the function to be continuous

$$dU = \left(\frac{\partial U}{\partial S}\right)_{V, n_1, n_2, \ldots n_K} dS + \left(\frac{\partial U}{\partial V}\right)_{S, n_1, n_2, \ldots, n_K} dV + $$
$$+ \left(\frac{\partial U}{\partial n_1}\right)_{S, V, n_2, \ldots, n_K} dn_1 + \left(\frac{\partial U}{\partial n_2}\right)_{S, V, n_1, n_3, \ldots, n_K} dn_2$$
$$+ \ldots + \left(\frac{\partial U}{\partial n_K}\right)_{S, V, n_1, n_2, \ldots, n_{K-1}} dn_K$$

or

$$dU = \left(\frac{\partial U}{\partial S}\right)_{V, n_i} dS + \left(\frac{\partial U}{\partial V}\right)_{S, n_i} dV + \sum_{i=1}^{K}\left(\frac{\partial U}{\partial n_i}\right)_{S, V, n_j} dn_i$$

where subscript i indicates any substance and subscript j any other substance except the one whose number of moles is changing.

If the composition does not change

$$dU = TdS - pdV$$

$$\therefore \quad \left(\frac{\partial U}{\partial S}\right)_{V, n_i} = T, \text{ and } \left(\frac{\partial U}{\partial V}\right)_{S, n_i} = -p$$

$$\therefore \quad dU = TdS - pdV + \sum_{i=1}^{K}\left(\frac{\partial V}{\partial n_i}\right)_{S, V, n_j} dn_i \qquad (11.18)$$

Molal *chemical potential*, μ, of component i is defined as

$$\mu_i = \left(\frac{\partial U}{\partial n_i}\right)_{S, V, nj}$$

signifying the change in internal energy per unit mole of component i when S, V, and the number of moles of all other components are constant.

$$\therefore \quad dU = TdS - pdV + \sum_{i=1}^{K} \mu_i \, dn_i$$

or

$$TdS = dU + pdV - \sum_{i=1}^{K} \mu_i \, dn_i \qquad (11.19)$$

This is known as *Gibbs entropy equation*.
In a similar manner

$$G = G(p, T, n_1, n_2, \ldots, n_K)$$

or $$dG = \left(\frac{\partial G}{\partial p}\right)_{T,\, n_i} dp + \left(\frac{\partial G}{\partial T}\right)_{p,\, n_i} dT + \sum_{i=1}^{K} \left(\frac{\partial G}{\partial n_i}\right)_{T,\, p,\, n_j} dn_i$$

$$= Vdp - SdT + \sum_{i=1}^{K} \left(\frac{\partial G}{\partial n_i}\right)_{T,\, p,\, n_j} dn_i \tag{11.20}$$

Since $G = U + pV - TS$

$$d(U + pV - TS) = VdP - SdT + \sum_{i=1}^{K} \left(\frac{\partial G}{\partial n_i}\right)_{T,\, p,\, n_j} dn_i$$

$$dU + pdV + Vdp - TdS - SdT$$

$$= Vdp - SdT + \sum_{i=1}^{K} \left(\frac{\partial G}{\partial n_i}\right)_{T,\, p,\, n_j} dn_i$$

or $$dU = TdS - pdV + \sum_{i=1}^{K} \left(\frac{\partial G}{\partial n_i}\right)_{T,\, p,\, n_j} dn_i$$

Comparing this equation with equation (11.18)

$$\left(\frac{\partial U}{\partial n_i}\right)_{S,\, V,\, n_j} = \left(\frac{\partial G}{\partial n_i}\right)_{T,\, p,\, n_j} = \mu_i$$

∴ Equation (11.20) becomes

$$dG = Vdp - SdT + \sum_{i=1}^{K} \mu_i \, dn_i$$

Similar equations can be obtained for changes in H and F.

Thus

$$dU = TdS - pdV + \sum_{i=1}^{k} \mu_i \, dn_i$$

$$dG = Vdp - SdT + \sum_{i=1}^{K} \mu_i \, dn_i$$

$$dH = TdS + Vdp + \sum_{i=1}^{K} \mu_i \, dn_i \tag{11.21}$$

$$dF = -SdT - pdV + \sum_{i=1}^{K} \mu_i \, dn_i$$

where

$$\mu_i = \left(\frac{\partial U}{\partial n_i}\right)_{S, V, n_j} = \left(\frac{\partial G}{\partial n_i}\right)_{T, p, n_j} = \left(\frac{\partial H}{\partial n_i}\right)_{S, p, n_j} = \left(\frac{\partial F}{\partial n_i}\right)_{T, V, n_j} \quad (11.22)$$

Chemical potential is an intensive property.

Let us consider a homogeneous phase of a multi-component system, for which

$$dU = TdS - pdV + \sum_{i=1}^{K} \mu_i \, dn_i$$

If the phase is enlarged in size, *U, S,* and *V* will increase, whereas *T, p,* and μ will remain the same. Thus

$$\Delta U = T \Delta S - p \Delta V + \Sigma \mu_i \Delta n_i$$

Let the system be enlarged to K-times the original size. Then

$$\Delta U = KU - U = (K-1)\, U$$

$$\Delta S = KS - S = (K-1)\, S$$

$$\Delta V = (K-1)\, V$$

$$\Delta n_i = (K-1)\, n_i$$

Substituting

$$(K-1)\, U = T\,(K-1)\, S - p\,(K-1)\, V + \Sigma \mu_i\,(K-1)\, n_i$$

$$\therefore \quad U = TS - pV + \Sigma \mu_i n_i$$

$$\therefore \quad G_{T,p} = \Sigma \mu_i n_i \quad (11.23)$$

Let us now find a relationship if there is a simultaneous change in intensive property. Differentiating equation (11.23)

$$dG = \Sigma n_i \, d\mu_i + \Sigma \mu_i \, dn_i \quad (11.24)$$

at constant *T* and *p,* with only μ changing.

When *T* and *p* change

$$dG = -SdT + Vdp + \Sigma \mu_i \, dn_i \quad (11.25)$$

Combining equations (11.24) and (11.25)

$$-SdT + Vdp - \Sigma n_i \, d\mu_i = 0 \quad (11.26)$$

This is known as *Gibbs-Duhem equation,* which shows the necessary relationship for simultaneous changes in *T, p,* and μ.

Since

$$G_{T,p} = \Sigma \mu_i n_i = \mu_1 n_1 + \mu_2 n_2 \ldots + \mu_K n_K$$

For a phase consisting of only one constituent

$$G = \mu \, n$$

or

$$\mu = \frac{G}{n} = g$$

i.e. the chemical potential is the molar Gibbs function and is a function of T and p only.

For a single phase, multi-component system, μ_i is a function of T, p, and the mole fraction x_i.

11.10 CONDITIONS OF EQUILIBRIUM OF A HETEROGENEOUS SYSTEM

Let us consider a heterogeneous system of volume V, in which several homogeneous phases ($\phi = a, b, \ldots, r$) exist in equilibrium. Let us suppose that each phase consists of i ($= 1, 2, \ldots, C$) constituents and that the number of constituents in any phase is different from the others.

Within each phase, a change in internal energy is accompanied by a change in entropy, volume and composition, according to

$$dU_\phi = T_\phi\, dS_\phi - p_\phi\, dV_\phi + \sum_{i=1}^{C} (\mu_i\, dn_i)_\phi$$

A change in the internal energy of the entire system can, therefore, be expressed as

$$\sum_{\phi=a}^{r} dU_\phi = \sum_{\phi=a}^{r} T_\phi\, dS_\phi - \sum_{\phi=a}^{r} p_\phi\, dV_\phi + \sum_{\phi=a}^{r} \sum_{i=1}^{C} (\mu_i\, dn_i)_\phi \tag{11.27}$$

Also, a change in the internal energy of the entire system involves changes in the internal energy of the constituent phases.

$$dU = dU_a + dU_b + \ldots + dU_r = \sum_{\phi=a}^{r} dU_\phi$$

Likewise, changes in the volume, entropy, or chemical composition of the entire system result from contributions from each of the phases

$$dV = dV_a + dV_b, + \ldots + dV_r = \sum_{\phi=a}^{r} dV_\phi$$

$$dS = dS_a + dS_b + \ldots + dS_r = \sum_{\phi=a}^{r} dS_\phi$$

$$dn = dn_a + dn_b + \ldots + dn_r = \sum_{\phi=a}^{r} dn_\phi$$

In a closed system in equilibrium, the internal energy, volume, entropy, and mass are constant.

$\therefore \quad dU = dV = dS = dn = 0$

or
$$\begin{aligned} dU_a &= -(dU_b + \dots + dU_r) = -\sum_j dU_j \\ dV_a &= -\sum_j dV_j \\ dS_a &= -\sum_j dS_j \\ dn_a &= -\sum_j dn_j \end{aligned} \tag{11.28}$$

where subscript j includes all phases except phase a.

Equation (11.27) can be written in terms of j independent variables and the dependent variable a (equation 11.28).

$$(T_a\, dS_a + \sum_j T_j\, dS_j) - (p_a\, dV_a + \sum_j p_j\, dV_j) + [\sum_j (\mu_i\, dn_i)_a + \sum_j \sum_j (\mu_i\, dn_i)_j] = 0$$

Substituting from equation (11.28)

$$(-T_a \sum_j dS_j + \sum_j T_j\, dS_j) - (-p_a \sum_j dV_j + \sum_j p_i\, dV_j) + [-\sum_j \sum_j \mu_{ia}\, dn_{ij} + \sum_j \sum_j (\mu_i\, dn_i)_j] = 0$$

where subscript $i\,a$ refers to component i of phase a.

Rearranging and combining the coefficients of the independent variables, dS_j, dV_j, and dn_j, gives

$$\sum_j (T_j - T_a)\, dS_j - \sum_j (p_j - p_a)\, dV_j + \sum_j \sum_j (\mu_{ij} - \mu_{ia})\, dn_{ij} = 0$$

But since dS_j, dV_j, and dn_j are independent, their coefficients must each be equal to zero.

$$\therefore \quad T_j = T_a, \; p_j = p_a, \; \mu_{ij} = \mu_{ia} \tag{11.29}$$

These equations represent conditions that exist when the system is in thermal, mechanical, and chemical equilibrium. The temperature and pressure of phase a must be equal to those of all other phases, and the chemical potential of the ith component in phase a must be equal to the chemical potential of the same component in all other phases.

11.11 GIBBS PHASE RULE

Let us consider a heterogeneous system of C chemical constituents which do not combine chemically with one another. Let us suppose that there are ϕ phases, and every constituent is present in each phase. The constituents are denoted by subscripts and the phases by superscripts. The Gibbs function of the whole heterogeneous system is

$$G_{T,\,p} = \sum_{i=1}^{C} n_i^{(1)}\, \mu_i^{(1)} + \sum_{i=1}^{C} n_i^{(2)}\, \mu_i^{(2)} + \dots + \sum_{i=1}^{C} n_i^{(\phi)}\, \mu_i^{(\phi)}$$

G is a function of T, p, and the n's of which there are $C\phi$ in number. Since there are no chemical reactions, the only way in which the n's may change is by the transport of the constituents from one phase to another. In this case the total number of moles of each constituent will remain constant.

$$n_1^{(1)} + n_1^{(2)} + \ldots + n_1^{(\phi)} = \text{constant}$$

$$n_2^{(1)} + n_2^{(3)} + \ldots + n_2^{(\phi)} = \text{constant}$$

$$\ldots\ldots\ldots\ldots\ldots\ldots\ldots\ldots\ldots\ldots$$

$$n_C^{(1)} + n_C^{(2)} + \ldots + n_C^{(\phi)} = \text{constant}$$

These are the *equations of constraint.*

At chemical equilibrium, G will be rendered a minimum at constant T and p, subject to these equations of constraint. At equilibrium, from equation (11.29).

$$\mu_{ij} = \mu_{ia}$$

$$\mu_1^{(1)} = \mu_1^{(2)} = \ldots\ldots = \mu_1^{(\phi)}$$

$$\mu_2^{(1)} = \mu_2^{(2)} = \ldots\ldots = \mu_2^{(\phi)}$$

$$\ldots\ldots\ldots\ldots\ldots\ldots \quad \ldots\ldots\ldots\ldots \tag{11.30}$$

$$\mu_C^{(1)} = \mu_C^{(2)} = \ldots\ldots = \mu_C^{(\phi)}$$

These are known as the *equations of phase equilibrium.* The equations of the phase equilibrium of one constituent are $(\phi - 1)$ in number. Therefore, for C constituents, there are $C\ (\phi - 1)$ such equations.

When equilibrium has been reached, there is no transport of matter from one phase to another. Therefore, in each phase, $\Sigma x = 1$. For ϕ phases, there are ϕ such equations available.

The state of the system at equilibrium is determined by the temperature, pressure, and $C\ \phi$ mole fractions. Therefore

Total number of variables $= C\phi + 2$.

Among these variables, there are $C\ (\phi - 1)$ equations of phase equilibrium and ϕ equations of $\Sigma x = 1$ type. Therefore

Total number of equations $= C\ (\phi - 1) + \phi$

If the number of variables is equal to the number of equations, the system is *nonvariant.* If the number of variables exceeds the number of equations by one, then the system is called *monovariant* and is said to have a variance of 1.

The excess of variables over equations is called the *variance, f.* This

$$f = (C\phi + 2) - [C\ (\phi - 1) + \phi]$$

or

$$f = C - \phi + 2$$

This is known as the *Gibbs Phase Rule.* The variance 'f' is also known as the *degree of freedom.*

For a pure substance existing in a single phase, $C = 1$, $\phi = 1$, and therefore, the variance is 2. There are two properties required to be known to fix up the state of the system at equilibrium.

If $C = 1$, $\phi = 2$, then $f = 1$, i.e. only one property is required to fix up the state of a single-component two-phase system.

If $C = 1$, $\phi = 3$, then $f = 0$. The state is thus unique for a substance, and refers to the triple point where all the three phases exist in equilibrium.

11.12 TYPES OF EQUILIBRIUM

The thermodynamic potential which controls equilibrium in a system depends on the particular *constraints* imposed on the system. Let $đQ$ be the amount of heat transfer involved between the system and the reservoir in an infinitesimal irreversible process (Fig. 11.8). Let dS denote the entropy change of the system and dS_0 the entropy change of the reservoir. Then, from the entropy principle

$$dS_0 + dS > 0$$

Since $\quad dS_0 = -\frac{đQ}{T}, \quad -\frac{đQ}{T} + dS > 0$

or $\quad đQ - TdS < 0$

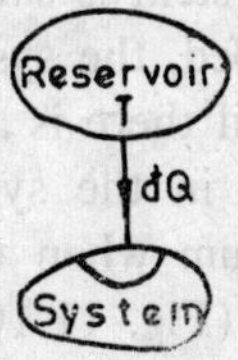

Fig. 11.8 Heat interaction between a system and its surroundings

During the infinitesimal process, the internal energy of the system changes by an amount dU, and an amount of work pdV is performed. So, by the first law

$$đQ = dU + pdV$$

Thus the inequality becomes

$$dU + pdV - TdS < 0 \tag{11.32}$$

If the constraints are constant U and V, then the equation (11.32) reduces to

$$dS > 0$$

The condition of constant U and V refers to an isolated system. Therefore, entropy is the critical parameter to determine the state of thermodynamic equilibrium of an isolated system. The entropy of an isolated system always increases and reaches a maximum value when equilibrium is reached.

If the constraints imposed on the system are constant T and V, the equation (11.32) reduces to

$$dU - d(TS) < 0$$

or $\quad d(U - TS) < 0$

$\therefore \quad dF < 0$

which expresses that the Helmholtz function decreases, becoming a minimum at the final equilibrium state.

If the constraints are constant T and p, the equation (11.32) becomes

$$dU + d(pV) - d(TS) < 0$$

$$d(U + pV - TS) < 0$$

$$\therefore \quad dG < 0$$

The Gibbs function of a system at constant T and p decreases during an irreversible process, becoming a minimum at the final equilibrium state. For a system constrained in a process to constant T and p, G is the critical parameter to determine the state of equilibrium.

The thermodynamic potential and the corresponding constrained variables are shown below.

$$\begin{array}{ccc} S & U & V \\ H & & F \\ P & G & T \end{array}$$

This trend of G, F, or S establishes four types of equilibrium, namely (a) stable, (b) neutral, (c) unstable, and (d) metastable.

A system is said to be in a state of *stable* equilibrium if, when the state is perturbed, the system returns to its original state. A system is not in equilibrium if there is a spontaneous change in the state. If there is a spontaneous change in the system, the entropy of the system increases and reaches a maximum when the equilibrium condition is reached (Fig. 11.9). Both A and B (Fig. 11.10) are assumed to be at the same temperature T. Let there

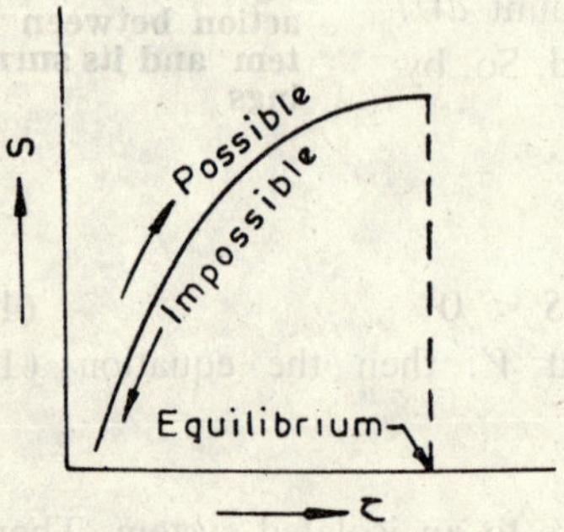

Fig. 11.9 Possible process for an isolated system.

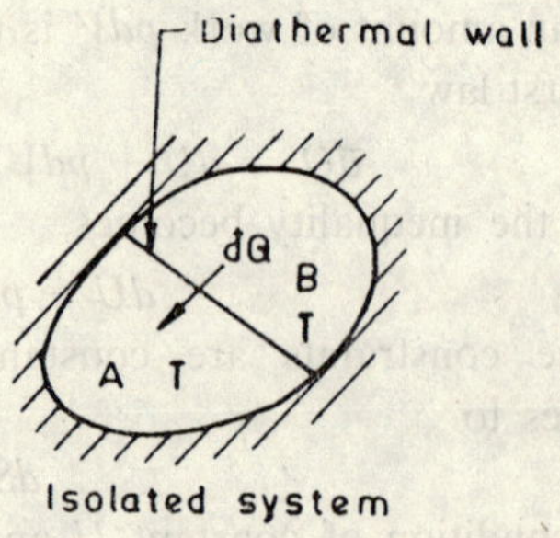

Fig. 11.10 Spontaneous changes in A and B due to heat interaction

be some spontaneous change; the temperature of A rises to $T + dT_1$, and that of B decreases to $T - dT_2$. For simplicity, let the heat capacities of the bodies be the same, so that $dT_1 = dT_2 = dT$. If $đQ$ is the heat interaction involved, then the entropy change

$$dS_A = \frac{đQ}{T + dT}, \quad dS_B = -\frac{đQ}{T - dT}$$

$$\therefore \quad dS = dS_A + dS_B = đQ \left[\frac{1}{T + dT} - \frac{1}{T - dT}\right]$$

$$= -\frac{2.dT}{T^2} \cdot đQ$$

So there is a decrease in entropy for the isolated system of A and B together. It is thus clear that the variation in temperature dT cannot take place. The system, therefore, exists in a stable equilibrium condition. Perturbation of the state produces an absurd situation and the system must revert to the original stable state. It may be observed:

If for all the possible variations in state of the isolated system, there is a negative change in entropy, then the system is in stable equilibrium.

$dS)_{U,\,V} > 0$	Spontaneous change
$dS)_{U,\,V} = 0$	Equilibrium
$dS)_{U,\,V} < 0$	Criterion of stability

Similarly

$dG)_{p,\,T} < 0,\ dF)_{T,\,V} < 0$	Spontaneous change
$dG)_{p,\,T} = 0,\ dF)_{T,\,V} = 0$	Equilibrium
$dG)_{p,\,T} > 0,\ dF)_{T,\,V} > 0$	Criterion of stability

A system is in a state of stable equilibrium if, for any finite variation of the system at constant T and p, G increases, i.e. the stable equilibrium state corresponds to the minimum value of G.

A system is said to be in a state of *neutral equilibrium* when the thermodynamic criterion of equilibrium (G, V, S, U, or H) remains at constant value for all possible variations of finite magnitude. If perturbed, the system does not revert to the original state.

For a system at constant T and p, the criterion of neutral equilibrium is

$$\delta G_{T,\,p} = 0$$

Similarly

$$\delta F_{T,\,V} = 0,\ \delta H_{S,\,p} = 0,\ \delta U_{S,\,V} = 0,\ \delta S_{U,V} = 0$$

A system is in a state of *unstable equilibrium* when the thermodynamic criterion is neither a minimum nor a constant value for all possible variations in the system. If the system is in an unstable equilibrium, there will be a spontaneous change accompanied by

$$\delta G_{T,\,p} < 0,\ \delta F_{T,\,V} < 0,\ \delta U_{S,\,V} < 0,\ \delta H_{S,\,p} < 0,\ \delta S_{U,\,V} > 0$$

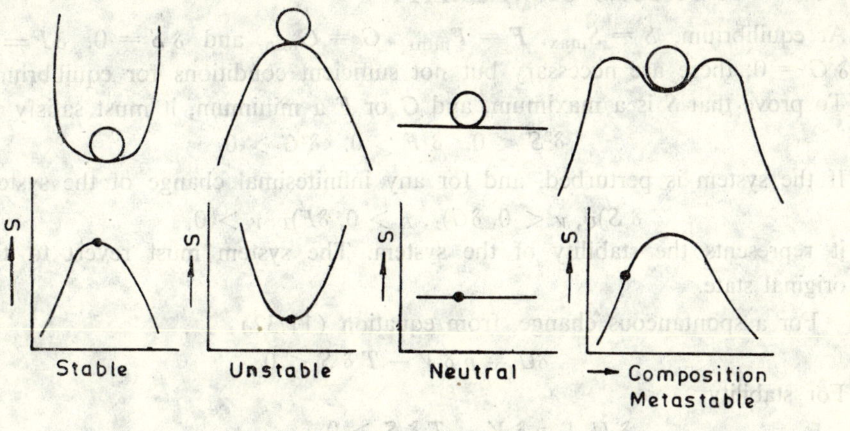

Fig. 11.11 Types of equilibrium

A system is in a state of *metastable equilibrium* if it is stable to small but not to large disturbances. A mixture of oxygen and hydrogen is in a metastable equilibrium. A little spark may start a chemical reaction. Such a mixture is not in its most stable state, even though in the absence of a spark it appears to be stable.

Figure 11.11 shows different types of equilibrium together with their mechanical analogies. S has been used as the criterion for equilibrium.

11.13 LOCAL EQUILIBRIUM CONDITIONS

Let an arbitrary division of an isolated system be considered, such that

$$S = S_1 + S_2,\; U = U_1 + U_2$$

Then for equilibrium, it must satisfy the condition

$$(\delta S)_{U,V} = 0$$

to first order in small displacements (otherwise δS could be made positive because of higher order terms). Now to the first order in a very small change

$$\delta_1 S = \left(\frac{\partial S}{\partial U_1}\right)_V \delta U_1 + \left(\frac{\partial S}{\partial U_2}\right)_V \delta U_2 + \left(\frac{\partial S}{\partial V_1}\right)_U \delta V_1 + \left(\frac{\partial S}{\partial V_2}\right)_U \delta V_2$$

Now $\quad TdS = dU + pdV$

$$\therefore \quad \left(\frac{\partial S}{\partial U}\right)_V = \frac{1}{T},\; \left(\frac{\partial S}{\partial V}\right)_U = \frac{p}{T}$$

$$\therefore \quad \delta_1 S = \frac{1}{T_1}\delta U_1 + \frac{1}{T_2}\delta U_2 + \frac{p_1}{T_1}\delta V_1 + \frac{p_2}{T_2}\delta V_2$$

Again $\quad \delta U_1 = -\delta U_2$ and $\delta V_1 = -\delta V_2$

$$\therefore \quad \delta_1 S = \left(\frac{1}{T_1} - \frac{1}{T_2}\right)\delta U_1 + \left(\frac{p_1}{T_1} - \frac{p_2}{T_2}\right)\delta V_1 + \text{Second order terms}$$

When $\quad \delta_1 S = 0$, at equilibrium

$$T_1 = T_2,\; p_1 = p_2$$

11.14 CONDITIONS OF STABILITY

At equilibrium, $S = S_{\max}$, $F = F_{\min}$, $G = G_{\min}$, and $\delta S = 0$, $\delta F = 0$; $\delta G = 0$; these are necessary but not sufficient conditions for equilibrium. To prove that S is a maximum, and G or F a minimum, it must satisfy

$$\delta^2 S < 0,\quad \delta^2 F > 0,\quad \delta^2 G > 0$$

If the system is perturbed, and for any infinitesimal change of the system

$$\delta S)_{U,V} < 0,\; \delta G)_{p,T} > 0,\; \delta F)_{T,V} > 0$$

it represents the stability of the system. The system must revert to the original state.

For a spontaneous change, from equation (11.32)

$$\delta U + p\,\delta V - T\,\delta S < 0$$

For stability

$$\delta U + p\,\delta V - T\,\delta S > 0$$

Let us choose $U = U\ (S, V)$ and expand δU in powers of δV and δS.

$$\delta U = \left(\frac{\partial U}{\partial S}\right)_V \delta S + \frac{1}{2}\left(\frac{\partial^2 U}{\partial S^2}\right)_V (\delta S)^2 + \left(\frac{\partial U}{\partial V}\right)_S \delta V$$

$$+ \frac{1}{2}\left(\frac{\partial^2 U}{\partial V^2}\right)_S (\delta V)^2 + \frac{\partial^2 U}{\partial V \cdot \partial S} \cdot \delta V \cdot \delta S + \ldots$$

$$= T\,\delta S - p\,\delta V + \frac{1}{2}\left(\frac{\partial^2 U}{\partial S^2}\right)_V (\delta S)^2$$

$$+ \frac{1}{2}\left(\frac{\partial^2 U}{\partial V^2}\right)_S (\delta V)^2 + \frac{\partial^2 U}{\partial V \cdot \partial S} \cdot \delta V \cdot \delta S + \ldots$$

The third order and higher terms are neglected.

Since $\delta U + p\,\delta v - T\,\delta S > 0$, it must satisfy the conditions given below

$$\left(\frac{\partial^2 U}{\partial S^2}\right)_V > 0,\ \left(\frac{\partial^2 U}{\partial V^2}\right)_S > 0,\ \frac{\partial^2 U}{\partial V \cdot \partial S} > 0$$

These inequalities indicate how the signs of some important physical quantities become restricted for a system to be stable.

Since $\left(\dfrac{\partial U}{\partial S}\right)_V = T$

$$\left(\frac{\partial^2 U}{\partial S^2}\right)_V = \left(\frac{\partial T}{\partial S}\right)_V = \frac{T}{C_v}$$

$\therefore$ $\dfrac{T}{C_v} > 0$

Since $T > 0\ K$

$\therefore$ $C_v > 0$

which is the condition of *thermal stability*.

Also $\left(\dfrac{\partial U}{\partial V}\right)_S = -p$

$$\left(\frac{\partial^2 U}{\partial V^2}\right)_S = -\left(\frac{\partial p}{\partial V}\right)_S$$

$\therefore$ $\left(\dfrac{\partial p}{\partial V}\right)_S < 0$

i.e. the adiabatic bulk modulus must be negative.

Similarly, if $F = F\ (T.\ V)$, then by Taylor's expansion, and using appropriate substitution

$$\delta F = -S\,\delta T - p\delta V + \frac{1}{2}\left(\frac{\partial^2 F}{\partial V^2}\right)_T (\delta V)^2 + \frac{1}{2}\left(\frac{\partial^2 F}{\partial T^2}\right)_S (\delta T)^2$$

$$+ \frac{\partial^2 \cdot F}{\partial V \cdot \partial T} \cdot \delta V \cdot \delta T + \ldots$$

For stability

$$\delta F + S\,\delta T + p\,\delta V > 0$$

$\therefore$ $\left(\dfrac{\partial^2 F}{\partial V^2}\right)_T > 0$

Again $\left(\frac{\partial F}{\partial V}\right)_T = -p$

$\therefore$ $\left(\frac{\partial^2 F}{\partial V^2}\right)_T = -\left(\frac{\partial p}{\partial V}\right)_T$

$\therefore$ $\left(\frac{\partial p}{\partial V}\right)_T < 0$

which is known as the condition of *mechanical stability*. The isothermal bulk modulus must also be negative.

EXAMPLE 11.1

(a) Derive the equation

$$\left(\frac{\partial C_p}{\partial p}\right)_T = -T\left(\frac{\partial^2 V}{\partial T^2}\right)_p$$

(b) Prove that C_p of an ideal gas is a function of T only.

(c) In the case of a gas obeying the equation of state

$$\frac{pv}{RT} = 1 + B'p$$

where B' is a function of T only, show that

$$C_p = -\bar{R}\,T\,p\,\frac{d^2}{dT^2}(B'T) + (C_p)_0$$

where $(C_p)_0$ is the value at very low pressures.

Solution

(a) $$C_p = T\left(\frac{\partial S}{\partial T}\right)_p$$

$$\left(\frac{\partial C_p}{\partial p}\right)_T = T\frac{\partial^2 S}{\partial T \cdot \partial p}$$

Now $\left(\frac{\partial S}{\partial p}\right)_T = -\left(\frac{\partial V}{\partial T}\right)_p$, by Maxwell's relation

$\therefore$ $$\frac{\partial^2 S}{\partial p \cdot \partial T} = -\left(\frac{\partial^2 V}{\partial T^2}\right)_p$$

$\therefore$ $$\left(\frac{\partial C_p}{\partial p}\right)_T = -T\left(\frac{\partial^2 V}{\partial T^2}\right)_p$$ Proved.

(b) For an ideal gas

$$V = \frac{n\bar{R}T}{p}$$

$$\left(\frac{\partial V}{\partial T}\right)_p = \frac{n\bar{R}}{p} \text{ and } \left(\frac{\partial^2 V}{\partial T^2}\right)_p = 0$$

$\therefore$ $\left(\frac{\partial C_p}{\partial p}\right)_T = 0$, i e. C_p is a function of T alone.

(c) $$\frac{pv}{\bar{R}T} = 1 + B'p$$

$$\therefore \quad B'p = \frac{pv}{\bar{R}T} - 1$$

or $$B'T = \frac{T}{p}\left(\frac{pv}{\bar{R}T} - 1\right) = \left(\frac{v}{\bar{R}} - \frac{T}{p}\right)$$

$$\left[\frac{\partial}{\partial T}(B'T)\right]_p = +\frac{1}{\bar{R}}\left(\frac{\partial v}{\partial T}\right)_p$$

$$\left[\frac{\partial^2(B'T)}{\partial T^2}\right]_p = \frac{1}{\bar{R}}\left(\frac{\partial^2 v}{\partial T^2}\right)_p = -\frac{1}{\bar{R}T}\left(\frac{\partial C_p}{\partial p}\right)_T$$

∴ On integration

$$C_p = -\bar{R}\,T\,p\frac{d^2}{dT^2}(B'T) + C_{p0}$$

where C_{po} (integration constant) is the value of C_p at very law values of pressure.

EXAMPLE 11.2

The Joule-Kelvin coefficient μ is a measure of the temperature change during a throttling process. A similar measure of the temperature change produced by an isentropic change of pressure is provided by the coefficient μ_S, where

$$\mu_S = \left(\frac{\partial T}{\partial p}\right)_S$$

Prove that

$$\mu_S - \mu = \frac{V}{C_p}$$

Solution

The Joule-Kelvin coefficient, μ, is given by

$$\mu = \frac{T\left(\frac{\partial V}{\partial T}\right)_p - V}{C_p}$$

Since $C_p = T\left(\frac{\partial S}{\partial T}\right)_p$ and by Maxwell's relation

$$\left(\frac{\partial V}{\partial T}\right)_p = -\left(\frac{\partial S}{\partial p}\right)_T$$

$$\mu = \frac{-T\left(\frac{\partial S}{\partial p}\right)_T}{T\left(\frac{\partial S}{\partial T}\right)_p} - \frac{V}{C_p}$$

$$\therefore \quad \mu = -\left(\frac{\partial S}{\partial p}\right)_T\left(\frac{\partial T}{\partial S}\right)_p - \frac{V}{C_p}$$

Since $$\left(\frac{\partial S}{\partial p}\right)_T \left(\frac{\partial T}{\partial S}\right)_p \left(\frac{\partial p}{\partial T}\right)_S = -1$$

$$\therefore \quad \mu = + \mu_S - \frac{V}{C_p}$$

$$\therefore \quad \mu_s - \mu = \frac{V}{C_p}$$ Proved.

Alternative method:

From the second *Tds* equation

$$Td_S = C_p \, dT - T\left(\frac{\partial V}{\partial T}\right)_p dp$$

$$\left(\frac{\partial T}{\partial p}\right)_S = \mu_s = \frac{T}{C_p}\left(\frac{\partial V}{\partial T}\right)_p$$

Now $$\mu = \frac{1}{C_p}\left[T\left(\frac{\partial V}{\partial T}\right)_p - V\right]$$

$$\therefore \quad \mu_s - \mu = \frac{V}{C_p}$$ Proved.

Example 11.3

The vapour pressure, in mm of mercury, of solid ammonia is given by

$$\ln p = 23.03 - \frac{3754}{T}$$

and that of liquid ammonia by

$$\ln p = 19.49 - \frac{3063}{T}$$

(a) What is the temperature of the triple point? What is the pressure? (b) What are the latent heats of sublimation and vaporization? (c) What is the latent heat of fusion at the triple point?

Solution

At the triple point, the saturated solid and saturated liquid lines meet.

$$23.03 - \frac{3754}{T} = 19.49 - \frac{3063}{T}$$

$$\therefore \quad T = 195.2 \text{ K}$$ Ans. (a).

$$\ln p = 23.03 - \frac{3754}{195.2}$$

$$\ln p = 3.80$$

$$p = 44.67 \text{ mm Hg}$$ Ans.

With the assumptions, $v''' \gg v'$ and $v''' \sim \frac{\bar{R}T}{p}$,

Clausius-Clapeyron equation reduces to

$$\frac{dp}{dT} = \frac{p}{\bar{R}\,T^2} \cdot l_{sub.}$$

where $l_{sub.}$ is the latent heat of sublimation.

The vapour pressure of solid ammonia is given by

$$\ln p = 23.03 - \frac{3754}{T}$$

$$\therefore \quad \frac{1}{p} \cdot \frac{dp}{dT} = \frac{3754}{T^2}$$

$$\therefore \quad \frac{dp}{dT} = 3754\,\frac{p}{T^2} = \frac{p}{\bar{R}T^2} \cdot l_{sub.}$$

$\therefore \quad l_{sub.} = 3754 \times 8.3143 = 31{,}200$ kJ/kgmol Ans. (b)

The vapour pressure of liquid ammonia is given by

$$\ln p = 19.49 - \frac{3063}{T}$$

$$\therefore \quad \frac{dp}{dT} = 3063\,\frac{p}{T^2} = \frac{p}{\bar{R}\,T^2}\,l_{vap.}$$

where $l_{vap.}$ is the latent heat of vaporization.

$\therefore \quad l_{vap.} = 3063 \times 8.3143 = 25{,}500$ kJ/kgmol Ans. (b)

At the triple point

$$l_{sub.} = l_{vap.} + l_{fu.}$$

where $l_{fu.}$ is the latent heat of fusion.

$\therefore \quad l_{fu.} = l_{sub.} - l_{vap.}$

$= 31{,}200 - 25{,}500 = 5{,}700$ kJ/kgmol Ans. (c)

EXAMPLE 11.4

Explain why the specific heat of a saturated vapour may be negative.

Solution

As seen in Fig. Ex. 11.4, if heat is transferred along the saturation line, there is a decrease in temperature. The slope of the saturated vapour line

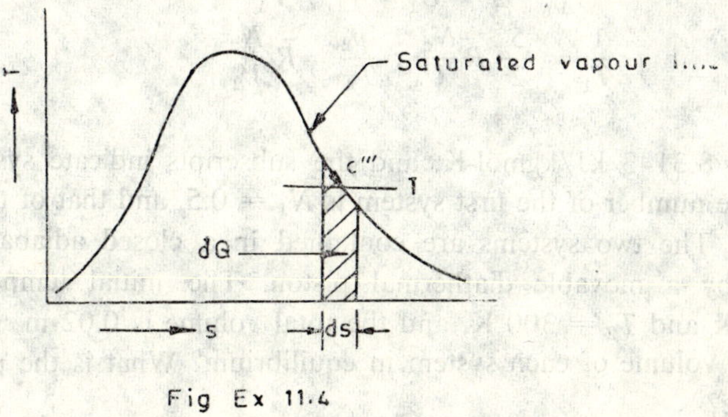

Fig Ex 11.4

is negative, i.e. when dS is positive, dT is negative. Therefore, the specific heat at constant saturation

$$C'''_{sat.} = T\left(\frac{dS'''}{dT}\right)$$

is negative. From the second TdS equation

$$TdS = C_p\, dT - T\left(\frac{\partial V}{\partial T}\right)_p dp$$

$$T\frac{dS'''}{dT} = C_p - T\left(\frac{\partial V'''}{\partial T}\right)_p \left(\frac{dp}{dT}\right)_{sat.}$$

$$= C_p - T \cdot \frac{n\bar{R}}{p} \cdot \frac{l_{vap.}}{T(V''' - V'')} \quad \text{[using } pV''' = n\bar{R}T \text{ and Clapeyron's equation]}$$

$$C'''_{sat.} = C_p - \frac{V'''}{T} \cdot \frac{l_{vap.}}{V'''} \quad [\because\ V''' \gg V'']$$

$$\therefore \quad C'''_{sat.} = C_p - \frac{l_{vap.}}{T}$$

Now the value of $l_{vap.}/T$ for common substances is about 83.74 J/gmol-K or 20 cal/gmol-K (*Trouton's rule*), where C_p is less than 41.87 J/gmol-K or 10 cal/gmol-K. Therefore, $C'''_{sat.}$ can be negative. Proved.

EXAMPLE 11.5

(a) Establish the condition of equilibrium of a closed composite system consisting of two simple systems separated by a movable diathermal wall that is impervious to the flow of matter.

(b) If the wall were rigid and diathermal, permeable to one type of material, and impermeable to all others, state the condition of equilibrium of the composite system.

(c) Two particular systems have the following equations of state

$$\frac{1}{T_1} = \frac{3}{2}\bar{R}\frac{N_1}{U_1}, \quad \frac{p_1}{T_1} = \frac{N_1}{V_1}\bar{R}$$

and

$$\frac{1}{T_1} = \frac{5}{2}\bar{R}\frac{N_2}{U_2}, \quad \frac{p_2}{T_2} = \bar{R}\frac{N_2}{V_2}$$

where $\bar{R} = 8.3143$ kJ/kgmol-K, and the subscripts indicate systems 1 and 2. The mole number of the first system is $N_1 = 0.5$, and that of the second is $N_2 = 0.75$. The two systems are contained in a closed adiabatic cylinder, separated by a movable diathermal piston. The initial temperatures are $T_1 = 200$ K and $T_2 = 300$ K, and the total volume is 0.02 m³. What is the energy and volume of each system in equilibrium? What is the pressure and temperature?

Solution

For the composite system, as shown in Fig. 11.5(a)

$$U_1 + U_2 = \text{constant}$$
$$V_1 + V_2 = \text{constant}$$

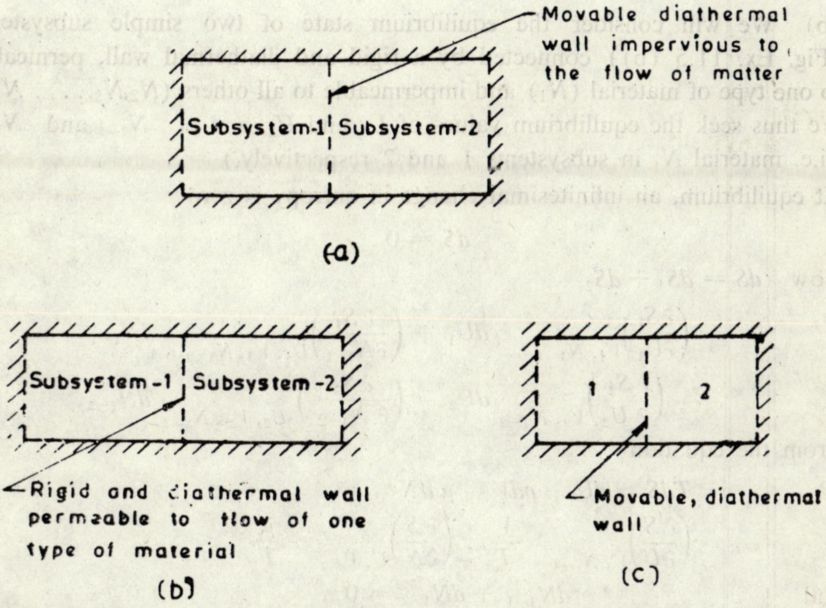

Fig. Ex. 11·5

The values of U_1, U_2, V_1, and V_2 would chnage in such a way as to maximize the value of entropy. Therefore, when the equilibrium condition is achieved

$$dS = 0$$

for the whole system. Since

$$S = S_1 + S_2$$
$$= S_1 (U_1, V_1, \ldots, N_{k1} \ldots) + S_2 (U_2, V_2, \ldots, N_{k2} \ldots)$$

$$\therefore \quad dS = \left(\frac{\partial S_1}{\partial U_1}\right)_{V_1, \ldots, N_{k1} \ldots} dU_1 + \left(\frac{\partial S_1}{\partial V_1}\right)_{U_1, \ldots, N_{k1} \ldots} dV_1$$

$$+ \left(\frac{\partial S_2}{\partial U_2}\right)_{V_2 \ldots, N_{k2} \ldots,} dU_2 + \left(\frac{\partial S_2}{\partial V_2}\right)_{V_2, \ldots N_{k2} \ldots} dV_2$$

$$= \frac{1}{T_1} dU_1 + \frac{p_1}{T_1} dV_1 + \frac{1}{T_2} dU_2 + \frac{p_2}{T_2} dV_2$$

Since $dU_1 + dU_2 = 0$ and $dV_1 + dV_2 = 0$

$$\therefore \quad dS = \left(\frac{1}{T_1} - \frac{1}{T_2}\right) dU_1 + \left(\frac{p_1}{T_1} - \frac{p_2}{T_2}\right) dV_1 = 0$$

Since the expression must vanish for arbitrary and independent values of dU_1 and dV_1

$$\frac{1}{T_1} - \frac{1}{T_2} = 0 \text{ and } \frac{p_1}{T_1} - \frac{p_2}{T_2} = 0$$

or $$p_1 = p_2 \text{ and } T_1 = T_2$$

∴ These are the conditions of mechanical and thermal equilibrium.

(b) We will consider the equilibrium state of two simple subsystems (Fig. Ex. 11.5 (b)) connected by a rigid and diathermal wall, permeable to one type of material (N_1) and impermeable to all others ($N_2, N_3, \ldots. N_r$). We thus seek the equilibrium values of U_1 and U_2, and of N_{1-1} and N_{1-2} (i.e. material N_1 in subsystems 1 and 2 respectively.)

At equilibrium, an infinitesimal change in entropy is zero

$$dS = 0$$

Now $dS = dS_1 + dS_2$

$$= \left(\frac{\partial S_1}{\partial U_1}\right)_{V_1, N_{1-1}, \ldots} dU_1 + \left(\frac{\partial S_1}{\partial N_{1-1}}\right)_{U_1, V_1, N_{1-2} \ldots} dN_{1-1}$$

$$+ \left(\frac{\partial S_2}{\partial U_2}\right)_{V_2, N_{1-2}} dU_2 + \left(\frac{\partial S_2}{\partial N_{1-2}}\right)_{U_2, V_2, N_{2-2}} dN_{1-2}$$

From the equation

$$TdS = dU + pdV - \mu\, dN$$

$$\therefore \quad \left(\frac{\partial S}{\partial U}\right)_{V, N, \ldots} = \frac{1}{T}, \quad \left(\frac{\partial S}{\partial N}\right)_{U, V} = -\frac{\mu}{T}$$

and $$dN_{1-1} + dN_{1-2} = 0$$

$$dU_1 + dU_2 = 0$$

$$dS = \left(\frac{1}{T_1} - \frac{1}{T_2}\right) dU_1 - \left(\frac{\mu_{1-1}}{T_1} - \frac{\mu_{1-2}}{T_2}\right) dN_{1-1} = 0$$

As dS must vanish for arbitrary values of both dU_1 and dN_{1-1}

$$T_1 = T_2$$

$$\mu_{1-1} = \mu_{1-2}$$

which are the conditions of thermal and chemical equilibrium.

(c) $N_1 = 0.5$ gmol, $N_2 = 0.75$ gmol

$$T_{i-1} = 200 \text{ K}, \quad T_{i-2} = 300 \text{ K}$$

$$V = V_1 + V_2 = 0.02 \text{ m}^3$$

$$U_1 + U_2 = \text{constant}$$

$$\Delta U_1 + \Delta U_2 = 0$$

Let T_f be the final temperature (Fig. Ex. 11.5 (c))

$$(U_{f-1} - U_{i-1}) = -(U_{f-2} - U_{i-2})$$

$$\frac{3}{2}\bar{R}N_1 (T_f - T_{i-1}) = -\frac{3}{2}\bar{R}N_2 (T_f - T_{i-2})$$

$$0.5\ (T_f - 200) = -0.75\ (T_f - 300)$$

$$\therefore \quad 1.25\ T_f = 325$$

$$\text{or } T_f = 260 \text{ K} \qquad \text{Ans.}$$

$$U_{f-1} = \frac{3}{2}\bar{R}N_1\ T_f = \frac{3}{2} \times 8.3143 \times 0.5 \times 10^{-3} \times 260 = 1.629 \text{ kJ}$$

$$U_{f-2} = \frac{3}{2} \times 8.3143 \times 0.75 \times 10^{-3} \times 260 = 2.430 \text{ kJ} \qquad \text{Ans.}$$

$$V_{f-2} = \frac{\bar{R}N_1T_{f-1}}{p_{f-1}} \qquad \text{At equilibrium}$$

$$p_{f-1} = p_{f-2} = p_f$$

$$V_{f-2} = \frac{\bar{R}N_2T_{f-2}}{p_{f-2}} \qquad T_{f-1} = T_{f-2} = T_f$$

$$V_{f-1} + V_{f-2} = \frac{\bar{R}T_f}{p_f}(N_1 + N_2) = 0.02 \text{ m}^3$$

$$\frac{8.3143 \times 260}{p_f} \times 1.25 \times 10^{-3} = 0.02 \text{ m}^3$$

$$\therefore \quad p_f = \frac{8.3143 \times 260 \times 1.25 \times 10^{-3}}{0.02} \text{ kN/m}^2$$

$$= 135 \text{ kN/m}^2 = 1.35 \text{ bar}$$

$$\therefore \quad V_{f-1} = \frac{8.3143 \times 0.5 \times 10^{-3} \times 260}{135} = 0.008 \text{ m}^3$$

$$\therefore \quad V_{f-2} = 0.02 - 0.008 = 0.012 \text{ m}^3 \qquad \text{Ans.}$$

EXAMPLE 11.6

Show that for a Van der Waals' gas

(a) $\left(\frac{\partial c_v}{\partial v}\right)_T = 0$

(b) $(s_2 - s_1)_T = R \ln \frac{v_2 - b}{v_1 - b}$

(c) $T\ (v - b)^{R/c_v} =$ constant, for an isentropic

(d) $c_p - c_v = \frac{T}{1 - 2a\ (v-b)^2/RTv^3}$

(e) $(h_2 - h_1)_T = (p_2\, v_2 - p_1\, v_1) + a \left(\frac{1}{v_1} - \frac{1}{v_2}\right)$

Solution

(a) From the energy equation (11.13)

$$\left(\frac{\partial U}{\partial V}\right)_T = T\left(\frac{\partial p}{\partial T}\right)_V - p$$

$$\frac{\partial^2 U}{\partial V \cdot \partial T} = T\left(\frac{\partial^2 p}{\partial T^2}\right) + \left(\frac{\partial p}{\partial T}\right)_V - \left(\frac{\partial p}{\partial T}\right)_V$$

$$\therefore \quad \frac{\partial^2 U}{\partial V \cdot \partial T} = T\left(\frac{\partial^2 p}{\partial T^2}\right)_V$$

$$C_v = \left(\frac{\partial^2 U}{\partial T}\right)_V$$

$$\therefore \quad \left(\frac{\partial C_v}{\partial V}\right)_T = \frac{\partial^2 U}{\partial T \cdot \partial V} = T\left(\frac{\partial^2 p}{\partial T^2}\right)_V = \left(\frac{\partial c_v}{\partial v}\right)_T$$

For a Van der Waals gas

$$\left(p + \frac{a}{v^2}\right)(v - b) = R\,T$$

$$p = \frac{R\,T}{v-b} - \frac{a}{v^2}$$

$$\because \quad \left(\frac{\partial^2 p}{\partial T^2}\right)_V = 0$$

$$\therefore \quad \left(\frac{\partial c_v}{\partial v}\right)_T = 0$$

Proved (a).

$\therefore$ c_v is independent of volume.

(b) From the first Tds equation (11.8)

$$Tds = c_v dT + T\left(\frac{\partial p}{\partial T}\right)_V dv$$

and energy equation (11.13) $\left(\frac{\partial U}{\partial V}\right)_T = T\left(\frac{\partial p}{\partial T}\right)_V - p$

$$ds = c_v \frac{dT}{T} + \frac{1}{T}\left[p + \left(\frac{\partial U}{\partial V}\right)_T\right] dv$$

For Van der Waals gas

$$\left(\frac{\partial U}{\partial V}\right)_T = \frac{a}{v^2}$$

$$\therefore \quad ds = c_v \frac{dT}{T} + \frac{1}{T}\left(p + \frac{a}{v^2}\right) dv$$

$$= c_v \frac{dT}{T} + \frac{R}{v-b} \cdot dv$$

$$\therefore \quad (s_2 - s_1)_T = R \ln \frac{v_2 - b}{v_1 - b}$$

Proved (b).

(c) At constant entropy

$$c_v \frac{dT}{T} + \frac{R}{v-b} dv = 0$$

or

$$\frac{dT}{T} + \frac{R}{c_v} \frac{dv}{v-b} = 0$$

by integration, $T\,(v - b)^{R/C_v}$ = constant

Proved (c).

(d) $c_p - c_v = T\left(\frac{\partial p}{\partial T}\right)_V \left(\frac{\partial v}{\partial T}\right)_p$

$$= \left[\left(\frac{\partial U}{\partial V}\right)_T + p\right] \left(\frac{\partial v}{\partial T}\right)_p$$

$$= \left(\frac{a}{v^2} + p\right)\left(\frac{\partial v}{\partial T}\right)_p$$

$$= \left(\frac{RT}{v-b}\right)\left(\frac{\partial v}{\partial T}\right)_p$$

From the equation

$$\left(p + \frac{a}{v^2}\right)(v - b) = RT$$

$$\therefore \quad (v - b)(-2av^{-3})\left(\frac{\partial v}{\partial T}\right)_p + \left(p + \frac{a}{v^2}\right)\left(\frac{\partial v}{\partial T}\right)_p = R$$

$$\left(\frac{\partial v}{\partial T}\right)_p = \frac{R/(v - b)}{\dfrac{RT}{(v-b)} - \dfrac{2a}{v^3}}$$

$$\therefore \quad c_p - c_v = \frac{R}{1 - 2a\,(v-b)^2/RTv^3}$$ Proved (d).

(e)
$$\left(\frac{\partial U}{\partial V}\right)_T = T\left(\frac{\partial p}{\partial T}\right)_V - p = \frac{a}{v^2}$$

$$du_T = \frac{a}{v^2} \cdot dv_T$$

$$\therefore \quad (u_2 - u_1)_T = a\left(\frac{1}{v_1} - \frac{1}{v_2}\right)$$

$$\therefore \quad (h_2 - h_1)_T = (p_2 v_2 - p_1 v_1) + a\left(\frac{1}{v_1} - \frac{1}{v_2}\right)$$ Proved (e).

PROBLEMS

11.1 Derive the following equations

(a) $U = F - T\left(\dfrac{\partial F}{\partial T}\right)_V = -T^2\left(\dfrac{\partial F/T}{\partial T}\right)_V$

(b) $C_v = -T\left(\dfrac{\partial^2 F}{\partial T^2}\right)_v$

(c) $H = G - T\left(\dfrac{\partial G}{\partial T}\right)_p = -T\left(\dfrac{\partial^2 G/T}{\partial T}\right)_p$

d) $C_p = -T\left(\dfrac{\partial^2 G}{\partial T^2}\right)_p$

11.2 (a) Derive the equation

$$\left(\frac{\partial c_v}{\partial v}\right)_T = T\left(\frac{\partial^2 p}{\partial T^2}\right)_V$$

(b) Prove that c_v of an ideal gas is a function of T only.

(c) In the case of a gas obeying the equation of state

$$\frac{pv}{RT} = 1 + \frac{B}{v}$$

where B'' is a function of T only, show that

$$c_v = -\frac{RT}{v}\frac{d^2}{dT^2}(B'' T) + c_{v0}$$

where $(c_v)_0$ is the value at very large volumes.

11.3 Derive the third TdS equation

$$TdS = C_v\left(\frac{\partial T}{\partial p}\right)_v dp + C_p\left(\frac{\partial T}{\partial V}\right)_p dV$$

and show that the three TdS equations may be written as

(a) $TdS = C_v\, dT + \frac{\beta T}{k} dV$

(b) $TdS = C_p\, dT - V\beta T\, dp$

(c) $TdS = \frac{C_v}{\beta} k\, dp + \frac{C_p}{\beta V} dV$

11.4 Derive the equations

(a) $C_p = T\left(\frac{\partial V}{\partial T}\right)_p \left(\frac{\partial p}{\partial T}\right)_s$

(b) $\left(\frac{\partial p}{\partial T}\right)_s = \frac{C_p}{V\beta T}$

(c) $\frac{(\partial p/\partial T)_s}{(\partial p/\partial T)_v} = \frac{\gamma}{\gamma - 1}$

11.5 Derive the equations

(a) $C_v = -T\left(\frac{\partial p}{\partial T}\right)_v \left(\frac{\partial V}{\partial T}\right)_s$

(b) $\left(\frac{\partial V}{\partial T}\right)_s = -\frac{C_v k}{\beta T}$

c) $\frac{(\partial V/\partial T)_s}{(\partial V/\partial T)_p} = \frac{1}{1-\gamma}$

11.6 (a) Prove that the slope of a curve on a Mollier diagram representing a reversible isothermal process is equal to

$$T - \frac{1}{\beta}$$

(b) Prove that the slope of a curve on a Mollier diagram representing a reversible isochoric process is equal to

$$T + \frac{\gamma - 1}{\beta}$$

11.7 (a) Show that

$$\mu\, c_p = T^2\left(\frac{\partial V/T}{\partial T}\right)_p$$

For 1 mole of a gas, in the region of moderate pressures, the equation of state may be written as

$$\frac{p v}{\bar{R} T} = 1 + B' p + C' p^2$$

where B' and C' are functions of temperature only.

(b) Show that as $p \to 0$

$$\mu\, c_p \to \bar{R}\, T^2 \frac{dB'}{dT}$$

(c) Show that the equation of the inversion curve is

$$p = -\frac{dB'/dT}{dC'/dT}$$

11.8 Prove the following function relationships of the reduced properties for the inversion curve of a Van der Waals gas

$$T_r = \frac{3(3v_r - 1)^2}{4v_r^2} \text{ and } p_r = \frac{9(2v_r - 1)}{v_r^2}$$

Hence, show that

$$\frac{\text{Maximum inversion temperature}}{\text{Critical temperature}} = 6.75$$

and
$$\frac{\text{Minimum inversion temperature}}{\text{Critical temperature}} = 0.75$$

11.9 Estimate the maximum inversion temperature of hydrogen if it is assumed to obey the equation of state

$$pV = RT + B_1 p + B_2 p^2 + B_3 p^3 + \ldots$$

For hydrogen, $B_1 \times 10^5 = a + 10^{-2}\, bT + 10^2\, c/T$

where $a = 166$, $b = -7.66$, $c = -172.33$.

11.10 The vapour pressure of mercury at 399 K and 401 K is found to be 0.988 mm and 1.084 mm of mercury respectively. Calculate the latent heat of vaporization of liquid mercury at 400 K.

Ans. 61,634.96 kJ/kgmol.

11.11 In the vicinity of the triple point, the vapour pressure of liquid ammonia (in atmospheres) is represented by

$$\ln p\ 15.16 - \frac{3063}{T}$$

This is the equation of the liquid-vapour boundary curve in a p-T diagram. Similarly, the vapour pressure of solid ammonia is

$$\ln p = 18.70 - \frac{3754}{T}$$

(a) What is the temperature and pressure at the triple point?
(b) What are the latent heats of sublimation and vaporization?
(c) What is the latent heat of fusion at the triple point?

Ans. 195.2 K, 0.585 atm., 1489 kJ/kg, 1836 kJ kg, 338 kJ/kg.

11.12 It is found that a certain liquid boils at a temperature of 95°C at the top of a hill, whereas it boils at a temperature of 105°C at the bottom. The latent heat is 4.187 kJ/g-mole. What is the approximate height of the hill? Ans. 321.12 m

11.13 Show that for an ideal gas in a mixture of ideal gases

$$d\mu_k = \frac{\mu_k - h_k}{T}\, dT + v_k\, dp + RTd \ln x_k$$

11.14 Compute μ for a gas whose equation of state is

$$p(v - b) = RT$$

Ans. $\mu = -b/c$

11.15 Show that

(a) $\left(\frac{\partial \beta}{\partial p}\right)_T = -\left(\frac{\partial k}{\partial T}\right)_p$

(b) $\left(\frac{\partial u}{\partial p}\right)_T = -T\left(\frac{\partial v}{\partial T}\right)_p - p\left(\frac{\partial v}{\partial p}\right)_T$

11.16 Two particular systems have the following equations of state

$$\frac{1}{T^{(1)}}=\frac{3}{2}\bar{R}\frac{N^{(1)}}{U^{(1)}} \text{ and } \frac{1}{T^{(2)}}=\frac{5}{2}\bar{R}\frac{N^{(2)}}{U^{(2)}}$$

where $\bar{R}=8.3143$ kJ/kgmol-K. The mole number of the first system is $N^{(1)}=2$, and that of the second is $N^{(2)}=3$. The two systems are separated by a diathermal wall, and the total energy in the composite system is 25.120 kJ. What is the internal energy of each system in equilibrium? Ans. 7.2 kJ, 17.92 kJ

11.17 Two systems with the equations of state given in Problem 11.16 are separated by a diathermal wall. The respective mole numbers are $N^{(1)}=2$ and $N^{(2)}=3$. The initial temperatures are $T^{(1)}=250$ K and $T^{(2)}=350$ K. What are the values of $U^{(1)}$ and $U^{(2)}$ after equilibrium has been established? What is the equilibrium temperature? Ans. 8.02 kJ, 20.04 kJ, 321.4 K.

11.18 Show that the change in latent heat L with temperature is given by the following relation

$$\left(\frac{dL}{dT}\right)=(C_p''' - C_p'') + \frac{L}{T} - \frac{v''' \beta''' - v'' \beta''}{v''' - v''}L.$$

12

Vapour Power Cycles

12.1 SIMPLE STEAM POWER CYCLE

A power cycle continuously converts heat (energy released by the burning of fuel) into work (shaft work), in which a working fluid repeatedly performs a succession of processes. In the vapour power cycle, the working fluid, which is water, undergoes a change of phase. Figure 12.1 gives the schematic of a

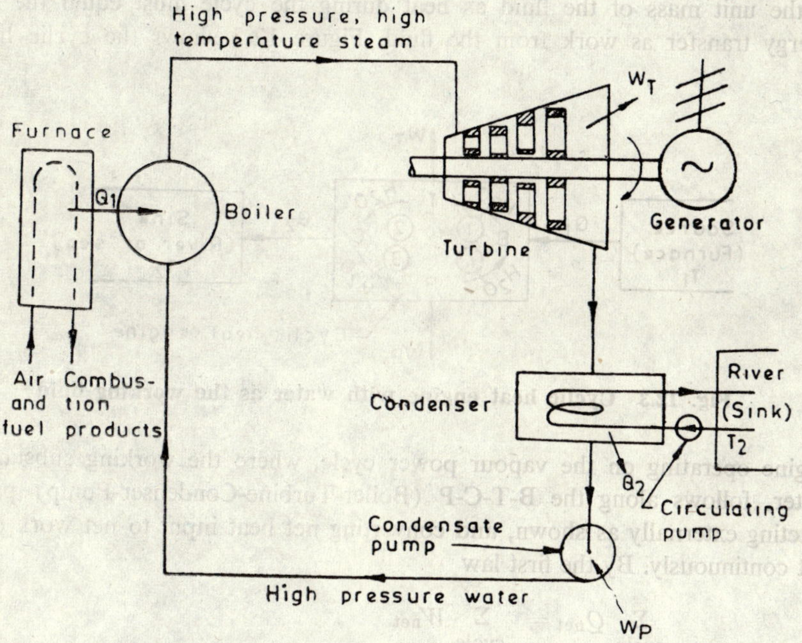

Fig. 12.1 Simple steam power plant

simple steam power plant working on the vapour power cycle. Heat is transferred to water in the boiler from an external source (furnace, where fuel is continuously burnt) to raise steam, the high pressure, high temperature steam leaving the boiler expands in the turbine to produce shaft work, the steam leaving the turbine condenses into water in the condenser (where cooling water circulates), rejecting heat, and then the water is pumped back to the boiler. Figure 12.2 shows how a unit mass of the working fluid,

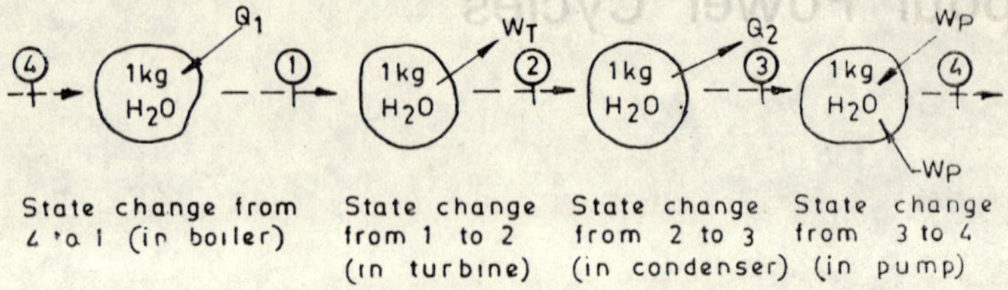

Fig. 12.2 One kg H_2O executing a heat engine cycle

sometimes in the liquid state and sometimes in the vapour state, undergoes various external heat and work interactions in executing a power cycle. Since the fluid is undergoing a cyclic process, there will be no net change in its internal energy over the cycle, and consequently the net energy transferred to the unit mass of the fluid as heat during the cycle must equal the net energy transfer as work from the fluid. Figure 12.3 shows the cyclic heat

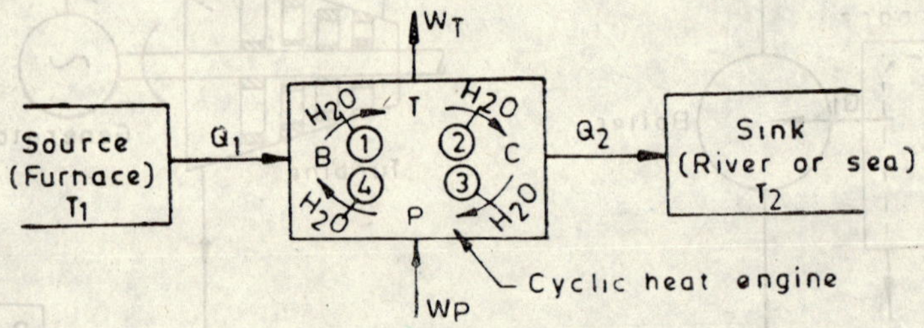

Fig. 12.3 Cyclic heat engine with water as the working fluid

engine operating on the vapour power cycle, where the working substance, water, follows along the B-T-C-P (Boiler-Turbine-Condenser-Pump) path, reacting externally as shown, and converting net heat input to net work output continuously. By the first law

$$\sum_{\text{cycle}} Q_{\text{net}} = \sum_{\text{cycle}} W_{\text{net}}$$

or
$$Q_1 - Q_2 = W_T - W_P$$

where Q_1 = heat transferred to the working fluid (kJ/kg)

Q_2 = heat rejected from the working fluid (kJ/kg)

W_T = work transferred from the working fluid (kJ/kg)

W_P = work transferred into the working fluid (kJ/kg).

The efficiency of the vapour power cycle would be given by

$$\eta_{\text{cycle}} = \frac{W_{\text{net}}}{Q_1} = \frac{W_T - W_P}{Q_1} = \frac{Q_1 - Q_2}{Q_1}$$

$$= 1 - \frac{Q_2}{Q_1}$$

12.2 RANKINE CYCLE

For each process in the vapour power cycle, it is possible to assume a hypothetical or ideal process which represents the basic intended operation and involves no extraneous effects. For the steam boiler, this would be a reversible constant pressure heating process of water to form steam, for the turbine the ideal process would be a reversible adiabatic expansion of steam, for the condenser it would be a reversible constant pressure heat rejection as the steam condenses till it becomes saturated liquid, and for the pump, the ideal process would be the reversible adiabatic compression of this liquid ending at the initial pressure. When all these four processes are ideal, the cycle is an ideal cycle, called a *Rankine cycle*. This is a reversible cycle. Figure 12.4

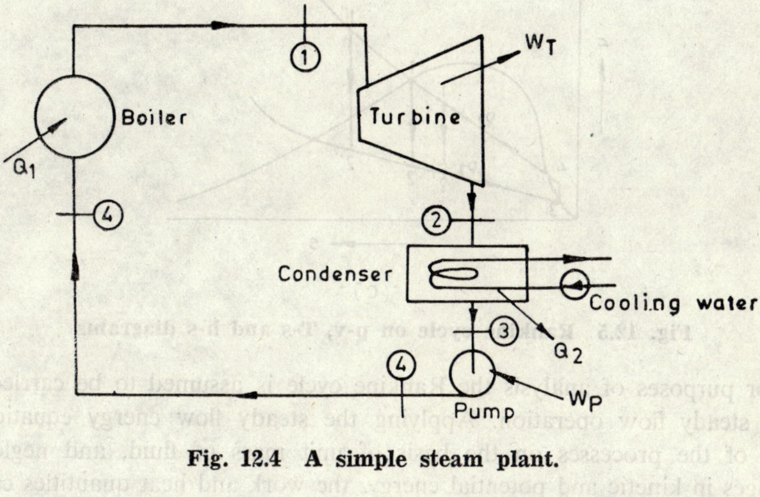

Fig. 12.4 A simple steam plant.

shows the flow diagram of the Rankine cycle, and in Fig. 12.5, the cycle has been plotted on the *p-v, T-s,* and *h-s* planes. The numbers on the plots correspond to the numbers on the flow diagram. For any given pressure, the steam approaching the turbine may be dry saturated (state 1), wet (state 1′), or superheated (state 1″), but the fluid approaching the pump is, in each case, saturated liquid (state 3). Steam expands reversibly and adiabatically in the turbine from state 1 to state 2 (or 1′ to 2′, or 1″ to 2″), the steam leaving the turbine condenses to water in the condenser reversibly at constant pressure from state 2 (or 2′, or 2″) to state 3, the water at state 3 is then pumped to the boiler at state 4 reversibly and adiabatically, and the water is heated in the boiler to form steam reversibly at constant pressure from state 4 to state 1 (or 1′ or 1″).

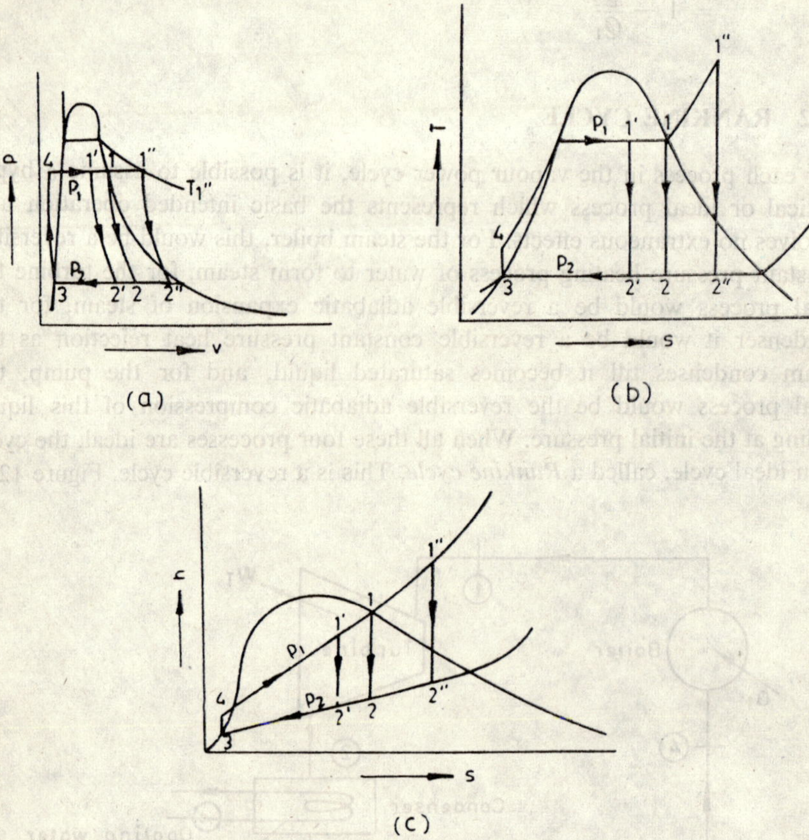

Fig. 12.5 Rankine cycle on p-v, T-s and h-s diagrams

For purposes of analysis the Rankine cycle is assumed to be carried out in a steady flow operation. Applying the steady flow energy equation to each of the processes on the basis of unit mass of fluid, and neglecting changes in kinetic and potential energy, the work and heat quantities can be evaluated in terms of the properties of the fluid.

For 1 kg fluid

The S.F.E.E. for the boiler (control volume) gives

$$h_4 + Q_1 = h_1$$

$$\therefore \quad Q_1 = h_1 - h_4 \tag{12.2}$$

The S.F.E.E. for the turbine as the control volume gives

$$h_1 = W_T + h_2$$

$$\therefore \quad W_T = h_1 - h_2 \tag{12.3}$$

Similarly, the S.F.E.E. for the condenser is

$$h_2 = Q_2 + h_3$$

$$\therefore \quad Q_2 = h_2 - h_3 \qquad (12.4)$$

and the S.F.E.E. for the pump gives

$$h_3 + W_P = h_4$$

$$\therefore \quad W_P = h_4 - h_3 \qquad (12.5)$$

The efficiency of the Rankine cycle is then given by

$$\eta = \frac{W_{net}}{Q_1} = \frac{W_T - W_P}{Q_1} = \frac{(h_1 - h_2) - (h_4 - h_3)}{h_1 - h_4} \qquad (12.6)$$

The pump handles liquid water which is incompressible, i.e., its density or specific volume undergoes little change with an increase in pressure. For reversible adiabatic compression, by the use of the general property relation

$$Tds = dh - vdp; \qquad ds = 0$$

and

$$dh = vdp$$

Since change in specific volume is negligible

$$\Delta h = v \, \Delta p$$

or

$$h_4 - h_3 = v_3 \ (p_1 - p_2)$$

If v is in m^3/kg and p is in bar

$$h_4 - h_3 = v_3 \ (p_1 - p_2) \times 10^5 \text{ J/kg} \qquad (12.7)$$

In MKS units

$$h_4 - h_3 = \frac{v_3 \, (p_1 - p_2) \times 10^4}{427} \text{ kcal/kg}$$

where p is in kgf/cm^2 abs.

Usually, the pump work is quite small compared to the turbine work and is sometimes neglected. Then $h_4 = h_3$, and the cycle efficiency approximately becomes

$$\eta \simeq \frac{h_1 - h_2}{h_1 - h_4}$$

The efficiency of the Rankine cycle is presented graphically in the *T-s* plot in Fig. 12.6. Thus Q_1 is proportional to area 1564, Q_2 is proportional to area 2563, and W_{net} ($= Q_1 - Q_2$) is proportional to area 1234 enclosed by the cycle.

The capacity of a steam plant is often expressed in terms of *steam rate,* which is defined as the rate of steam flow (kg/hr) required to produce unit shaft output (1 kW or 1 PS). Therefore

$$\text{Steam rate} = \frac{860}{W_{net}} = \frac{860}{W_T - W_P} \text{ kg/kW hr}$$

or

$$\text{Steam rate} = \frac{632}{W_T - W_P} \text{ kg/PS hr} \qquad (12.8)$$

where $(W_T - W_P)$ is in kcal/kg.

In S.I. units the steam rate is given by

$$\text{Steam rate} = \frac{1}{W_T - W_P}\ \frac{\text{kg}}{\text{kJ}} \cdot \frac{1\ \text{kJ/sec}}{1\ \text{kW}}$$

$$= \frac{1}{W_T - W_P}\ \frac{\text{kg}}{\text{kW sec}} = \frac{3600}{W_T - W_P}\ \frac{\text{kg}}{\text{kW hr}}$$

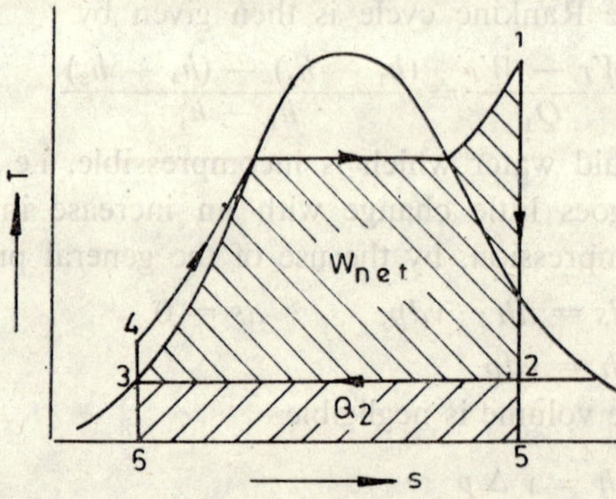

Fig. 12.6 Q_1, W_{net} **and** Q_2 **are proportional to areas**

The cycle efficiency is sometimes expressed alternatively as *heat rate* which is the rate of heat input (Q_1) required to produce unit work output (1 kW or 1 PS)

$$\therefore \qquad \text{Heat rate} = \frac{860\ Q_1}{W_T - W_P}\ \text{kcal/kW hr}$$

or

$$\text{Heat rate} = \frac{632\ Q_1}{W_T - W_P}\ \text{kcal/PS hr} \qquad (12.9)$$

In S.I. units

$$\text{Heat rate} = \frac{3600\ Q_1}{W_T - W_P}\ \text{kJ/kW hr}$$

12.3 ACTUAL VAPOUR CYCLE PROCESSES

The processes of an actual cycle differ from those of the ideal cycle. In the actual cycle conditions might be as indicated in Figs. 12.7 and 12.8, showing the various losses. The thermal efficiency of the actual cycle is

$$\mu_{th} = \frac{W_{net}}{Q_1}$$

where the work and heat quantities are the measured values for the actual cycle, which are different from the corresponding quantities of the ideal cycle.

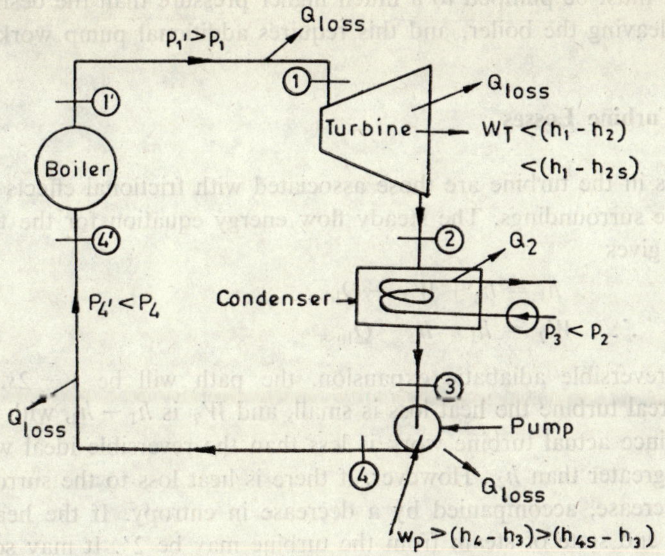

Fig. 12.7 Various losses in a steam plant.

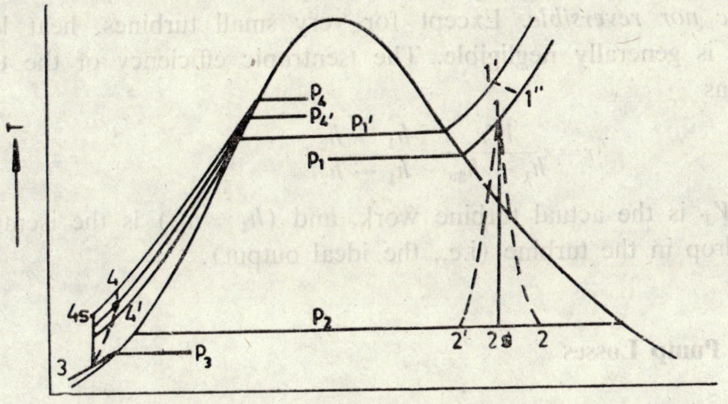

Fig. 12.8 Various losses on T-s plot

12.3.1 Piping Losses

Pressure drop due to friction and heat loss to the surroundings are the most important piping losses. States 1′ and 1 (Fig. 12.8) represent the states of the steam leaving the boiler and entering the turbine respectively, 1′ − 1″ represents the frictional losses, and 1″ − 1 shows the constant pressure heat loss to the surroundings. Both the pressure drop and heat transfer reduce the availability of steam entering the turbine.

A similar loss is the pressure drop in the boiler and also in the pipeline from the pump to the boiler. Due to this pressure drop, the water entering the boiler must be pumped to a much higher pressure than the desired steam pressure leaving the boiler, and this requires additional pump work.

12.3.2 Turbine Losses

The losses in the turbine are those associated with frictional effects and heat loss to the surroundings. The steady flow energy equation for the turbine in Fig. 12.7 gives

$$h_1 = h_2 + W_T + Q_{\text{loss}}$$

$$\therefore \quad W_T = h_1 - h_2 - Q_{\text{loss}} \tag{12.10}$$

For the reversible adiabatic expansion, the path will be $1 - 2s$. For an ordinary real turbine the heat loss is small, and W_T is $h_1 - h_2$, with Q_2 equal to zero. Since actual turbine work is less than the reversible ideal work output, h_2 is greater than h_{2s}. However, if there is heat loss to the surroundings, h_2 will decrease, accompanied by a decrease in entropy. If the heat loss is large, the end state of steam from the turbine may be $2'$. It may so happen that the entropy increase due to frictional effects just balances the entropy decrease due to heat loss, with the result that the initial and final entropies of steam in the expansion process are equal, *but the expansion is neither adiabatic nor reversible*. Except for very small turbines, heat loss from turbines is generally negligible. The isentropic efficiency of the turbine is defined as

$$\eta_T = \frac{W_T}{h_1 - h_{2s}} = \frac{h_1 - h_2}{h_1 - h_{2s}} \tag{12.11}$$

where W_T is the actual turbine work, and $(h_1 - h_{2s})$ is the isentropic enthalpy drop in the turbine (i.e., the ideal output).

12.3.3. Pump Losses

The losses in the pump are similar to those of the turbine, and are primarily due to the irreversibilities associated with fluid friction. Heat transfer is usually negligible. The pump efficiency is defined as

$$\eta_P = \frac{h_{4s} - h_3}{W_P} = \frac{h_{4s} - h_3}{h_4 - h_3} \tag{12.12}$$

where W_P is the actual pump work.

12.3.4 Condenser Losses

The losses in the condenser are usually small. These include the loss of pressure and the cooling of condensate below the saturation temperature.

12.4 COMPARISON OF RANKINE AND CARNOT CYCLES

Although the Carnot cycle has the maximum possible efficiency for the given limits of temperature, it is not suitable in steam power plants. Figure 12.9 shows the Rankine and Carnot cycles on the *T-s* diagram. The reversible adiabatic expansion in the turbine, the constant temperature heat rejection in the condenser, and the reversible adiabatic compression in the pump, are similar characteristic features of both the Rankine and Carnot cycles. But whereas the heat addition process in the Rankine cycle is reversible and at constant pressure, in the Carnot cycle it is reversible and isothermal. In Figs. 12.9(a) and 12.9(c), Q_2 is the same in both the cycles, but since Q_1

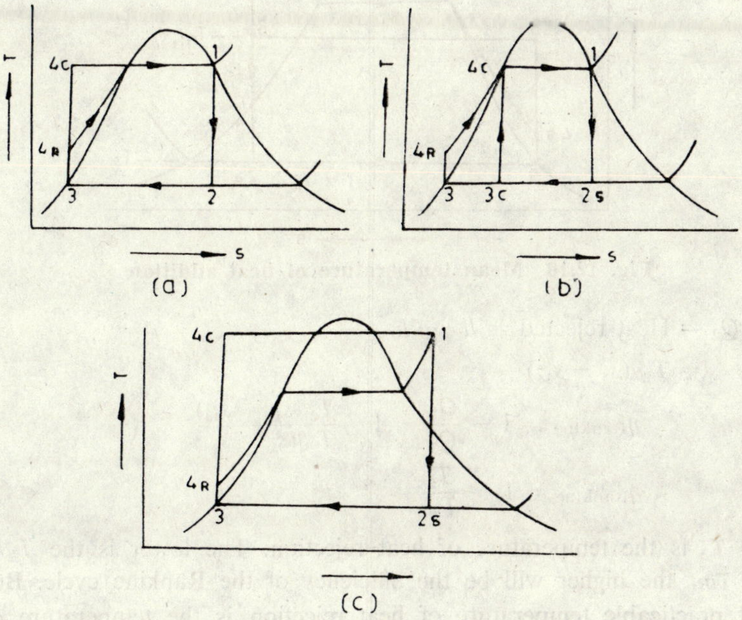

Fig. 12.9 Comparison of Carnot and Rankine cycles

is more, η_{Carnot} is greater than η_{Rankine}. The two Carnot cycles in Fig. 12.9(a) and 12.9(b) have the same thermal efficiency. Therefore, in Fig. 12.9(b) also, $\eta_{\text{Carnot}} > \eta_{\text{Rankine}}$. But the Carnot cycle cannot be realized in practice because the pump work [in all the three cycles (a), (b), and (c)] is very large. Whereas in (a) and (c) it is impossible to add heat at infinite pressures and at constant temperature from state 4c to state 1, in (b), it is difficult to control the quality at 3c, so that isentropic compression leads to a saturated liquid state.

12.5 MEAN TEMPERATURE OF HEAT ADDITION

In the Rankine cycle, heat is added reversibly at a constant pressure, but at infinite temperatures. If T_{m1} is the mean temperature of heat addition, as shown in Fig. 12.10, so that the area under 4*s* and 1 is equal to the area under 5 and 6, then heat added

$$Q_1 = h_1 - h_{4s} = T_{m1}(s_1 - s_{4s})$$

$\therefore$ T_{m_1} = Mean temperature of heat addition

$$= \frac{h_1 - h_{4s}}{s_1 - s_{4s}}$$

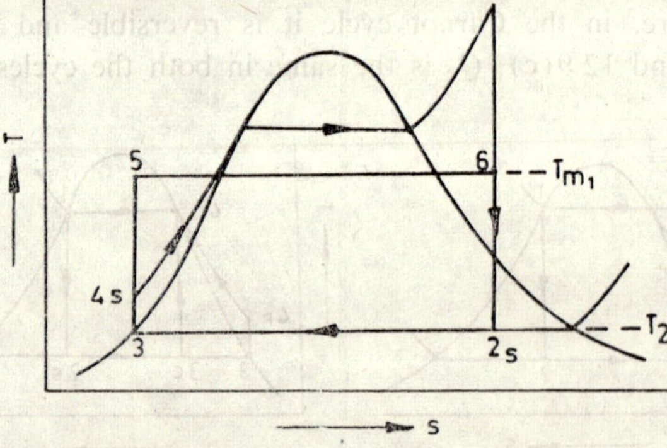

Fig. 12.10 Mean temperature of heat addition

Since Q_2 = Heat rejected $= h_{2s} - h_3$

$$= T_2\,(s_1 - s_{4s})$$

$$\eta_{\text{Rankine}} = 1 - \frac{Q_2}{Q_1} = 1 - \frac{T_2\,(s_1 - s_{4s})}{T_{m1}(s_1 - s_{4s})}$$

$$\therefore \qquad \eta_{\text{Rankine}} = 1 - \frac{T_2}{T_{m1}} \qquad (12.13)$$

where T_2 is the temperature of heat rejection. The lower is the T_2 for a given Tm_1 the higher will be the efficiency of the Rankine cycle. But the lowest practicable temperature of heat rejection is the temperature of the surroundings (T_0). This is being fixed

$$\eta_{\text{Rankine}} = f\,(T_{m_1})\text{ only} \qquad (12.14)$$

The higher the mean temperature of heat addition, the higher will be the cycle efficiency.

The effect of increasing the initial temperature at constant pressure on cycle efficiency is shown in Fig. 12.11. When the initial state changes from 1 to 1′, Tm_1 between 1 and 1′ is higher than Tm_1 between 4*s* and 1. So an increase in the superheat at constant pressure increases the mean temperature of heat addition and hence the cycle efficiency.

The maximum temperature of steam that can be used is *fixed from metallurgical considerations* (i.e., the materials used for the manufacture of the components which are subjected to high-pressure, high-temperature steam like the superheaters, valves, pipelines, inlet stages of turbine, etc.). When

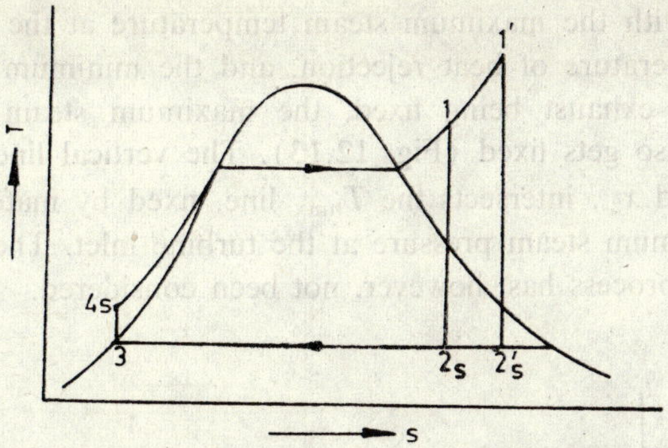

Fig. 12.11 Effect of superheat on mean temperature of heat addition

the maximum temperature is fixed, as the operating steam pressure at which heat is added in the boiler increases from p_1 to p_2 (Fig. 12.12), the mean temperature of heat addition increases, since T_{m1} between $7s$ and 5 is

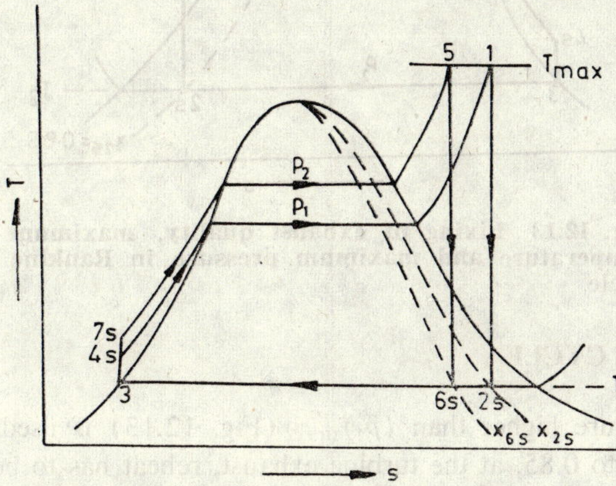

Fig. 12.12 Effect of increase of pressure on Rankine cycle.

temperature of heat addition increases, since Tm_1 between $7s$ and 5 is creases from p_1 to p_2, the ideal expansion line shifts to the left and the moisture content at the turbine exhaust increases (because $x_{6s} < x_{2s}$). If the moisture content of steam in the later stages of the turbine is high, the entrained water particles along with the vapour coming out from the nozzles with high velocity strike the blades and erode their surfaces, as a result of which the longevity of the blades decreases. From a consideration of the erosion of blades in the later stages of the turbine, the maximum moisture content at the turbine exhaust is not allowed to exceed 15%, or the quality to fall below 85%. It is desirable that most of the turbine expansion should take place in the single phase or vapour region.

Therefore, with the maximum steam temperature at the turbine inlet, the minimum temperature of heat rejection, and the minimum quality of steam at the turbine exhaust being fixed, the maximum steam pressure at the turbine inlet also gets fixed (Fig. 12.13). The vertical line drawn from 2*s*, fixed by T_2 and x_{2s}, intersects the T_{max} line, fixed by material, at 1, which gives the maximum steam pressure at the turbine inlet. The irreversibility in the expansion process has, however, not been considered.

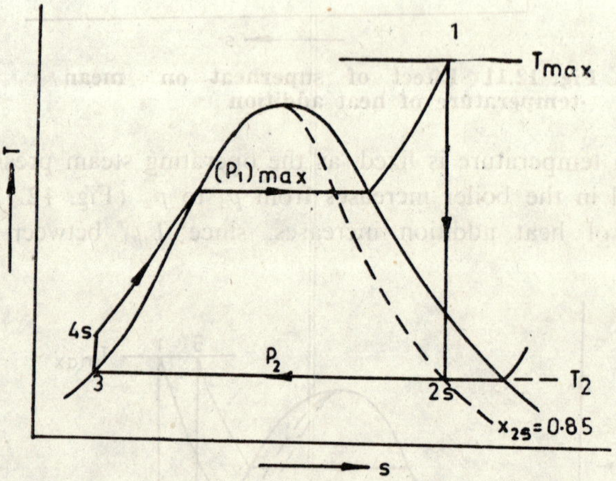

Fig. 12.13 Fixing of exhaust quality, maximum temperature and maximum pressure in Rankine cycle

12.6 REHEAT CYCLE

If a steam pressure higher than $(p_1)_{max}$ (Fig. 12.13) is used, in order to limit the quality to 0.85, at the turbine exhaust, reheat has to be adopted. In that case all the steam after partial expansion in the turbine is brought back to the boiler, reheated by combustion gases and then fed back to the turbine for further expansion. The flow, *T-s*, and *h-s* diagrams for the ideal Rankine cycle with reheat are shown in Fig. 12.14. In the reheat cycle the expansion of steam from the initial state 1 to the condenser pressure is carried out in two or more steps, depending upon the number of reheats used. In the first step, steam expands in the high pressure (H.P.) turbine from the initial state to approximately the saturated vapour line (process 1-2*s* in Fig. 12.14). The steam is then resuperheated (or reheated) at constant pressure in the boiler (process 2*s*-3) and the remaining expansion (process 3-4*s*) is carried out in the low pressure (L.P.) turbine. In the case of use of two reheats, steam is resuperheated twice at two different constant pressures. To protect the reheater tubes, steam is not allowed to expand deep into the two-phase region before it is taken for reheating, because in that case the moisture particles in steam while evaporating would leave behind solid deposits in the form of scale which is difficult to remove. Also, a low reheat pressure may bring down T_{m1} and hence, cycle efficiency. Again, a high reheat pressure increases the moisture content at turbine exhaust. Thus, the reheat pressure is optimized. For the cycle in Fig. 12.14, for 1 kg of steam

$$Q_1 = h_1 - h_{6s} + h_3 - h_{2s}$$

$$Q_2 = h_{4s} - h_5$$

$$W_T = h_1 - h_{2s} + h_3 - h_{4s}$$

$$W_P = h_{6s} - h_5$$

$$\eta = \frac{W_T - W_P}{Q_1} = \frac{(h_1 - h_{2s} + h_3 - h_{4s}) - (h_{6s} - h_5)}{h_1 - h_{6s} + h_3 - h_{2s}} \tag{12.15}$$

$$\underset{\text{(SI)}}{\text{Steam rate}} = \frac{3600}{(h_1 - h_{2s} + h_3 - h_{4s}) - (h_{6s} - h_5)} \text{ kg/kW hr}$$

where enthalpy is in kJ/kg.

$$\underset{\text{(MKS)}}{\text{Steam rate}} = \frac{860}{(h_1 - h_{2s} + h_3 - h_{4s}) - (h_{6s} - h_5)} \text{ kg/kW hr} \tag{12.16}$$

where enthalpy is in kcal/kg.

Since higher pressures are used in a reheat cycle, pump work may be appreciable.

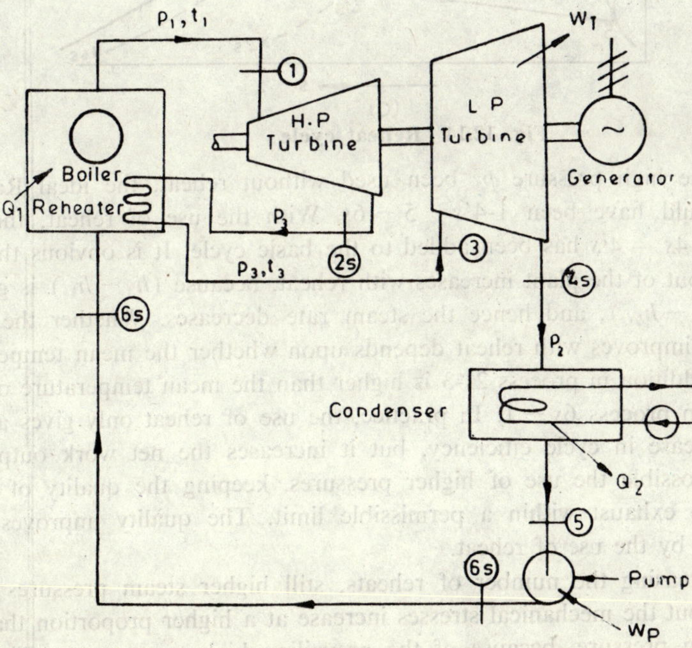

(a)

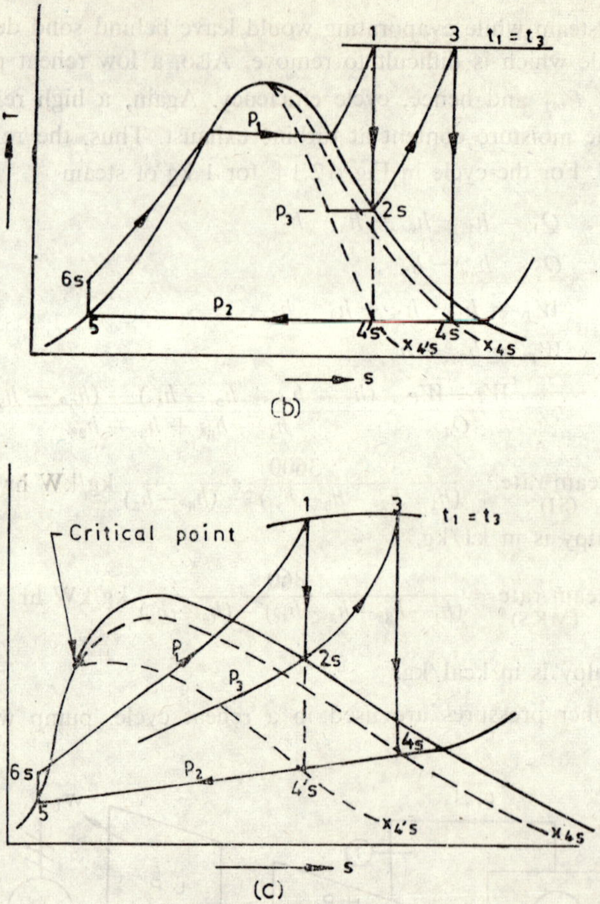

Fig. 12.14 Reheat cycle

Had the high pressure p_1 been used without reheat, the ideal Rankine cycle would have been $1\text{-}4's - 5 - 6s$. With the use of reheat, the area $2s - 3 - 4s - 4's$ has been added to the basic cycle. It is obvious that net work output of the plant increases with reheat, because $(h_3 - h_{4s})$ is greater than $(h_{2s} - h_{4's})$, and hence the steam rate decreases. Whether the cycle efficiency improves with reheat depends upon whether the mean temperature of heat addition in process $2s$-3 is higher than the mean temperature of heat addition in process $6s - 1$. In practice, the use of reheat only gives a marginal increase in cycle efficiency, but it increases the net work output by making possible the use of higher pressures, keeping the quality of steam at turbine exhaust within a permissible limit. The quality improves from $x_{4's}$ to x_{4s} by the use of reheat.

By increasing the number of reheats, still higher steam pressures could be used, but the mechanical stresses increase at a higher proportion than the increase in pressure, because of the prevailing high temperature. The cost and fabrication difficulties will also increase. In that way, the maximum steam pressure gets fixed, and more than two reheats have not yet been used so far.

In Fig. 12.14, only ideal processes have been considered. The irreversibilities in the expansion and compression processes have been considered in the example given later.

12.7 IDEAL REGENERATIVE CYCLE

In order to increase the mean temperature of heat addition (T_{m1}), attention was so far confined to increasing the amount of heat supplied at high temperatures, such as increasing superheat, using higher pressure and temperature of steam, and using reheat. The mean temperature of heat addition can also be increased by decreasing the amount of heat added at low temperatures. In a saturated steam Rankine cycle (Fig. 12.15), a considerable part

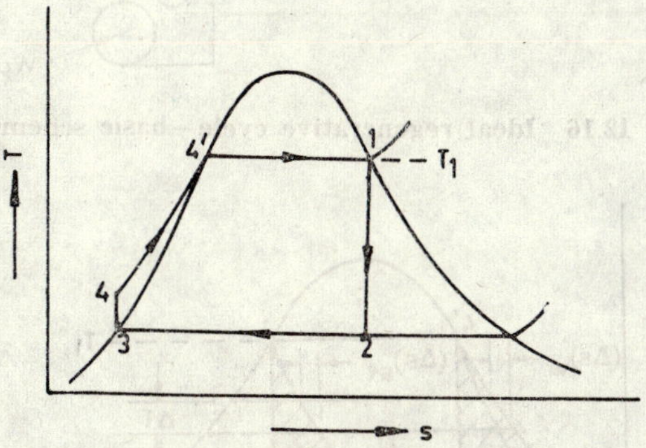

Fig. 12.15 Simple Rankine cycle

of the total heat supplied is in the liquid state when heating up water from 4 to 4′, at a temperature lower than T_1, the maximum temperature of the cycle. For maximum efficiency, all heat should be supplied at T_1, and feedwater should enter the boiler at state 4′. This may be accomplished in what is known as an ideal regenerative cycle, the flow diagram of which is shown in Fig. 12.16 and the corresponding *T*-*s* diagram in Fig. 12.17.

The unique feature of the ideal regenerative cycle is that the condensate, after leaving the pump circulates around the turbine casing, counterflow to the direction of vapour flow in the turbine (Fig. 12.16). Thus it is possible to transfer heat from the vapour as it flows through the turbine to the liquid flowing around the turbine. Let us assume that this is a reversible heat transfer, i.e., at each point the temperature of the vapour is only infinitesimally higher than the temperature of the liquid. The process 1-2 (Fig. 12.17) thus represents reversible expansion of steam in the turbine with reversible heat rejection. For any small step in the process of heating the water,

$$\Delta T \text{ (water)} = -\Delta T \text{ (steam)}$$

and
$$\Delta s \text{ (water)} = -\Delta s \text{ (steam)}$$

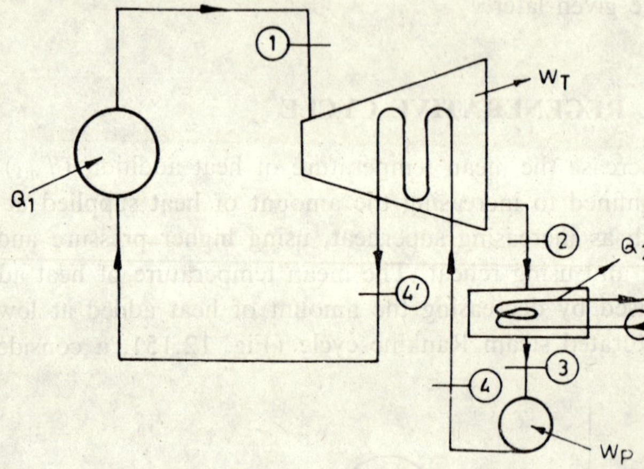

Fig. 12.16 Ideal regenerative cycle—basic scheme

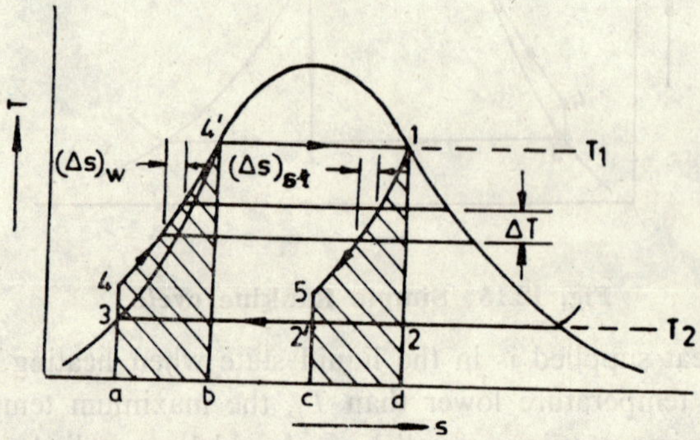

Fig. 12.17 Ideal regenerative cycle on T-s plot

Then the slopes of lines 1-2′ and 4′-3 (Fig. 12.17) will be identical at every temperature and the lines will be identical in contour. Areas 4-4′-b-a-4 and 2′-1-d-c-2′ are not only equal but congruous. Therefore, all the heat added from an external source (Q_1) is at the constant temperature T_1, and all the heat rejected (Q_2) is at the constant temperature T_2, both being reversible. Then

$$Q_1 = h_1 - h_{4'} = T_1 (s_1 - s_{4'})$$

$$Q_2 = h_{2'} - h_3 = T_2 (s_{2'} - s_3)$$

Since

$$s_{4'} - s_3 = s_1 - s_2$$

or

$$s_1 - s_{4'} = s_{2'} - s_3$$

$$\therefore \qquad \eta = 1 - \frac{Q_2}{Q_1} = 1 - \frac{T_2}{T_1}$$

The efficiency of the ideal regenerative cycle is thus equal to the Carnot cycle efficiency.

Writing the steady flow energy equation for the turbine

$$h_1 - W_T - h_{2'} + h_4 - h_{4'} = 0$$

or

$$W_T = (h_1 - h_{2'}) - (h_{4'} - h_4) \tag{12.17}$$

The pump work remains the same as in the Rankine cycle, i.e.

$$W_P = h_4 - h_3$$

The net work output of the ideal regenerative cycle is thus less, and hence its steam rate will be more, although it is more efficient, when compared with the Rankine cycle. However, the cycle is not practicable for the following reasons:

(a) Reversible heat transfer cannot be obtained in finite time.

(b) Heat exchanger in the turbine is mechanically impracticable.

(c) The moisture content of the steam in the turbine will be high.

12.8 REGENERATIVE CYCLE

In the practical regenerative cycle, the feedwater enters the boiler at a temperature between 4 and 4′ (Fig. 12.17), and it is heated by steam extracted from intermediate stages of the turbine. The flow diagram of the regenerative cycle with saturated steam at the inlet to the turbine, and the corresponding *T-s* diagram are shown in Figs. 12.18 and 12.19 respectively. For every kg

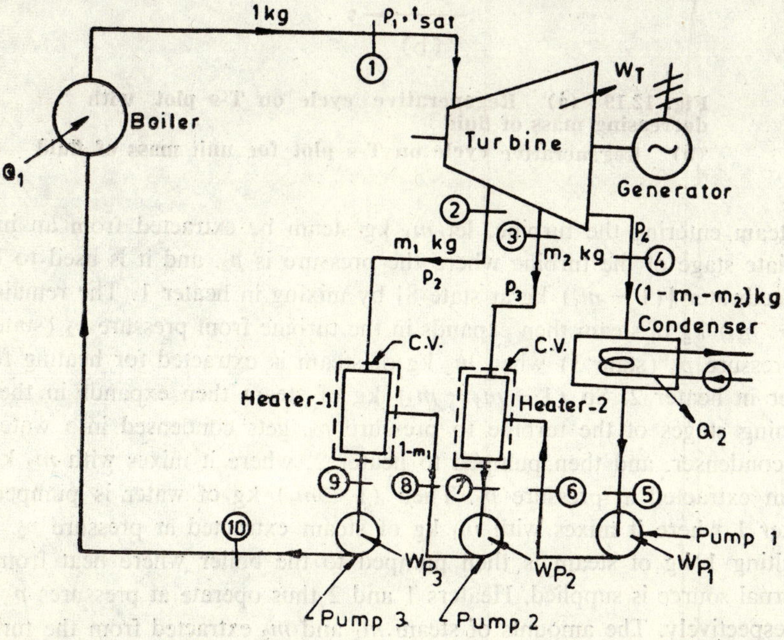

Fig. 12.18 Regenerative cycle flow diagram with two feedwater heaters.

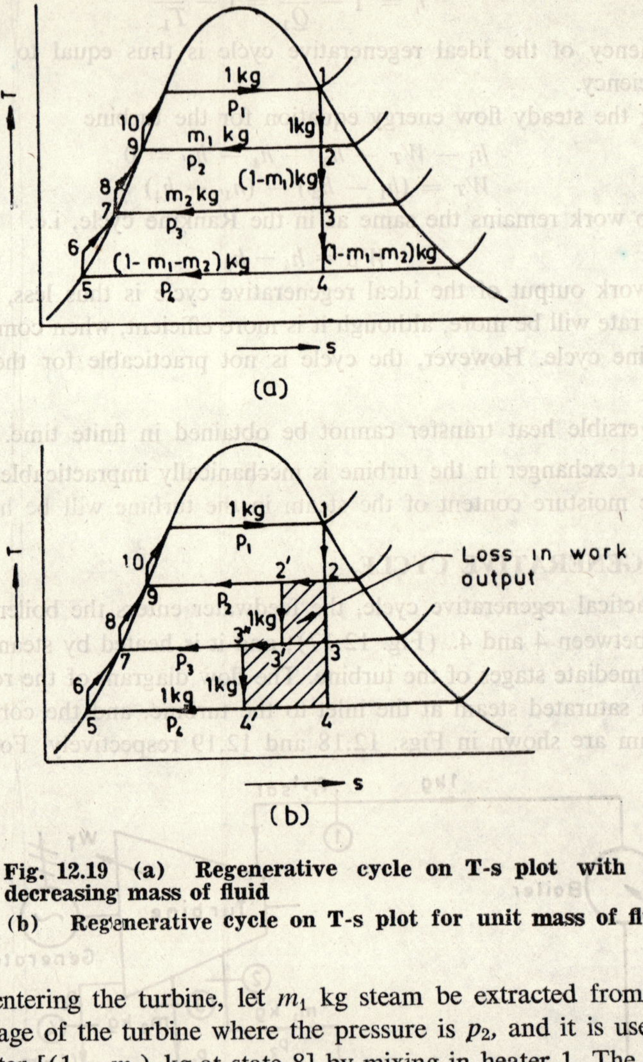

Fig. 12.19 (a) Regenerative cycle on T-s plot with decreasing mass of fluid
(b) Regenerative cycle on T-s plot for unit mass of fluid

of steam entering the turbine, let m_1 kg steam be extracted from an intermediate stage of the turbine where the pressure is p_2, and it is used to heat up feedwater [$(1 - m_1)$ kg at state 8] by mixing in heater 1. The remaining $(1 - m_1)$ kg of steam then expands in the turbine from pressure p_2 (state 2) to pressure p_3 (state 3) when m_2 kg of steam is extracted for heating feedwater in heater 2. So $(1 - m_1 - m_2)$ kg of steam then expands in the remaining stages of the turbine to pressure p_4, gets condensed into water in the condenser, and then pumped to heater 2, where it mixes with m_2 kg of steam extracted at pressure p_3. Then $(1 - m_1)$ kg of water is pumped to heater 1 where it mixes with m_1 kg of steam extracted at pressure p_2. The resulting 1 kg of steam is then pumped to the boiler where heat from an external source is supplied. Heaters 1 and 2 thus operate at pressures p_2 and p_3 respectively. The amounts of steam m_1 and m_2 extracted from the turbine are such that at the exit from each of the heaters, the state is saturated liquid at the respective pressures. The heat and work transfer quantities of the cycle are

$$W_T = 1\,(h_1-h_2) + (1-m_1)\,(h_2-h_3) + (1-m_1-m_2)\,(h_3-h_4) \text{ kJ/kg} \tag{12.18}$$

$$W_P = W_{P1} + W_{P2} + W_{P3}$$

$$= (1-m_1-m_2)\,(h_6-h_5) + (1-m_1)\,(h_8-h_7) + 1\,(h_{10}-h_9) \text{ kJ/kg} \tag{12.19}$$

$$Q_1 = 1\,(h_1-h_{10}) \text{ kJ/kg} \tag{12.20}$$

$$Q_2 = (1-m_1-m_2)\,(h_4-h_5) \text{ kJ/kg} \tag{12.21}$$

$$\text{Cycle efficiency, } \eta = \frac{Q_1-Q_2}{Q_1} = \frac{W_T-W_P}{Q_1}$$

$$\text{Steam rate (S.I.)} = \frac{3600}{W_T-W_P} \text{ kg/kW hr}$$

In the Rankine cycle operating at the given pressures, p_1 and p_4, the heat addition would have been from state 6 to state 1. By using two stages of regenerative feedwater heating, feedwater enters the boiler at state 10, instead of state 6, and heat addition is, therefore, from state 10 to state 1. Therefore

$$(T_{m_1})_{\text{with regeneration}} = \frac{h_1-h_{10}}{s_1-s_{10}} \tag{12.22}$$

and

$$(T_{m_1})_{\text{without regeneration}} = \frac{h_1-h_6}{s_1-s_6} \tag{12.23}$$

Since $(T_{m_1})_{\text{with regeneration}} > (T_{m_1})_{\text{without regeneration}}$

the efficiency of the regenerative cycle will be higher than that of the Rankine cycle.

The energy balance for heater 1 gives

$$m_1 h_2 + (1-m_1)\,h_8 = 1\,h_9$$

$$\therefore \quad m_1 = \frac{h_9-h_8}{h_2-h_8} \tag{12.24}$$

The energy balance for heater 1 gives

$$m_2 h_3 + (1-m_1-m_2)\,h_6 = (1-m_1)\,h_7$$

or

$$m_2 = (1-m_1)\frac{h_7-h_6}{h_3-h_6} \tag{12.25}$$

From equations (12.24) and (12.25), m_1 and m_2 can be evaluated. Equations (12.24) and (12.25) can also be written alternatively as

$$(1-m_1)\,(h_9-h_8) = m_1\,(h_2-h_9)$$

$$(1-m_1-m_2)\,(h_7-h_6) = m_2\,(h_3-h_7)$$

Energy gain of feedwater = Energy given off by vapour in condensation

Heaters have been assumed to be adequately insulated, and there is no heat gain from, or heat loss to, the surroundings.

Path 1-2-3-4 in Fig. 12.19 represents the states of a decreasing mass of fluid. For 1 kg of steam, the states would be represented by the dotted path 1-2′-3′-4′. From equation (12.18),

$$W_T = (h_1 - h_2) + (1 - m_1)(h_2 - h_3) + (1 - m_1 - m_2)(h_3 - h_4)$$
$$= (h_1 - h_2) + (h_{2'} - h_{3'}) + (h_{3''} - h_{4'}) \quad (12.26)$$

where
$$(1 - m_1)(h_2 - h_3) = 1\,(h_{2'} - h_{3'}) \quad (12.27)$$
$$(1 - m_1 - m_2)(h_3 - h_4) = 1\,(h_3 - h_{4'}) \quad (12.28)$$

The cycle 1-2-2′-3′-3″-4′-5-6-7-8-9-10-1 represents 1 kg of working fluid. The heat released by steam condensing from 2 to 2′ utilized in heating up the water from 8 to 9.

$$\therefore \quad 1\,(h_2 - h_{2'}) = 1\,(h_9 - h_8) \quad (12.29)$$

Similarly

$$1\,(h_{3'} - h_{3''}) = 1\,(h_7 - h_6) \quad (12.30)$$

From equations (12.26), (12.29) and (12.30)

$$W_T = (h_1 - h_{4'}) - (h_2 - h_{2'}) - (h_{3'} - h_{3''})$$
$$= (h_1 - h_{4'}) - (h_9 - h_8) - (h_7 - h_6) \quad (12.31)$$

The similarity of equations (12.17) and 12.31) can be noticed. It is seen that the stepped cycle 1-2′-3′-4′-5-6-7-8-9-10 approximates the ideal regenerative cycle in Fig. 12.17, and that a greater number of stages would give a closer approximation (Fig. 12.20). Thus the heating of feedwater by steam 'bled' from the turbine, known as regeneration, *carnotizes* the Rankine cycle.

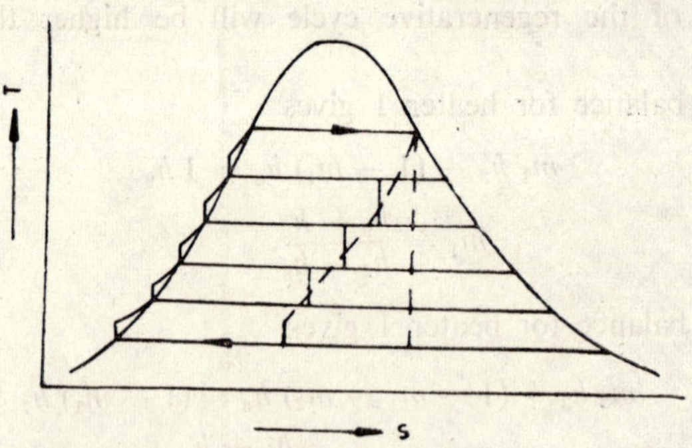

Fig. 12.20 Regenerative cycle with many stages of feed water heating

The heat rejected Q_2 in the cycle decreases from $(h_4 - h_5)$ to $(h_{4'} - h_5)$. There is also loss in work output by the amount (Area under 2-2′ + Area under 3′ − 3″ + Area under 4-4′), as shown by the hatched area in Fig. 12.19(b). So the steam rate increases by regeneration, i.e., more steam has to circulate per hour to produce unit shaft output.

The enthalpy-entropy diagram of a regenerative cycle is shown in Fig. 12.21.

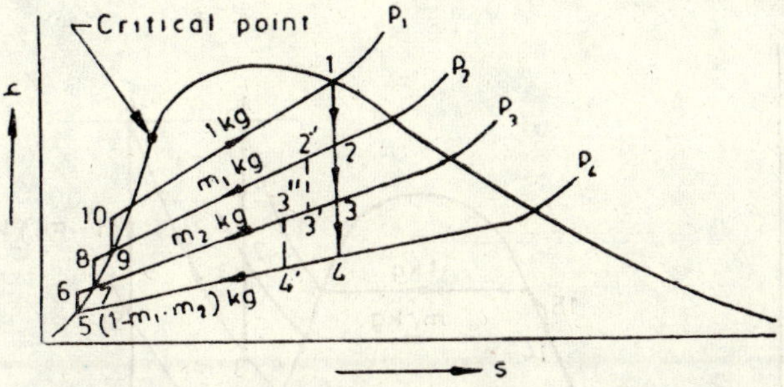

Fig. 12.21 Regenerative cycle on h-s diagram

12.9 REHEAT-REGENERATIVE CYCLE

The reheating of steam is adopted when the vaporization pressure is high. The effect of reheat alone on the thermal efficiency of the cycle is very small. Regeneration or the heating up of feedwater by steam extracted from the turbine has a marked effect on cycle efficiency. A modern steam power plant is equipped with both. Figures 12.22 and 12.23 give the flow and *T-s*

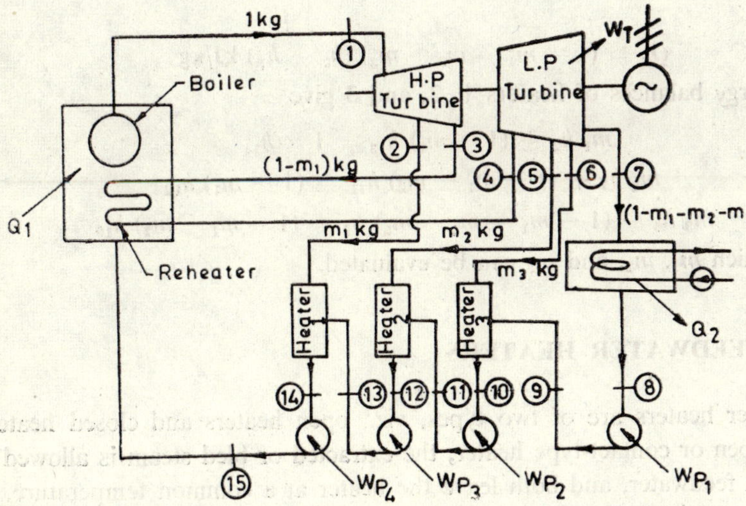

Fig. 12.22 Reheat-regenerative cycle flow diagram

diagrams of a steam plant with reheat and three stages of feedwater heating. Here

$$W_T = (h_1 - h_2) + (1 - m_1)(h_2 - h_3) + (1 - m_1)(h_4 - h_5) + (1 - m_1 - m_2)(h_5 - h_6) + (1 - m_1 - m_2 - m_3)(h_6 - h_7) \text{ kJ/kg}$$

$$W_P = (1 - m_1 - m_2 - m_3)(h_9 - h_8) + (1 - m_1 - m_2)(h_{11} - h_{10}) + (1 - m_1)(h_{13} - h_{12}) + 1\,(h_{15} - h_{14}) \text{ kJ/kg}$$

$$Q_1 = (h_1 - h_{15}) + (1 - m_1)(h_4 - h_3) \text{ kJ/kg}$$

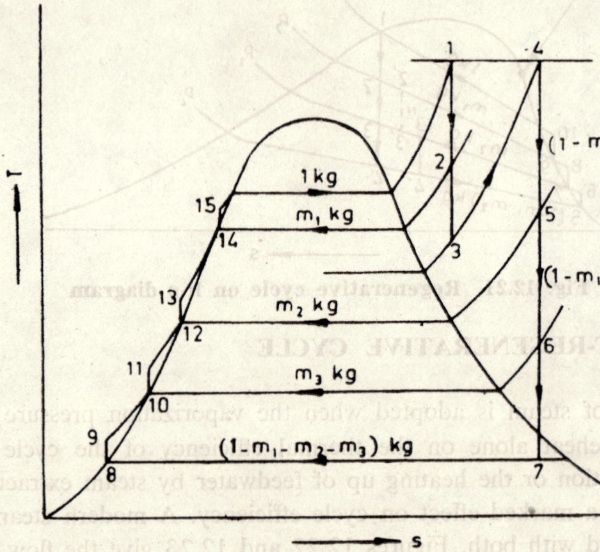

Fig. 12.23 T-s diagram of reheat-regenerative cycle

and

$$Q_2 = (1 - m_1 - m_2 - m_3)(h_7 - h_8) \text{ kJ/kg}$$

The energy balances of heaters 1, 2, and 3 give

$$m_1 h_2 + (1 - m_1) h_{13} = 1 \times h_{14}$$

$$m_2 h_5 + (1 - m_1 - m_2) h_{11} = (1 - m_1) h_{12}$$

$$m_3 h_6 + (1 - m_1 - m_2 - m_3) h_9 = (1 - m_1 - m_2) h_{10}$$

from which m_1, m_2, and m_3 can be evaluated.

12.10 FEEDWATER HEATERS

Feedwater heaters are of two types, viz., open heaters and closed heaters. In an open or contact-type heater, the extracted or bled steam is allowed to mix with feedwater, and both leave the heater at a common temperature, as shown in Figs. 12.18 and 12.22. In a closed heater, the fluids are kept separate, and not allowed to mix together (Fig. 12.24). The feedwater flows through the tubes in the heater and the extracted steam condenses on the outside of the tubes in the shell. The heat released by condensation is transferred to the feedwater through the walls of the tubes. The condensate (saturated water at the steam extraction pressure), sometimes called the heater-drip, then passes through a trap into the next lower pressure heater. This, to some extent, reduces the steam required by that heater. The trap passes only liquid and no vapour. The drip from the lowest pressure heater could similarly be trapped to the condenser, but this would be throwing away energy to the condenser cooling water. To avoid this waste, a drip pump feeds the drip directly into the feedwater stream.

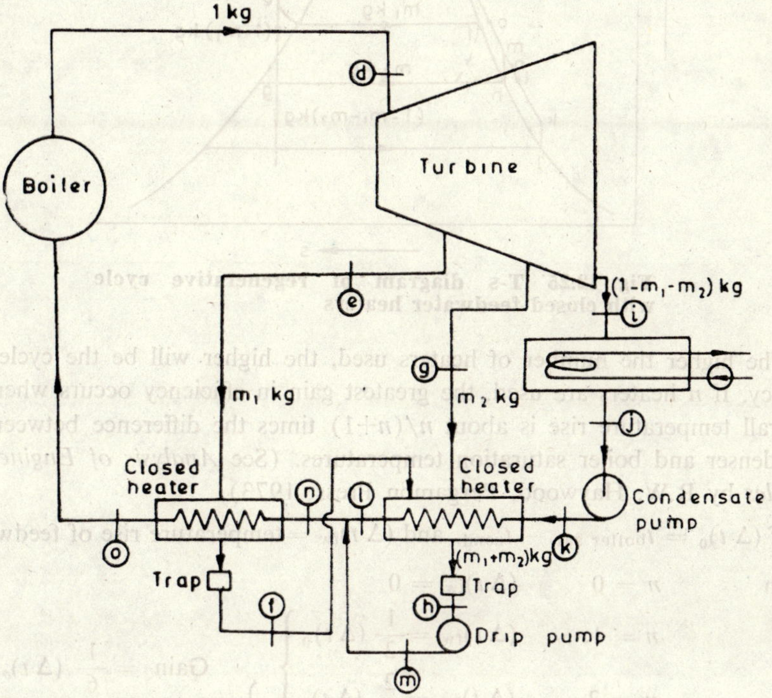

Fig. 12.24 Regenerative cycle flow diagram with closed feed-water heaters

Figure 12.25 shows the *T-s* plot corresponding to the flow diagram in Fig. 12.24. The temperature of the feedwater (at '*l*' or 'o') leaving a particular heater is always less than the saturation temperature at the steam extraction pressure (at *e* or *g*), the difference being known as the *terminal temperature difference* of the heater.

The advantages of the open heater are simplicity, lower cost, and high heat transfer capacity. The disadvantage is the necessity of a pump at each heater to handle the large feedwater stream.

A closed heater system requires only a single pump for the main feedwater stream regardless of the number of heaters. The drip pump, if used, is relatively small. Closed heaters are costly and may not give as high a feedwater temperature as do open heaters. In most steam power plants, closed heaters are favoured, but at least one open heater is used, primarily for the purpose of feedwater deaeration. The open heater in such a system is called the *deaerator*.

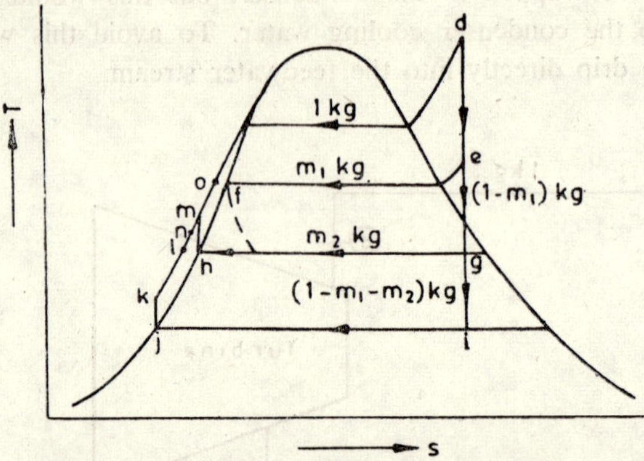

Fig. 12.25 T-s diagram of regenerative cycle with closed feedwater heaters

The higher the number of heaters used, the higher will be the cycle efficiency. If n heaters are used, the greatest gain in efficiency occurs when the overall temperature rise is about $n/(n+1)$ times the difference between the condenser and boiler saturation temperatures. (See *Analysis of Engineering Cycles* by R.W. Haywood, Pergamon Press, 1973).

If $(\Delta t)_0 = t_{\text{boiler sat.}} - t_{\text{cond.}}$ and $(\Delta t)_{\text{fw}}$ = temperature rise of feedwaters

with $n = 0$ $\quad (\Delta t)_{\text{fw}} = 0$

$n = 1, \quad (\Delta t)_{\text{fw}} = \frac{1}{2} (\Delta t)_0$

Gain $= \frac{1}{6} (\Delta t)_0$

$n = 2, \quad (\Delta t)_{\text{fw}} = \frac{2}{3} (\Delta t)_0$

Gain $= \frac{1}{12} (\Delta t)_0$

$n = 3, \quad (\Delta t)_{\text{fw}} = \frac{3}{4} (\Delta t)_0$

Gain $= \frac{1}{20} (\Delta t)_0$

$n = 4, \quad (\Delta t)_{\text{fw}} = \frac{4}{5} (\Delta t)_0$

and so on. Since the cycle efficiency is proportional to $(\Delta t)_{fw}$, the efficiency gain follows *the law of diminishing returns* with the increase in the number of heaters. The greatest increment in efficiency occurs by the use of the first heater. The increments for each additional heater thereafter successively diminish. The number of heaters is fixed up by the energy balance of the whole plant when it is found that the cost of adding another does not justify the saving in Q_1 or the marginal increase in cycle efficiency. An increase in feedwater temperature may, in some cases, cause a reduction in boiler efficiency. So the number of heaters gets optimized. Five points of extraction are often used in practice. Some cycles use as many as nine.

12.11 CHARACTERISTICS OF AN IDEAL WORKING FLUID IN VAPOUR POWER CYCLES

There are certain drawbacks with steam as the working substance in a power cycle. The maximum temperature that can be used in steam cycles consistent with the best available material is about 600°C, while the critical temperature of steam is 375°C, which necessitates large superheating and permits the addition of only an infinitesimal amount of heat at the highest temperature.

High moisture content is involved in going to higher steam pressures in order to obtain higher mean temperature of heat addition (Tm_1). The use of reheat is thus necessitated. Since reheater tubes are costly, the use of more than two reheats is hardly recommended. Also, as pressure increases, the metal stresses increase, and the thicknesses of the walls of boiler drums, tubes, pipe lines, etc., increase not in proportion to pressure increase, but much faster, because of the prevalence of high temperature.

It may be noted that high T_{m1} is only desired for high cycle efficiency. High pressures are only forced by the characteristics (weak) of steam.

If the lower limit is now considered, it is seen that at the heat rejection temperature of 40°C, the saturation pressure of steam is 0.075 bar, which is considerably lower than atmospheric pressure. The temperature of heat rejection can be still lowered by using some refrigerant as a coolant in the condenser. The corresponding vacuum will be still higher, and to maintain such low vacuum in the condenser is a big problem.

It is the low temperature of heat rejection that is of real interest. The necessity of a vacuum is a disagreeable characteristic of steam.

The saturated vapour line in the *T-s* diagram of steam is sufficiently inclined, so that when steam is expanded to lower pressures (for higher turbine output as well as cycle efficiency), it involves more moisture content, which is not desired from the consideration of the erosion of turbine blades in later stages.

The desirable characteristics of the working fluid in a vapour power cycle to obtain best thermal efficiency are given below.

(a) The fluid should have a high critical temperature so that the saturation pressure at the maximum permissible temperature (metallurgical limit) is relatively low. It should have a large enthalpy of evaporation at that pressure.

(b) The saturation pressure at the temperature of heat rejection should be above atmospheric pressure so as to avoid the necessity of maintaining vacuum in the condenser.

(c) The specific heat of liquid should be small so that little heat transfer is required to raise the liquid to the boiling point.

(d) The saturated vapour line of the *T-s* diagram should be steep, very close to the turbine expansion process so that excessive moisture does not appear during expansion.

(e) The freezing point of the fluid should be below room temperature, so that it does not get solidified while flowing through the pipelines.

(f) The fluid should be chemically stable and should not contaminate the materials of construction at any temperature.

(g) The fluid should be nontoxic, noncorrosive, not excessively viscous, and low in cost.

The characteristics of such an ideal fluid are approximated in the *T-s* diagram as shown in Fig. 12.26. Some superheat is desired to reduce piping losses and improve turbine efficiency. The thermal efficiency of the cycle is very close to the Carnot efficiency.

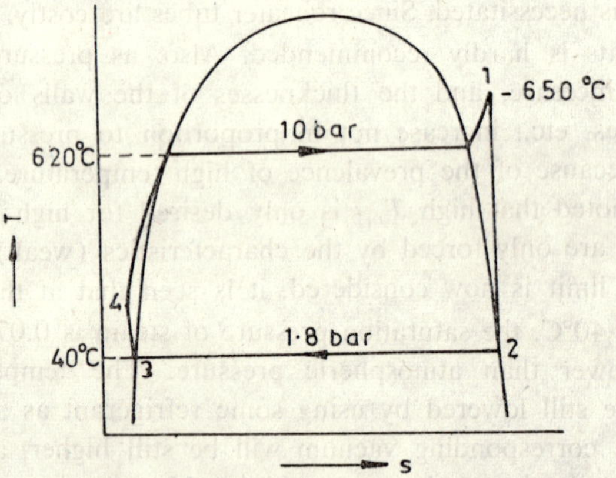

Fig. 12.26 T-s diagram of an ideal working fluid for a vapour power cycle

12.12 BINARY VAPOUR CYCLES

No single fluid can meet all the requirements as mentioned above. Although in the overall evaluation, water is better than any other working fluid, however, in the high temperature range, there are a few better fluids, and notable among them are (a) diphenyl ether, $(C_6H_5)_2O$, (b) aluminium bromide, Al_2Br_6, and (c) mercury and other liquid metals like sodium or potassium. From among these, only mercury has actually been used in practice. Diphenyl ether could be considered, but it has not yet been used because, like most organic substances, it decomposes gradually at high temperatures. Aluminium bromide is a possibility and yet to be considered.

When $p = 12$ bar, the saturation temperature for water, aluminium bromide, and mercury are 187°C, 482.5°C, and 560°C respectively. Mercury is thus a better fluid in the high temperature range, because at high temperature, its vaporization pressure is relatively low. Its critical pressure and temperature are 1080 bar and 1460°C respectively.

But in the low temperature range, mercury is unsuitable, because its saturation pressure becomes exceedingly low and it would be impractical to maintain such a high vacuum in the condenser. At 30°C, the saturation pressure of mercury is only 2.7×10^{-4} cm Hg. Its specific volume at such a low pressure is very large, and it would be difficult to accommodate such a large volume flow.

For this reason, mercury vapour leaving the mercury turbine is condensed at a higher temperature, and the heat released during the condensation of mercury is utilized in evaporating water to form steam to operate on a conventional turbine.

Thus in the binary (or two-fluid) cycle, two cycles with different working fluids are coupled in series, the heat rejected by one being utilized in the other.

The flow diagram of mercury-steam binary cycle and the corresponding *T-s* diagram are given in Figs. 12.27 and 12.28 respectively. The mercury

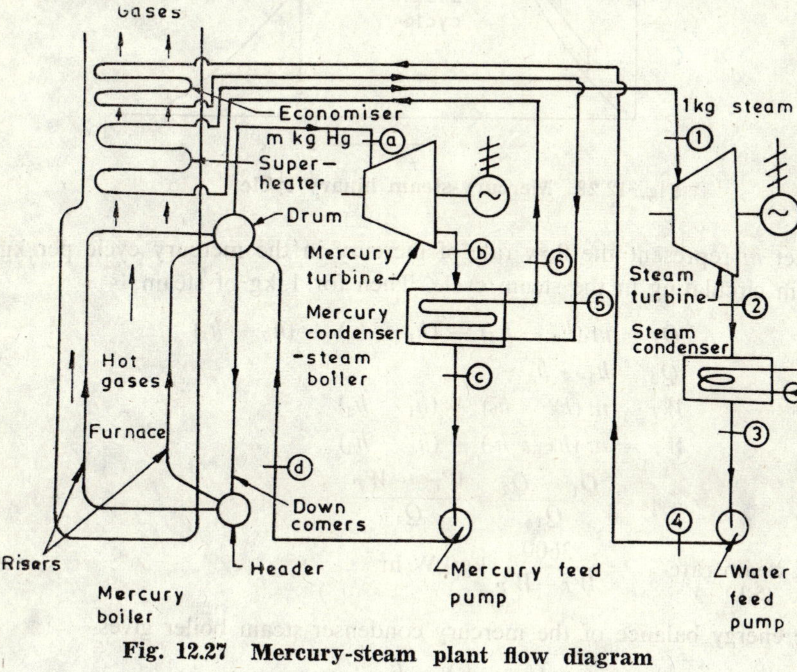

Fig. 12.27 Mercury-steam plant flow diagram

cycle, *a-b-c-d,* is a simple Rankine type of cycle using saturated vapour. Heat is supplied to the mercury in process *d-a*. The mercury expands in a turbine (process *a-b*) and is then condensed in process *b-c*. The feed pump process, *c-d,* completes the cycle.

The heat rejected by mercury during condensation is transferred to boil water and form saturated vapour (process 5-6). The saturated vapour is heated from the external source (furnace) in the superheater (process 6-1). Superheated steam expands in the turbine (process 1-2) and is then condensed (process 2-3). The feedwater (condensate) is then pumped (process 3-4), heated till it is saturated liquid in the economizer (process 4-5) before going to the mercury condenser-steam boiler, where the latent heat is absorbed. In an actual plant the steam cycle is always a regenerative cycle, but for the sake of simplicity, this complication has been omitted.

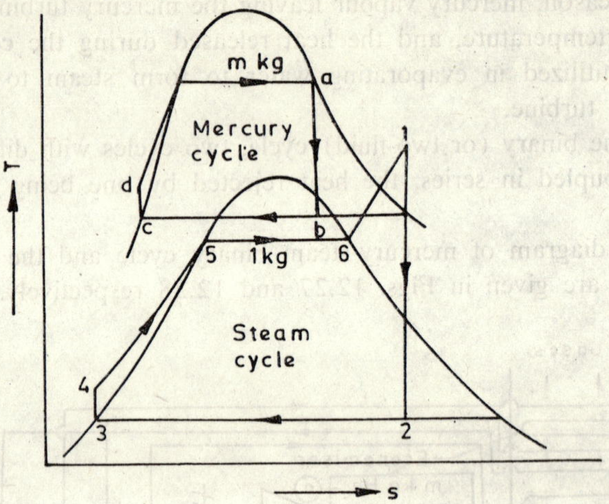

Fig. 12.28 Mercury-steam binary cycle

Let m represent the flow rate of mercury in the mercury cycle per kg of steam circulating in the steam cycle. Then for 1 kg of steam

$$Q_1 = m\,(h_a - h_d) + (h_1 - h_6) + (h_5 - h_4)$$

$$Q_2 = h_2 - h_3$$

$$W_T = m\,(h_a - h_b) + (h_1 - h_2)$$

$$W_P = m\,(h_d - h_c) + (h_4 - h_3)$$

$$\eta_{\text{cycle}} = \frac{Q_1 - Q_2}{Q_1} = \frac{W_T - W_P}{Q_1}$$

and steam rate (S.I.) $= \dfrac{3600}{W_T - W_P}$ kg/kW hr

The energy balance of the mercury condenser-steam boiler gives

$$m\,(h_b - h_c) = h_6 - h_5$$

$$\therefore \quad m = \frac{h_6 - h_5}{h_b - h_c} \text{ kg Hg/kg } H_2O$$

To vaporize one kg of water, seven to eight kg of mercury must condense.

The addition of the mercury cycle to the steam cycle results in a marked increase in the mean temperature of heat addition to the plant as a whole and consequently the efficiency is increased. The maximum pressure is relatively low.

It may be interesting to note that the concept of the binary vapour cycle evolved from the need of improving the efficiency of the reciprocating steam engine. When steam expands upto, say, atmospheric temperature, the resultant volume flow rate of steam becomes too large for the steam engine cylinder to accommodate. So most of the early steam engines are found to be non-condensing. The binary cycle with steam in the high temperature and ammonia or sulphur dioxide in the low temperature range, was first suggested by Professor Josse of Germany in the middle of the nineteenth century. Steam exhausted from the engine at a relatively higher pressure and temperature was used to evaporate ammonia or sulphur dioxide which operated on another cycle. But with the progress in steam turbine design, such a cycle was found to be of not much utility, since modern turbines can cope efficiently with a large volume flow of steam.

The mercury-steam cycle has been in actual commercial use for more than three decades. One such plant is the Schiller Station in the USA. But it has never attained wide acceptance because there has always been the possibility of improving steam cycles by increasing pressure and temperature, and by using reheat and regeneration. Over and above, mercury is expensive, limited in supply, and highly toxic.

The mercury-steam cycle represents the two-fluid cycles. The mercury cycle is called the *topping cycle* and the steam cycle is called the *bottoming cycle*. If a sulphur dioxide cycle is added to it in the low temperature range, so that the heat released during the condensation of steam is utilized in forming sulphur dioxide vapour which expands in another turbine, then the mercury-steam-sulphur dioxide cycle is a three-fluid or tertiary cycle. Similarly, other liquid metals, apart from mercury, like sodium or potassium, may be considered for a working fluid in the topping cycle. Apart from SO_2, other refrigerants (ammonia, freons etc.) may be considered as working fluids for the bottoming cycle.

Since the possibilities of improving steam cycles are diminishing, and the incentives to reduce fuel cost are very much increasing, coupled cycles, like the mercury-steam cycle, may receive more favourable consideration in the near future.

12.13 THERMODYNAMICS OF COUPLED CYCLES

If two cycles are coupled in series where heat lost by one is absorbed by the other (Fig. 12.29), as in the mercury-steam binary cycle, let η_1 and η_2 be the efficiencies of the topping and bottom cycles respectively, and η be the overall efficiency of the combined cycle.

$$\eta_1 = 1 - \frac{Q_2}{Q_1} \text{ and } \eta_2 = 1 - \frac{Q_3}{Q_2}$$

or

$$Q_2 = Q_1 (1 - \eta_1) \text{ and } Q_3 = Q_2 (1 - \eta_2)$$

Now

$$\eta = 1 - \frac{Q_3}{Q_1} = 1 - \frac{Q_2 (1 - \eta_2)}{Q_1}$$

$$= 1 - \frac{Q_1 (1 - \eta_1)(1 - \eta_2)}{Q_1}$$

$$= 1 - (1 - \eta_1)(1 - \eta_2)$$

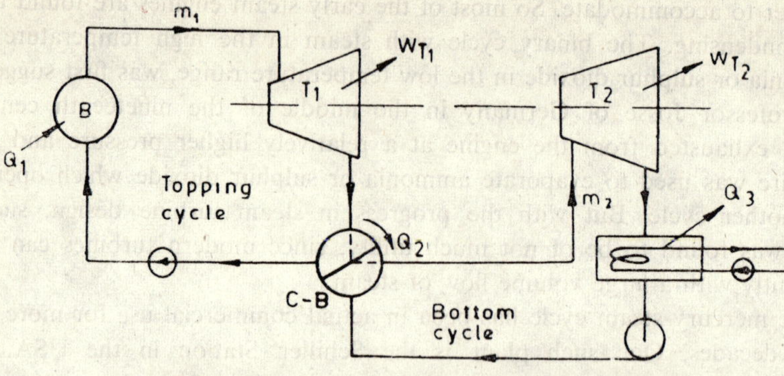

Fig. 12.29 Two vapour cycles coupled in series

If there are n cycles coupled in series, the overall efficiency would be given by

$$\eta = 1 - \pi_{i=1}^{n} (1 - \eta_i)$$

i.e. $$\eta = 1 - (1 - \eta_1)(1 - \eta_2)(1 - \eta_3) \ldots\ldots (1 - \eta_n)$$

or $$1 - \eta = (1 - \eta_1)(1 - \eta_2)(1 - \eta_3) \ldots\ldots (1 - \eta_n)$$

$\therefore$ Total loss = Product of losses in all the cycles. For two cycles coupled in series

$$\eta = 1 - (1 - \eta_1)(1 - \eta_2)$$
$$= 1 - (1 - \eta_1 - \eta_2 + \eta_1 \eta_2)$$
$$= \eta_1 + \eta_2 - \eta_1 \eta_2$$

or $$\eta = \eta_1 + \eta_2 - \eta_1 \eta_2$$

This shows that the overall efficiency of two cycles coupled in series equals the sum of the individual efficiencies minus their product.

By combining two cycles in series, even if individual efficiencies are low, it is possible to have a fairly high combined efficiency, which cannot be attained by a single cycle.

For example, if $\eta_1 = 0.50$ and $\eta_2 = 0.40$

$$\eta = 0.5 + 0.4 - 0.5 \times 0.4 = 0.70$$

It is almost impossible to achieve such a high efficiency in a single cycle.

12.14 PROCESS HEAT AND BY-PRODUCT POWER

There are several industries, such as paper mills, textile mills, chemical factories, dying plants, rubber manufacturing plants, sugar factories, etc., where saturated steam at the desired temperature is required for heating, drying, etc. For constant temperature heating (or drying), steam is a very good medium, since isothermal condition can be maintained by allowing saturated steam to condense at that temperature and utilizing the latent heat released for heating purposes. Apart from the process heat, the factory also needs power to drive various machines, for lighting, and for other purposes.

Formerly it was the practice to generate steam for power purposes at a moderate pressure and to generate separately saturated steam for process work at a pressure which gave the desired heating temperature. Having two separate units for process heat and power is wasteful, for of the total heat supplied to the steam for power purposes, a greater part will normally be carried away by the cooling water in the condenser.

By modifying the initial steam pressure and exhaust pressure, it is possible to generate the required power and make available for process work the required quantity of exhaust steam at the desired temperature. In Fig. 12.30,

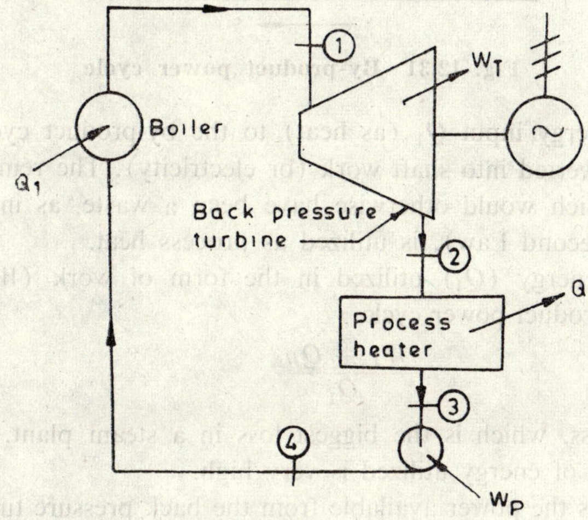

Fig. 12.30 Back pressure turbine

the exhaust steam from the turbine is utilized for process heating, the process heater replacing the condenser of the ordinary Rankine cycle. The pressure at exhaust from the turbine is the saturation pressure corresponding to the temperature desired in the process heater. Such a turbine is called *a back pressure turbine.* A plant processing both power and process heat is sometimes known as a *co-generating plant.* When the process steam is the basic need, and the power is produced incidentally as a by-product, the cycle is sometimes called a by-product power cycle. Figure 12.31 shows the *T-s* plot of such a cycle. If W_T is the turbine output in kW, Q_H the process heat required in kJ/hour, and w is the steam flow rate in kg/hour

$$W_T \times 3600 = w(h_1 - h_2)$$

nd

$$w(h_2 - h_3) = Q_H$$

$$\therefore \quad W_T \times 3600 = \frac{Q_H}{h_2 - h_3}(h_1 - h_2)$$

$$\text{or} \quad Q_H = \frac{W_T \times 3600 \times (h_2 - h_3)}{h_1 - h_2} \text{ kJ/hr}$$

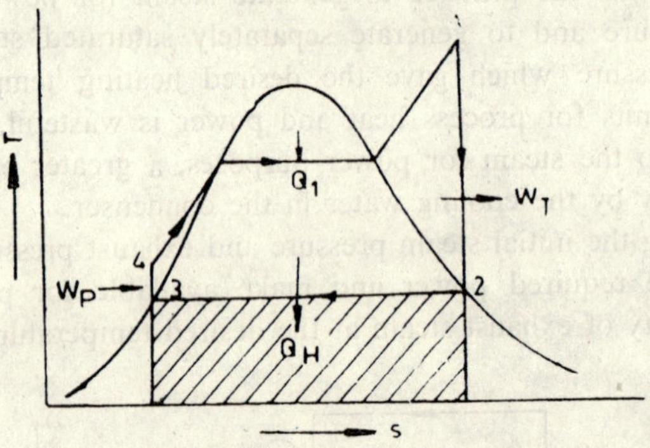

Fig. 12.31 By-product power cycle

Of the total energy input Q_1 (as heat) to the by-product cycle, W_T, only part of it is converted into shaft work (or electricity). The remaining energy $(Q_1 - W_T)$, which would otherwise have been a waste, as in the Rankine cycle (by the Second Law), is utilized as process heat.

Fraction of energy (Q_1) utilized in the form of work (W_T) and heat (Q_H) in a by-product power cycle

$$= \frac{W_T + Q_H}{Q_1}$$

Condenser loss, which is the biggest loss in a steam plant, is here zero, and the fraction of energy utilized is very high.

In many cases the power available from the back pressure turbine through which the whole of the heating steam flows is appreciably less than that required in the factory. This may be due to relatively high back pressure, or small heating requirement, or both. *Pass-out turbines* are employed in these cases, where a certain quantity of steam is continuously extracted for heating purposes at the desired temperature and pressure. (Figs. 12.32 and 12.33).

$$Q_1 = w\,(h_1 - h_8) \text{ kJ/hr}$$
$$Q_2 = (w - w_1)\,(h_3 - h_4) \text{ kJ/hr}$$
$$Q_H = w_1\,(h_2 - h_6) \text{ kJ/hr}$$
$$W_T = w\,(h_1 - h_2) + (w - w_1)\,(h_2 - h_3) \text{ kJ/hr}$$
$$W_P = (w - w_1)\,(h_5 - h_4) + w_1\,(h_7 - h_6) \text{ kJ/hr}$$
$$w_1 h_7 + (w - w_1)\,h_5 = w \times h_8$$

where w is the boiler capacity (kg/hour) and w_1 is the steam flow rate required (kg/hr) at the desired temperature for process heating.

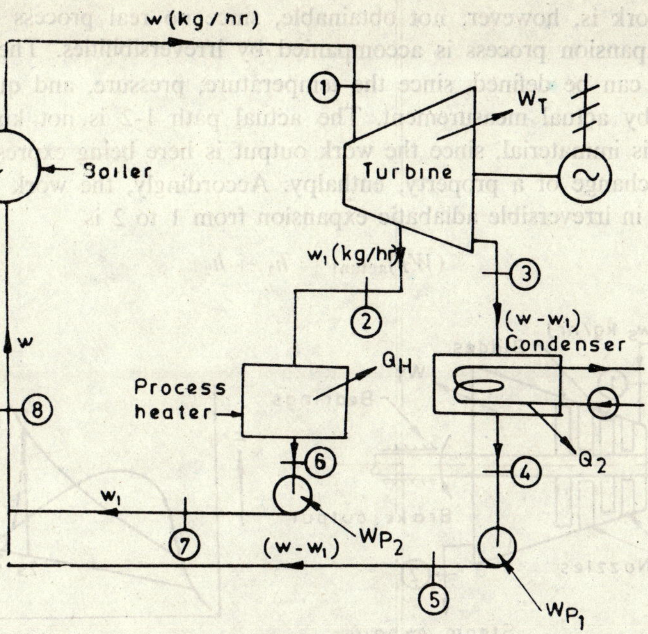

Fig. 12.32 Pass-out turbine

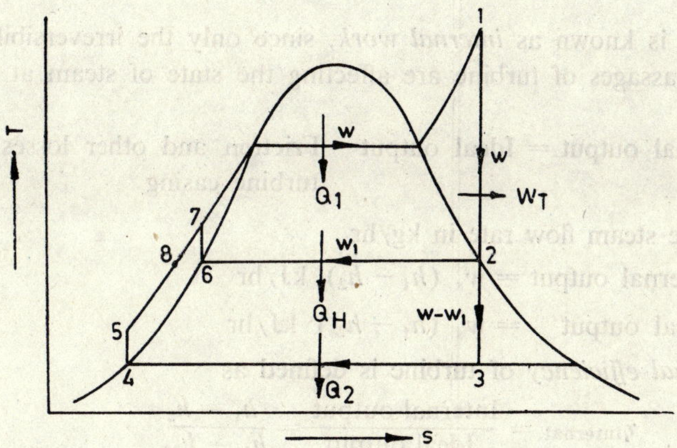

Fig. 12.33 T-s diagram of power and process heat plant

12.15 EFFICIENCIES IN STEAM POWER PLANT

For the steady flow operation of a turbine, neglecting changes in K.E. and P.E. (Figs. 12.34 and 12.35).

Maximum or ideal work output per unit mass of steam

$$(W_T)_{\max} = (W_T)_{\text{ideal}} = h_1 - h_{2s}$$

= Reversible and adiabatic enthalpy drop in turbine

This work is, however, not obtainable, since no real process is reversible. The expansion process is accompanied by irreversibilities. The actual final state 2 can be defined, since the temperature, pressure, and quality can be found by actual measurement. The actual path 1-2 is not known and its nature is immaterial, since the work output is here being expressed in terms of the change of a property, enthalpy. Accordingly, the work done by the turbine in irreversible adiabatic expansion from 1 to 2 is

$$(W_T)_{\text{actual}} = h_1 - h_2$$

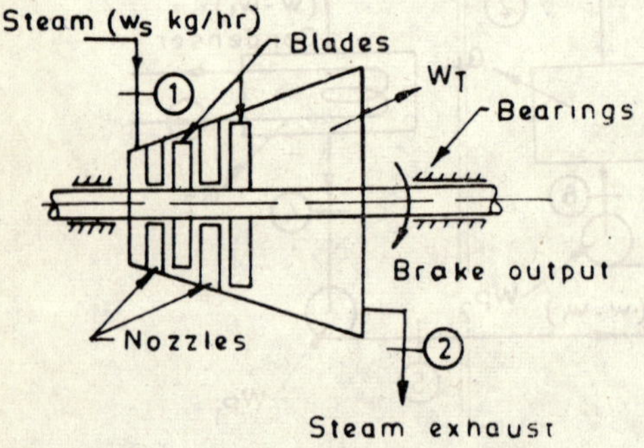

Fig. 12.34 Efficiencies in a steam turbine

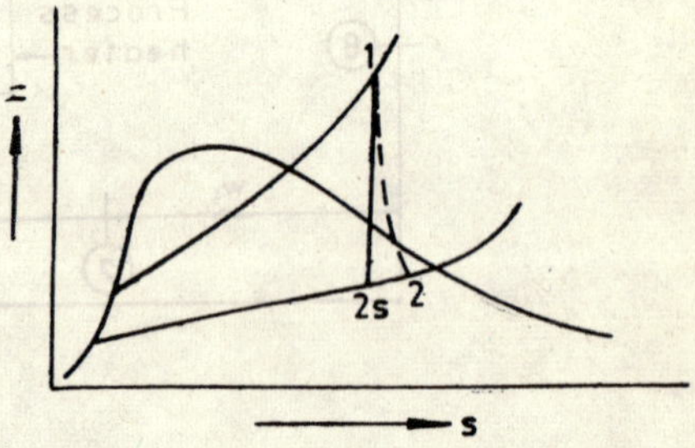

Fig. 12.35 Internal efficiency of a steam turbine

This work is known as *internal work,* since only the irreversibilities within the flow passages of turbine are affecting the state of steam at the turbine exhaust.

∴ Internal output = Ideal output − Friction and other losses within the turbine casing

If w_s is the steam flow rate in kg/hr

$$\text{Internal output} = w_s\,(h_1 - h_2)\ \text{kJ/hr}$$

$$\text{Ideal output} = w_s\,(h_1 - h_{2s})\ \text{kJ/hr}$$

The *internal efficiency* of turbine is defined as

$$\eta_{\text{internal}} = \frac{\text{Internal output}}{\text{Ideal output}} = \frac{h_1 - h_2}{h_1 - h_{2s}}$$

Work output available at the shaft is less than the internal output because of the external losses in the bearings, etc.

∴ Brake output or shaft output

= Internal output − External losses

= Ideal output − Internal and External losses

= (kW × 3600 kJ/hr)

or (kW × 860 kcal/hr)

or (BHP × 632 kcal/hr)

The *brake efficiency* of turbine is defined as

$$\eta_{\text{brake}} = \frac{\text{Brake output}}{\text{Ideal output}}$$

$$= \frac{\text{kW} \times 3600}{w_s\,(h_1 - h_{2s})}$$

In MKS units

$$\eta_{\text{brake}} = \frac{(\text{BHP} \times 632) \text{ or } (\text{kW} \times 860)}{w_s\,(h_1 - h_{2s})}$$

The *mechanical efficiency* of turbine is defined as

$$\eta_{\text{mech.}} = \frac{\text{Brake output}}{\text{Internal output}}$$

$$= \frac{\text{kW} \times 3600}{w_s\,(h_1 - h_2)} \text{ or } \frac{(\text{BHP} \times 632) \text{ or } (\text{kW} \times 860)}{w_s\,(h_1 - h_2)}$$

$$\therefore \quad \eta_{\text{brake}} = \eta_{\text{internal}} \times \eta_{\text{mech.}}$$

While the internal efficiency takes into consideration the internal losses, and the mechanical efficiency considers only the external losses, the brake efficiency takes into account both the internal and external losses (with respect to turbine casing).

The generator (or *alternator*) efficiency is defined as

$$\eta_{\text{generator}} = \frac{\text{Output at generator terminals}}{\text{Brake output of turbine}}$$

The *efficiency of the boiler* is defined as

$$\eta_{\text{boiler}} = \frac{\text{Energy utilized}}{\text{Enregy supplied}} = \frac{w_s\,(h_1 - h_4)}{w_f \times \text{C.V.}}$$

where w_f is the fuel burning rate in the boiler (kg/hr) and C.V. is the calorific value of the fuel (kJ/kg), i.e., the heat energy released by the complete combustion of unit mass of fuel.

The power plant is an *energy converter* from fuel to electricity (Fig. 12.36), and the *overall efficiency* of the plant is defined as

$$\eta_{\text{overall}} = \eta_{\text{plant}} = \frac{\text{kW} \times 3600}{w_f \times \text{C.V.}}$$

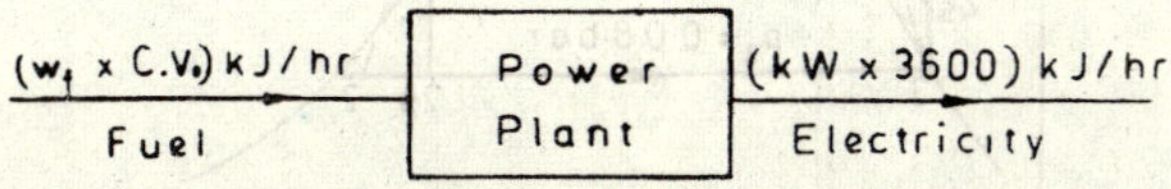

Fig. 12.36 Power plant—an energy converter from fuel to electricity

This may be expressed as

$$\eta_{\text{overall}} = \frac{\text{kW} \times 3600}{w_f \times \text{C.V.}} = \frac{w_s (h_1 - h_4)}{w_f \times \text{C.V.}} \times \frac{w_s (h_1 - h_2)}{w_s (h_1 - h_4)} \times \frac{\text{Brake output}}{w_s (h_1 - h_2)} \times \frac{\text{kW} \times 3600}{\text{Brake output}}$$

or

$$\eta_{\text{overall}} = \eta_{\text{boiler}} \times \eta_{\text{cycle}} \times \eta_{\text{turbine (mech.)}} \times \eta_{\text{generator}}$$

where pump work has been neglected in the expression for cycle efficiency.

EXAMPLE 12.1

Steam at 20 bar, 360°C is expanded in a steam turbine to 0.08 bar. It then enters a condenser, where it is condensed to saturated liquid water. The pump feeds back the water into the boiler. (a) Assuming ideal processes, find per kg of steam the net work and the cycle efficiency. (b) If the turbine and the pump have each 80% efficiency, find the percentage reduction in the net work and cycle efficiency.

Solution

The property values at different state points (Fig. Ex. 12.1) found from the steam tables are given below.

$$h_1 = 3159.3 \text{ kJ/kg} \qquad s_1 = 6.9917 \text{ kJ/kg-K}$$

$$h_3 = h_{f_{p_2}} = 173.88 \text{ kJ/kg} \quad s_3 = s_{f_{p_2}} = 0.5926 \text{ kJ/kg-K}$$

$$h_{fg_{p_2}} = 2403.1 \text{ kJ/kg} \qquad s_{g_{p_2}} = 8.2287 \text{ kJ/kg-K}$$

$$v_{f_{p_2}} = 0.001008 \text{ m}^3\text{/kg} \quad \therefore s_{fg_{p_2}} = 7.6361 \text{ kJ/kg-K}$$

Now

$$s_1 = s_{2s} = 6.9917 = s_{f_{p_2}} + x_{2s}\, s_{fg_{p_2}} = 0.5926 + x_2 \cdot 7.6361$$

$$\therefore \quad x_{2s} = \frac{6.3991}{7.6361} = 0.838$$

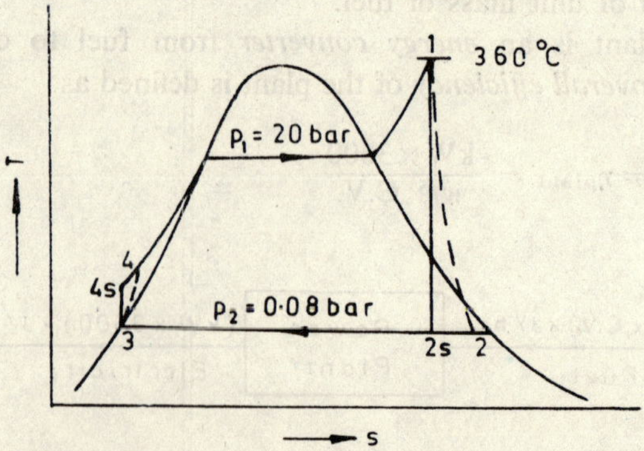

Fig. Ex. 12·1

$$\therefore \quad h_{2s} = h_{f_{p_2}} + x_{2s}\, h_{fg_{p_2}} = 173.88 + 0.838 \times 2403.1$$

$$= 2187.68 \text{ kJ/kg}$$

(a) $W_P = h_{4s} - h_3 = v_{fp_2}(p_1 - p_2) = 0.001008 \dfrac{\text{m}^3}{\text{kg}} \times 19.92 \times 100 \dfrac{\text{kN}}{\text{m}^2}$

$$= 2.008 \text{ kJ/kg}$$

$$h_{4s} = 175.89 \text{ kJ/kg}$$

$$W_T = h_1 - h_{2s}$$

$$= 3159.3 - 2187.68 = 971.62 \text{ kJ/kg}$$

$$\therefore \quad W_{\text{net}} = W_T - W_P = 969.61 \text{ kJ/kg} \qquad \text{Ans.}$$

$$Q_1 = h_1 - h_{4s} = 3159.3 - 175.89$$

$$= 2983.41 \text{ kJ/kg}$$

$$\therefore \quad \eta_{\text{cycle}} = \frac{W_{\text{net}}}{Q_1} = \frac{969.61}{2983.41} = 0.325, \text{ or } 32.5\% \qquad \text{Ans.}$$

(b) If $\eta_P = 80\%$, and $\eta_T = 80\%$

$$W_P = \frac{2.008}{0.8} = 2.51 \text{ kJ/kg}$$

$$W_T = 0.8 \times 971.62 = 777.3 \text{ kJ/kg}$$

$$\therefore \quad W_{\text{net}} = W_T - W_P = 774.8 \text{ kJ/kg}$$

$\therefore$ % Reduction in work output

$$= \frac{969.61 - 774.8}{969.61} \times 100 = 20.1\% \qquad \text{Ans.}$$

$$h_{4s} = 173.88 + 2.51 = 176.39 \text{ kJ/kg}$$

$$\therefore \quad Q_1 = 3159.3 - 176.39 = 2982.91 \text{ kJ/kg}$$

$$\therefore \quad \eta_{\text{cycle}} = \frac{774.8}{2982.91} = 0.2597, \text{ or } 25.97\%$$

$\therefore$ % Reduction in cycle efficiency

$$= \frac{0.325 - 0.2597}{0.325} \times 100 = 20.1\% \qquad \text{Ans.}$$

EXAMPLE 12.2

A cyclic steam power plant is to be designed for a steam temperature at a turbine inlet of 360°C and an exhaust pressure of 0.08 bar. After isentropic expansion of steam in the turbine, the moisture content at the turbine exhaust is not to exceed 15%. Determine the greatest allowable steam pressure at the turbine inlet, and calculate the Rankine cycle efficiency for these steam conditions. Estimate also the mean temperature of heat addition.

Solution

At state 2*s* (Fig. Ex. 12.2), the quality and pressure are known.

$$\therefore \quad s_{2s} = s_f + x_{2s}\, s_{fg} = 0.5926 + 0.85\ (8.2287 - 0.5926)$$

$$= 7.0833 \text{ kJ/kg-K}$$

Since $s_1 = s_{2s}$

$\therefore \quad s_1 = 7.0833$ kJ/kg-K

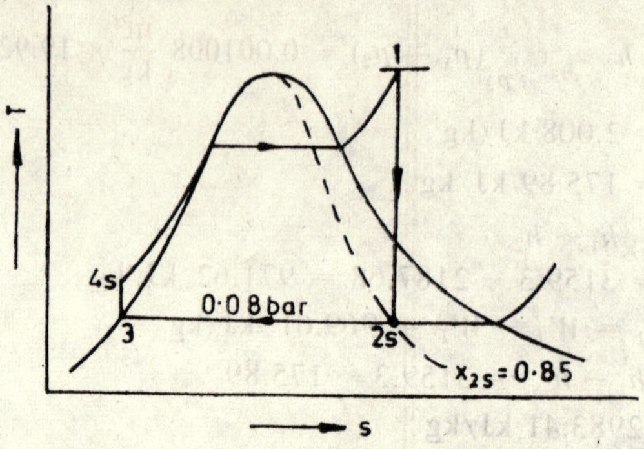

Fig Ex 12 2

At state 1, the temperature and entropy are thus known. At 360°C, $s_g = 5.0526$ kJ/kg-K, which is less than s_1. So from the table of superheated steam, at $t_1 = 360$°C and $s_1 = 7.0833$ kJ/kg-K, the pressure is found to be 16.832 bar (by interpolation).

$\therefore$ The greatest allowable steam pressure is

$$p_1 = 16.832 \text{ bar} \qquad \text{Ans.}$$

$$h_1 = 3165.54 \text{ kJ/kg}$$

$$h_{2s} = 173.88 + 0.85 \times 2403.1 = 2216.52 \text{ kJ/kg}$$

$$h_3 = 173.88 \text{ kJ/kg}$$

$$h_{4s} - h_3 = 0.001 \times (16.83 - 0.08) \times 100 = 1.675 \text{ kJ/kg}$$

$$\therefore \quad h_{4s} = 175.56 \text{ kJ/kg}$$

$$Q_1 = h_1 - h_{4s} = 3165.54 - 175.56$$

$$= 2990 \text{ kJ/kg}$$

$$W_T = h_1 - h_{2s} = 3165.54 - 2216.52 = 949 \text{ kJ/kg}$$

$$W_P = 1.675 \text{ kJ/kg}$$

$$\eta_{\text{cycle}} = \frac{W_{\text{net}}}{Q_1} = \frac{947.32}{2990} = 0.3168 \text{ or } 31.68\% \qquad \text{Ans.}$$

Mean temperature of heat addition

$$T_{m1} = \frac{h_1 - h_{4s}}{s_1 - s_{4s}} = \frac{2990}{7.0833 - 0.5926}$$

$$= 460.66 \text{ K} = 187.51 \text{ °C}$$

EXAMPLE 12.3

A steam power station uses the following cycle:

Steam at boiler outlet—150 bar, 550°C

Reheat at 40 bar to 550°C

Condenser at 0.1 bar.

Using the Mollier chart and assuming ideal processes, find the (a) quality at turbine exhaust, (b) cycle efficiency, and (c) steam rate.

Solution

The property values at different states (Fig. Ex. 12.3) are read from the Mollier chart.

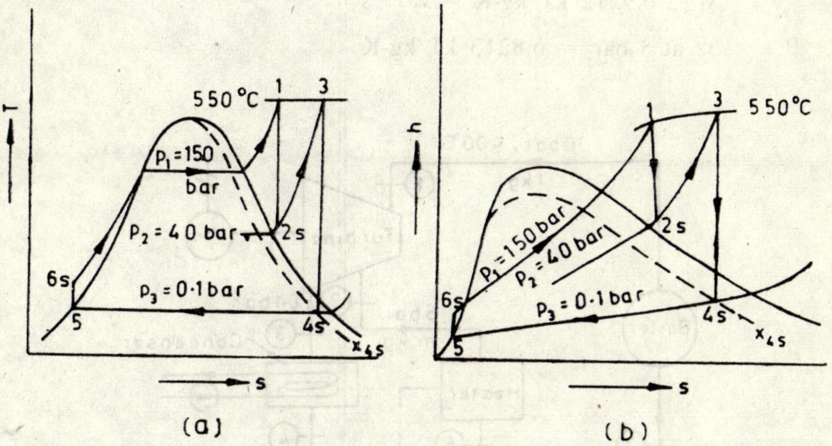

Fig Ex 12.3

$h_1 = 3465$, $h_{2s} = 3065$, $h_3 = 3565$,

$h_{4s} = 2300$ kJ/kg $x_{4s} = 0.88$, h_5 (steam table) $= 191.83$ kJ/kg

Quality at turbine exhaust = 0.88 Ans. (a).

$W_P = v\,\Delta p = 10^{-3} \times 150 \times 10^2 = 15$ kJ/kg

$\therefore\; h_{6s} = 206.83$ kJ/kg

$Q_1 = (h_1 - h_{6s}) + (h_3 - h_{2s})$

$= (3465 - 206.83) + (3565 - 3065) = 3758.17$ kJ/kg

$W_T = (h_1 - h_{2s}) + (h_3 - h_{4s})$

$= (3465 - 3065) + (3565 - 2300) = 1665$ kJ/kg

$\therefore\; W_{net} = W_T - W_P = 1665 - 15 = 1650$ kJ/kg

$\therefore\; \eta_{cycle} = \dfrac{W_{net}}{Q_1} = \dfrac{1650}{3758.17} = 0.4390$, or 43.9% Ans. (b).

Steam rate $= \dfrac{3600}{1650} = 2.18$ kg/kW hr Ans. (c).

EXAMPLE 12.4

In a single-heater regenerative cycle the steam enters the turbine at 30 bar, 400°C and the exhaust pressure is 0.10 bar. The feedwater heater is a direct-contact type which operates at 5 bar. Find (a) the efficiency and the steam rate of the cycle, and (b) the increase in mean temperature of heat addition, efficiency and steam rate, as compared to the Rankine cycle (without regeneration). Neglect pump work.

Solution

Figure Ex. 12.4 gives the flow, *T-s,* and *h-s* diagrams. From the steam tables, the property values at various states have been obtained.

$$h_1 = 3230.9 \text{ kJ/kg}$$

$$s_1 = 6.9212 \text{ kJ/kg-K} = s_2 = s_3$$

$$s_g \text{ at 5 bar} = 6.8213 \text{ kJ/kg-K}$$

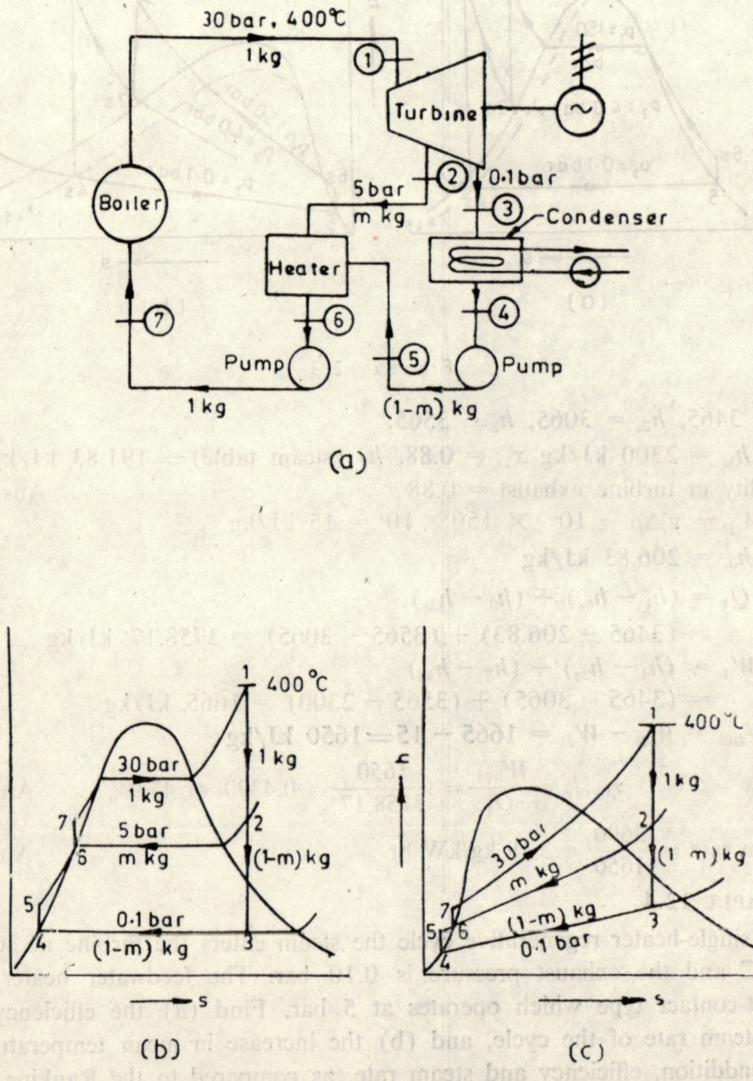

Fig. Ex 12.4

Since $s_2 > s_g$, the state 2 must lie in the superheated region. From the table for superheated steam $t_2 = 172°C$, $h_2 = 2796$ kJ/kg

$$s_3 = 6.9212 = s_{f_{0.1\text{ bar}}} + x_3\, s_{fg_{0.1\text{ bar}}}$$
$$= 0.6493 + x_3\, 7.5009$$

$$\therefore \quad x_3 = \frac{6.2719}{7.5009} = 0.836$$

$$\therefore \quad h_3 = 191.83 + 0.836 \times 2392.8 = 2192.2 \text{ kJ/kg}$$

Since pump work is neglected

$$h_4 = 191.83 \text{ kJ/kg} = h_5$$
$$h_6 = 640.23 \text{ kJ/kg} = h_7$$

Energy balance for the heater gives

$$m(h_2 - h_6) = (1-m)\ (h_6 - h_5)$$
$$m(2796 - 640.23) = (1-m)\ (640.23 - 191.83)$$
$$2155.77\, m = 548.4 - 548.4\, m$$

$$\therefore \quad m = \frac{548.4}{2704.17} = 0.203 \text{ kg}$$

$$\therefore \quad W_T = (h_1 - h_2) + (1-m)\ (h_2 - h_3)$$
$$= (3230.9 - 2796) + 0.797\ (2796 - 2192.2)$$
$$= 916.13 \text{ kJ/kg}$$

$$Q_1 = h_1 - h_6 = 3230.9 - 640.23 = 2590.67 \text{ kJ/kg}$$

$$\eta_{\text{cycle}} = \frac{916.13}{2590.67} = 0.3536, \text{ or } 35.36\%$$ Ans. (a).

Steam rate $= \dfrac{3600}{916.13} = 3.93$ kg/kW hr Ans. (a).

$$T_{m1} = \frac{h_1 - h_7}{s_1 - s_7} = \frac{2590.67}{6.9212 - 1.8607} = 511.95 \text{ K}$$
$$= 238.8\ °C$$

$$T_{m1} \text{ (without regeneration)} = \frac{h_1 - h_4}{s_1 - s_4}$$
$$= \frac{3039.07}{6.9212 - 0.6493}$$
$$= 484.55 \text{ K}$$
$$= 211.4 \text{ K}$$

Increase in T_{m1} due to regeneration

$$= 238.8 - 211.4 = 27.4°C$$ Ans. (b).

$$W_T \text{ (without regeneration)} = h_1 - h_3$$
$$= 3230.9 - 2192.2 = 1038.7 \text{ kJ/kg}$$

Steam rate (without regeneration) $= \dfrac{3600}{1038.7} = 3.46$ kg/kW hr

$\therefore$ Increase in steam rate due to regeneration

$$= 3.93 - 3.46 = 0.47 \text{ kg/kW hr}$$ Ans. (b).

$$\eta_{\text{cycle}} \text{ (without regeneration)} = \frac{h_1 - h_3}{h_1 - h_4} = \frac{1038.7}{3039.07}$$

$$= 0.3418 \text{ or } 34.18\%$$

$\therefore$ Increase in cycle efficiency due to regeneration

$$= 35.36 - 34.18 = 1.18\%$$ Ans. (c).

EXAMPLE 12.5

A certain chemical plant requires heat from process steam at 120°C at the rate of 5.83 MJ/s and power at the rate of 1000 kW from the generator terminals. Both the heat and power requirements are met by a back pressure turbine of 80% brake and 85% internal efficiency, which exhausts steam at 120°C dry saturated. All the latent heat released during condensation is utilized in the process heater. Find the pressure and temperature of steam at the inlet to the turbine. Assume 90% efficiency for the generator.

Solution

At 120°C, $h_{fg} = 2202.6$ kJ/kg $= h_2 - h_3$ (Fig. Ex. 12.5)

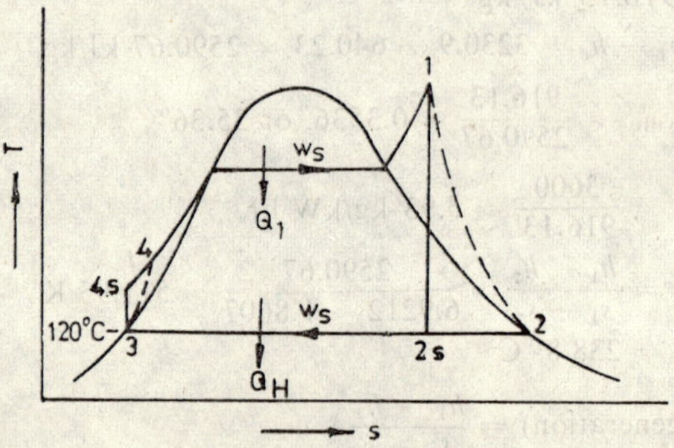

Fig. Ex 12.5

$$Q_H = w_s\,(h_2 - h_3) = 5.83 \text{ MJ/s}$$

$\therefore$ $$w_s = \frac{5830}{2202.6} = 2.647 \text{ kg/s}$$

$$W_{\text{net}} = \frac{1000}{0.9} \text{ kJ/s} = \text{Brake output}$$

Now $$\eta_{\text{brake}} = \frac{\text{Brake output}}{\text{Ideal output}} = \frac{(1000)/0.9}{w_s\,(h_1 - h_{2s})} = 0.80$$

$\therefore$ $$h_1 - h_{2s} = \frac{1000}{0.9 \times 0.8 \times 2.647} = 524.7 \text{ kJ/kg}$$

Again $\eta_{\text{internal}} = \dfrac{h_1 - h_2}{h_1 - h_{2s}} = 0.85$

$\therefore$ $h_1 - h_2 = 0.85 \times 524.7 = 446$ kJ/kg

$h_2 = h_g$ at 120°C$=2706.3$ kJ/kg

$\therefore$ $h_1 = 3152.3$ kJ/kg

$h_{2s} = h_1 - 524.7 = 2627.6$ kJ/kg

$= h_f + x_{2s} h_{fg}$

$= 503.71 + x_{2s} \times 2202.6$

$\therefore$ $x_{2s} = \dfrac{2123.89}{2202.6} = 0.964$

$\therefore$ $s_{2s} = s_f + x_{2s}\, s_{fg} = 1.5276 + 0.964 \times 5.6020$

$= 6.928$ kJ/kg-K

$= s_1$

At state 1, $h_1 = 3152.3$ kJ/kg

$s_1 = 6.928$ kJ/kg-K

From the Mollier chart

$p_1 = 22.5$ bar

$t_1 = 360$°C Ans.

EXAMPLE 12.6

A certain factory has an average electrical load of 1500 kW and requires 3.5 MJ/s for heating purposes. It is proposed to install a single-extraction pass-out steam turbine to operate under the following conditions:

Initial pressure 15 bar.

Initial temperature 300°C.

Condenser pressure 0.1 bar.

Steam is extracted between the two turbine sections at 3 bar, 0.96 dry, and is isobarically cooled without subcooling in heaters to supply the heating load. The internal efficiency of the turbine (in the L.P. Section) is 0.80 and the efficiency of the boiler is 0.85 when using oil of calorific value 44 MJ/kg.

If 10% of boiler steam is used for auxiliaries calculate the oil consumption per day. Assume that the condensate from the heaters (at 3 bar) and that from the condenser (at 0.1 bar) mix freely in a separate vessel (hot well) before being pumped to the boiler. Neglect extraneous losses.

Solution

Let w_s be the flow rate of steam (kg/hr) entering the turbine, and w the amount of steam extracted per hour for process heat (Fig. Ex. 12.6).

$h_1 = 3037.3$ kJ/kg

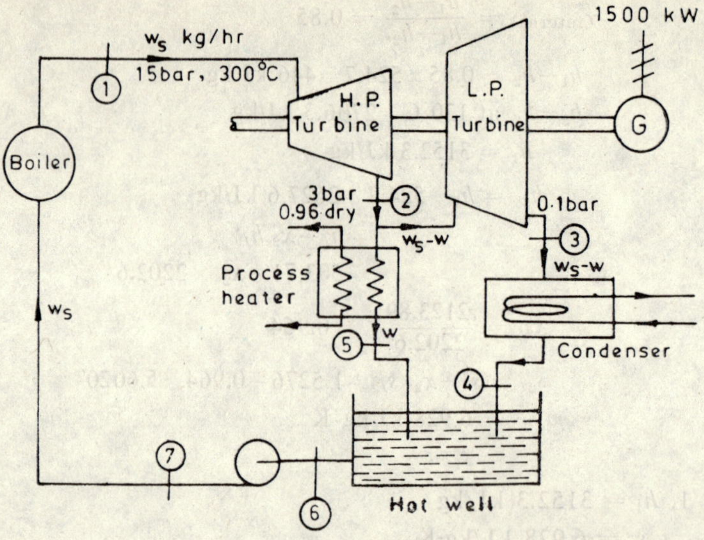

Fig. Ex. 12·6 (a)

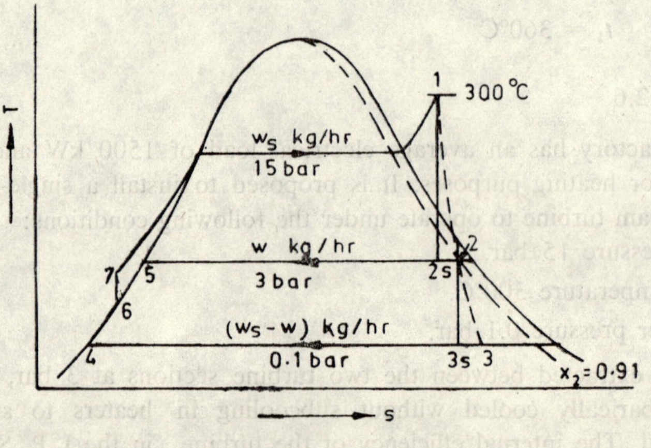

Fig Ex. 12·6 (b)

$$h_2 = 561.47 + 0.96 \times 2163.8$$
$$= 2638.7 \text{ kJ/kg}$$
$$s_2 = 1.6718 + 0.96 \times 5.3201$$
$$= 6.7791 \text{ kJ/kg-K}$$
$$= s_{3s}$$
$$s_{3s} = 6.7791 = 0.6493 + x_{3s} \times 7.5009$$
$$x_{3s} = \frac{6.1298}{7.5009} = 0.817$$
$$h_{3s} = 191.83 + 0.817 \times 2392.8 = 2146.75 \text{ kJ/kg}$$

$$h_2 - h_{3s} = 2638.7 - 2146.75 = 491.95 \text{ kJ/kg}$$

$$h_2 - h_3 = 0.8 \times 491.95 = 393.56 \text{ kJ/kg}$$

$$\therefore \quad h_3 = 2638.7 - 393.56 = 2245.14 \text{ kJ/kg}$$

$$h_5 = 561.47 \text{ kJ/kg}, \quad h_4 = 191.83 \text{ kJ/kg}$$

$$Q_H = w\,(h_2 - h_5) = w\,(2638.7 - 561.47) = 3.5 \text{ MJ/s}$$

$$\therefore \quad w = \frac{3500}{2077.23} = 1.685 \text{ kg/s}$$

Now

$$W_T = w_s\,(h_1 - h_2) + (w_s - w)\,(h_2 - h_3)$$

$$= w_s\,(3037.3 - 2638.7) + (w_s - 1.685) \times 393.56$$

$$= w_s \times 398.6 + w_s \times 393.56 - 663.15$$

$$= 792.16\, w_s - 663.15$$

Neglecting pump work

$$W_T = 1500 \text{ kJ/s} = 792.16\, w_s - 663.15$$

$$\therefore \quad w_s = \frac{2163.15}{792.16} = 2.73 \text{ kg/s} = 9828 \text{ kg/hr}$$

By making energy balance for the hot well

$$(w_s - w)\, h_4 + w h_5 = w_s\, h_6$$

$$(2.73 - 1.685)\, 191.83 + 1.685 \times 561.47 = 2.73 \times h_6$$

$$200.46 + 946.08 = 2.73\, h_6$$

$$\therefore \quad h_6 = 419.98 \text{ kJ/kg} \simeq h_7$$

Steam raising capacity of the boiler = 1.1 w_s kg/hr, since 10% of boiler steam is used for auxiliaries.

$$\therefore \quad \eta_{\text{boiler}} = \frac{1.1\, w_s\,(h_1 - h_7)}{w_f \times \text{C.V.}}$$

where w_f = fuel burning rate (kg/hr)
and C.V. = calorific value of fuel = 44 MJ/kg

$$\therefore \quad 0.85 = \frac{1.1 \times 9828 \times (3037.3 - 419.98)}{w_f \times 44000}$$

or

$$w_f = \frac{1.1 \times 9828 \times 2617.32}{0.85 \times 44000} = 756.56 \text{ kg/hr}$$

$$= \frac{756.56 \times 24}{1000} = 18.16 \text{ tonnes/day}$$

Ans.

EXAMPLE 12.7

A steam turbine gets its supply of steam at 70 bar and 450°C. After expanding to 25 bar in high pressure stages, it is reheated to 420°C at the constant pressure. Next, it is expanded in intermediate pressure stages to an appropriate minimum pressure such that part of the steam bled at this pressure heats the feedwater to a temperature of 180°C. The remaining steam expands from this pressure to a condenser pressure of 0.07 bar in the low pressure stage. The isentropic efficiency of the h.p. stage is 78.5%, while that of the intermediate and l.p. stages is 83% each. From the above data (a) determine the minimum pressure at which bleeding is necessary, and sketch a line diagram of the arrangement of the plant, (b) sketch on the *T-s* diagram all the processes, (c) determine the quantity of steam bled per kg of flow at the turbine inlet, and (d) calculate the cycle efficiency. Neglect pump work.

Solution

Figure Ex. 12.7 gives the flow and *T-s* diagrams of the plant. It would be assumed that the feedwater heater is an open heater. Feedwater is heated to 180°C. So $p_{sat.}$ at 180°C ≅ 10 bar is the pressure at which the heater operates.

Therefore, the pressure at which bleeding is necessary is 10 bar. Ans.(a).

From the Mollier chart

$$h_1 = 3285,\ h_{2s} = 3010,\ h_3 = 3280,\ h_{4s} = 3030 \text{ kJ/kg}$$
$$h_3 - h_4 = 0.83\ (h_3 - h_{4s}) = 0.83 \times 250 = 207.5 \text{ kJ/kg}$$
$$h_4 = 3280 - 207.5 = 3072.5 \text{ kJ/kg}$$
$$h_{5s} = 2225 \text{ kJ/kg}$$
$$h_4 - h_5 = 0.83\ (h_4 - h_{5s}) = 0.83 \times 847.5 = 703.4 \text{ kJ/kg}$$
$$\therefore \quad h_5 = 3072.5 - 703.4 = 2369.1 \text{ kJ/kg}$$
$$h_6 = 162.7 \text{ kJ/kg}$$
$$h_8 = 762.81 \text{ kJ/kg}$$
$$h_1 - h_2 = 0.785\ (h_1 - h_{2s}) = 0.785 \times 275 = 215.9 \text{ kJ/kg}$$
$$h_2 = 3285 - 215.9 = 3069.1 \text{ kJ/kg}$$

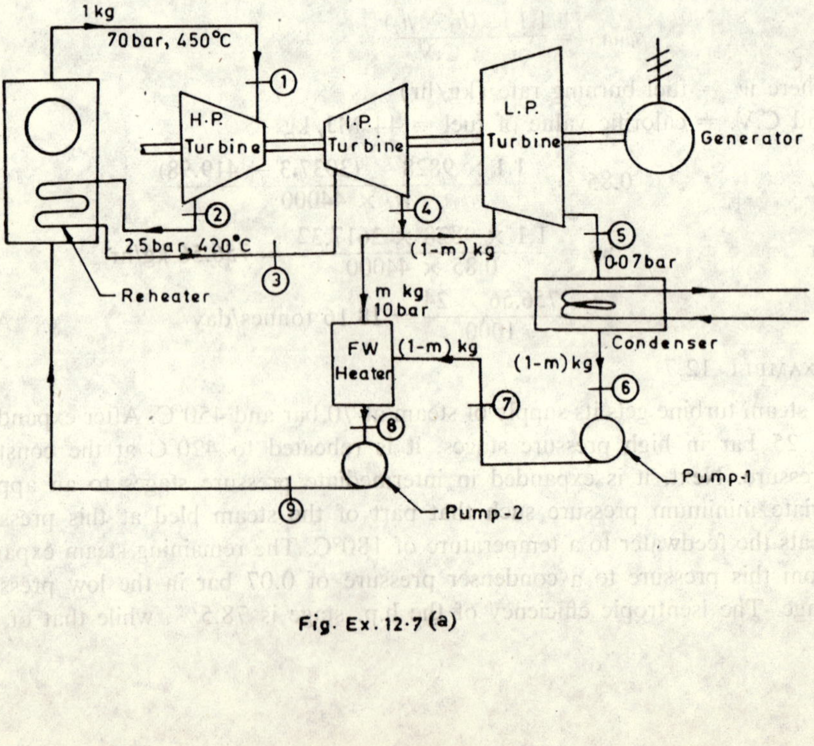

Fig. Ex. 12·7 (a)

Energy balance for the heater gives

$$m \times h_4 + (1 - m)\, h_7 = 1 \times h_8$$

$$m \times 3072.5 + (1 - m) \times 162.7 = 1 \times 762.81$$

$$m = \frac{600.11}{2909.8} = 0.206 \text{ kg/kg steam flow at turbine inlet}$$

Ans. (c).

$$\therefore \quad \eta_{\text{cycle}} = \frac{(h_1 - h_2) + (h_3 - h_4) + (1 - m)(h_4 - h_5)}{(h_1 - h_8) + (h_3 - h_2)}$$

$$= \frac{215.9 + 207.5 + 0.794 \times 703.4}{2522.2 + 210.9}$$

$$= \frac{981.9}{2733.1} = 0.3592 \text{ or } 35.92\%$$

Ans. (d).

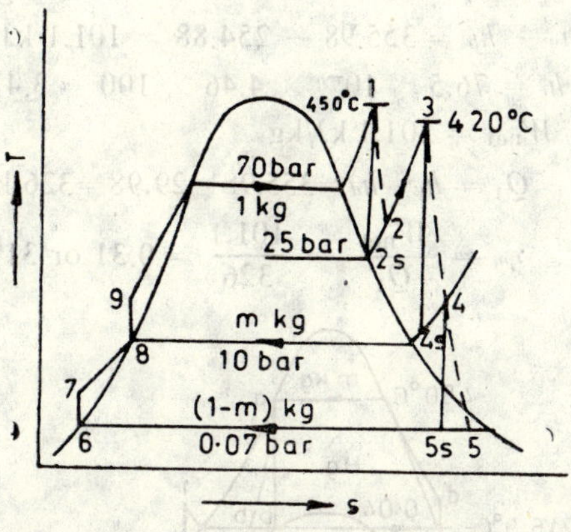

Fig. Ex. 12.7 (b)

Example 12.8

A binary-vapour cycle operates on mercury and steam. Saturated mercury vapour at 4.5 bar is supplied to the mercury turbine, from which it exhausts at 0.04 bar. The mercury condenser generates saturated steam at 15 bar which is expanded in a steam turbine to 0.04 bar. (a) Find the overall efficiency of the cycle. (b) If 50,000 kg/hr of steam flows through the steam turbine, what is the flow through the mercury turbine? (c) Assuming that all processes are reversible, what is the useful work done in the binary vapour cycle for the specified steam flow? (d) If the steam leaving the mercury condenser is superheated to a temperature of 300°C in a superheater located in the mercury boiler, and if the internal efficiencies of the mercury and steam turbines are 0.85 and 0.87 respectively, calculate the overall efficiency of the cycle. The properties of saturated mercury are given below

p (bar)	t(°C)	h_f (kJ/kg)	h_g	s_f (kJ/kg-K)	s_g	v_f (m³/kg)	v_g
4.5	450	62.93	355.98	0.1352	0.5397	79.9×10^{-6}	0.068
0.04	216.9	29.98	329.85	0.0808	0.6925	76.5×10^{-6}	5.178

Solution

The cycle is shown in Fig. Ex. 12.8.

For the mercury cycle, $h_a = 355.98$ kJ/kg

$$s_a = 0.5397 \text{ kJ/kg-K} = s_b = s_f + x_b\, s_{fg}$$

$$= 0.0808 + x_b\,(0.6925 - 0.0808)$$

$$\therefore \quad x_b = \frac{0.4589}{0.6117} = 0.75$$

$$h_b = 29.98 + 0.75 \times 299.87 = 254.88 \text{ kJ/kg}$$

$$(W_T)_m = h_a - h_b = 355.98 - 254.88 = 101.1 \text{ kJ/kg}$$

$$(W_P)_m = h_d - h_c = 76.5 \times 10^{-6} \times 4.46 \times 100 = 3.41 \times 10^{-2} \text{ kJ/kg}$$

$$\therefore \quad W_{\text{net}} = 101.1 \text{ kJ/kg}$$

$$Q_1 = h_a - h_d = 355.98 - 29.98 = 326 \text{ kJ/kg}$$

$$\therefore \quad \eta_m = \frac{W_{\text{net}}}{Q_1} = \frac{101.1}{326} = 0.31 \text{ or } 31\%$$

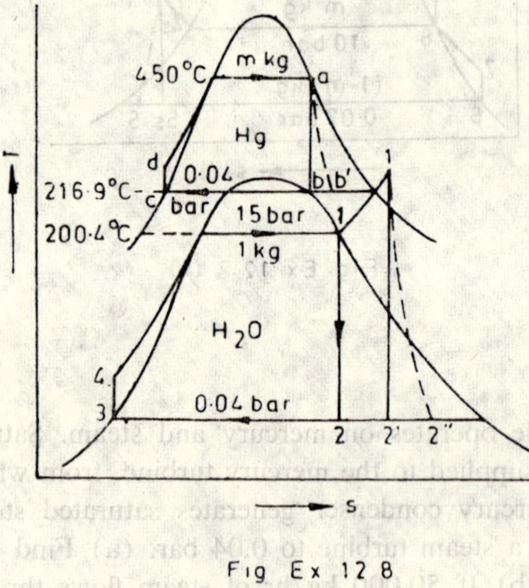

Fig Ex 12.8

For the steam cycle

$$h_1 = 2792.2 \text{ kJ/kg}$$

$$s_1 = 6.4448 \text{ kJ/kg-K} = s_2 = s_f + x_2\, s_{fg2}$$

$$= 0.4226 + x_2\,(8.4746 - 0.4226)$$

$$x_2 = \frac{6.0222}{8.0520} = 0.748$$

$$h_2 = 121.46 + 0.748 \times 2432.9 = 1941.27 \text{ kJ/kg}$$

$$(W_T)_{St.} = h_1 - h_2 = 2792.2 - 1941.27$$
$$= 850.93 \text{ kJ/kg}$$

$$(W_P)_{St.} = h_4 - h_3 = 0.001 \times 14.96 \times 100 = 1.496 \text{ kJ/kg}$$
$$\simeq 1.5 \text{ kJ/kg}$$

$$h_4 = 121.46 + 1.5 = 122.96 \text{ kJ/kg}$$

$$Q_1 = h_1 - h_4 = 2792.2 - 122.96 = 2669.24 \text{ kJ/kg}$$

$$(W_{\text{net}})_{St} = 850.93 - 1.5 = 849.43 \text{ kJ/kg}$$

$$\therefore \quad \eta_{\text{st.}} = \frac{W_{\text{net}}}{Q_1} = \frac{849.43}{2669.24} = 0.318 \text{ or } 31.8\%$$

Overall efficiency of the binary cycle would be

$$\eta_{\text{overall}} = \eta_{\text{m}} + \eta_{\text{st.}} - \eta_{\text{m}} \cdot \eta_{\text{st.}}$$
$$= 0.31 + 0.318 - 0.31 \times 0.318$$
$$= 0.5294 \text{ or } 52.94\% \qquad \text{Ans. (a).}$$

η_{overall} can also be determined in the following way:

By writing the energy balance for a mercury condenser-steam boiler

$$m\,(h_b - h_c) = 1\,(h_1 - h_4)$$

where m is the amount of mercury circulating for 1 kg of steam in the bottom cycle.

$$\therefore \quad m = \frac{h_1 - h_4}{h_b - h_c} = \frac{2669.24}{254.88 - 29.98} = \frac{2669.24}{224.90} = 11.87 \text{ kg}$$

$$(Q_1)_{\text{total}} = m\,(h_a - h_d) = 11.87 \times 326 = 3869.6 \text{ kJ/kg}$$

$$(W_T)_{\text{total}} = m\,(h_a - h_b) + (h_1 - h_2)$$
$$= 11.87 \times 101.1 + 850.93 = 2051 \text{ kJ/kg}$$

$(W_P)_{\text{total}}$ may be neglected.

$$\therefore \quad \eta_{\text{overall}} = \frac{W_T}{Q_1} = \frac{2051}{3869.6} = 0.53 \text{ or } 53\%$$

If 50,000 kg/hr of steam flows through the steam turbine, the flow rate of mercury w_m would be

$$w_m = 50{,}000 \times 11.87 = 59.35 \times 10^4 \text{ kg/hr} \qquad \text{Ans. (b)}$$

$$(W_T)_{\text{total}} = 2051 \times 50{,}000 = 10255 \times 10^4 \text{ kJ/hr}$$
$$= 0.2849 \times 10^5 \text{ kW} = 28.49 \text{ MW} \qquad \text{Ans (c).}$$

Considering the efficiencies of turbines

$$(W_T)_m = h_a - h_b' = 0.85 \times 101.1 = 85.94 \text{ kJ/kg}$$

$$\therefore \quad h_b' = 355.98 - 85.94 = 270.04 \text{ kJ/kg}$$

$$\therefore \quad m'\,(h_{b'} - h_{c'} = 1\,(h_1) - h_4)$$

$$\therefore \quad m' = \frac{2669.24}{240.06} = 11.12 \text{ kg}$$

$$(Q_1)_{\text{total}} = m'\,(h_a - h_d) + 1\,(h_1' - h_1)$$
$$= 11.12 \times 326 + (3037.3 + 2792.2)$$
$$= 3870.22 \text{ kJ/kg}$$

$$s_1' = 6.9160 = 0.4226 + x_2'\,(8.4746 - 0.4226)$$

$$\therefore \quad x_2' = \frac{6.4934}{8.0520} = 0.806$$

$$h_2' = 121.46 + 0.806 \times 2432.9 = 2082.38 \text{ kJ/kg}$$

$$(W_T)_{st.} = h_1' - h_2'' = 0.87\ (3037.3 - 2082.38)$$

$$= 830.78 \text{ kJ/kg}$$

$$(W_T)_{total} = 11.12 \times 85.94 + 830.78$$

$$= 1786.43 \text{ kJ/kg}$$

Pump work is neglected.

$$\therefore \quad \eta_{overall} = \frac{1786.43}{3870.22} = 0.462 \text{ or } 46.2\%$$

Ans. (d).

PROBLEMS

12.1 For the following steam cycles, find (a) W_T in kJ/kg, (b) W_P in kJ/kg, (c) Q_1 in kJ/kg, (d) cycle efficiency, (e) steam rate in kg/kW hr, and (f) moisture at the end of the turbine process. Show the results in tabular form with your comments.

	Boiler Outlet	*Condenser Pressure*	*Type of Cycle*
(a)	10 bar, saturated	1 bar	Ideal Rankine Cycle
(b)	—do—	—do—	Neglect W_P
(c)	—do—	—do—	Assume 75% pump and turbine efficiency
(d)	—do—	0.1 bar	Ideal Rankine Cycle
(e)	10 bar, 300°C	—do—	—do—
(f)	160 bar, 600°C	—do—	—do—
(g)	—do—	—do—	Reheat to 600°C at maximum intermediate pressure to limit end moisture to 15%
(h)	—do—	—do—	—do— but with 85% turbine efficiency
(i)	10 bar, saturated	0.1 bar	Isentropic pump process ends on saturated liquid line
(j)	—do—	—do—	—do— but with 80% machine efficiencies
(k)	—do—	—do—	Ideal regenerative cycle
(l)	—do—	—do—	Single open heater at 110°C
(m)	—do—	—do—	Two open heaters at 90°C and 135°C.
(n)	—do—	—do—	—do— but the heaters are closed heaters

12.2 A geothermal power plant utilizes steam produced by natural means underground. Steam wells are drilled to tap this steam supply which is available at 4.5 bar and 175°C. The steam leaves the turbine at 100 mm Hg absolute pressure. The turbine isentropic efficiency is 0.75. Calculate the efficiency of the plant. If the unit produces 12.5 MW, what is the steam flow rate ?

12.3 A simple steam power cycle uses solar energy for the heat input. Water in the cycle enters the pump as a saturated liquid at 40°C, and is pumped to 2 bar. It then evaporates in the boiler at this pressure, and enters the turbine as saturated vapour. At the turbine exhaust the conditions are 40°C and 10% moisture. The flow rate is 150 kg/hr. Determine (a) the turbine isentropic efficiency, (b) the net work output, (c) the cycle efficiency, and (b) the area of solar collector needed if the collectors pick up 0.58 kW/m^2. Ans. 12.78%, 18.2 m^2.

12.4 In a reheat cycle, the initial steam pressure and the maximum temperature are 150 bar and 550°C respectively. If the condenser pressure is 0.1 bar and the moisture at the condenser inlet is 5%, and assuming ideal processes, determine (a) the reheat pressure, (b) the cycle efficiency, and (c) the steam rate.

Ans. 13.5 bar, 43.6%, 205 kg/kW hr.

12.5 In a nuclear power plant heat is transferred in the reactor to liquid sodium. The liquid sodium is then pumped to a heat exchanger where heat is transferred to steam. The steam leaves this heat exchanger as saturated vapour at 55 bar, and is then superheated in an external gas-fired superheater to 650°C. The steam then enters the turbine, which has one extraction point at 4 bar, where steam flows to an open feedwater heater. The turbine efficiency is 75% and the condenser temperature is 40°C. Determine the heat transfer in the reactor and in the superheater to produce a power output of 80 MW.

12.6 In a reheat cycle, steam at 500°C expands in a h.p. turbine till it is saturated vapour. It is reheated at constant pressure to 400°C and then expands in a l.p. turbine to 40°C. If the maximum moisture content at the turbine exhaust is limited to 15%, find (a) the reheat pressure, (b) the pressure of steam at the inlet to the h.p. turbine, (c) the net specific work output, (d) the cycle efficiency, and (e) the steam rate. Assume all ideal processes.

What would have been the quality, the work output, and the cycle efficiency without the reheating of steam ? Assume that the other conditions remain the same.

12.7 A regenerative cycle operates with steam supplied at 30 bar and 300°C, and condenser pressure of 0.08 bar. The extraction points for two heaters (one closed and one open) are at 3.5 bar and 0.7 bar respectively. Calculate the thermal efficiency of the plant, neglecting pump work.

12.8 The net power output of the turbine in an ideal reheat-regenerative cycle is 100 MW. Steam enters the high-pressure (H.P.) turbine at 90 bar, 550°C. After expansion to 7 bar, some of the steam goes to an open heater and the balance is reheated to 400°C, after which it expands to 0.07 bar. (a) What is the steam flow rate to the H.P. turbine ? (b) What is the total pump work ? (c) Calculate the cycle efficiency. (d) If there is a 10°C rise in the temperature of the cooling water, what is the rate of flow of the cooling water in the condenser ? (e) If the velocity of the steam flowing from the turbine to the condenser is limited to a maximum of 130 m/s, find the diameter of the connecting pipe.

12.9 A mercury cycle is superposed on the steam cycle operating between the boiler outlet condition of 40 bar, 400°C and the condenser temperature of 40°C. The heat released by mercury condensing at 0.2 bar is used to impart the latent heat of vaporization to the water in the steam cycle. Mercury enters the mercury turbine as saturated vapour at 10 bar. Compute (a) the kg of mercury circulated per kg of water, and (b) the efficiency of the combined cycle.

The property values of saturated mercury are given below

p (bar)	t (°C)	h_f (kJ/kg)	h_g	s_f (kJ/kg-K)	s_g	v_f (m³/kg)	v_g
10	515.5	72.23	363.0	0.1478	0.5167	80.9×10^{-6}	0.0333
0.2	277.3	38.35	336.55	0·0967	0.6385	77.4×10^{-6}	1.163

12.10 In an electric generating station, using a binary vapour cycle with mercury in the upper cycle and steam in the lower, the ratio of mercury flow to steam flow is 10:1 on a mass basis. At an evaporation rate of 1,000,000 kg/hr for the mercury, its specific enthalpy rises by 356 kJ/kg in passing through the boiler. Superheating the steam in the boiler furnace adds 586 kJ to the steam specific enthalpy. The mercury gives up 251.2 kJ/kg during condensation, and the steam gives up 2003 kJ/kg in its condenser. The overall boiler efficiency is 85%. The combined turbine mechanical and generator efficiencies are each 95% for the mercury and steam units. The steam auxiliaries require 5% of the energy generated by the units. Find the overall efficiency of the plant.

12.11 A sodium-mercury-steam cycle operates between 1000°C and 40°C. Sodium rejects heat at 670°C to mercury. Mercury boils at 24.6 bar and rejects heat at 0.141 bar. Both the sodium and mercury cycles are saturated. Steam is formed at 30 bar and is superheated in the sodium boiler to 350°C. It rejects heat at 0.08 bar. Assume isentropic expansions, no heat losses, and no regeneration and neglect pumping work. Find (a) the amounts of sodium and mercury used per kg of steam, (b) the heat added and rejected in the composite cycle per kg steam, (c) the total work done per kg steam. (d) the efficiency of the composite cycle, (e) the efficiency of the corresponding Carnot cycle, and (f) the work, heat added, and efficiency of a supercritical pressure steam (single fluid) cycle operating at 250 bar and between the same temperature limits.

For mercury, at 24.6 bar, $h_g = 366.78$ kJ/kg

$s_g = 0.48$ kJ/kg-K, and at 0.141 bar, $s_f = 0.09$

and $s_g = 0.64$ kJ/kg-K, $h_f = 36.01$ and $h_g = 330.77$ kJ/kg

For sodium, at 1000°C, $h_g = 4982.53$ kJ/kg

At turbine exhaust, $h = 3914.85$ kJ/kg

At 670°C, $h_f = 745.29$ kJ/kg.

For a supercritical steam cycle, the specific enthalpy and entropy at the turbine inlet may be computed by extrapolation from the steam tables.

12.13 A 10,000 kW steam turbine operates with steam at the inlet at 40 bar, 400°C saturated and 1000 kW of power, for which a back pressure turbine of 7% internal efficiency is to be used. Find the steam condition required at the inlet to the turbine.

12.13 A 10,000 kW steam turbine operates with steam at the inlet at 40 bar, 400°C and exhausts at 0.1 bar. Ten thousand kg/hr of steam at 3 bar are to be extracted for process work. The turbine has 75% isentropic efficiency throughout. Find the boiler capacity required.

12.14 A 50 MW steam plant built in 1935 operates with steam at the inlet at 60 bar, 450°C and exhausts at 0.1 bar, with 80% turbine efficiency. It is proposed to scrap the old boiler and put in a new boiler and a topping turbine of efficiency 85% operating with inlet steam at 180 bar, 500°C. The exhaust from the topping turbine at 60 bar is reheated to 450°C and admitted to the old turbine. The flow rate is just sufficient to produce the rated output from the old turbine. Find the improvement in efficiency with the new set up. What is the additional power developed?

12.15 A steam plant operates with an initial pressure at 20 bar and temperature 400°C, and exhausts to a heating system at 2 bar. The condensate from the heating system is returned to the boiler plant at 65°C, and the heating system utilizes for its intended purpose 90% of the energy transferred from the steam it receives. The turbine efficiency is 70%. (a) What fraction of the energy supplied to the steam plant serves a useful purpose ? (b) If two separate steam plants had been set up to produce the same useful energy, one to generate heating steam at 2 bar, and the other to generate power through a cycle working between 20 bar, 400°C and 0.07 bar, what fraction of the energy supplied would have served a useful purpose ? Ans. 91.2%, 64.5%.

12.16 In a nuclear power plant saturated steam at 30 bar enters a h.p. turbine and expands isentropically to a pressure at which its quality is 0.841. At this pressure the steam is passed through a moisture separator which removes all the liquid. Saturated vapour leaves the separator and is expanded isentropically to 0.04 bar in l.p. turbine, while the saturated liquid leaving the separator is returned via a feed pump to the boiler. The condensate leaving the condenser at 0.04 bar is also returned to the boiler via a second feed pump. Calculate the cycle efficiency and turbine outlet quality taking into account the feed pump term. Recalculate the same quantities for a cycle with the same boiler and condenser pressures but without moisture separation.

Ans. 35.5%, 0.824; 35%; 0.716.

12.17 The net power output of an ideal regenerative-reheat steam cycle is 80 MW. Steam enters the h.p. turbine at 80 bar, 500°C and expands till it becomes saturated vapour. Some of the steam then goes to an open feedwater heater and the balance is reheated to 400°C, after which it expands in the l.p. turbine to 0.07 bar. Compute (a) the reheat pressure, (b) the steam flow rate to the h.p. turbine, and (c) the cycle efficiency. Neglect pump work. Ans. 6.5 bar, 58.4 kg/s, 43.7%.

12.18 Figure P. 12.18 shows the arrangement of a steam plant in which steam is also required for an industrial heating process. The steam leaves boiler B at 30 bar, 320°C and expands in the H.P. turbine to 2 bar, the efficiency of the H.P. turbine being 75%. At this point one half of the steam passes to the process heater P and the remainder enters separator S which removes all the moisture. The dry steam enters the L.P. turbine at 2 bar and expands to the condenser pressure 0.07 bar, the efficiency

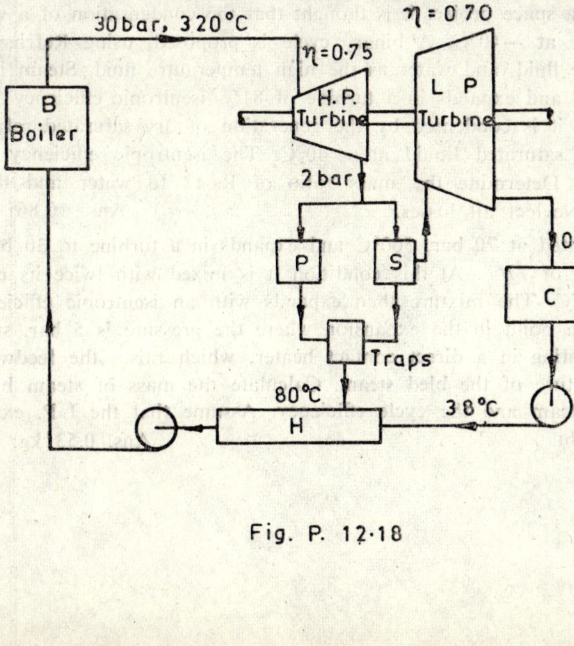

Fig. P. 12·18

of the L.P. turbine being 70%. The drainage from the separator mixes with the condensate from the process heater and the combined flow enters the hotwell *H* at 80°C. Traps are provided at the exits from *P* and *S*. A pump extracts the condensate from condenser *C* and this enters the hotwell at 38°C. Neglecting the feed pump work and radiation loss, estimate the temperature of water leaving the hotwell which is at atmospheric pressure. Also calculate, as percentage of heat transferred in the boiler, (a) the heat transferred in the process heater, and (b) the work done in the turbines.

12.19 In a combined power and process plant the boiler generates 21,000 kg/hr of steam at a pressure of 17 bar, and temperature 230°C. A part of the steam goes to a process heater which consumes 132.56 kW, the steam leaving the process heater 0.957 dry at 17 bar being throttled to 3.5 bar. The remaining steam flows through a H.P. turbine which exhausts at a pressure of 3.5 bar. The exhaust steam mixes with the process steam before entering the L.P. turbine which develops 1337.5 kW. At the exhaust the pressure is 0.3 bar, and the steam is 0.912 dry. Draw a line diagram of the plant and determine (a) the steam quality at the exhaust from the H.P. turbine, (b) the power developed by the H.P. turbine, and (c) the isentropic efficiency of the H.P. turbine. Ans. (a) 0.96, (b) 1125 kW, (c) 77%.

12.20 Steam is supplied to a pass-out turbine at 35 bar, 350°C and dry saturated process steam is required at 3.5 bar. The low pressure stage exhausts at 0.07 bar and the condition line may be assumed to be straight (the condition line is the locus passing through the states of steam leaving the various stages of the turbine). If the power required is 1 MW and the maximum process load is 1.4 kW, estimate the maximum steam flow through the high and low pressure stages. Assume that the steam just condenses in the process plant. Ans. 1.543 and 1.182 kg/s.

12.21 Geothermal energy from a natural geyser can be obtained as a continuous supply of steam 0.87 dry at 2 bar and at a flow rate of 2700 kg/hr. This is utilized in a mixed-pressure cycle to augment the superheated exhaust from a high pressure turbine of 83% internal efficiency, which is supplied with 5500 kg/hr of steam at 40 bar and 500°C. The mixing process is adiabatic and the mixture is expanded to a condenser pressure of 0.10 bar in a low pressure turbine of 78% internal efficiency. Determine the power output and the thermal efficiency of the plant.

Ans. 1745 kW, 35%

12.22 In a study for a space project it is thought that the condensation of a working fluid might be possible at —40°C. A binary cycle is proposed, using Refrigerant-12 as the low temperature fluid, and water as the high temperautre fluid. Steam is generated at 80 bar, 500°C and expands in a turbine of 81% isentropic efficiency to 0.06 bar, at which pressure it is condensed by the generation of dry saturated refrigerant vapour at 30°C from saturated liquid at —40°C. The isentropic efficiency of the R-12 turbine is 83%. Determine the mass ratio of Rs-12 to water and the efficiency of the cycle. Neglect all losses. Ans. 10.86; 44.4%.

12.23 Steam is generated at 70 bar, 500°C and expands in a turbine to 30 bar with an isentropic efficiency of 77%. At this condition it is mixed with twice its mass of steam at 30 bar, 400°C. The mixture then expands with an isentropic efficiency of 80% to 0.06 bar. At a point in the expansion where the pressure is 5 bar, steam is bled for feedwater heating in a direct contact heater, which raises the feedwater to the saturation temperature of the bled steam. Calculate the mass of steam bled per kg of high pressure steam and the cycle efficiency. Assume that the L.P. expansion condition line is straight. Ans. 0.53 kg; 31.9%.

13

Gas Power Cycles

Here gas is the working fluid. It does not undergo any phase change. Engines operating on gas cycles may be either cyclic or non-cyclic. Hot air engines using air as the working fluid operate on a closed cycle. Internal combustion engines where the combustion of fuel takes place inside the engine cylinder are non-cyclic heat engines.

13.1 CARNOT CYCLE (1824)

The Carnot cycle (Fig. 13.1) has been discussed in Chapters 6 and 7. It consists of:

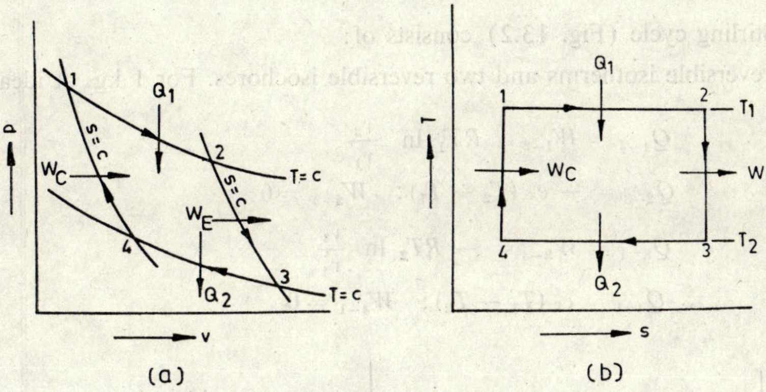

Fig. 13.1 Carnot cycle

Two reversibile isotherms and two reversible adiabatics. If an ideal gas is assumed as the working fluid, then for 1 kg of gas

$$Q_{1-2} = RT_1 \ln \frac{v_2}{v_1}; \qquad W_{1-2} = RT_1 \ln \frac{v_2}{v_1}$$

$$Q_{2-3} = 0; \qquad W_{2-3} = -c_v\,(T_3 - T_2)$$

$$Q_{3-4} = RT_2 \ln \frac{v_4}{v_3}; \qquad W_{3-4} = RT_2 \ln \frac{v_4}{v_3}$$

$$Q_{4-1} = 0; \qquad W_{4-1} = -c_v\,(T_1 - T_4)$$

$$\therefore \qquad \sum_{\text{cycle}} đQ = \sum_{\text{cycle}} đW$$

Now

$$\frac{v_2}{v_3} = \left(\frac{T_2}{T_1}\right)^{1/(\gamma-1)}$$

and

$$\frac{v_1}{v_4} = \left(\frac{T_2}{T_1}\right)^{1/(\gamma-1)}$$

$$\therefore \quad \frac{v_2}{v_3} = \frac{v_1}{v_4} \text{ or } \frac{v_2}{v_1} = \frac{v_3}{v_4}$$

Therefore

$$Q_1 = \text{Heat added} = RT_1 \ln \frac{v_2}{v_1}$$

$$W_{\text{net}} = Q_1 - Q_2 = R \ln \frac{v_2}{v_1} \cdot (T_1 - T_2)$$

$$\therefore \quad \eta_{\text{cycle}} = \frac{W_{\text{net}}}{Q_1} = \frac{T_1 - T_2}{T_1} \tag{13.1}$$

The large back work ($W_c = W_{4-1}$) is a big drawback for the Carnot gas cycle, as in the case of the Carnot vapour cycle.

13.2 STIRLING CYCLE (1827)

The Stirling cycle (Fig. 13.2) consists of:

Two reversible isotherms and two reversible isochores. For 1 kg of ideal gas

$$Q_{1-2} = W_{1-2} = RT_1 \ln \frac{v_2}{v_1}$$

$$Q_{2-3} = -c_v (T_2 - T_1); \quad W_{2-3} = 0$$

$$Q_{3-4} = W_{3-4} = -RT_2 \ln \frac{v_3}{v_4}$$

$$Q_{4-1} = c_v (T_1 - T_2); \quad W_{4-1} = 0$$

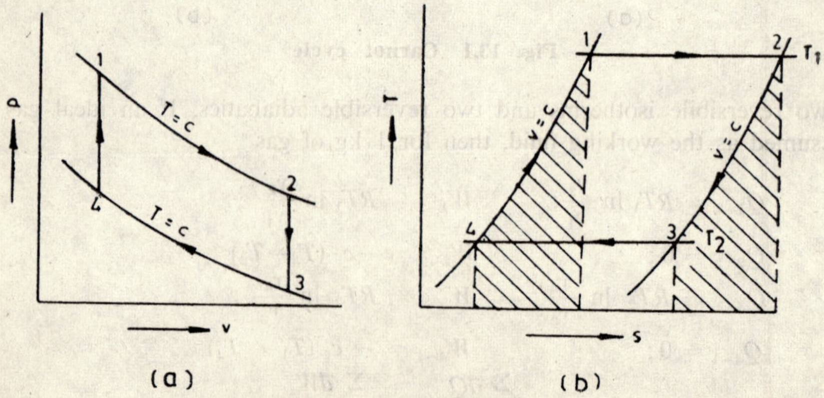

Fig. 13.2 Stirling cycle

Due to heat transfers at constant volume processes, the efficiency of the Stirling cycle is less than that of the Carnot cycle. However, if a regenerative arrangement is used such that

$Q_{2-3} = Q_{4-1}$, i.e., the area under 2-3 is equal to the area under 4-1, then the cycle efficiency becomes

$$\eta = \frac{RT_1 \ln \frac{v_2}{v_1} - RT_2 \ln \frac{v_3}{v_4}}{RT_1 \ln \frac{v_2}{v_1}} = \frac{T_1 - T_2}{T_1} \tag{13.2}$$

So, the regenerative Stirling cycle has the same efficiency as the Carnot cycle

13.3 ERICSSON CYCLE (1850)

The Ericsson cycle (Fig. 13.3) is made up of:
Two reversible isotherms and two reversible isobars.
For 1 kg of ideal gas

$$Q_{1-2} = W_{1-2} = RT_1 \ln \frac{p_1}{p_2}$$

$$Q_{2-3} = c_p (T_2 - T_1); \quad W_{2-3} = p_2 (v_3 - v_2) = R (T_2 - T_1)$$

$$Q_{3-4} = W_{3-4} = - RT_2 \ln \frac{p_1}{p_2}$$

$$Q_{4-1} = c_p (T_1 - T_4); \quad W_{4-1} = p_1 (v_1 - v_4) = R (T_1 - T_2)$$

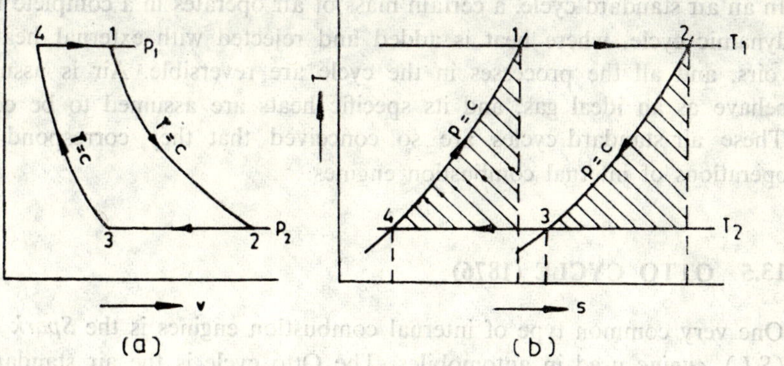

Fig. 13.3 Ericsson cycle

Since part of the heat is transferred at constant pressure and part at constant temperature, the efficiency of the Ericsson cycle is less than that of the Carnot cycle. But with ideal regeneration, $Q_{2-3} = {}_{4-1}$. so that all the heat is added from the external source at T_1 and all the heat is rejected to an external sink at T_2, the efficiency of the cycle becomes equal to the Carnot cycle efficiency, since

$$\eta = 1 - \frac{Q_2}{Q_1} = 1 - \frac{R\,T_2 \ln \dfrac{p_1}{p_2}}{R\,T_1 \ln \dfrac{p_1}{p_2}} = 1 - \frac{T_2}{T_1} \quad (13.3)$$

The regenerative Stirling and Ericsson cycles have the same efficiency as the Carnot cycle, but much less back work. Hot air engines working on these cycles have been successfully operated. But it is difficult to transfer heat to a gas at high rates, since the gas film has a very low thermal conductivity. So there has not been much progress in the development of hot air engines. However, since the cost of internal combustion engine fuels is getting excessive, these may find a field of use in the near future.

13.4 AIR STANDARD CYCLES

Internal combustion engines in which the combustion of fuel occurs in the engine cylinder itself are non-cyclic heat engines. The temperature due to the evolution of heat because of the combustion of fuel inside the cylinder is so high that the cylinder is cooled by water circulation around it to avoid rapid deterioration. The working fluid, the fuel-air mixture, undergoes permanent chemical change due to combustion, and the products of combustion after doing work are thrown out of the engine, and a fresh charge is taken. So the working fluid does not undergo a complete thermodynamic cycle.

To simplify the analysis of I.C. engines, *air standard cycle* are conceived. In an air standard cycle, a certain mass of air operates in a complete thermodynamic cycle, where heat is added and rejected with external heat reservoirs, and all the processes in the cycle are reversible. Air is assumed to behave as an ideal gas, and its specific heats are assumed to be constant. These air standard cycles are so conceived that they correspond to the operations of internal combustion engines.

13.5 OTTO CYCLE (1876)

One very common type of internal combustion engines is the *Spark Ignition (S.I.) engine* used in automobiles. The Otto cycle is the air standard cycle of such an engine. The sequence of processes in the elementary operation of the S.I. engine is given below, with reference to Fig. 13.4, where the sketches of the engine and the indicator diagram are given.

Process 1-2, Intake. The inlet valve is open, the piston moves to the right, admitting fuel-air mixture into the cylinder at constant pressure.

Process 2-3, Compression. Both the valves are closed, the piston compresses the combustible mixture to the minimum volume.

Process 3-4, Combustion. The mixture is then ignited by means of a spark, combustion takes place, and there is an increase in temperature and pressure.
Process 4-5, Expansion. The products of combustion do work on the piston which moves to the right, and the pressure and temperature of the gases decrease.

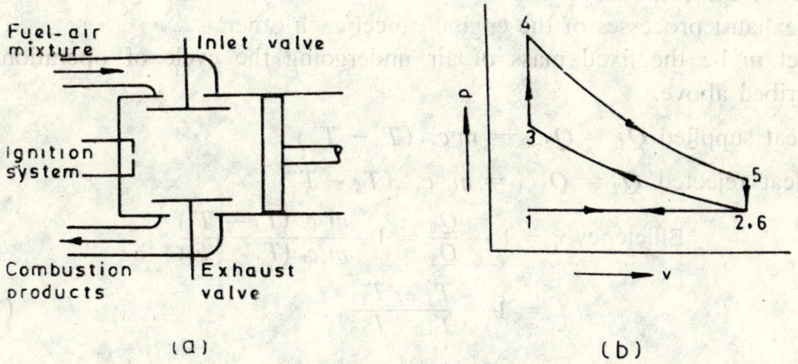

Fig. 13.4 (a) S.I. engine. (b) Indicator diagram

Process 5-6, Blow-down. The exhaust valve opens, and the pressure drops to the initial pressure.
Process 6-1, Exhaust. With the exhaust valve open, the piston moves inwards to expel the combustion products from the cylinder at constant pressure.

The series of process as described above constitute a *mechanical cycle,* and not a thermodynamic cycle. The cycle is completed in four strokes of the piston.

Figure 13.5 shows the air standard cycle (Otto cycle) corresponding to the above engine. It consists of:

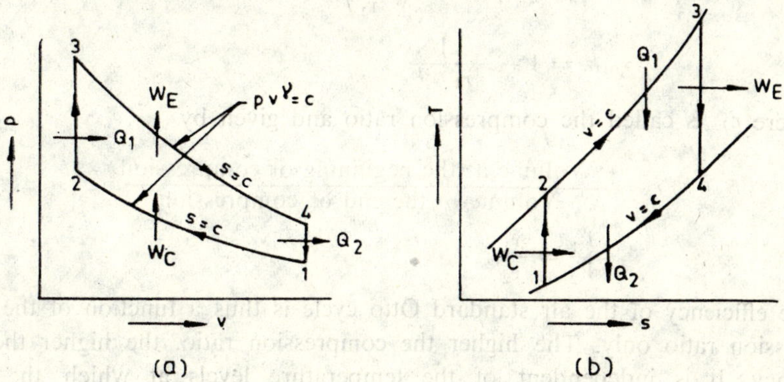

Fig. 13.5 Otto cycle

Two reversible adiabatics and two reversible isochores.

Air is compressed in process 1-2 reversibly and adiabatically. Heat is then added to air reversibly at constant volume in process 2-3. Work is done by air in expanding reversibly and adiabatically in process 3-4. Heat is then rejected by air reversibly at constant volume in process 4-1, and the system (air) comes back to its initial state. Heat transfer processes have been substituted for the combustion and blow-down processes of the engine. The intake and exhaust processes of the engine cancel each other.

Let m be the fixed mass of air undergoing the cycle of operations as described above.

Heat supplied $Q_1 = Q_{2-3} = m\, c_v\, (T_3 - T_2)$

Heat rejected $Q_2 = Q_{4-1} = m\, c_v\, (T_4 - T_1)$

$$\text{Efficiency } \eta = 1 - \frac{Q_2}{Q_1} = 1 - \frac{m\, c_v\, (T_4 - T_1)}{m\, c_v\, (T_3 - T_2)}$$

$$= 1 - \frac{T_4 - T_1}{T_3 - T_2} \tag{13.4}$$

Process 1-2, $\dfrac{T_2}{T_1} = \left(\dfrac{v_1}{v_2}\right)^{\gamma-1}$

Process 3-4, $\dfrac{T_3}{T_4} = \left(\dfrac{v_4}{v_3}\right)^{\gamma-1} = \left(\dfrac{v_1}{v_2}\right)^{\gamma-1}$

$$\therefore \quad \frac{T_2}{T_1} = \frac{T_3}{T_4}$$

or

$$\frac{T_3}{T_2} = \frac{T_4}{T_1}$$

$$\frac{T_3}{T_2} - 1 = \frac{T_4}{T_1} - 1$$

$$\therefore \quad \frac{T_4 - T_1}{T_3 - T_2} = \frac{T_1}{T_2} = \left(\frac{v_2}{v_1}\right)^{\gamma-1}$$

$\therefore$ From equation (13.4), $\eta = 1 - \left(\dfrac{v_2}{v_1}\right)^{\gamma-1}$

or

$$\eta_{\text{otto}} = 1 - \frac{1}{r_k^{\gamma-1}} \tag{13.5}$$

where r_k is called the compression ratio and given by

$$r_k = \frac{\text{Volume at the beginning of compression}}{\text{Volume of the end of compression}}$$

$$= \frac{V_1}{V_2} = \frac{v_1}{v_2}$$

The efficiency of the air standard Otto cycle is thus a function of the compression ratio only. The higher the compression ratio, the higher the efficiency. It is independent of the temperature levels at which the cycle operates. The compression ratio cannot, however, be increased beyond a certain limit, because of a noisy and destructive combustion phenomenon, known as detonation. It also depends upon the fuel, the engine design, and the operating conditions.

13.6 DIESEL CYCLE (1892)

The limitation on compression ratio in the S.I. engine can be overcome by compressing air alone, instead of the fuel-air mixture, and then injecting the fuel into the cylinder in spray form when combustion is desired. The temperature of air after compression must be high enough so that the fuel sprayed into the hot air burns spontaneously. The rate of burning can, to some extent, be controlled by the rate of injection of fuel. An engine operating in this way is called a *compression ignition* (*C.I.*) *engine*. The sequence of processes in the elementary operation of a C.I. engine, shown in Fig. 13.6, is given below.

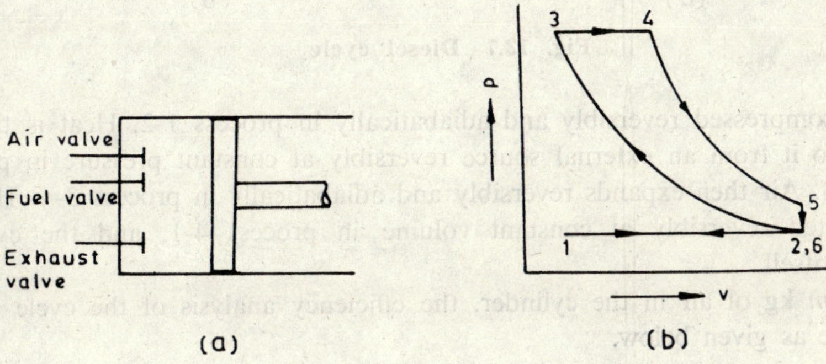

Fig. 13.6 (a) C.I. engine (b) Indicator diagram

Process 1-2, Intake. The air valve is open. The piston moves out admitting air into the cylinder at constant pressure.
Process 2-3, Compression. The air is then compressed by the piston to the minimum volume with all the valves closed.
Process 3-4, Fuel injection and combustion. The fuel valve is open, fuel is sprayed into the hot air, and combustion takes place at constant pressure.
Process 4-5, Expansion. The combustion products expand, doing work on the piston which moves out to the maximum volume.
Process 5-6, Blow-down. The exhaust valve opens, and the pressure drops to the initial pressure.
Pressure 6-1, Exhaust. With the exhaust valve open, the piston moves towards the cylinder cover driving away the combustion products from the cylinder at constant pressure.

The above processes constitute an engine cycle, which is completed in four strokes of the piston or two revolutions of the crank shaft.

Figure 13.7 shows the air standard cycle, called the *Diesel cycle,* corresponding to the C.I. engine, as described above. The cycle is composed of:

Two reversible adiabatics, one reversible isobar, and one reversible isochore.

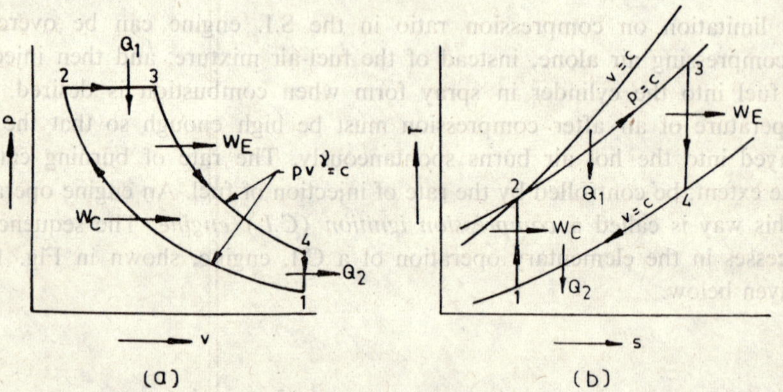

Fig. 13.7 Diesel cycle

Air is compressed reversibly and adiabatically in process 1-2. Heat is then added to it from an external source reversibly at constant pressure in process 2-3. Air then expands reversibly and adiabatically in process 3-4. Heat is rejected reversibly at constant volume in process 4-1, and the cycle repeats itself.

For m kg of air in the cylinder, the efficiency analysis of the cycle can be made as given below.

Heat supplied, $Q_1 = Q_{2-3} = m\, c_p\, (T_3 - T_2)$

Heat rejected, $Q_2 = Q_{4-1} = m\, c_v\, (T_4 - T_1)$

Efficiency $\eta = 1 - \dfrac{Q_2}{Q_1} = 1 - \dfrac{m\, c_v\, (T_4 - T_1)}{m\, c_p\, (T_3 - T_2)}$

$$\therefore \quad \eta = 1 - \frac{T_4 - T_1}{\gamma\, (T_3 - T_2)} \qquad (13.6)$$

The efficiency may be expressed in terms of any two of the following three ratios

Compression ratio, $r_k = \dfrac{V_1}{V_2} = \dfrac{v_1}{v_2}$

Expansion ratio, $r_e = \dfrac{V_4}{V_3} = \dfrac{v_4}{v_3}$

Cut-off ratio, $r_c = \dfrac{V_3}{V_2} = \dfrac{v_3}{v_2}$

It is seen that

$$r_k = r_e \cdot r_c$$

Process 3-4

$$\frac{T_4}{T_3} = \left(\frac{v_3}{v_4}\right)^{\gamma-1} = \frac{1}{r_e^{\gamma-1}}$$

$$\therefore \quad T_4 = T_3 \cdot = \frac{r_c^{\gamma-1}}{r_k^{\gamma-1}}$$

Process 2-3

$$\frac{T_2}{T_3} = \frac{p_2 v_2}{p_3 v_3} = \frac{v_2}{v_3} = \frac{1}{r_c}$$

$$\therefore \quad T_2 = T_3 \cdot \frac{1}{r_c}$$

Process 1-2

$$\frac{T_1}{T_2} = \left(\frac{v_2}{v_1}\right)^{\gamma-1} = \frac{1}{r_k^{\gamma-1}}$$

$$\therefore \quad T_1 = T_2 \cdot \frac{1}{r_k^{\gamma-1}} = \frac{T_3}{r_c} \cdot \frac{1}{r_k^{\gamma-1}}$$

Substituting the values of T_1, T_2, and T_4 in the expression of efficiency (equation 13.6)

$$\eta = 1 - \frac{T_3 \cdot \frac{r_c^{\gamma-1}}{r_k^{\gamma-1}} - \frac{T_3}{r_k} \frac{1}{r_k^{\gamma-1}}}{\gamma\left(T_3 - T_3 \cdot \frac{1}{r_c}\right)}$$

$$\therefore \quad \eta_{\text{Diesel}} = 1 - \frac{1}{\gamma} \cdot \frac{1}{r_k^{\gamma-1}} \cdot \frac{r_c^{\gamma-1}}{r_c - 1} \qquad (13.7)$$

As $r_c > 1$, $\frac{1}{\gamma}\left(\frac{r_c^{\gamma-1}}{r_c-1}\right)$ is also greater than unity. Therefore, the efficiency of the Diesel cycle is less than that of the Otto cycle for the same compression ratio.

13.7 LIMITED PRESSURE CYCLE, MIXED CYCLE OR DUAL CYCLE

The air standard Diesel cycle does not simulate exactly the pressure-volume variation in an actual compression ignition engine, where the fuel injection is started before the end of compression stroke. A closer approximation is the limited pressure cycle in which some part of heat is added to air at constant volume, and the remainder at constant pressure.

Figure 13.8 shows the *p-v* and *T-s* diagrams of the dual cycle. Heat is added reversibly, partly at constant volume (2-3) and partly at constant pressure (3-4).

Heat supplied $Q_1 = m\, c_v\, (T_3 - T_2) + m\, c_p\, (T_4 - T_3)$

Heat rejected $Q_2 = m\, c_v\, (T_5 - T_1)$

Efficiency $\eta = 1 - \frac{Q_2}{Q_1}$

$$= 1 - \frac{m\, c_v\, (T_5 - T_1)}{m\, c_v\, (T_3 - T_2) + m\, c_p\, (T_4 - T_3)}$$

$$= 1 - \frac{T_5 - T_1}{(T_3 - T_2) + \gamma\, (T_4 - T_3)} \qquad (13.8)$$

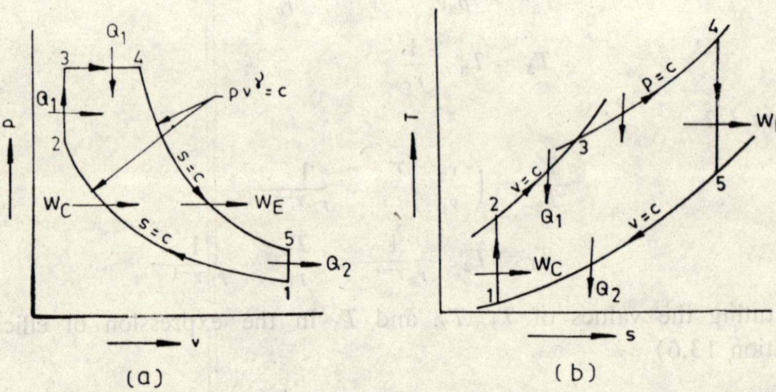

Fig. 13.8 Limited pressure cycle, Mixed cycle or Dual cycle

The efficiency of the cycle can be expressed in terms of the following ratios

Compression ratio, $r_k = \frac{V_1}{V_2}$

Expansion ratio, $r_e = \frac{V_5}{V_4}$

Cut-off ratio, $r_c = \frac{V_4}{V_3}$

Constant volume pressure ratio, $r_p = \frac{p_3}{p_2}$

It is seen, as before that

$$r_k = r_c \cdot r_e$$

or $$r_e = \frac{r_k}{r_c}$$

Process 3-4

$$r_c = \frac{V_4}{V_3} = \frac{T_4\, p_3}{p_4\, T_3} = \frac{T_4}{T_2}$$

$$\therefore \quad T_3 = \frac{T_4}{r_3}$$

Process 2-3

$$\frac{p_2 V_2}{T_2} = \frac{p_3 V_3}{T_3}$$

$$T_2 = T_3 \cdot \frac{p_2}{p_3} = \frac{T_4}{r_p \cdot r_c}$$

Process 1-2

$$\frac{T_1}{T_2} = \left(\frac{v_2}{v_1}\right)^{\gamma-1} = r_k^{1/(\gamma-1)}$$

$$\therefore \quad T_1 = \frac{T_4}{r_p \cdot r_c \cdot r_k^{\gamma-1}}$$

Process 4-5

$$\frac{T_5}{T_4} = \left(\frac{v_4}{v_5}\right)^{\gamma-1} = r_e^{1/(\gamma-1)}$$

$$\therefore \quad T_5 = T_4 \cdot \frac{r_c^{\gamma-1}}{r_k^{\gamma-1}}$$

Substituting the values of T_1, T_2, T_3, and T_5 in the expression of efficiency (equation 13.8),

$$\eta = 1 - \frac{T_4 \cdot \frac{r_c^{\gamma-1}}{r_k^{\gamma-1}} - \frac{T_4}{r_p \cdot r_c \cdot r_k^{\gamma-1}}}{\left(\frac{T_4}{r_c} - \frac{T_4}{r_p \cdot r_c}\right) + \gamma\left(T_4 - \frac{T_4}{r_c}\right)}$$

$$\therefore \quad \eta_{\text{Dual}} = 1 - \frac{1}{r_k^{\gamma-1}} \cdot \frac{r_p \cdot r_c^{\gamma} - 1}{r_p - 1 + \gamma r_p (r_c - 1)} \tag{13.9}$$

13.8 COMPARISON OF OTTO, DIESEL, AND DUAL CYCLES

The three cycles can be compared on the basis of either the same compression ratio or the same maximum pressure and temperature.

Figure 13.9 shows the comparison of Otto, Diesel, and Dual cycles for the same compression ratio and heat rejection. Here

1-2-6-5 —Otto cycle
1-2-7-5 —Diesel cycle
1-2-3-4-5—Dual cycle

For the same Q_2, the higher the Q_1, the higher is the cycle efficiency. In the *T-s* diagram, the area under 2-6 represents Q_1 for the Otto cycle, the area under 2-7 represents Q_1 for the Diesel cycle, and the area under 2-3-4 represents Q_1 for the Dual cycle. Therefore, for the same r_k and Q_2

$$\eta_{\text{Otto}} > \eta_{\text{Dual}} > \eta_{\text{Diesel}}$$

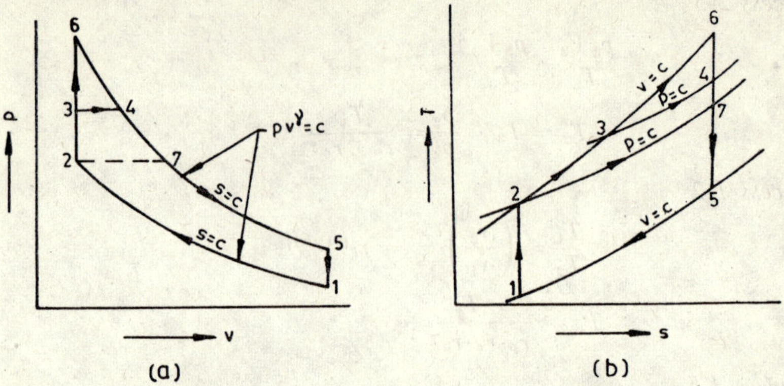

Fig. 13.9 Comparison of Otto, Diesel and Dual cycles for the same compression ratio

Figure 13.10 shows a comparison of the three air standard cycles for the same maximum pressure and temperature (state 4), the heat rejection being also the same. Here

1-6-4-5 —Otto cycle
1-7-4-5 —Diesel cycle
1-2-3-4-5—Dual cycle

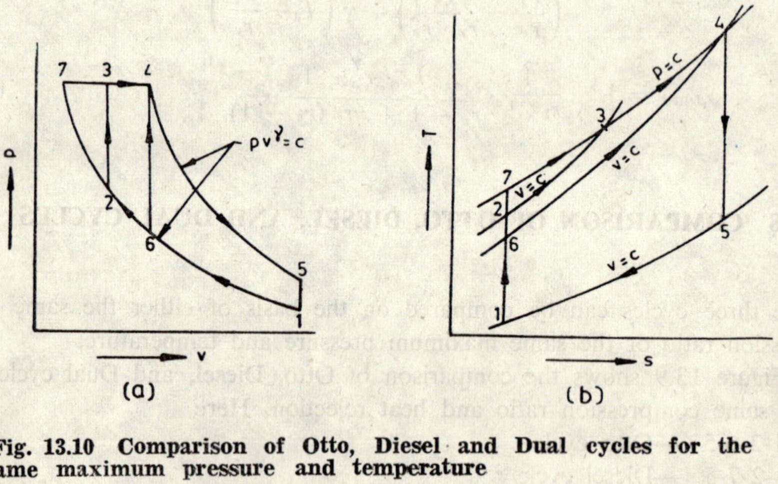

Fig. 13.10 Comparison of Otto, Diesel and Dual cycles for the same maximum pressure and temperature

Q_1 is represented by the area under 6-4 for the Otto cycle, by the area under 7-4 for the Diesel cycle and by the area under 2-3-4 for the Dual cycle in the *T-s* plot, Q_2 being the same.

$$\therefore \qquad \eta_{\text{Deisel}} > \eta_{\text{Dual}} > \eta_{\text{Otto}}$$

This comparison is of greater significance, since the Diesel cycle would definitely have a higher compression ratio than the Otto cycle.

13.9 BRAYTON CYCLE

A simple gas turbine power plant is shown in Fig. 13.11. Air is first compressed adiabatically in process *a-b*, it then enters the combustion chamber where fuel is injected and burned essentially at constant pressure in process *b-c*, and then the products of combustion expand in the turbine to the ambient pressure in process *c-d* and are thrown out to the surroundings. The cycle is open. The state diagram on the *p-v* coordinates is shown in Fig. 13.12.

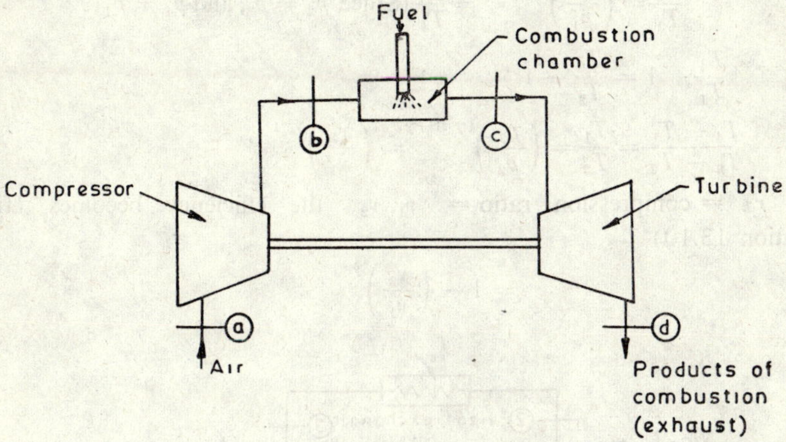

Fig. 13.11 A simple gas turbine plant

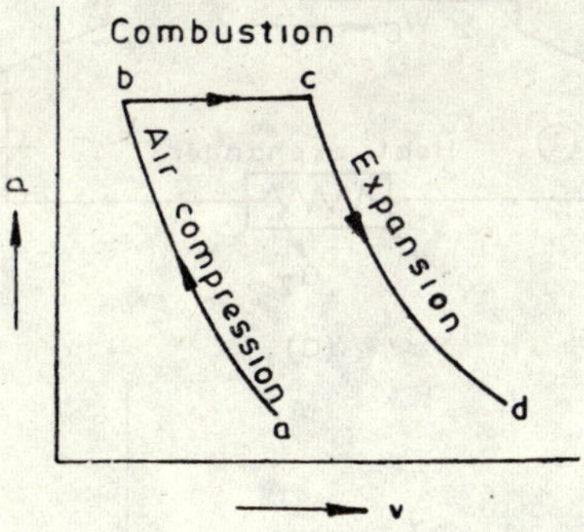

Fig. 13.12 State diagram of a gas turbine plant on p-v plot

The Brayton cycle is the air standard cycle for the gas turbine power plant. Here air is first compressed reversibly and adiabatically, heat is added to it reversibly at constant pressure, air expands in the turbine reversibly and adiabatically, and heat is then rejected from the air reversibly at constant pressure to bring it to the initial state. The Brayton cycle, therefore, consists of:

Two reversible isobars and two reversible adiabatics.

The flow, *p-v*, and *T-s* diagrams are shown in Fig. 13.13. For *m* kg of air

$$Q_1 = \text{heat supplied} = m\,c_p\,(T_3 - T_2)$$
$$Q_2 = \text{heat rejected} = m\,c_p\,(T_4 - T_1)$$

$\therefore$ Cycle efficiency, $\eta = 1 - \dfrac{Q_2}{Q_1}$

$$= 1 - \frac{T_4 - T_1}{T_3 - T_2} \quad (13.10)$$

Now

$$\frac{T_2}{T_1} = \left(\frac{p_2}{p_1}\right)^{(\gamma-1)/\gamma} = \frac{T_3}{T_4} \text{ (Since } p_2 = p_3, \text{ and } p_4 = p_1)$$

$$\therefore \quad \frac{T_4}{T_1} - 1 = \frac{T_3}{T_2} - 1$$

or
$$\frac{T_4 - T_1}{T_3 - T_2} = \frac{T_1}{T_2} = \left(\frac{p_1}{p_2}\right)^{(\gamma-1)/\gamma} = \left(\frac{v_2}{v_1}\right)^{\gamma-1}$$

If r_k = compression ratio = v_1/v_2 the efficiency becomes (from equation 13.10)

$$\eta = 1 - \left(\frac{v_2}{v_1}\right)^{\gamma-1}$$

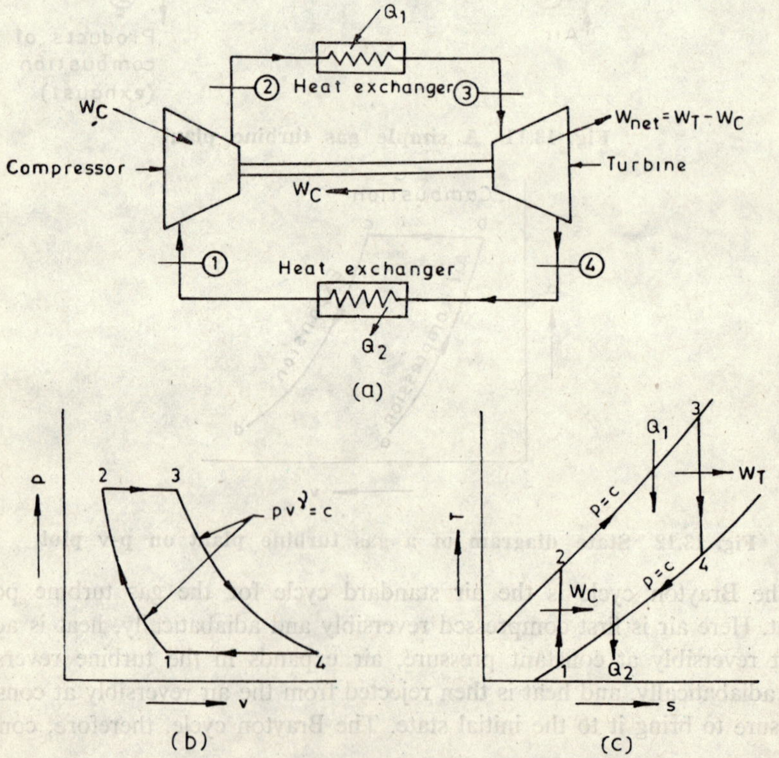

Fig. 13.13 Brayton cycle

or $$\eta_{\text{Brayton}} = 1 - \frac{1}{r_k^{\gamma-1}} \tag{13.11}$$

If r_p = pressure ratio = p_2/p_1 the efficiency may be expressed in the following form also

$$\eta = 1 - \left(\frac{p_1}{p_2}\right)^{(\gamma-1)/\gamma}$$

or $$\eta_{\text{Brayton}} = 1 - \frac{1}{(r_p)^{(\gamma-1)/\gamma}} \tag{13.12}$$

The efficiency of the Brayton cycle, therefore, depends upon either the compression ratio or the pressure ratio. For the same compression ratio, the Brayton cycle efficiency is equal to the Otto cycle efficiency.

13.9.1 Comparison between Brayton Cycle and Otto Cycle

Brayton and Otto cycles are shown superimposed on the *p-v* and *T-s* diagrams in Fig. 13.14. For the same r_k and work capacity, the Brayton cycle (1-2-5-6) handles a larger range of volume and a smaller range of pressure and temperature than does the Otto cycle (1-2-3-4).

In the reciprocating engine field, the Brayton cycle is not suitable. A reciprocating engine cannot efficiently handle a large volume flow of low pressure gas, for which the engine size ($\pi/4$ D^2L) becomes large, and the friction losses also become more. So the Otto cycle is more suitable in the reciprocating engine field.

In turbine plants, however, the Brayton cycle is more suitable than the Otto cycle. An internal combustion engine is exposed to the highest temperature (after the combustion of fuel) only for a short while, and it gets

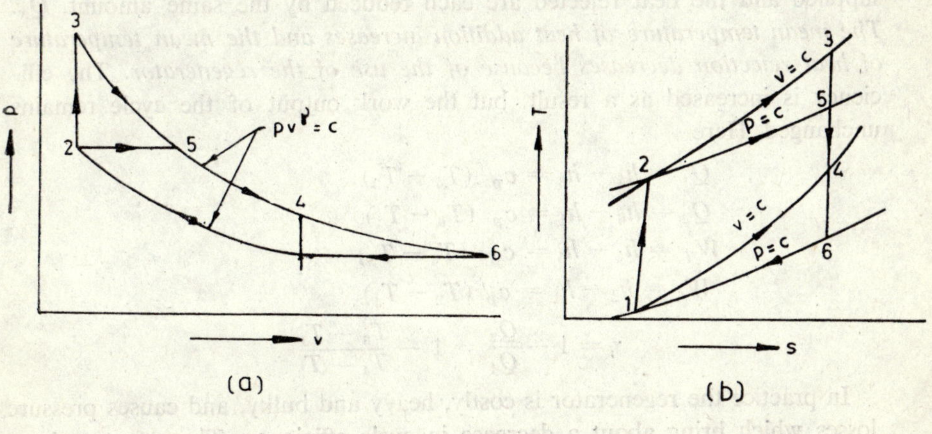

Fig. 13.14 Comparison of Otto and Brayton cycles.

time to become cool in the other processes of the cycle. On the other hand, a gas turbine plant, a seady flow device, is always exposed to the highest temperature used. So to protect material, the maximum temperature of gas that can be used in a gas turbine plant cannot be as high as in an internal combustion engine. Also, in the steady flow machinery, it is more difficult to carry out heat transfer at constant volume than at constant pressure. Moreover, a gas turbine can handle a large volume flow of gas quite efficiently. So we find that the Brayton cycle is the basic air standard cycle for all modern gas turbine plants.

13.9.2 Effect of Regeneration on Brayton Cycle Efficiency

The efficiency of the Brayton cycle can be increased by utilizing part of the energy of the exhaust gas from the turbine in heating up the air leaving the compressor in a heat exchanger called a *regenerator,* thereby reducing the amount of heat supplied from an external source and also the amount of head rejected. Such a cycle is illustrated in Fig. 13.15. The temperature of air leaving the turbine at 5 is higher than that of air leaving the compressor at 2. In the regenerator, the temperature of air leaving the compressor is raised by heat transfer from the turbine exhaust. The maximum temperature to which the cold air at 2 could be heated is the temperature of the hot air leaving the turbine at 5. This is possible only in an infinite heat exchanger. In the real case, the temperature at 3 is less than that at 5. The ratio of the actual temperature rise of air to the maximum possible rise is called the *effectiveness* of the regenerator. For the case illustrated

$$\text{Effectiveness} = \frac{t_3 - t_2}{t_5 - t_2}$$

When the regenerator is used in the idealized cycle (Fig. 13.15), the heat supplied and the heat rejected are each reduced by the same amount, Q_x. *The mean temperature of heat addition increases and the mean temperature of heat rejection decreases because of the use of the regenerator.* The efficiency is increased as a result, but the work output of the cycle remains unchanged. Here

$$Q_1 = h_4 - h_3 = c_p \ (T_4 - T_3)$$

$$Q_2 = h_6 - h_1 = c_p \ (T_6 - T_1)$$

$$W_T = h_4 - h_5 = c_p \ (T_4 - T_5)$$

$$W_c = h_2 - h_1 = c_p \ (T_2 - T_1)$$

$$\eta = 1 - \frac{Q_2}{Q_1} = 1 - \frac{T_6 - T_1}{T_4 - T_3}$$

In practice the regenerator is costly, heavy and bulky, and causes pressure losses which bring about a decrease in cycle efficiency. These factors have to be balanced against the gain in efficiency to decide whether it is worthwhile to use the regenerator.

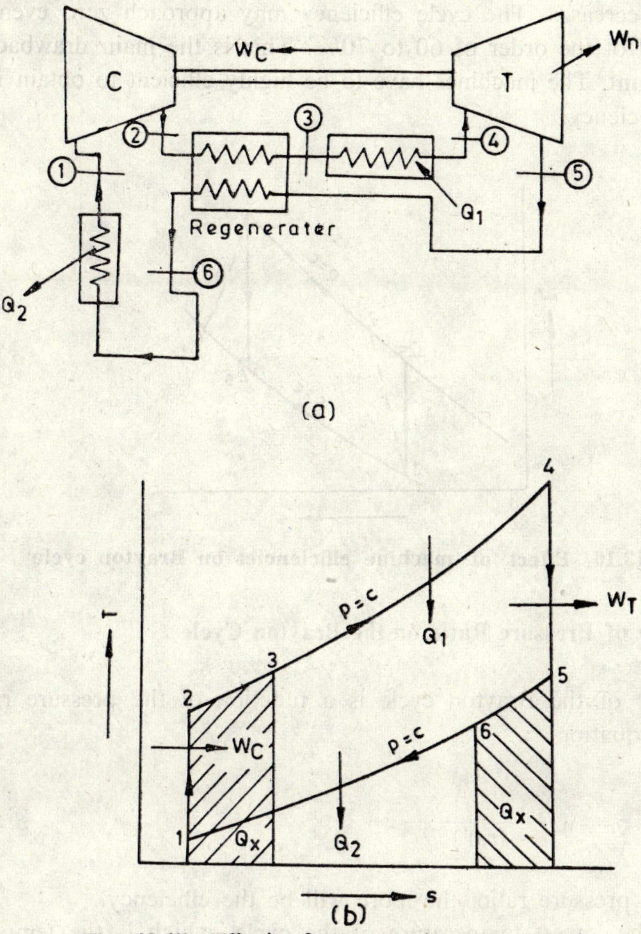

Fig. 13.15 Effect of regeneration on Brayton cycle

13.9.3 Effect of Irreversibilities in Turbine and Compressor

The Brayton cycle is highly sensitive to the real machine efficiencies of the turbine and the compressor. Figure 13.16 shows the actual and ideal expansion and compression processes.

$$\text{Turbine efficiency, } \eta_T = \frac{h_3 - h_4}{h_3 - h_{4s}} = \frac{T_3 - T_4}{T_3 - T_{4s}}$$

$$\text{Compressor efficiency, } \eta_C = \frac{h_{2s} - h_1}{h_2 - h_1} = \frac{T_{2s} - T_1}{T_2 - T_1}$$

$$W_{\text{net}} = W_T - W_C = (h_3 - h_4) - (h_2 - h_1)$$

$$Q_1 = h_3 - h_2 \text{ and } Q_2 = h_4 - h_1$$

The net output is reduced by the amount $[(h_4 - h_{4s}) + (h_2 - h_{2s})]$, and the heat supplied is reduced by the amount $(h_2 - h_{2s})$. The efficiency of the cycle will thus be less than that of the ideal cycle. As η_T and η_C decrease, η_{cycle} also decreases. The cycle efficiency may approach zero even when η_T and η_C are of the order of 60 to 70%. This is the main drawback of a gas turbine plant. The machines have to be highly efficient to obtain reasonable cycle efficiency.

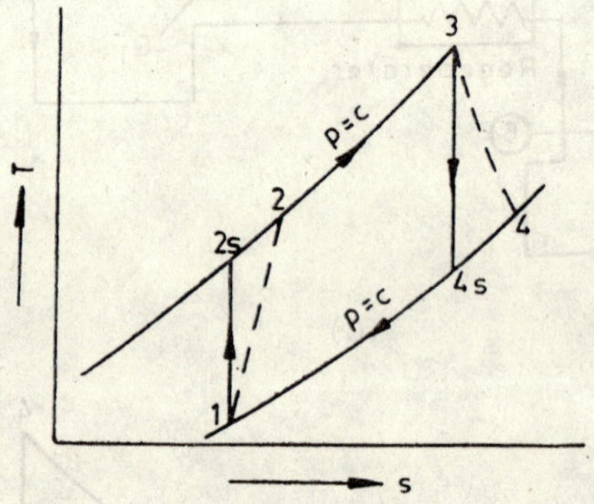

Fig. 13.16 Effect of machine efficiencies on Brayton cycle

13.9.4 Effect of Pressure Ratio on the Brayton Cycle

The efficiency of the Brayton cycle is a function of the pressure ratio as given by the equation

$$\eta = 1 - \frac{1}{(r_p)^{(\gamma-1)/\gamma}}$$

The more the pressure ratio, the more will be the efficiency.

Let T_1 = the lowest temperature of the cycle, which is the temperature of the surroundings (T_{min}), and T_3 = the maximum or the highest temperature of the cycle limited by the characteristics of the material available for burner and turbine construction (T_{max}).

Since the turbine, a steady-flow machine, is always exposed to the highest temperature gas, the maximum temperature of gas at the inlet to the turbine is limited to about 800°C by using a high air-fuel ratio.

Figure 13.17 shows the Brayton cycles operating between the same T_{max} and T_{min} at various pressure ratios. As the pressure ratio changes, the cycle shape also changes. For the cycle 1-2-3-4 of low pressure ratio r_p, since the average temperature of heat addition

$$T_{m1} = \frac{h_3 - h_2}{s_3 - s_2}$$

is only a little greater than the average temperature of heat rejection

$$T_{m2} = \frac{h_4 - h_1}{s_4 - s_1}$$

the efficiency will be low. *At the lower limit of unity pressure ratio, both work output and efficiency will be zero.*

As the pressure ratio is increased, the efficiency steadily increases, because T_{m1} increases and T_{m2} decreases. The mean temperature of heat addition T_{m1} approaches T_{max} and the mean temperature of heat rejection T_{m2} approaches T_{min}, with the increase in r_p. *In the limit when the compression process ends at T_{max}, the Carnot efficiency is reached, r_p has the maximum value (r_{pmax}), but the work capacity again becomes zero.*

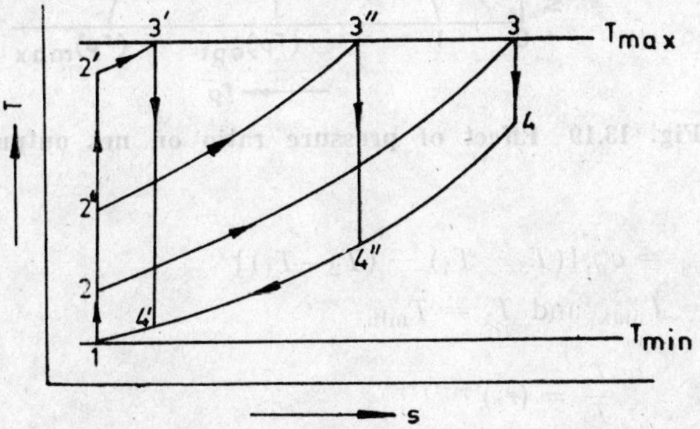

Fig. 13.17 Effect of pressure ratio on Brayton cycle

Figure 13.18 shows how the cycle efficiency varies with the pressure ratio, with r_p varying between two limiting values of 1 and $(r_p)_{max}$ when the Carnot efficiency is reached. When $r_p = (r_p)_{max}$, the cycle efficiency is given by

$$\eta = 1 - \frac{1}{((r_p)_{max})^{(\gamma-1)/\gamma}} = \eta_{Carnot}$$

$$= 1 - \frac{T_{min}}{T_{max}}$$

$$\therefore ((r_p)_{max})^{(\gamma-1)/\gamma} = \frac{T_{max}}{T_{min}}$$

$$\therefore (r_p)_{max} = \left(\frac{T_{max}}{T_{min}}\right)^{\gamma/(\gamma-1)} \quad (13.13)$$

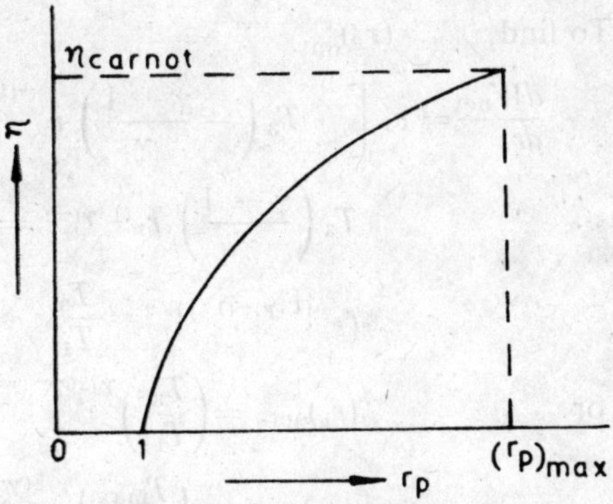

Fig. 13.18 Effect of pressure ratio on Brayton cycle efficiency

From Fig. 13.17 it is seen that the work capacity of the cycle, operating between T_{max} and T_{min}, is zero when $r_p = 1$ passes through a maximum, and then again becomes zero when the Carnot efficiency is reached. There is an optimum value of pressure ratio $(r_p)_{opt}$. at which work capacity becomes a maximum, as shown in Fig. 13.19.

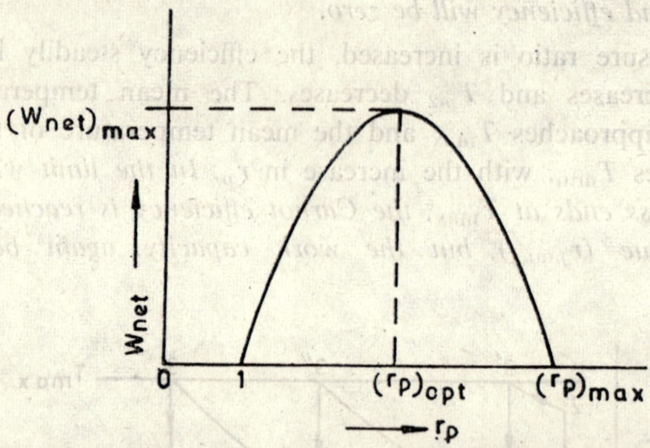

Fig. 13.19 Effect of pressure ratio on net output

For 1 kg,

$$W_{net} = c_p\ [(T_3 - T_4) - (T_2 - T_1)]$$

where $T_3 = T_{max}$ and $T_1 = T_{min}$

Now $$\frac{T_3}{T_4} = (r_p)^{(\gamma-1)/\gamma}$$

$\therefore$ $$T_4 = T_3 \cdot r_p^{-(\gamma-1)/\gamma}$$

Similarly $$T_2 = T_1 \cdot r_p^{(\gamma-1)/\gamma}$$

Substituting in the expression for W_{net}

$$W_{net} = c_p\,[T_3 - T_3 \cdot (r_p)^{-(\gamma-1)/\gamma} - T_1\ r_p^{(\gamma-1)/\gamma} + T^1] \quad (13.14)$$

To find $(r_p)_{opt.}$,

$$\frac{dW_{net}}{dr_p} = c_p \left[- T_3 \left(- \frac{\gamma - 1}{\gamma} \right) r_p^{-(1+\gamma)/-1} - T_1 \left(\frac{\gamma - 1}{\gamma} \right) r_p^{1-(1-\gamma)-1} \right] = 0$$

$\therefore$ $$T_3 \left(\frac{\gamma - 1}{\gamma} \right) r_p^{(1/\gamma)-2} = T_1 \left(\frac{\gamma - 1}{\gamma} \right) \cdot r_p^{-1/\gamma}$$

$\therefore$ $$r_p^{-(1/\gamma)-(1/\gamma)+2} = \frac{T_3}{T_1}$$

or $$(r_p)_{opt.} = \left(\frac{T_3}{T_1} \right)^{\gamma/2(\gamma-1)}$$

$\therefore$ $$(r_p)_{opt.} = \left(\frac{T_{max}}{T_{min}} \right)^{\gamma/2(\gamma-1)} \quad (13.15)$$

From equations (13.13) and (13.15)

$$(r_p)_{opt.} = \sqrt{(r_p)_{max}} \quad (13.16)$$

Substituting the values of $(r_p)_{opt.}$ in equation (13.14)

$$W_{net} = (W_{net})_{max} = c_p \left[T_3 - T_3 \left(\frac{T_1}{T_3}\right)^{\gamma/2(\gamma-1)\cdot((\gamma-1)/\gamma)} - T_1 \cdot \left(\frac{T_3}{T_1}\right)^{\gamma/2(\gamma-1)\cdot((\gamma-1)/\gamma)} + T_1 \right]$$

$$= c_p \left[T_3 - 2\sqrt{T_1 T_3} + T_1\right]$$

or $$(W_{net})_{max} = c_p \left(\sqrt{T_{max}} - \sqrt{T_{min}}\right)^2 \quad (13.17)$$

Considering the cycles 1-2′-3′-4′ and 1-2″-3″-4″ (Fig. 13.17), it is obvious that to obtain a reasonable work capacity, a certain reduction in efficiency must be accepted.

13.9.5 Effect of Intercooling and Reheating on Brayton Cycle

The efficiency of the Brayton cycle may often be increased by the use of staged compression with intercooling, or by using staged heat supply, called reheat.

Let the compression process be divided into two stages. Air, after being compressed in the first stage, is cooled to the initial temperature in a heat exchanger, called an intercooler, and then compressed further in the second stage (Fig. 13.20). 1-2′-5-6 is the ideal cycle without intercooling, having a

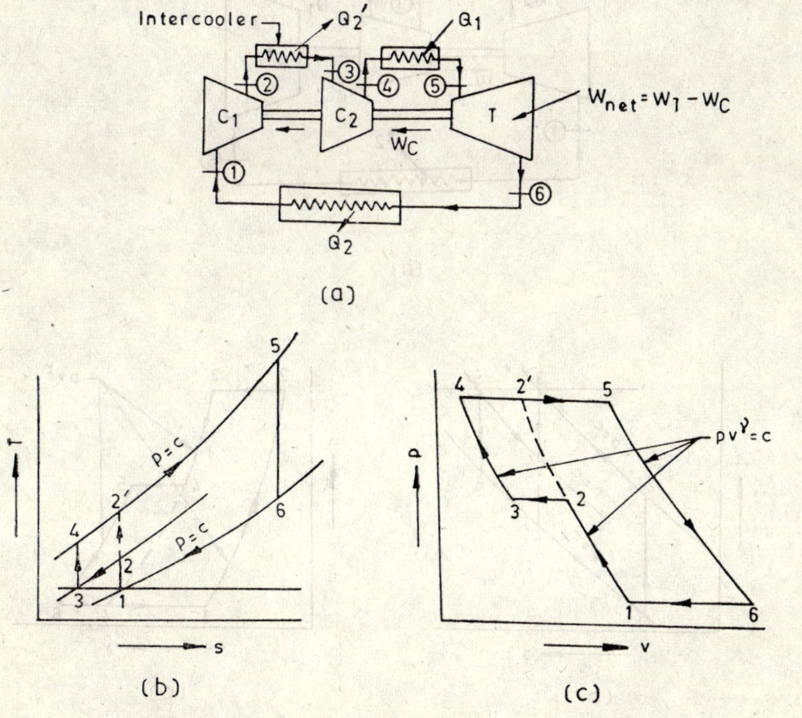

Fig. 13.20 Effect of intercooling on Brayton cycle

single-stage compression. 1-2-3-4-6 is the cycle with intercooling, having a two-stage compression. The cycle 2-3-4-2′ is thus added to the basic cycle 1-2′-5-6. There is more work capacity, since the included area is more. There is more heat supply also. For the cycle 4-2′-2-3, T_{m1} is lower and T_{m2} higher (lower r_p) than those of the basic cycle 1-2′-5-6. So the efficiency of the cycle reduces by staging the compression and intercooling. But if a regenerator is used, the low temperature heat addition (4-2′) may be obtained by recovering energy from the exhaust gases from the turbine. So there may be a net gain in efficiency when intercooling is adopted in conjunction with a regenerator.

Similarly, let the total heat supply be given in two stages and the expansion process be divided in stages in two turbines (T_1 and T_2) with intermediate reheat, as shown in Fig. 13.21. 1-2-3-4′ is the cycle with a single-stage heat supply having no reheat, with total expansion in one turbine only. 1-2-3-4-5-6 is the cycle with a single-stage reheat, having the expansion divided into two stages. With the basic cycle, the cycle 4-5-6-4′ is added because of reheat. The work capacity increases, but the heat supply also increases.

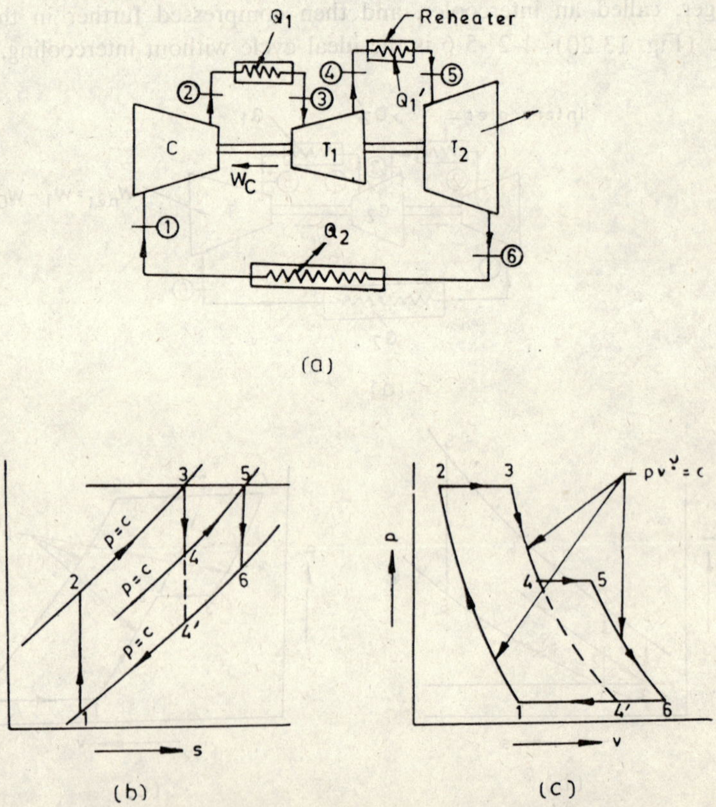

Fig. 13.21 Effect of reheat on Brayton cycle.

In the cycle 4-5-6-4′, r_p is lower than in the basic cycle 1-2-3-4′, so its efficiency is lower. Therefore, the efficiency of the cycle decreases with the use of reheat. But T_6 is greater than T_4'. Therefore, if regeneration is employed, these is more energy that can be recovered from the turbine exhaust gases. So when regeneration is employed in conjunction with reheat, there may be a net gain in cycle efficiency.

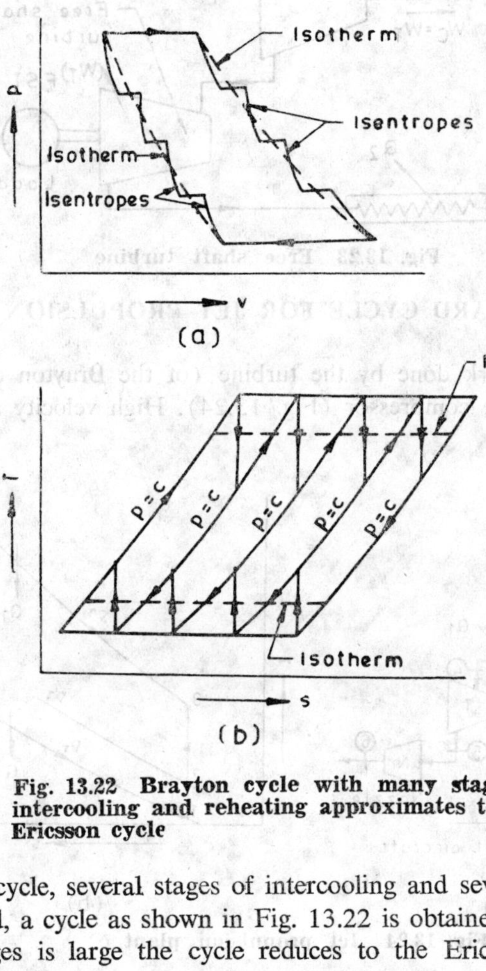

Fig. 13.22 Brayton cycle with many stages of intercooling and reheating approximates to Ericsson cycle

If in one cycle, several stages of intercooling and several stages of reheat are employed, a cycle as shown in Fig. 13.22 is obtained. When the number of such stages is large the cycle reduces to the Ericsson cycle with two reversible isobars and two reversible isotherms. With ideal regeneration the cycle efficiency becomes equal to the Carnot efficiency.

13.9.6 Free-shaft Turbine

So far only a single shaft has been shown in flow diagrams, on which were mounted all the compressors and the turbines. Sometimes, for operating convenience and part-load efficiency, one turbine is used for driving the compressor only on one shaft, and a separate turbine is used on another shaft, known as free-shaft, for supplying the load, as shown in Fig. 13.23.

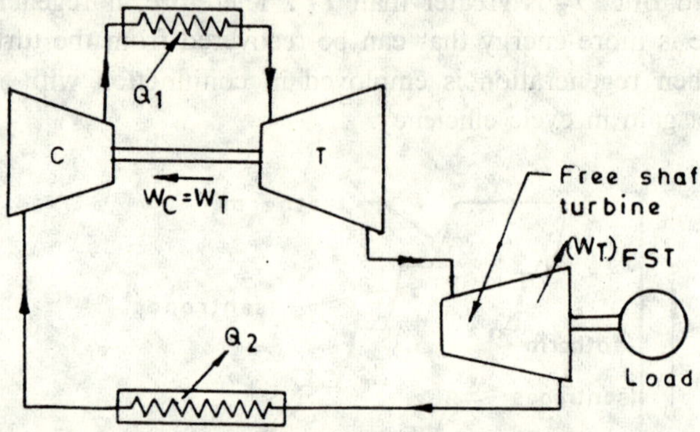

Fig. 13.23 Free shaft turbine

13.10 AIR STANDARD CYCLE FOR JET PROPULSION

In this cycle the work done by the turbine (of the Brayton cycle) is just sufficient to drive the compressor (Fig. 13.24). High velocity air first flows

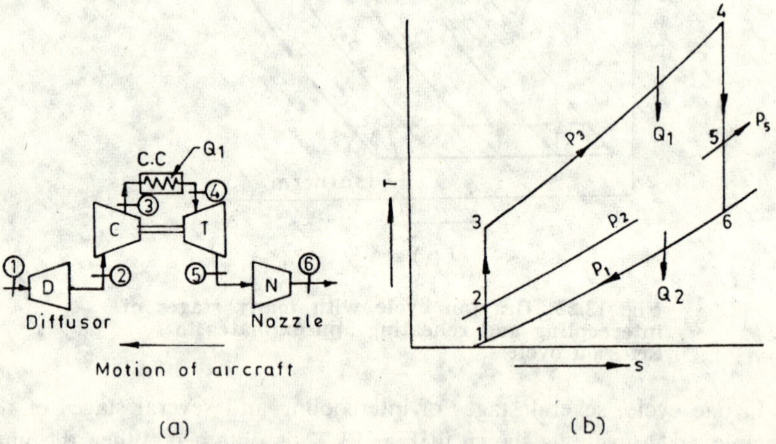

Fig. 13.24 Jet propulsion plant

through a diffusor (where the pressure of air increases at the expense of its K.E.) where it is decelerated, and then it is compressed reversibly and adiabatically in compressor *C*. Air is then heated reversibly at constant pressure from an external source (in the actual engine, fuel is burnt in the combustion chamber, *CC*), after which it expands in turbine *T* to a pressure so as to produce work output equal to the compressor work. The exhaust from the turbine is then expanded in a nozzle to the ambient pressure. Since air (in the actual plant, combustion gases) leaves at high velocity, the change in the momentum of the fluid results in a thrust upon the aircraft (to propel it forward) in which the engine is installed.

EXAMPLE 13.1

An engine working on the Otto cycle is supplied with air at 0.1 MPa, 35°C. The compression ratio is 8. Heat supplied is 2100 kJ/kg. Calculate the maximum pressure and temperature of the cycle, the cycle efficiency, and the mean effective pressure. (For air, $c_p = 1.005$, $c_v = 0.718$, and $R = 0.287$ kJ/kg-K).

Solution

From Fig. Ex. 13.1

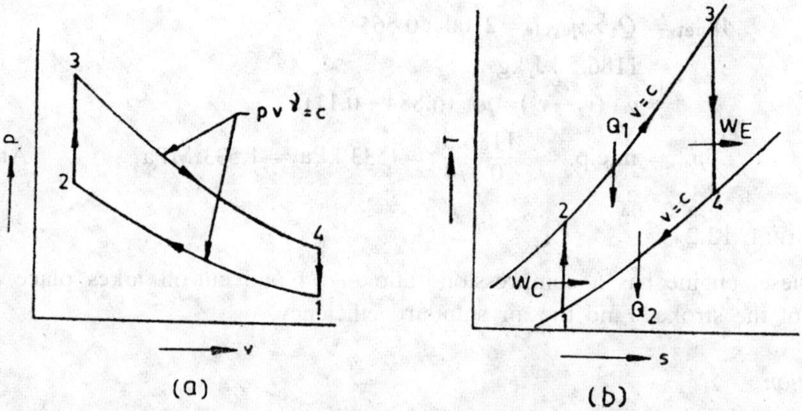

Fig. Ex. 13.1

$$T_1 = 273 + 35 = 308 \text{ K}$$

$$p_1 = 0.1 \text{ MPa} = 100 \text{ kN/m}^2$$

$$Q_1 = 2100 \text{ kJ/kg}$$

$$r_k = 8, \ \gamma = 1.4$$

$$\therefore \quad \eta_{\text{cycle}} = 1 - \frac{1}{r_k^{\gamma-1}} = 1 - \frac{1}{8^{0.4}} = 1 - \frac{1}{2.3}$$

$$= 0.565 \text{ or } 56.5\% \qquad \text{Ans.}$$

$$\frac{v_1}{v_2} = 8, \ v_1 = \frac{RT_1}{p_1} = \frac{0.287 \times 308}{100} = 0.884 \text{ m}^3/\text{kg}$$

$$\therefore \quad v_2 = \frac{0.884}{8} = 0.11 \text{ m}^3/\text{kg}$$

$$\frac{T_2}{T_1} = \left(\frac{v_1}{v_2}\right)^{\gamma-1} = (8)^{0.4} = 2.3$$

$\therefore \quad T_2 = 2.3 \times 308 = 708.4$ K

$$Q_1 = c_v\,(T_3 - T_2) = 2100 \text{ kJ/kg}$$

$$\therefore \quad T_3 - 708.4 = \frac{2100}{0.718} = 2925 \text{ K}$$

or $\quad T_3 = T_{max} = 3633$ K **Ans**

$$\frac{p_2}{p_1} = \left(\frac{v_1}{v_2}\right)^{\gamma} = (8)^{1.4} = 18.37$$

$\therefore \quad p_2 = 1.873$ MPa

Again

$$\frac{p_3 v_3}{T_3} = \frac{p_2 v_2}{T_2}$$

$$\therefore \quad p_3 = p_{max} = 1.837 \times \frac{363}{708} = 9.426 \text{ MPa}$$ **Ans.**

$$W_{net} = Q_1 \times \eta_{cycle} = 2100 \times 0.565$$
$$= 1186.5 \text{ kJ/kg}$$
$$= p_m\,(v_1 - v_2) = p_m\,(0.884 - 0.11)$$

$$\therefore \quad p_m = \text{m.e.p.} = \frac{1186.5}{0.774} = 1533 \text{ kPa} = 1.533 \text{ MPa}$$ **Ans.**

EXAMPLE 13.2

A Diesel engine has a compression ratio of 14 and cut-off takes place at 6% of the stroke. Find the air standard efficiency.

Solution

From Fig. Ex. 13.2

$$r_k = \frac{v_1}{v_2} = 14$$

$$v_3 - v_2 = 0.06\,(v_1 - v_2)$$
$$= 0.06\,(14 v_2 - v_2)$$
$$= 0.78\, v_2$$
$$v_3 = 1.78\, v_2$$

$$\therefore \quad \text{Cut-off ratio, } r_c = \frac{v_3}{v_2} = 1.78$$

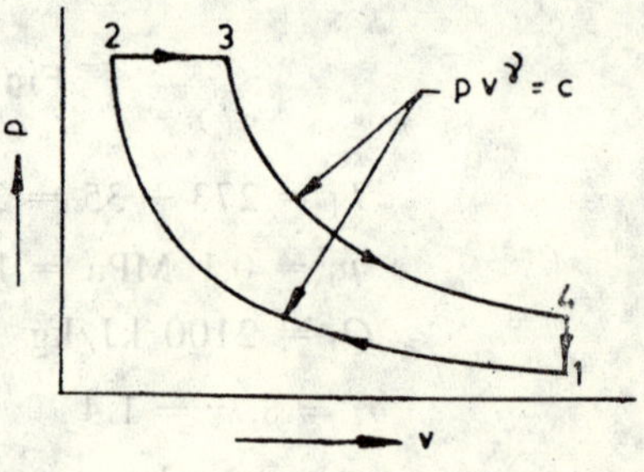

Fig. Ex. 13.2

$$\eta_{Diesel} = 1 - \frac{1}{\gamma} \cdot \frac{1}{r_k^{\gamma - 1}} \cdot \frac{r_c^{\gamma} - 1}{r_c - 1}$$

$$= 1 = \frac{1}{1.4} \cdot \frac{1}{(14)^{0.4}} \cdot \frac{(1.78)^{1.4} - 1}{1.78 - 1}$$

$$= 1 - 0.248 \cdot \frac{1.24}{0.78} = 0.605, \text{ i.e., } 60.5\%$$ **Ans.**

EXAMPLE 13.3

In an air standard Diesel cycle, the compression ratio is 16, and at the beginning of isentropic compression, the temperature is 15°C and the pressure is 0.1 MPa. Heat is added until the temperature at the end of the constant pressure process is 1480°C. Calculate (a) the cut-off ratio, (b) the heat supplied per kg of air, (c) the cycle efficiency, and (d) the m.e.p.

Solution

From Fig. Ex. 13.3

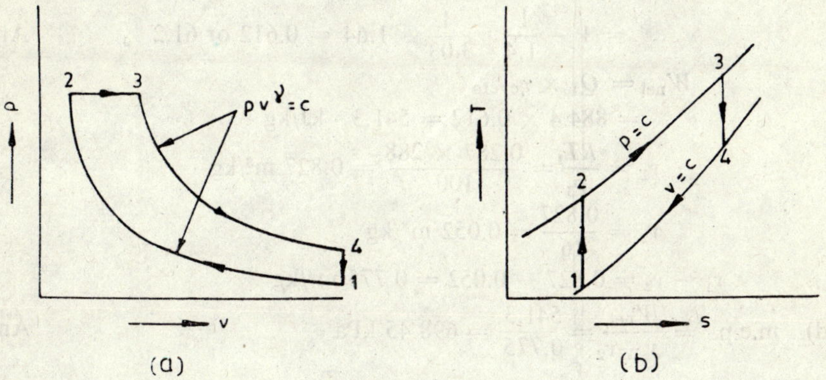

Fig. Ex 13·3

$$r_k = \frac{v_1}{v_2} = 16$$

$$T_1 = 273 + 15 = 288 \text{ K}$$

$$p_1 = 0.1 \text{ MPa} = 100 \text{ kN/m}^2$$

$$T_3 = 1480 + 273 = 1753 \text{ K}$$

$$\frac{T_2}{T_1} = \left(\frac{v_1}{v_2}\right)^{\gamma-1} = (16)^{0.4} = 3.03$$

$$\therefore \quad T_2 = 288 \times 3.03 = 873 \text{ K}$$

$$\frac{p_2 v_2}{T_2} = \frac{p_3 v_3}{T_3}$$

(a) Cut-off ratio, $r_c = \frac{v_3}{v_2} = \frac{T_3}{T_2} = \frac{1753}{273} = 2.01$ Ans.

(b) Heat supplied, $Q_1 = c_p \, (T_3 - T_2)$

$$= 1.005 \, (1753 - 873)$$

$$= 884.4 \text{ kJ/kg}$$

$$\frac{T_3}{T_4} = \left(\frac{v_4}{v_3}\right)^{\gamma-1} = \left(\frac{v_1}{v_2} \times \frac{v_2}{v_3}\right)^{\gamma-1} = \left(\frac{16}{2.01}\right)^{0.4} = 2.29$$

$$\therefore \quad T_4 = \frac{1753}{2.29} = 766 \text{ K}$$

Heat rejected, $Q_2 = c_v\ (T_4 - T_1) \quad = 0.718\ (766 - 288) = 343.2$ kJ/kg

$\therefore$ Cycle efficiency $= 1 - \dfrac{Q_2}{Q_1}$

$$= 1 - \frac{343.2}{884.4} = 0.612 \text{ or } 61.2\% \qquad \text{Ans.}$$

It may also be estimated from the equation

$$\eta_{\text{cycle}} = 1 - \frac{1}{\gamma} \cdot \frac{1}{r_k^{\gamma-1}} \cdot \frac{r_c^{\gamma} - 1}{r_c - 1}$$

$$= 1 - \frac{1}{1.4}\ \frac{1}{(16)^{0.4}} \cdot \frac{(2.01)^{1.4} - 1}{2.01 - 1}$$

$$= 1 - \frac{1}{1.4} \cdot \frac{1}{3.03} \cdot 1.64 = 0.612 \text{ or } 61.2\ \% \qquad \text{Ans.}$$

$$W_{\text{net}} = Q_1 \times \eta_{\text{cycle}}$$

$$= 884.4 \times 0.612 = 541.3 \quad \text{kJ/kg}$$

$$v_1 = \frac{RT_1}{p_1} = \frac{0.287 \times 288}{100} = 0.827 \text{ m}^3\text{/kg}$$

$$v_2 = \frac{0.827}{16} = 0.052 \text{ m}^3\text{/kg}$$

$\therefore \quad v_1 - v_2 = 0.827 - 0.052 = 0.775$ m^3/kg

(d) m.e.p. $= \dfrac{W_{\text{net}}}{v - v_2} = \dfrac{541.3}{0.775} = 698.45$ kPa Ans.

EXAMPLE 13.4

An air standard dual cycle has a compression ratio of 16, and compression begins at 1 bar, 50°C. The maximum pressure is 70 bar. The heat transferred to air at constant pressure is equal to that at constant volume. Estimate (a) the pressures and temperatures at the cardinal points of the cycle, (b) the cycle efficiency, and (c) the m.e.p. of the cycle. $c_v = 0.718$ kJ/kg-K, $c_p = 1.005$ kJ/kg-K.

Solution

Given : (Fig. Ex. 13.4)

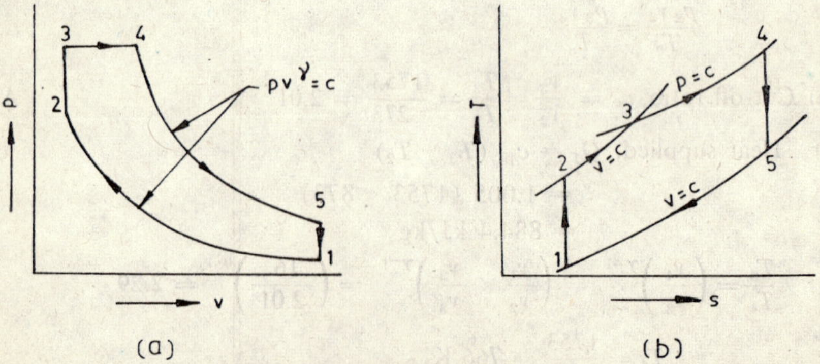

Fig. Ex. 13·4

$$T_1 = 273 + 50 = 323 \text{ K}$$

$$\frac{T_2}{T_1} = \left(\frac{v_1}{v_2}\right)^{\gamma-1} = (16)^{0.4}$$

$$\therefore \quad T_2 = 979 \text{ K}$$

$$p_2 = p_1\left(\frac{v_1}{v_2}\right)^{\gamma} = 1.0 \times (16)^{1.4} = 48.5 \text{ bar}$$

$$T_3 = T_2 \cdot \frac{p_3}{p_2} = 979 \times \frac{70}{48.5} = 1413 \text{ K}$$

$$Q_{2-3} = c_v\,(T_3 - T_2) = 0.718\,(1413 - 979) = 312 \text{ kJ/kg}$$

Now

$$Q_{2-3} = Q_{3-4} = c_p\,(T_4 - T_3)$$

$$\therefore \quad T_4 = \frac{312}{1.005} + 1413 = 1723 \text{ K}$$

$$\frac{v_4}{v_3} = \frac{T_4}{T_3} = \frac{1723}{1413} = 1.22$$

$$\therefore \quad \frac{v_5}{v_4} = \frac{v_1}{v_2} \times \frac{v_3}{v_4} = \frac{16}{1.22} = 13.1$$

$$\therefore \quad T_5 = T_4\left(\frac{v_4}{v_5}\right)^{\gamma-1} = 1723 \times \frac{1}{(13.1)^{0.4}} = 615 \text{ K}$$

$$p_5 = p_1\left(\frac{T_5}{T_1}\right) = 1.0 \times \frac{615}{323} = 1.9 \text{ bar}$$

$$\eta_{\text{cycle}} = 1 - \frac{Q_2}{Q_1} = 1 - \frac{c_v\,(T_5 - T_1)}{c_v\,(T_3 - T_2) + c_p\,(T_4 - T_3)}$$

$$= 1 - \frac{0.718\,(615 - 323)}{312 + 312}$$

$$= 1 - \frac{0.718 \times 292}{624} = 0.665 \text{ or } 66.5\%$$ Ans.(b).

$$v_1 = \frac{RT_1}{p_1} = \frac{0.287 \text{ kJ/kg-K} \times 323 \text{ K}}{10^2 \text{ kN/m}^2} = 0.927 \text{ m}^3\text{/kg}$$

$$v_1 - v_2 = v_1 - \frac{v_1}{16} = \frac{15}{16} v_1$$

$$W_{\text{net}} = Q_1 \times \eta_{\text{cycle}} = 0.665 \times 624 \text{ kJ/kg}$$

$$\therefore \quad \text{m.e.p.} = \frac{W_{\text{net}}}{v_1 - v_2} = \frac{0.665 \times 624 \text{ kJ/kg}}{\frac{15}{16} \times 0.927 \text{ m}^3\text{/kg}}$$

$$= 476 \text{ kN/m}^2 = 4.76 \text{ bar}$$ Ans.(c)

EXAMPLE 13.5

In a gas turbine plant, working on the Brayton cycle with a regenerator of 75% effectiveness, the air at the inlet to the compressor is at 0.1 MPa, 30°C, the pressure ratio is 6, and the maximum cycle temperature is 900°C. If the turbine and compressor have each an efficiency of 80%, find the percentage increase in the cycle efficiency due to regeneration.

Solution

Given : (Fig. Ex. 13.5)

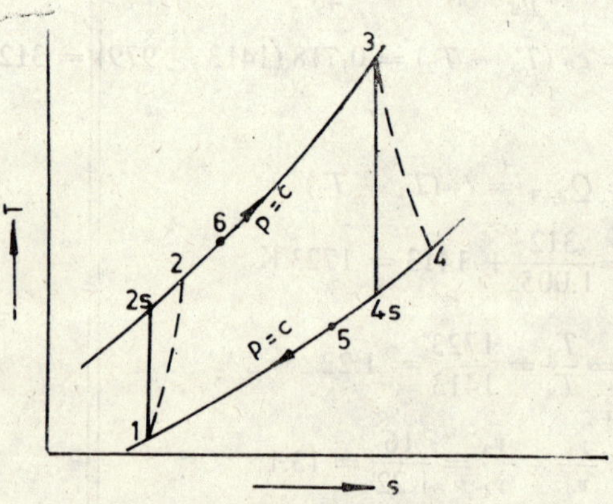

Fig. Ex. 13.5

$$p_1 = 0.1 \text{ MPa}$$
$$T_1 = 303 \text{ K}$$
$$T_3 = 1173 \text{ K}$$
$$r_p = 6, \eta_T = \eta_C = 0.8$$

Wihout a regenerator

$$\frac{T_{2s}}{T_1} = \left(\frac{p_2}{p_1}\right)^{(\gamma-1)/\gamma} = \frac{T_3}{T_{4s}} = (6)^{0.4/1.4} = 1.668$$

$$T_{2s} = 303 \times 1.668 = 505 \text{ K}$$

$$T_{4s} = \frac{1173}{1.668} = 705 \text{ K}$$

$$T_2 - T_1 = \frac{T_{2s} - T_1}{\eta_C} = \frac{505 - 303}{0.8} = 252 \text{ K}$$

$$T_3 - T_4 = \eta_T (T_3 - T_{4s}) = 0.8\,(1173 - 705) = 375 \text{ K}$$

$$W_T = h_3 - h_4 = c_p (T_3 - T_4) = 1.005 \times 375 = 376.88 \text{ kJ/kg}$$

$$W_C = h_2 - h_1 = c_p (T_2 - T_1) = 1.005 \times 252 = 253.26 \text{ kJ/kg}$$

$$T_2 = 252 + 303 = 555 \text{ K}$$

$$Q_1 = h_3 - h_2 = c_p\,(T_3 - T_2) = 1.005\,(1173 - 555) = 621.09 \text{ kJ/kg}$$

$$\therefore \quad \eta = \frac{W_T - W_C}{Q_1} = \frac{376.88 - 253.26}{621.09} = 0.199 \text{ or } 19.9\%$$

With regenerator

$$T_4 = T_3 - 375 = 1173 - 375 = 798 \text{ K}$$

$$\text{Regenerator effectiveness} = \frac{T_6 - T_2}{T_4 - T_2} = 0.75$$

$$\therefore \quad T_6 - 555 = 0.75\ (798 - 555)$$

$$\text{or} \quad T_6 = 737.3 \text{ K}$$

$$\therefore \quad Q_1 = h_3 - h_6 = c_p\ (T_3 - T_6)$$

$$= 1.005\ (1173 - 737.3)$$

$$= 437.88 \text{ kJ/kg}$$

W_{net} remains the same.

$$\therefore \quad \eta = \frac{W_{\text{net}}}{Q_1} = \frac{123.62}{435.7} = 0.2837 \text{ or } 28.37\%$$

$\therefore$ Percentage increase due to regeneration

$$= \frac{0.2837 - 0.199}{0.199} = 0.4256, \text{ or } 42.56\%$$

EXAMPLE 13.6

A gas turbine plant operates on the Brayton cycle between $T_{\text{min}} = 300$ K and $T_{\text{max}} = 1073$ K. Find the maximum work done per kg of air, and the corresponding cycle efficiency. How does this efficiency compare with the Carnot cycle efficiency operating between the same two temperatures?

Solution

$$(W_{\text{net}})_{\text{max}} = c_p\ (\sqrt{T_{\text{max}}} - \sqrt{T_{\text{min}}})^2$$

$$= 1.005\ (\sqrt{1073} - \sqrt{300})^2$$

$$= 1.005\ (15.43)^2 = 239.28 \text{ kJ/kg} \qquad \text{Ans.}$$

$$\eta_{\text{cycle}} = 1 - \frac{1}{(r_p)^{(\gamma-1)/\gamma}} = 1 - \sqrt{\frac{T_{\text{min}}}{T_{\text{max}}}}$$

$$= 1 - \sqrt{\frac{300}{1073}} = 0.47 \text{ or } 47\% \qquad \text{Ans.}$$

$$\eta_{\text{Carnot}} = 1 - \frac{T_{\text{min}}}{T_{\text{max}}} = 1 - \frac{300}{1073} = 0.721 \text{ or } 72.1\%$$

$$\therefore \quad \frac{\eta_{\text{Brayton}}}{\eta_{\text{Carnot}}} = \frac{0.47}{0.721} = 0.652 \qquad \text{Ans.}$$

EXAMPLE 13.7

In a gas turbine plant the ratio of $T_{\text{max}}/T_{\text{min}}$ is fixed. Two arrangements of components are to be investigated: (a) single-stage compression followed by expansion in two turbines of equal pressure ratios with reheat to the maximum cycle temperature, and (b) compression in two compressors of equal pressure ratios, with intercooling to the minimum cycle temperature, followed by single-stage expansion. If η_C and η_T are the compressor and turbine efficiencies, show that the optimum specific output is obtained at the same overall pressure ratio for each arrangement.

If η_C is 0.85 and η_T is 0.9, and T_{max}/T_{min} is 3.5, determine the above pressure ratio for optimum specific output and show that with arrangement (a) the optimum output exceeds that of arrangement (b) by about 11%.

Solution

(a) With reference to Fig. Ex. 13.7(a)

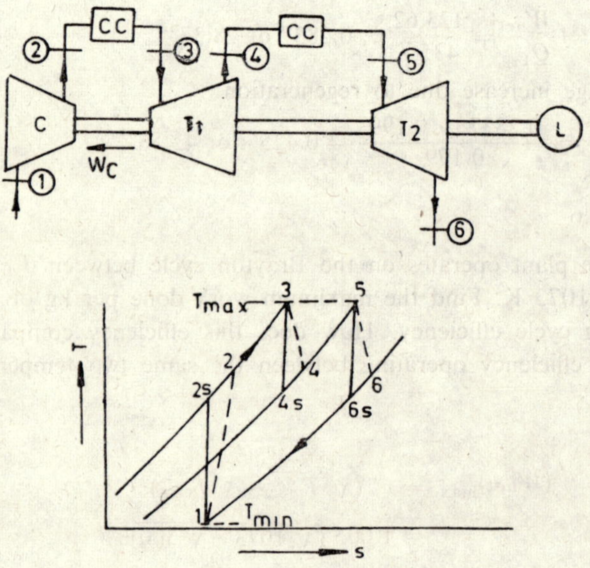

Fig. Ex. 13·7(a)

$$T_1 = T_{min},\ T_3 = T_5 = T_{max},\ \frac{p_2}{p_4} = \frac{p_4}{p_1}$$

$$\therefore \quad p_4 = \sqrt{p_1 p_2}$$

$$\frac{p_{2s}}{p_1} = r, \text{ pressure ratio}$$

$$\therefore \quad p_{2s} = p_2 = rp_1$$

$$\therefore \quad p_4 = \sqrt{r} \cdot p_1$$

$$\frac{T_{2s}}{T_1} = \left(\frac{p_2}{p_1}\right)^{(\gamma-1)/\gamma} = r^x$$

where $$x = \frac{\gamma - 1}{\gamma}$$

$$\therefore \quad T_{2s} = T_{min}\, r^x$$

$$(\Delta T_s)_{comp.} = T_{2s} - T_1 = T_{min}\, r^x - T_{min} = T_{min}\,(r^x - 1)$$

$$\therefore \quad (\Delta T)_{\text{comp.}} = \frac{T_{\min}(r^x - 1)}{\eta_C}$$

$$\frac{T_3}{T_{4s}} = \left(\frac{p_3}{p_4}\right)^{(\gamma-1)/\gamma} = \left(\frac{rp_1}{\sqrt{r}\cdot p_1}\right)^x = r^{x/2}$$

$$\therefore \quad T_{4s} = T_3\, r^{-x/2} = T_{\max}\, r^{-x/2}$$

$$(\Delta T_s)_{\text{turb.}} = T_3 - T_{4s} = T_{\max} - T_{\max}\cdot r^{-x/2}$$

$$= T_{\max}\,(1 - r^{-x/2})$$

$$\therefore \quad (\Delta T)_{\text{turb.1}} = \eta_T\, T_{\max}\,(1 - r^{-x/2}) = (\Delta T)_{\text{turb.2}}$$

$$\therefore \quad W_{\text{net}} = c_p\left[2\eta_T\, T_{\max}\,(1 - r^{-x/2}) - \frac{T_{\min}}{\eta_C}(r^x - 1)\right]$$

$$\frac{dW_{\text{net}}}{dr} = c_p\left[2\eta_T\, T_{\max}\,\frac{x}{2}\cdot r^{-x/2-1} - \frac{T_{\min}}{\eta_C}\,x\cdot r^{x-1}\right] = 0$$

On simplification

$$r^{3x/2} = \eta_T \eta_C \frac{T_{\max}}{T_{\min}}$$

$$\therefore \quad r_{\text{opt.}} = \left(\eta_T\, \eta_C \frac{T_{\max}}{T_{\min}}\right)^{2/3(\gamma/(\gamma-1))}$$

(b) With reference to Fig. Ex. 13.7(b)

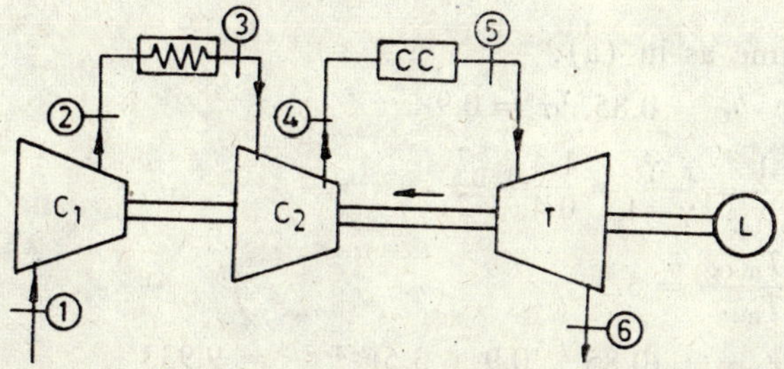

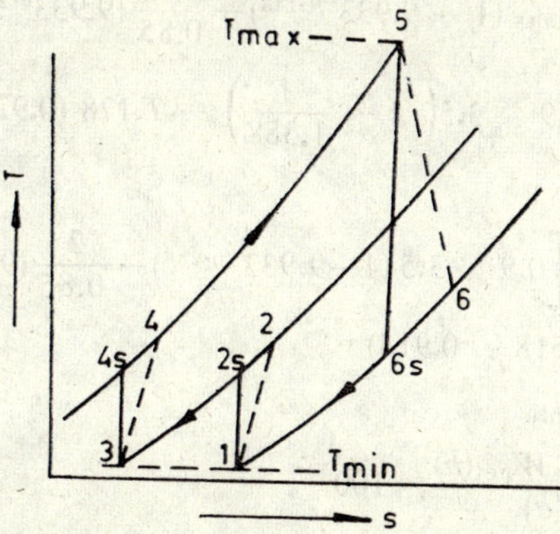

Fig. Ex. 13.7(b)

$$\frac{T_{2s}}{T_{\min}} = \left(\frac{p_2}{p_1}\right)^{(\gamma-1)/\gamma} = (\sqrt{r}\,)^x = r^{x/2}$$

$$(\Delta T_s)_{\text{comp}\cdot 1} = T_{2s} - T_1 = T_{\min}(r^{x/2} - 1)$$

$$(\Delta T)_{\text{comp}\cdot 1} = \frac{T_{\min}(r^{x/2} - 1)}{\eta_c} = (\Delta T)_{\text{comp}\cdot 2}$$

$$\frac{T_{\max}}{T_{6s}} = \left(\frac{p_5}{p_6}\right)^{(\gamma-1)/\gamma} = r^x$$

$$\therefore \quad T_{6s} = T_{\max} \cdot r^{-x}$$

$$(\Delta T_s)_{\text{turb}\cdot} = T_{\max} - T_{6s} = T_{\max}(1 - r^{-x})$$

$$(\Delta T)_{\text{turb}\cdot} = \eta_T T_{\max}(1 - r^{-x})$$

$$W_{\text{net}} = c_p \left[\eta_T\, T_{\max}(1 - r^{-x}) - \frac{2T_{\min}(r^{x/2} - 1)}{\eta_C}\right]$$

$$\frac{dW_{\text{net}}}{dr} = c_p \left[\eta_T T_{\max}\, x \cdot r^{-(x+1)} - \frac{2T_{\min}}{\eta_C} \frac{x}{2} \cdot r^{(x/2)-1}\right] = 0$$

On simplification

$$r_{\text{opt}\cdot} = \left(\eta_T\, \eta_C \frac{T_{\max}}{T_{\min}}\right)^{3\gamma/3(\gamma-1)}$$

This is the same as in (a).

If $\eta_C = 0.85, \eta_T = 0.9$

$$\frac{1}{x} = \frac{\gamma}{\gamma-1} = \frac{1.4}{0.4} = \frac{7}{2}$$

$$\frac{T_{\max}}{T_{\min}} = 3.5$$

$$\therefore \quad r_{\text{opt}\cdot} = (0.85 \times 0.9 \times 3.5)^{2/3 \times 7/2} = 9.933 \qquad \textbf{Ans.}$$

$$W_{\text{net}}(a) = c_p\left[2\eta_T\, T_{\max}(1 - r^{-x/2}) - \frac{T_{\min}}{\eta_C}(r^x - 1)\right]$$

$$= c_p\,[2 \times 0.9 \times T_{\max}(1 - 9.933^{-0.143}) - \frac{T_{\min}}{0.85}(9.933^{0.286} - 1)]$$

$$= c_p \cdot T_{\min}\left[2 \times 0.9 \times 3.5\left(1 - \frac{1}{1.388}\right) - T.178\,(0.928)\right]$$

$$= 0.670\, c_p \cdot T_{\min}$$

$$W_{\text{net}}(b) = c_p\, T_{\min}\left[0.9 \times 3.5\,(1 - 9.933^{-0.286}) - \frac{2}{0.85}(9.933^{0.143} - 1)\right]$$

$$= c_p\, T_{\min}\,(1.518 - 0.914)$$

$$= 0.604\, c_p\, T_{\min}$$

$$\therefore \quad \frac{W_{\text{net}}(a) - W_{\text{net}}(b)}{W_{\text{net}}(a)} \times 100$$

$$= \frac{0.670 - 0.604}{0.670} \times 100 = 10.9\,\% \qquad \textbf{Proved.}$$

EXAMPLE 13.8

Consider an ideal jet propulsion cycle in which air enters the compressor at 1 atm., 15°C. The pressure leaving the compressor is 5 atm. and the maximum temperature is 870°C. The air expands in the turbine to such a pressure that the turbine work is just equal to the compressor work. On leaving the turbine the air expands reversibly and adiabatically in a nozzle to 1 atm. Determine the velocity of the air leaving the nozzle.

Solution

With reference to Fig. 13.24

$$p_1 = 1 \text{ atm.}, \ T_1 = 288 \text{ K}, \ p_3 = 5 \text{ atm.}$$

$$T_4 = 870 + 273 = 1143 \text{ K}$$

$$\frac{T_3}{T_1} = \left(\frac{p_3}{p_1}\right)^{(\gamma-1)/\gamma} = (5)^{0.4/1.4} = 1.58$$

$$T_3 = 1.58 \times 288 = 455 \text{ K}$$

Since $W_C = W_T$

$$c_p\,(T_3 - T_1) = c_p\,(T_4 - T_5)$$

$$\therefore \quad 455 - 288 = 1143 - T_5$$

$$\therefore \quad T_5 = 976 \text{ K}$$

$$\frac{T_4}{T_6} = \left(\frac{p_4}{p_6}\right)^{(\gamma-1)/\gamma} = (5)^{0.4/1.4} = 1.58$$

$$\therefore \quad T_6 = \frac{1143}{1.58} = 725 \text{ K}$$

Assuming that the velocity of air entering the nozzle is very small, then by the first law,

$$h_5 = h_6 + \frac{V_6^2}{2}$$

$$\frac{V_6^2}{2} = h_5 - h_6 = c_p\,(T_5 - T_6) = 1.005\,(976 - 725)$$

$$= 252.26 \text{ kJ/kg}$$

$$V_6 = \sqrt{50.45 \times 10^4} = 710.3 \text{ m/s}$$ Ans.

PROBLEMS

13.1 In a Stirling cycle the volume varies between 0.03 and 0.06 m^3, the maximum pressure is 0.2 MPa, and the temperature varies between 540°C and 270°C. The working fluid is air (an ideal gas). (a) Find the efficiency and the work done per cycle for the simple cycle. (b) Find the efficiency and the work done per cycle for the cycle with an ideal regenerator, and compare with the Carnot cycle having the same isothermal heat supply process and the same temperature range.

Ans. (a) 27.7%, 53.7 kJ/kg, (b) 33.2%.

13.2 An Ericsson cycle operating with an ideal regenerator works between 1100 K and 288 K. The pressure at the beginning of isothermal compression is 1.013 bar. Determine (a) the compressor and turbine work per kg of air, and (b) the cycle efficiency. Ans. (a) $W_T = 465$ kJ/kg, $W_C = 121.8$ kJ/kg, (b) 0.738.

13.3 Plot the efficiency of the air standard Otto cycle as a function of the compression ratio for compression ratios from 4 to 16.

13.4 Find the air standard efficiencies for Otto cycles with a compression ratio of 6 using ideal gases having specific heat ratios 1.3, 1.4, and 1.67. What are the advantages and disadvantages of using helium as the working fluid?

13.5 An engine equipped with a cylinder having a bore of 15 cm and a stroke of 45 cm operates on an Otto cycle. If the clearance volume is 2000 cm^3, compute the air standard efficiency. Ans. 47.4%

13.6 In an air standard Otto cycle the compression ratio is 7, and compression begins at 35°C, 0.1 MPa. The maximum temperature of the cycle is 1100°C. Find (a) the temperature and pressure at the cardinal points of the cycle, (b) the heat supplied per kg of air, (c) the work done per kg of air, (d) the cycle efficiency, and (e) the m.e.p. of the cycle.

13.7 An engine working on the Otto cycle has an air standard cycle efficiency of 56% and rejects 544 kJ/kg of air. The pressure and temperature of air at the beginning of compression are 0.1 MPa and 60°C respectively. Compute (a) the compression ratio of the engine, (b) the work done per kg of air, (c) the pressure and temperature at the end of compression, and (d) the maximum pressure in the cycle.

13.8 For an air standard Diesel cycle with a compression ratio of 15 plot the efficiency as a function of the cut-off ratio for cut-off ratios from 1 to 4. Compare with the results of Problem 13.3.

13.9 In an air standard Diesel cycle, the compression ratio is 15. Compression begins at 0.1 MPa, 40°C. The heat added is 1.675 MJ/kg. Find (a) the maximum temperature of the cycle, (b) the work done per kg of air, (c) the cycle efficiency, (d) the temperature at the end of the isentropic expansion, (e) the cut-off ratio, (f) the maximum pressure of the cycle, and (g) the m.e.p. of the cycle.

13.10 Two engines are to operate on Otto and Diesel cycles with the following data:

Maximum temperature 1400 K, exhaust temperature 700 K. State of air at the beginning of compression 0.1 MPa, 300 K.

Estimate the compression ratios, the maximum pressures, efficiencies, and rate of work outputs (for 1 kg/min of air) of the respective cycles.

13.11 An air standard limited pressure cycle has a compression ratio of 15 and compression begins at 0.1 MPa, 40°C. The maximum pressure is limited to 6 MPa and the heat added is 1.675 MJ/kg. Compute (a) the heat supplied at constant volume per kg of air, (b) the heat supplied at constant pressure per kg of air, (c) the work done per kg of air, (d) the cycle efficiency, (e) the temperature at the end of the constant volume heating process, (f) the cut-off ratio, and (g) the m.e.p. of the cycle.

13.12 In an ideal cycle for an internal combustion engine the pressure and temperature at the beginning of adiabatic compression are respectively 0.11 MPa and 115°C, the compression ratio being 16. At the end of compression heat is added to the working fluid, first, at constant volume, and then at constant pressure reversibly. The working fluid is then expanded adiabatically and reversibly to the original volume.

If the working fluid is air and the maximum pressure and temperature are respectively 6 MPa and 2000°C, determine, per kg of air (a) the pressure, temperature, volume, and entropy of the air at the five cardinal points of the cycle (take s_1 as the entropy of air at the beginning of compression), and (b) the work output and efficiency of the cycle.

13.13 Show that the air standard efficiency for a cycle comprising two constant pressure processes and two isothermal processes (all reversible) is given by

$$\eta = \frac{(T_1 - T_2)\ln (r_p)^{(\gamma-1)/\gamma}}{T_1\,[1 + \ln(r_p)^{(\gamma-1)/\gamma}] - T_2}$$

where T_1 and T_2 are the maximum and minimum temperatures of the cycle, and r_p is the pressure ratio.

13.14 Obtain an expression for the specific work done by an engine working on the Otto cycle in terms of the maximum and minimum temperatures of the cycle, the compression ratio r_k, and constants of the working fluid (assumed to be an ideal gas).

Hence show that the compression ratio for maximum specific work output is given by

$$r_k = \left(\frac{T_{min}}{T_{max}}\right)^{1/2(1-\gamma)}$$

13.15 A dual combustion cycle operates with a volumetric compression ratio $r_k = 12$, and with a cut-off ratio 1.615. The maximum pressure is given by $p_{max} = 54p_1$, where p_1 is the pressure before compression. Assuming indices of compression and expansion of 1.35, show that the m.e.p. of the cycle

$$p_m = 10\ p_1$$

Hence evaluate (a) temperatures at cardinal points with $T_1 = 335$ K, and (b) cycle efficiency.

Ans. (a) $T_2 = 805$ K, $p_2 = 29.2\ p_1$, $T_3 = 1490$ K, $T_4 = 2410$ K, $T_5 = 1200$ K, (b) $\eta = 0.67$.

13.16 Recalculate (a) the temperatures at the cardinal points, (b) the m.e.p., and (c) the cycle efficiency when the cycle of Problem 13.15 is a Diesel cycle with the same compression ratio and a cut-off ratio such as to give an expansion curve coincident with the lower part of that of the dual cycle of Problem 13.15.

Ans. (a) $T_2 = 805$ K, $T_3 = 1970$ K, $T_4 = 1142$ K, (b) $6.82\ p_1$, (c) $\eta = 0.513$.

13.17 In an air standard Brayton cycle the compression ratio is 7 and the maximum temperature of the cycle is 800°C. The compression begins at 0.1 MPa, 35°C. Compare the maximum specific volume and the maximum pressure with the Otto cycle of Problem 13.6. Find (a) the heat supplied per kg of air, (b) the net work done per kg of air, (c) the cycle efficiency; and (d) the temperature at the end of the expansion process.

13.18 A gas turbine plant operates on the Brayton cycle between the temperatures 27°C and 800°C. (a) Find the pressure ratio at which the cycle efficiency approaches the Carnot cycle efficiency, (b) find the pressure ratio at which the work done per kg of air is maximum, and (c) compare the efficiency at this pressure ratio with the Carnot efficiency for the given temperatures.

13.19 In a gas turbine plant working on the Brayton cycle the air at the inlet is at 27°C, 0.1 MPa. The pressure ratio is 6.25 and the maximum temperature is 800°C. The turbine and compressor efficiencies are each 80%. Find (a) the compressor work per kg of air, (b) the turbine work per kg of air, (c) the heat supplied per kg of air, (d) the cycle efficiency, and (e) the turbine exhaust temperaure.

13.20 Solve Problem 13.19 if a regenerator of 75% effectiveness is added to the plant.

13.21 Solve Problem 13.19 if the compression is divided into two stages, each of pressure ratio 2.5 and efficiency 80%, with intercooling to 27°C.

13.22 Solve Problem 13.21 if a regenerator of 75% effectiveness is added to the plant.

13.23 Solve Problem 13.19 if a reheat cycle is used. The turbine expansion is divided into two stages, each of pressure ratio 2.5 and efficiency 80%, with reheat to 800°C.

13.24 Solve Problem 13.23 if a regenerator of 75% effectiveness is added to the plant.

13.25 Solve Problem 13.24 if the staged compression of Problem 13.21 is used in the plant.

13.25 Find the inlet condition for the free-shaft turbine if a two-shaft arrangement is used in Problem 13.19.

13.27 A simple gas turbine plant operating on the Brayton cycle has air inlet temperature 27°C, pressure ratio 9, and maximum cycle temperature 727°C.

What will be the improvement in cycle efficiency and output if the turbine process is divided into two stages each of pressure ratio 3, with intermediate reheating to 727°C ? Ans. — 18.3%, 30.6%.

13.28 Obtain an expression for the specific work output of a gas turbine unit in terms of pressure ratio, isentropic efficiencies of the compressor and turbine, and the maximum and minimum temperatures, T_3 and T_1. Hence show that the pressure ratio r_p for maximum power is given by

$$r_p = \left(\eta_T \, \eta_C \frac{T_3}{T_1}\right)^{\gamma/2(\gamma-1)}$$

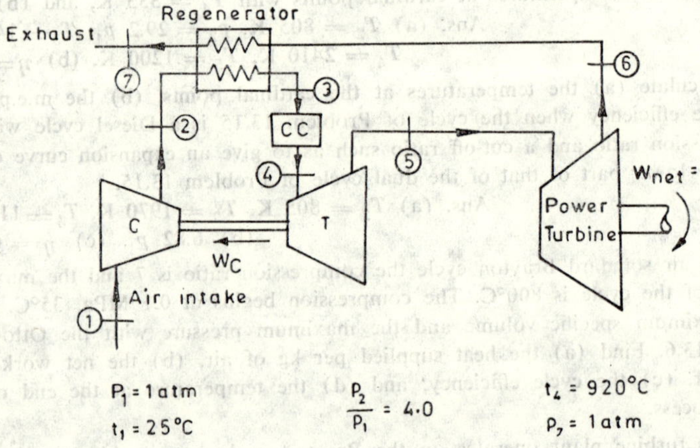

Fig. P 13·30

If $T_3 = 1073$ K, $T_1 = 300$ K, $\eta_C = 0.8$, $\eta_T = 0.8$, and $\gamma = 1.4$, compute the optimum value of pressure ratio, the maximum net work output per kg of air, and corresponding cycle efficiency. Ans. 4.263, 100 kJ/kg, 17.2%.

13.29 A gas turbine plant draws in air at 1.013 bar, 10°C and has a pressure ratio of 5.5. The maximum temperature in the cycle is limited to 750°C.

Compression is conducted in an uncooled rotary compressor having an isentropic efficiency of 82%, and expansion takes place in a turbine with as isentropic efficiency of 85%. A heat exchanger with an efficiency of 70% is fitted between the compressor outlet and combustion chamber. For an air flow of 40 kg/s, find (a) the overall cycle efficiency, (b) the turbine output, and (c) the air-fuel ratio if the calorific value of the fuel used is 45.22 MJ/kg. Ans. (a) 30.4%, (b) 4272 kW, (c) 115.

13.30 A gas turbine for use as an automotive engine is shown in Fig. P. 13.30. In the first turbine, the gas expands to just a low enough pressure p_5, for the turbine to drive the compressor. The gas is then expanded through a second turbine connected to the drive wheels. Consider air as the working fluid, and assume that all processes are ideal. Determine (a) pressure p_5, (b) the net work per kg and mass flow rate, (c) temperature T_3 and cycle thermal efficiency, and (d) the T-s diagram for the cycle.

13.31 Repeat Problem 13.30 assuming that the compressor has an efficiency of 80%, both the turbines have efficiencies of 85%, and the regenerator has an efficiency of 72%.

13.32 An ideal air cycle consists of isentropic compression, constant volume heat transfer, isothermal expansion to the original pressure, and constant pressure heat transfer to the original temperature. Deduce an expression for the cycle efficiency in terms of volumetric compression ratio, r_k, and isothermal expansion ratio, r_e. In such a cycle, the pressure and temperature at the start of compression are 1 bar and 40°C, the compression ratio is 8, and the maximum pressure is 100 bar. Determine the cycle efficiency and the m.e.p. Ans. 51.5%, 3.45 bar.

13.33 For a gas turbine jet propulsion unit, shown in Fig. 13.24, the pressure and temperature entering the compressor are 1 atm. and 15°C respectively. The pressure ratio across the compressor is 6 to 1 and the temperature at the turbine inlet is 1000°C. On leaving the turbine the air enters the nozzle and expands to 1 atm. Determine the pressure at the nozzle inlet and the velocity of the air leaving the nozzle.

13.34 Repeat Problem 13.33, assuming that the efficiency of the compressor and turbine are both 85%, and that the nozzle efficiency is 95%.

13.35 Develop expressions for work output per kg and the efficiency of an ideal Brayton cycle with regeneration, assuming maximum possible regeneration. For fixed maximum and minimum temperatures, how do the efficiency and work outputs vary with the pressure ratio? What is the optimum pressure ratio?

13.36 For an air standard Otto cycle with fixed intake and maximum temperatures, T_1 and T_3, find the compression ratio that renders the net work per cycle a maximum. Derive the expression for cycle efficiency at this compression ratio.

If the air intake temperature, T_1, is 300 K and the maximum cycle temperature, T_3, is 1200 K, compute the compression ratio for maximum net work, maximum work output per kg in a cycle, and the corresponding cycle efficiency.

Find the changes in work output and cycle efficiency when the compression ratio is increased from this optimum value to 8. Take $c_v = 0.718$ kJ/kg-K.

Ans. 5.65, 215 kJ/kg, 50%, $\Delta W_{net} = -9$ kJ/kg, $\Delta\eta = +6.4\%$.

14

Refrigeration Cycles

14.1 REFRIGERATION BY NON-CYCLIC PROCESSES

Refrigeration is the cooling of a system below the temperature of its surroundings.

The melting of ice or snow was one of the earliest methods of refrigeration and is still employed. Ice melts at 0°C. So when ice is placed in a given space warmer than 0°C, heat flows into the ice and the space is cooled or refrigerated. The latent heat of fusion of ice is supplied from the surroundings, and the ice changes its state from solid to liquid.

Another medium of refrigeration is solid carbon dioxide or dry ice. At atmospheric pressure CO_2 cannot exist in a liquid state, and consequently, when solid CO_2 is exposed to atmosphere, it sublimates, i.e., it goes directly from solid to vapour, by absorbing the latent heat of sublimation (620 kJ/kg at 1 atm., – 78.5°C) from the surroundings (Fig. 14.1). Thus dry ice is suitable for low temperature refrigeration.

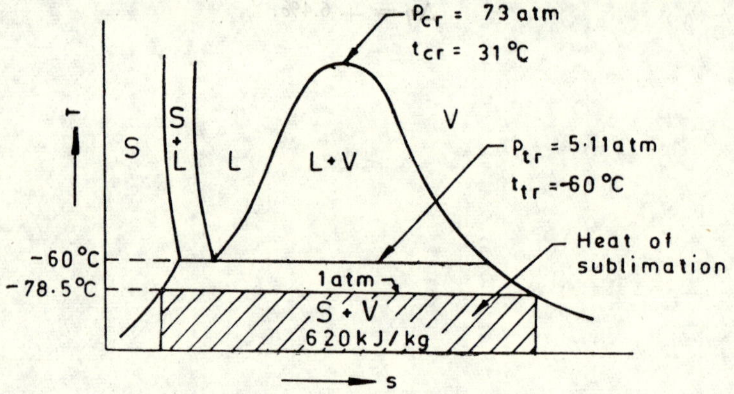

Fig. 14.1 *T-s* **diagram of CO_2**

In these two examples it is observed that the refrigeration effect has been accomplished by non-cyclic processes. Of greater importance, however, are the methods in which the cooling substance is not consumed and discarded, but used again and again in a thermodynamic cycle.

14.2 REVERSED HEAT ENGINE CYCLE

A reversed heat engine cycle, as explained in Sec. 6.12, is visualized as an engine operating in the reverse way, i.e., receiving heat from a low temperature region, discharging heat to a high temperature region, and receiving a net inflow of work (Fig. 14.2). Under such conditions the cycle is called a *heat pump cycle* or a *refrigeration cycle* (see Sec. 6.6). For a heat pump

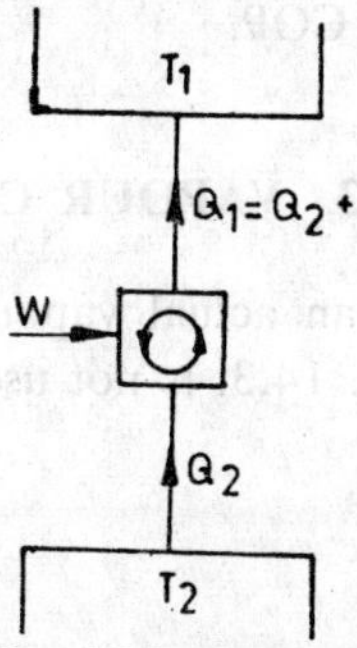

Fig. 14.2 Reversed heat engine cycle.

$$(\mathrm{COP})_{\mathrm{H.P.}} = \frac{Q_1}{W} = \frac{Q_1}{Q_1 - Q_2}$$

and for a refrigerator

$$(\mathrm{COP})_{\mathrm{ref.}} = \frac{Q_2}{W} = \frac{Q_2}{Q_1 - Q_2}$$

The working fluid in a refrigeration cycle is called a *refrigerant*. In the reversed Carnot cycle (Fig. 14.3), the refrigerant is first compressed reversibly and adiabatically in process 1-2 where the work input per kg of refrigerant is W_c, then it is condensed reversibly in process 2-3 where the heat

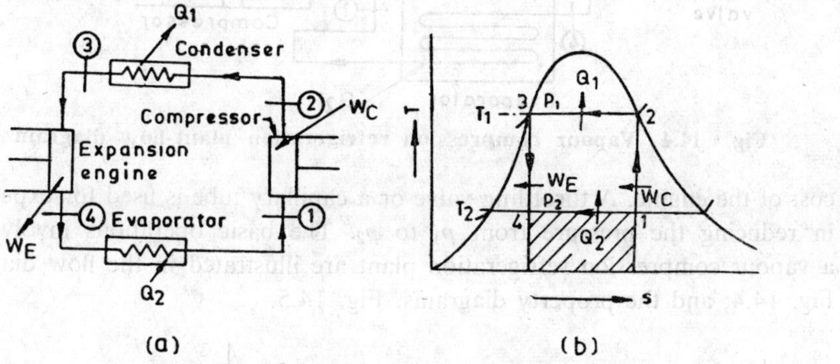

Fig. 14.3 Reversed Carnot cycle

rejection is Q_1, the refrigerant then expands reversibly and adiabatically in process 3-4 where the work output is W_E, and finally it absorbs heat Q_2 reversibly by evaporation from the surroundings in process 4-1.

Here $\quad Q_1 = T_1 (s_2 - s_3),\ Q_2 = T_2 (s_1 - s_4)$

and $\quad W_{\text{net}} = W_C - W_E = Q_1 - Q_2 = (T_1 - T_2)(s_1 - s_4)$

where T_1 is the temperature of heat rejection and T_2 the temperature of heat absorption.

$$(\mathrm{COP}_{\mathrm{ref.}})_{\mathrm{rev.}} = \frac{Q_2}{W_{\text{net}}} = \frac{T_2}{T_1 - T_2}$$

$$\text{and} \quad (\mathrm{COP}_{\mathrm{H.P.}})_{\mathrm{rev.}} = \frac{Q_1}{W_{\text{net}}} = \frac{T_1}{T_1 - T_2} \tag{14.1}$$

As shown in Sec. 6.16, these are the maximum values for any refrigerator or heat pump operating between T_1 and T_2. It is important to note that for the same T_2 or T_1, the COP increases with the decrease in the temperature difference $(T_1 - T_2)$, i.e., the closer the temperatures T_1 and T_2, the higher the COP.

14.3 VAPOUR COMPRESSION REFRIGERATION CYCLE

In an actual vapour refrigeration cycle, an expansion engine, as shown in Fig. 14.3, is not used, since power recovery is small and does not justify the

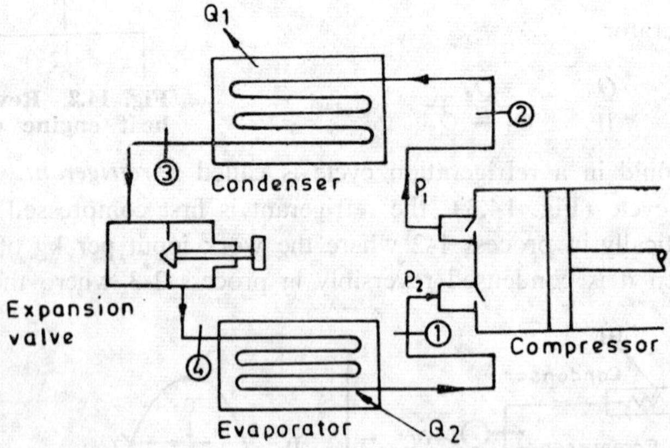

Fig. 14.4 Vapour compression refrigeration plant-flow diagram

cost of the engine. A throttling valve or a capillary tube is used for expansion in reducing the pressure from p_1 to p_2. The basic operations involved in a vapour compression refrigeration plant are illustrated in the flow diagram, Fig. 14.4, and the property diagrams, Fig. 14.5.

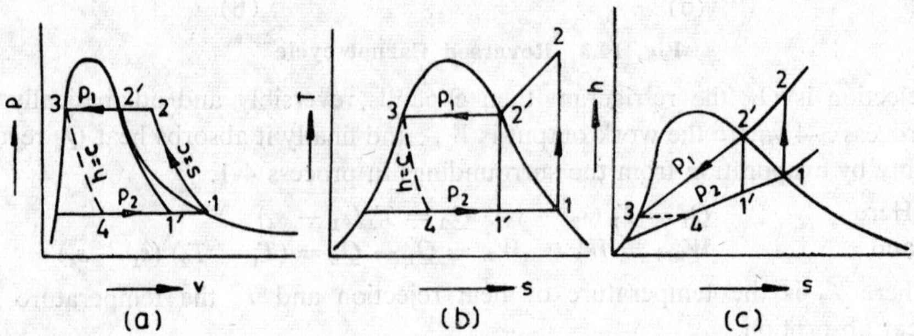

Fig. 14.5 Vapour compression refrigeration cycle—property diagrams

The operations represented are as follows for an idealized plant:

1. *Compression.* A reversible adiabatic process 1-2 or 1′-2′ either

starting with saturated vapour (state 1), called *dry compression,* or starting with wet vapour (state 1′), called *wet compression.* Dry compression (1-2) is always preferred to wet compression (1-′2′), because with wet compression there is a danger of the liquid refrigerant being tapped in the head of the cylinder by the rising piston which may damage the valves or the cylinder head, and the droplets of liquid refrigerant may wash away the lubricating oil from the walls of the cylinder, thus accelerating wear.

2. *Cooling and Condensing.* A reversible constant pressure process, 2-3, first desuperheated and then condensed, ending with saturated liquid. Heat Q_1 is transferred out.

3. *Expansion.* An adiabatic throttling process 3-4, for which enthalpy remains unchanged. States 3 and 4 are equilibrium points. Process 3-4 is adiabatic (then only $h_3 = h_4$ by S.F.E.E.), but not isentropic.

$$Tds = dh - vdp, \text{ or } s_4 - s_3 = -\int_{p_1}^{p_2} \frac{vdp}{T}.$$

Hence it is irreversible and cannot be shown in property diagrams. States 3 and 4 have simply been joined by a dotted line.

4. *Evaporation.* A constant pressure reversible process, 4-1, which completes the cycle. The refrigerant is throttled by the expansion valve to a pressure, the saturation temperature at this pressure being below the temperature of the surroundings. Heat then flows, by virtue of temperature difference, from the surroundings, which gets cooled or refrigerated, to the refrigerant, which then evaporates, absorbing the heat of evaporation. The evaporator thus produces the cooling or the *refrigerating effect,* absorbing heat Q_2 from the surroundings by evaporation.

In refrigeration practice, enthalpy is the most sought-after property. The diagram in *p-h* coordinates is found to be the most convenient. The constant property lines in the *p-h* diagram are shown in Fig. 14.6, and the vapour compression cycle in Fig. 14.7.

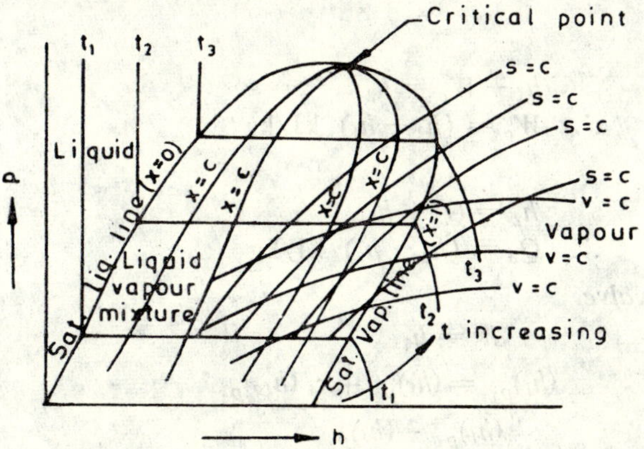

Fig. 14.6 Phase diagram with constant property lines on *p-h* plot

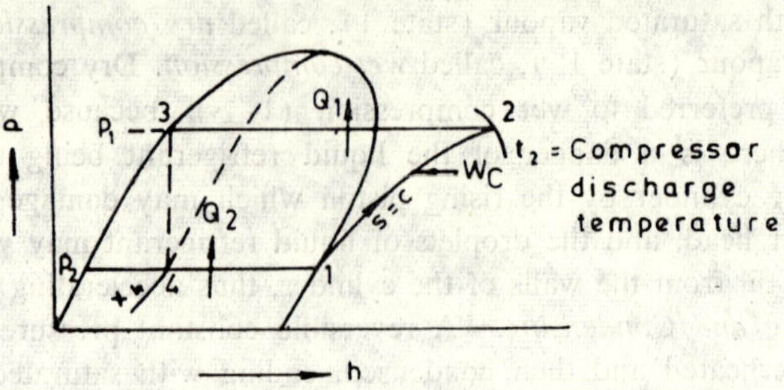

Fig. 14.7 Vapour compression cycle on *p-h* diagram

14.3.1 Performance and Capacity of a Vapour Compression Plant

Figure 14.8 shows the simplified diagram of a vapour compression refrigeration plant.

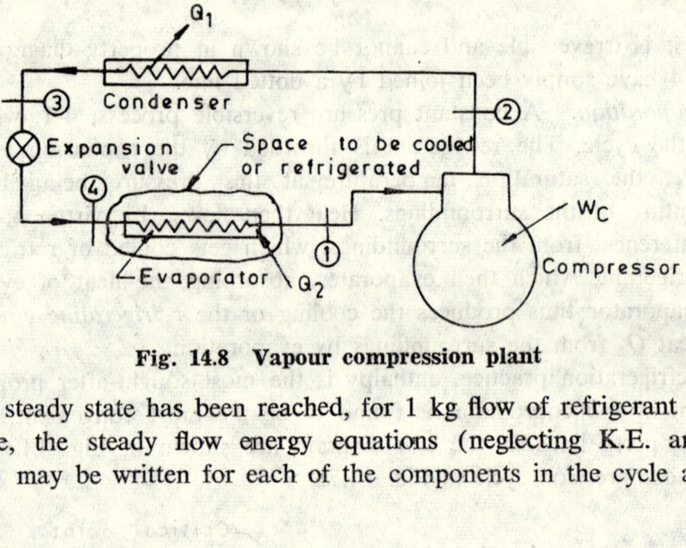

Fig. 14.8 Vapour compression plant

When steady state has been reached, for 1 kg flow of refrigerant through the cycle, the steady flow energy equations (neglecting K.E. and P.E. changes) may be written for each of the components in the cycle as given below.

Compressor

$$h_1 + W_c = h_2$$

$$\therefore \quad W_c = (h_2 - h_1) \text{ kJ/kg}$$

Condenser

$$h_2 = Q_3 + h_3$$

$$\therefore \quad Q_3 = (h_2 - h_3) \text{ kJ/kg}$$

Expansion valve

$$h_3 = h_4$$

or

$$(h_f)_{p_1} = (h_f)_{p_2} + x_4 (h_{fg})_{p_2}$$

$$\therefore \quad x_4 = \frac{(h_f)_{p_1} - (h_f)_{p_2}}{(h_{fg})_{p_2}}$$

This is the quality of the refrigerant at the inlet to the evaporator (mass fraction of vapour in liquid-vapour mixture).

Evaporator

$$h_4 + Q_2 = h_1$$

$$\therefore \quad Q_2 = (h_1 - h_4) \ \text{kJ/kg}$$

This is known as the refrigerating effect, the amount of heat removed from the surroundings per unit mass flow of refrigerant.

If the *p-h* chart for a particular refrigerant is available with the given parameters, it is possible to obtain from the chart the values of enthalpy at all the cardinal points of the cycle. Then for the cycle

$$\text{COP} = \frac{Q_2}{W_c} = \frac{h_1 - h_4}{h_2 - h_1} \qquad (14.2)$$

If w is the mass flow of refrigerant in kg/sec, then the rate of heat removal from the surroundings

$$= w\,(h_1 - h_4) \ \text{kJ/sec} = w\,(h_1 - h_4) \times 3600 \ \text{kJ/hr}$$

One tonne of refrigeration is defined as the rate of heat removal from the surroundings equivalent to the heat required for melting 1 tonne of ice in one day. If the latent heat of fusion of ice is taken as 336 kJ/kg, then 1 tonne is equivalent to heat removal at the rate of $(1000 \times 336)/24$ or 14,000 kJ/hr.

$\therefore$ Capacity of the refrigerating plant

$$\frac{w\,(h_1 - h_4) \times 3600}{14{,}000} \ \text{tonnes}$$

The rate of heat removal in the condenser

$$Q_1 = w\,(h_2 - h_3) \ \text{kJ/s}$$

If the condenser is water-cooled, $\dot{m}_c$ the flow-rate of cooling water in kg/s, and $(t_{c_2} - t_{c_1})$ the rise in temperature of water, then

$$Q_1 = w\,(h_2 - h_3) = \dot{m}_c\,(t_{c_2} - t_{c_1}) \ \text{kJ/s}$$

provided the heat transfer is confined only between the refrigerant and water, and there is no heat interaction with the surroundings.

The rate of work input to the compressor

$$W_c = w\,(h_2 - h_1) \ \text{kJ/s}$$

$$= w\,(h_2 - h_1) \ \text{kW}$$

or

$$W_c = \frac{w\,(h_2 - h_1) \times 3600}{632} \ \text{P.S., in MKS units.}$$

where h_1 and h_2 are in kcal/kg.

14.3.2 Actual Vapour Compression Cycle

In order to ascertain that there is no droplet of liquid refrigerant being carried over into the compressor, some superheating of vapour is recommended after the evaporator.

A small degree of subcooling of the liquid refrigerant after the condenser is also used to reduce the mass of vapour formed during expansion, so that too many vapour bubbles do not impede the flow of liquid refrigerant through the expansion valve.

Both the superheating of vapour at the evaporator outlet and the subcooling of liquid at the condenser outlet contribute to an increase in the refrigerating effect, as shown in Fig. 14.9. The compressor discharge temperature, however, increases, due to superheat, from t'_2 to t_2, and the load on the condenser also increases.

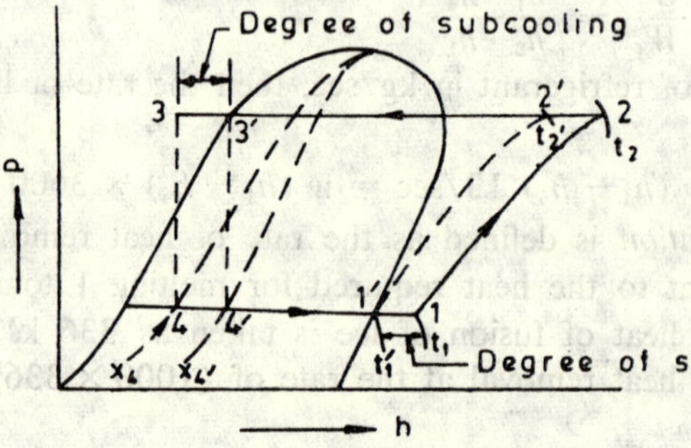

Fig. 14.9 Superheat and subcooling in a vapour compression cycle

Sometimes, a liquid-line heat exchanger is used in the plant, as shown in Fig. 14.10. The liquid is subcooled in the heat exchanger, reducing the load on the condenser and improving the COP. For 1 kg flow

$$Q_2 = h_6 - h_5,\ Q_1 = h_2 - h_3$$

$$W_c = h_2 - h_1 \text{ and } h_1 - h_6 = h_3 - h_4$$

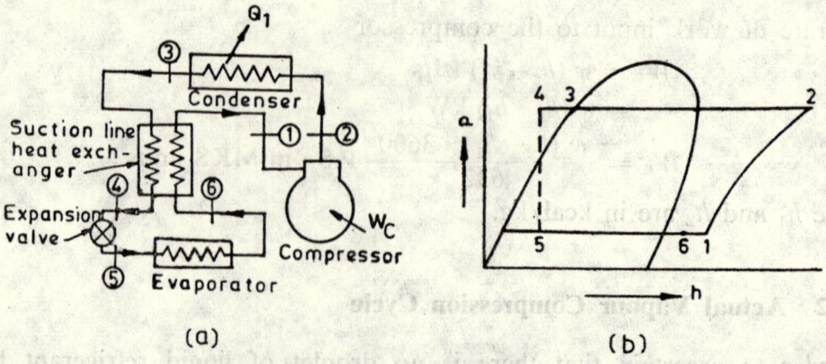

Fig. 14.10 Vapour compression cycle with a suction-line heat exchanger.

14.3.3 Components in a Vapour Compression Plant

Condenser. It must desuperheat and then condense the compressed refrigerant. Condensers may be either air-cooled or water-cooled. An air-cooled condenser is used in small self-contained units. Water-cooled condensers are used in larger installations.

Expansion device. It reduces the pressure of the refrigerant, and also regulates the flow of the refrigerant to the evaporator. Two widely used types of expansion devices are: capillary tubes and throttle valves (thermostatic expansion valves). Capillary tubes are used only for small units. Once the size and length are fixed, the evaporator pressure, etc., gets fixed. No modification in operating conditions is possible. Throttle valves are used in larger units. These regulate the flow of the refrigerant according to the load on the evaporator.

Compressor. Compressors may be of three types: (a) reciprocating, (b) rotary, and (c) centrifugal. When the volume flow rate of the refrigerant is large, centrifugal compressors are used. Rotary compressors are used for small units. Reciprocating compressors are used in plants up to 100 tonnes capacity. For plants of higher capacities, centrifugal compressors are employed.

In reciprocating compressors, which may be single-cylinder or multi-cylinder ones, because of clearance, leakage past the piston and valves, and throttling effects at the suction and discharge valves, the actual volume of gas drawn into the cylinder is less than the volume displaced by the piston. This is accounted for in the term *volumetric efficiency,* which is defined as

$$\eta_{vol.} = \frac{\text{Actual volume of gas drawn at evaporator pressure and temperature}}{\text{Piston displacement}}$$

$\therefore$ Volume of gas handled by the compressor

$$= w.v_1 \text{ (m}^3\text{/s)} = \left(\frac{\pi}{4} D^2 L \frac{N}{60} n\right) \times \eta_{vol.}$$

where w is the refrigerant flow rate,

v_1 is the specific volume of the refrigerant at the compressor inlet,

D and L are the diameter and stroke of the compressor,

n is the number of cylinders in the compressor, and

N is the r.p.m.

Evaporator. A common type of evaporator is a coil brazed on to a plate, called a plate evaporator. In a 'flooded evaporator' the coil is filled only with a liquid refrigerant. In an indirect expansion coil, water (up to 0°C) or brine (for temperatures between 0 and −21°C) may be chilled in the evaporator, and the chilled water or brine may then be used to cool some other medium.

14.3.4 Refrigerants

The most widely used refrigerants now-a-days are a group of halogenated hydrocarbons, marketed under the various proprietary names of freon, genetron, arcton, isotron, frigen, and, in India, mafron. These are either methane-based or ethane-based, where the hydrogen atoms are replaced by chlorine or fluorine atoms. Methane-based compounds are denoted by a number of two digits, where the first digit minus one is the number of hydrogen atoms and the second digit indicates the number of fluorine atmos, while the other atoms are chlorine. For Refrigerant-12 (R-12), e.g., the number of hydrogen atoms is zero, the number of fluorine atoms is two, and hence the other two atoms must be chlorine. Therefore, the compound is CCl_2F_2, dichloro-difluoro methane. Similarly, for R-22, it is $CHClF_2$, monochloro-difluoro methane; for R-50, it is methane, CH_4; for R-10, it is CCl_4, carbon tetrachloride, and so on. If the compound is ethane-based, a three-digit number is assigned to the refrigerant, where the first digit is always 1, the second digit minus one is the number of hydrogen atoms, and the third digit indicates the number of fluorine atoms, all other atoms in the hydrocarbon being chlorine. For example, R-110 is C_2Cl_6, R-113 is $C_2Cl_3F_3$, R-142 is $C_2H_3ClF_2$, and so on.

Apart from the halocarbon group, ammonia (NH_3) is also a popular refrigerant. But it is toxic and inflammable. It is used in the absorption refrigeration system, which has been discussed in the next section. Other fluids used as refrigerants are sulphur dioxide SO_2, methylchloride CH_3Cl, carbon dioxide CO_2, air, water, and various hydrocarbons like propane, C_3H_8 and butane, C_4H_{10}.

14.4 ABSORPTION REFRIGERATION CYCLE

The absorption refrigeration system is a *heat operated unit* which uses a refrigerant that is *alternately absorbed and liberated from the absorbent*. In the basic absorption system, the compressor in the vapour compression cycle *is replaced by an absorber-generator assembly* involving less mechanical work. Figure 14.11 gives the basic absorption refrigeration cycle, in which *ammonia is the refrigerant* and *water is the absorbent*. This is known as the *aqua-ammonia absorption* system.

Ammonia vapour is vigorously absorbed in water. So when low-pressure ammonia vapour from the evaporator comes in contact in the absorber with the weak solution (the concentration of ammonia in water is low) coming from the generator, it is readily absorbed, releasing the latent heat of condensation. The temperature of the solution tends to rise, while the absorber is cooled by the circulating water, absorbing the *heat of solution* (Q_A), and maintaining a constant temperature. Strong solution, rich in ammonia, is pumped to

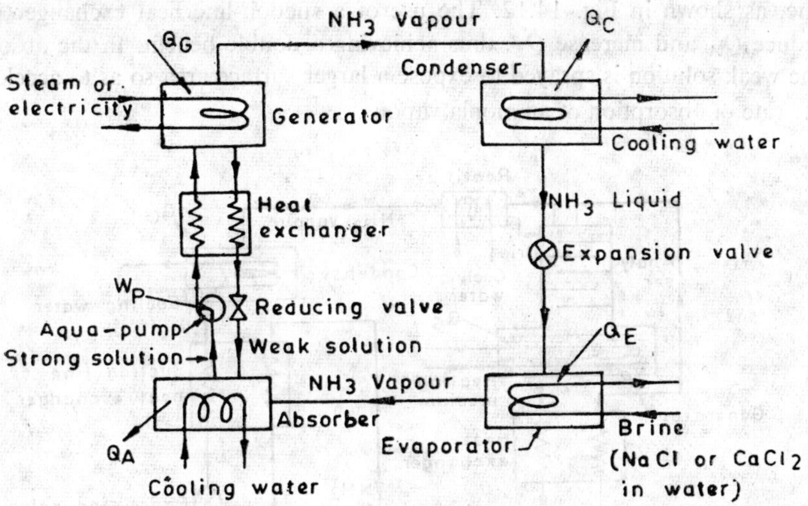

Fig. 14.11 Vapour absorption refrigeration plant-flow diagram

the generator where heat (Q_G) is supplied from an external source (steam, electricity, gas flame, etc.). Since the boiling point of ammonia is less than that of water, the ammonia vapour is given off from the aqua-ammonia solution at high pressure, and the weak solution returns to the absorber through a pressure reducing valve. The heat exchanger preheats the strong solution and precools the weak solution, reducing both Q_G and Q_A, the heat to be supplied in the generator and the heat to be removed in the absorber respectively. The ammonia vapour then condenses in the condenser, is throttled by the expansion valve, and then evaporates, absorbing the heat of evaporation from the surroundings or the brine to be chilled.

In driving the ammonia vapour out of the solution in the generator, it is impossible to avoid evaporating some of the water. This water vapour going to the condenser along with the ammonia vapour, after condensation, may get frozen to ice and block the expansion valve. So an *analyzer-rectifier combination* (Fig. 14.12) is used to eliminate water vapour from the ammonia vapour going into the condenser.

The analyzer is a direct-contact heat exchanger consisting of a series of trays mounted above the generator. The strong solution from the absorber flows downward over the trays to cool the outgoing vapours. Since the saturation temperature of water is higher than that of ammonia at a given pressure, it is the water vapour which condenses first. As the vapour passes upward through the analyzer, it is cooled and enriched by ammonia, and the liquid is heated. Thus the vapour going to the condenser is lower in temperature and richer in ammonia, and the heat input to the generator is decreased.

The final reduction in the percentage of water vapour in the ammonia going to the condenser occurs in the rectifier which is a water-cooled heat exchanger which condenses water vapour and returns it to the generator through the drip line, as shown in Fig. 14.12. The use of a suction-line heat exchanger is to reduce Q_A and increase Q_E, thus achieving a double benefit. In the absorber the weak solution is sprayed to expose a larger surface area so as to accelerate the rate of absorption of ammonia vapour.

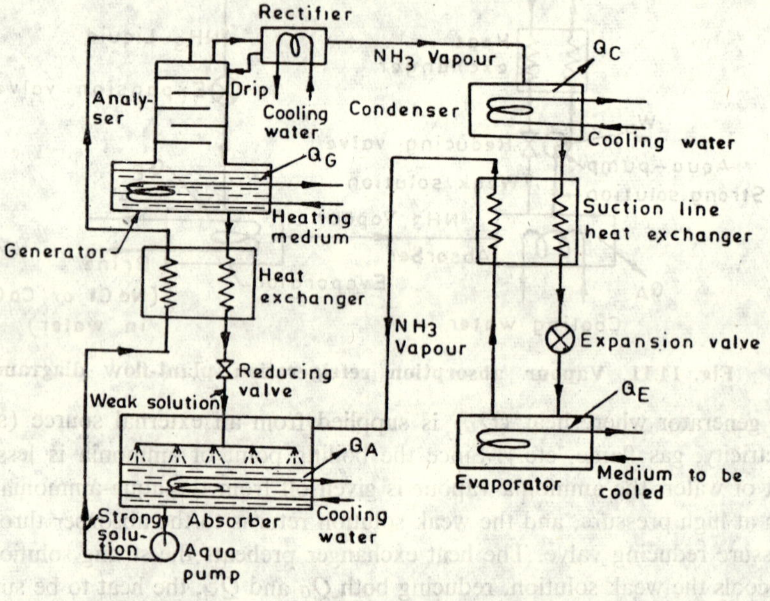

Fig. 14.12 Actual vapour absorption refrigeration plant with analyzer and rectifier

There is another absorption refrigeration system, namely, lithium bromide-water vapour absorption (Fig. 14.13). Here the refrigerant is water and the

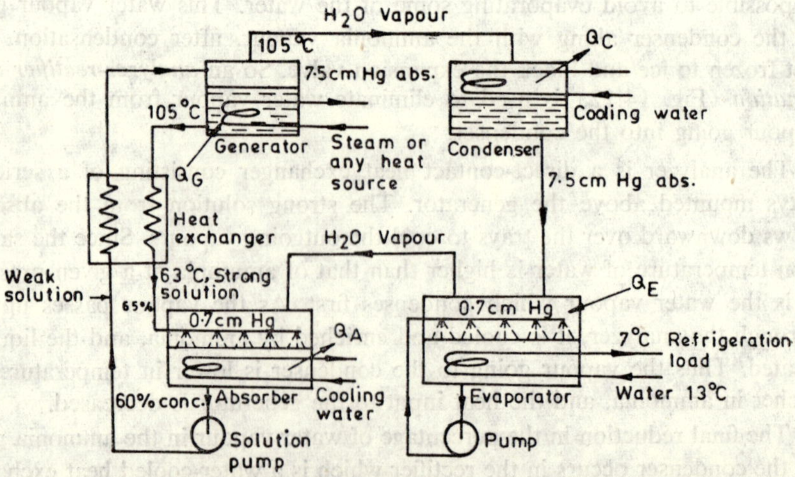

Fig. 14.13 Lithium bromide-water absorption refrigeration plant

absorbent is the solution of lithium bromide salt in water. Since water cannot be cooled below 0°C, it can be used as a refrigerant in air conditioning units. Lithium bromide solution has a strong affinity for water vapour because of its very low vapour pressure. It absorbs water vapour as fast as it is released in the evaporator.

While the vapour compression refrigeration system requires the expenditure of 'high-grade' energy in the form of shaft work to drive the compressor with the concommitant disadvantage of vibration and noise, the absorption refrigeration system requires only 'low-grade' energy in the form of heat to drive it, and it is relatively silent in operation and subject to little wear. Although the COP $= Q_E/Q_G$ is low, the absorption units are usually built when waste heat is available, and they are built in relatively bigger sizes. One current application of the absorption system that may grow in importance is the *utilization of solar energy* for the generator heat source of a refrigerator for food preservation and perhaps for comfort cooling.

14.4.1 Theoretical COP of an Absorption System

Let us assume that an absorption refrigeration plant uses heat Q_G from a source at T_1, provides refrigeration Q_E for a region at T_R, and rejects heat $(Q_A + Q_C)$ to a sink (atmosphere) at T_2, as shown in Fig. 14.14.

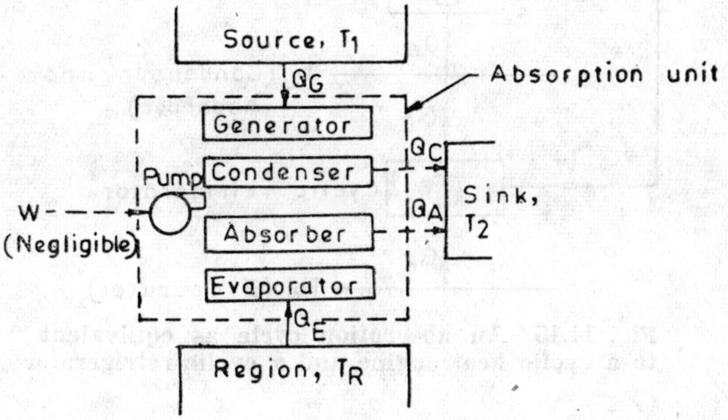

Fig. 14.14 Energy fluxes in vapour absorption plant

By the first law

$$Q_E + Q_G = Q_C + Q_A \tag{14.3}$$

By the second law

$$(\Delta S)_{\text{source}} + (\Delta S)_{\text{sink}} + (\Delta S)_{\text{region}} \geqslant 0$$

$$\therefore \quad -\frac{Q_G}{T_1} + \frac{Q_C + Q_A}{T_2} - \frac{Q_E}{T_R} \geqslant 0$$

From equation (14.3)

$$-\frac{Q_G}{T_1} + \frac{Q_E + Q_G}{T_2} - \frac{Q_E}{T_R} \geqslant 0$$

$$\therefore \quad \frac{T_1-T_2}{T_1T_2}\,Q_G+\frac{T_R-T_2}{T_2T_R}\,Q_E \geqslant 0$$

$$\therefore \quad \frac{T_2-T_R}{T_2T_R}\,Q_E \leqslant \frac{T_1-T_2}{T_1T_2}\,Q_G$$

$$\therefore \quad \frac{Q_E}{Q_G} \leqslant \frac{(T_1-T_2)\,T_R}{(T_2-T_R)\,T_1}$$

or
$$\text{COP} \leqslant \frac{(T_1-T_2)\,T_R}{(T_2-T_R)\,T_1}$$

$$\therefore \quad (\text{COP})_{\max} = \frac{(T_1-T_2)\,T_R}{(T_2-T_R)\,T_1} \tag{14.4}$$

or
$$(\text{COP})_{\max} = \frac{T_R}{T_2-T_R} \times \frac{T_1-T_2}{T_1}$$

Therefore, the maximum possible COP is the product of the ideal COP of a refrigerator working between T_R and T_2, and the ideal thermal efficiency of an engine working between T_1 and T_2. The cyclic heat engine and the refrigerator together, as shown in Fig. 14.15, is equivalent to the absorption cycle.

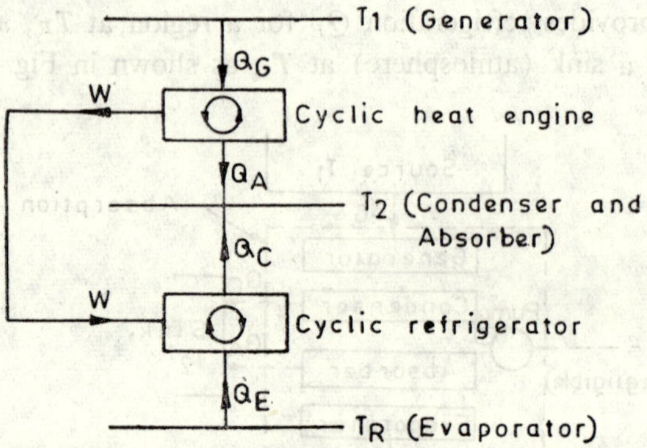

Fig. 14.15 An absorption cycle as equivalent to a cyclic heat engine and a cyclic refrigerator

14.5 GAS CYCLE REFRIGERATION

Refrigeration can also be accomplished by means of a gas cycle. In the gas cycle, an expander replaces the throttle valve of a vapour compression system, because the drop in temperature by throttling a real gas is very small. For an ideal gas, enthalpy is a function of temperature only, and since in throttling enthalpy remains unchanged, there would not be any change in temperature also. Work output obtained from the expander is used as an aid in compression, thus decreasing the net work input. The ideal gas-refrigeration cycle is the same as the *reversed Brayton cycle*. The flow and *T-s* diagrams of the cycle are shown in Fig. 14.16. *Since there is no* phase change, the condenser and evaporator in a vapour compression system are here called the cooler and refrigerator respectively. The COP of the refrigeration cycle, assuming the gas to be ideal.

$$\text{COP} = \frac{Q_2}{W_{\text{net}}} = \frac{h_1 - h_4}{(h_2 - h_1) - (h_3 - h_4)}$$

$$= \frac{T_1 - T_4}{(T_2 - T_1) - (T_3 - T_4)} = \frac{T_1 - T_4}{T_1\left(\dfrac{T_2}{T_1} - 1\right) - T_4\left(\dfrac{T_3}{T_4} - 1\right)}$$

For isentropic compression and expansion

$$\frac{T_2}{T_1} = \left(\frac{p_1}{p_2}\right)^{(\gamma-1)/\gamma} = \frac{T_3}{T_4}$$

$$\therefore \quad \text{COP} = \frac{T_1 - T_4}{(T_1 - T_4)\left(\dfrac{T_3}{T_4} - 1\right)} = \frac{T_4}{T_3 - T_4}$$

Also $\quad \text{COP} = \dfrac{1}{\left(\dfrac{p_1}{p_2}\right)^{(\gamma-1)/\gamma} - 1}$, (14.6)

where p_1 is the pressure after compression and p_2 is the pressure before compression.

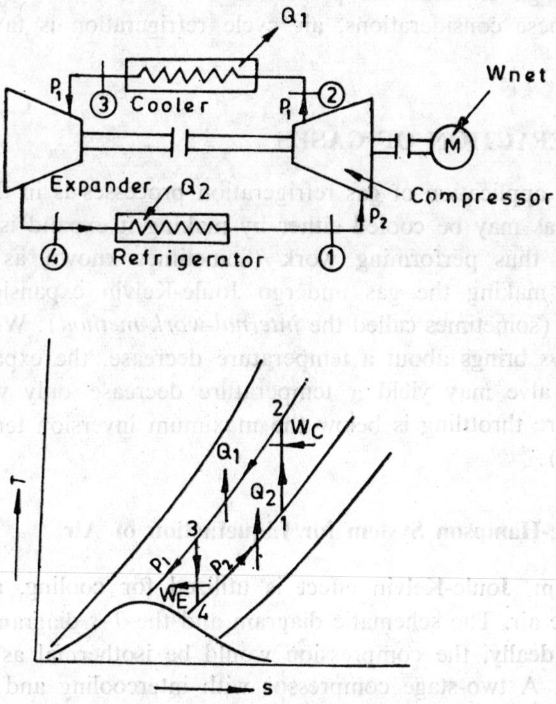

Fig. 14.16 Gas refrigeration cycle

The COP of a gas-cycle refrigeration system is low. The power requirement per unit capacity is high. Its prominent application is in aircrafts missiles, where the vapour compression refrigeration system becomes heavy and bulky. Figure 14.17 shows the open cycle aircraft cabin cooling.

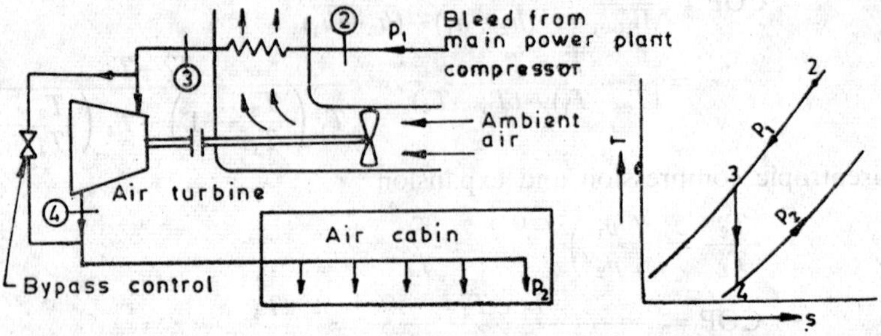

Fig. 14.17 Open cycle aircraft cabin cooling

The compressed air is available and is a small percentage of the amount handled by the compressor of a turbojet or a supercharged aircraft engine. Large amounts of cool ambient air are available for cooling the compressed air. In addition to cooling, the replacement of stale air in the cabin is possible. At high altitudes the pressurization of cabin air is also possible. Because of these considerations, air cycle refrigeration is favoured in air-crafts.

14.6 LIQUEFACTION OF GASES

An important application of gas refrigeration processes is in the liquefaction of gases. A gas may be cooled either by making it expand isentropically in an expander, thus performing work (sometimes known as *external-work method*), or making the gas undergo Joule-Kelvin expansion through a throttle valve (sometimes called the *internal-work method*). While the former method always brings about a temperature decrease, the expansion through the throttle valve may yield a temperature decrease only when the temperature before throttling is below the maximum inversion temperature (see Section 13.7).

14.6.1 Linde-Hampson System for Liquefaction of Air

In this system, Joule-Kelvin effect is utilized for cooling, and ultimately, liquefying the air. The schematic diagram and the T-s diagram are shown in Fig. 14.18. Ideally, the compression would be isothermal as shown on the T-s diagram. A two-stage compressor with intercooling and aftercooling is shown. The *yield, Y,* of the system is defined as the ratio of the mass of liquid produced to the mass of gas compressed. The energy required per unit mass of liquid produced is known as the *specific work* consumption, W.

The theoretical yield, assuming perfect insulation, can be determined by a mass and energy balance for the control volume (Joule-Kelvin refrigeration system) as shown in Fig. 14.18. Let $\dot{m}_f$ be the rate at which liquid air is

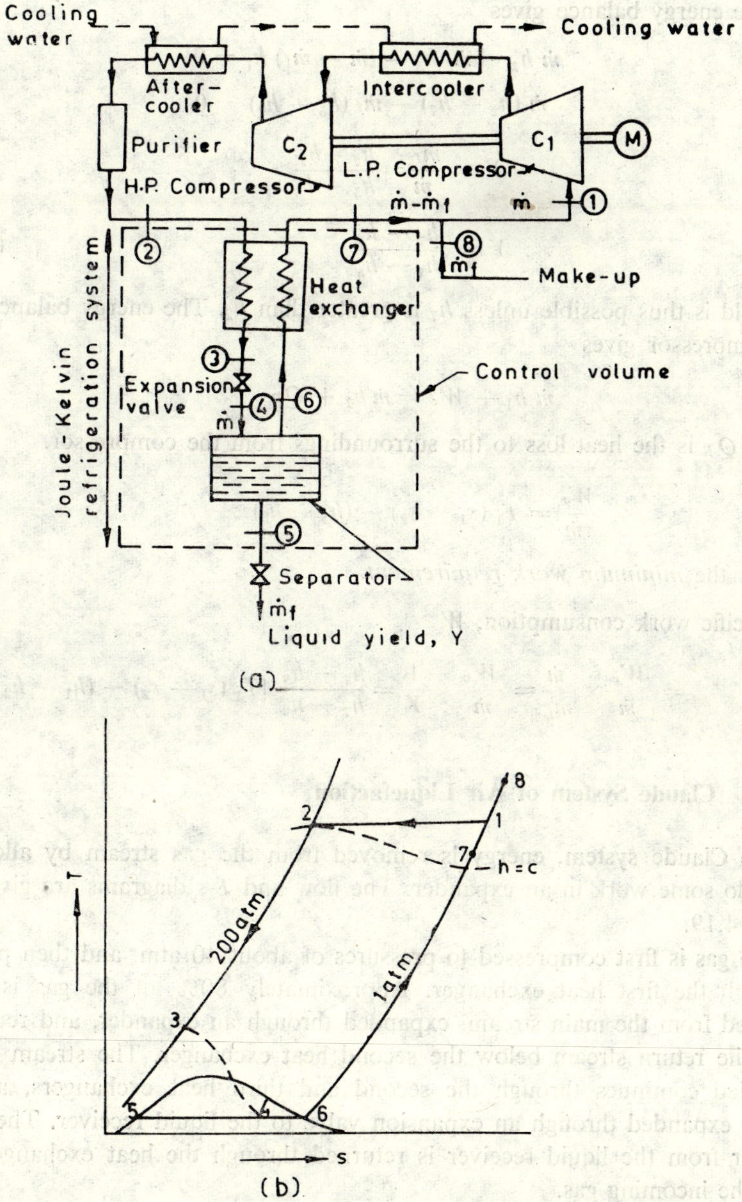

Fig. 14.18 Linde-Hampson cycle for air liquefaction

produced (the same given to the system as make-up), and $\dot{m}$ the rate at which air is compressed and then expanded. Then the yield is

$$Y = \frac{\dot{m}_f}{\dot{m}}$$

and the energy balance gives

$$\dot{m}\, h_2 - \dot{m}_f\, h_5 - (\dot{m} - \dot{m}_f)\, h_7 = 0$$

$$\therefore \quad \dot{m}\,(h_2 - h_7) - \dot{m}_f\,(h_5 - h_7) = 0$$

$$Y = \frac{\dot{m}_f}{\dot{m}} = \frac{h_2 - h_7}{h_5 - h_7}$$

$$\therefore \quad Y = \frac{h_7 - h_2}{h_7 - h_5} \tag{14.6}$$

No yield is thus possible unless h_7 is greater than h_2. The energy balance for the compressor gives

$$\dot{m}\, h_1 + W_c = \dot{m}\, h_2 + Q_R$$

where Q_R is the heat loss to the surroundings from the compressor.

$$\therefore \quad \frac{W_c}{\dot{m}} = T_1\,(s_1 - s_2) - (h_1 - h_2)$$

This is the *minimum work requirement.*

Specific work consumption, W

$$= \frac{W_c}{\dot{m}} \times \frac{\dot{m}}{\dot{m}_f} = \frac{W_c}{\dot{m}} \cdot \frac{1}{Y} = \frac{h_7 - h_5}{h_7 - h_2}\,[T_1\,(s_1 - s_2) - (h_1 - h_2)]$$

14.6.2. Claude System of Air Liquefaction

In the Claude system, energy is removed from the gas stream by allowing it to do some work in an expander. The flow and T-s diagrams are given in Fig. 14.19.

The gas is first compressed to pressures of about 40 atm. and then passed through the first heat exchanger. Approximately 80% of the gas is then diverted from the main stream, expanded through an expander, and reunited with the return stream below the second heat exchanger. The stream to be liquefied continues through the second and third heat exchangers, and is finally expanded through an expansion valve to the liquid receiver. The cold vapour from the liquid receiver is returned through the heat exchangers to cool the incoming gas.

The yield and the specific work consumption may be computed by making the mass and energy balance as in the Linde-Hampson system.

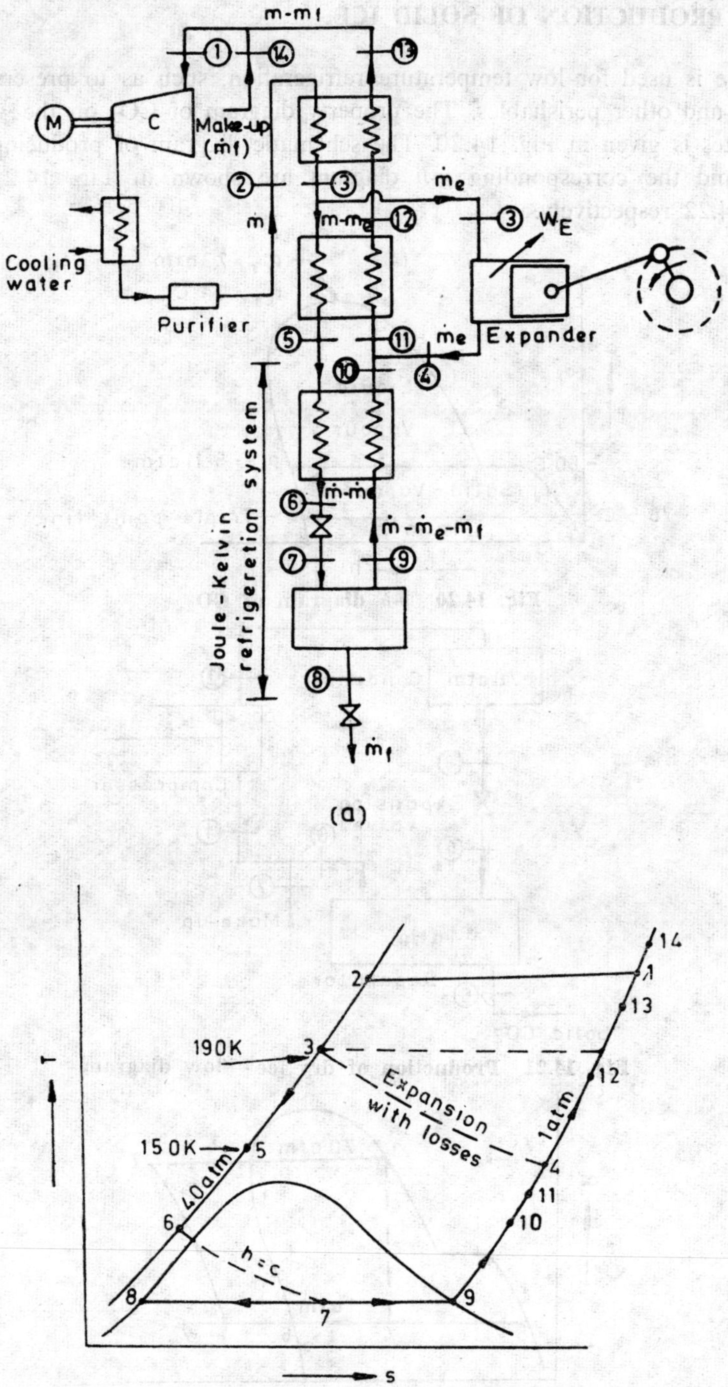

Fig. 14.19 Claude cycle for air liquefaction

14.7 PRODUCTION OF SOLID ICE

Dry ice is used for low temperature refrigeration, such as to preserve ice-cream and other perishables. The property diagram of CO_2 on the *p-h* co-ordinates is given in Fig. 14.20. The schematic diagram of producing solid CO_2 and the corresponding *p-h* diagram are shown in Fig. 14.21 and Fig. 14.22 respectively.

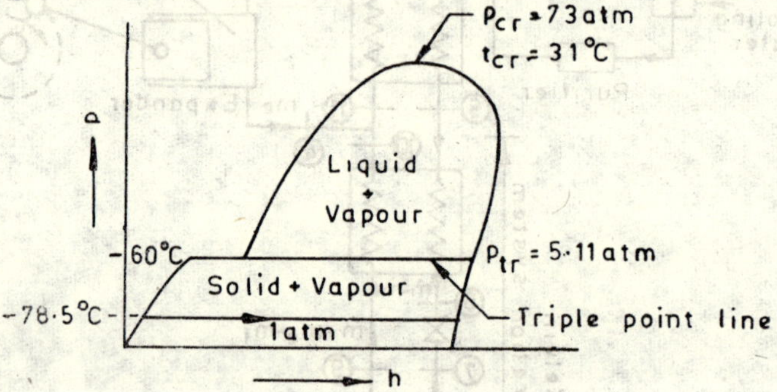

Fig. 14.20 ***p-h*** **diagram of CO_2**

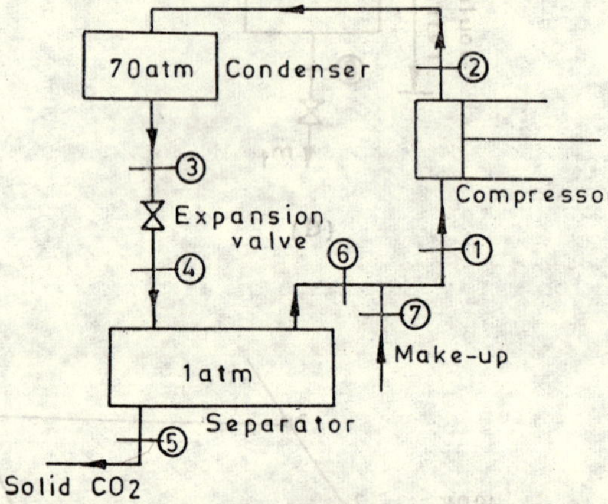

Fig. 14.21 Production of dry ice—flow diagram

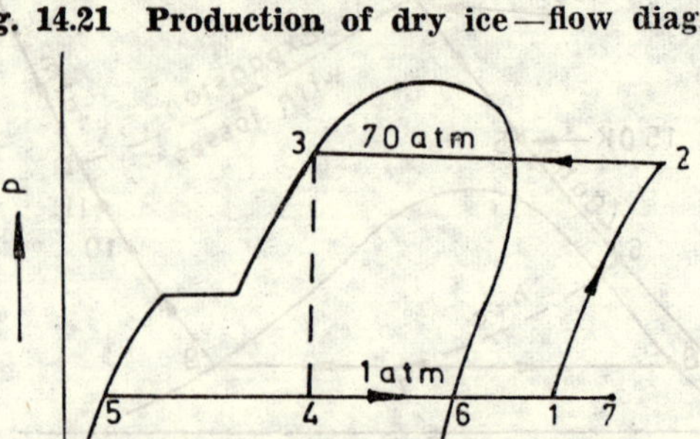

Fig. 14.22 Refrigeration cycle of a dry ice plant on *p-h* plot

EXAMPLE 14.1

A cold storage is to be maintained at −5°C while the surroundings are at 35°C. The heat leakage from the surroundings into the cold storage is estimated to be 29 kW. The actual COP of the refrigeration plant used is one-third that of an ideal plant working between the same temperatures. Find the power required (in kW) to drive the plant.

Solution

$$\text{COP (Ideal)} = \frac{T_2}{T_1 - T_2} = \frac{268}{308 - 268} = 6.7$$

$\therefore$ Actual COP $= \frac{1}{3} \times 6.7$

$$= 2.23 = \frac{Q_2}{W}$$

$\therefore$ Power required to drive the plant (Fig. Ex. 14.1)

$$W = \frac{Q_2}{2.23} = \frac{29}{2.23}$$

$$= 13 \text{ kW} \quad \text{Ans.}$$

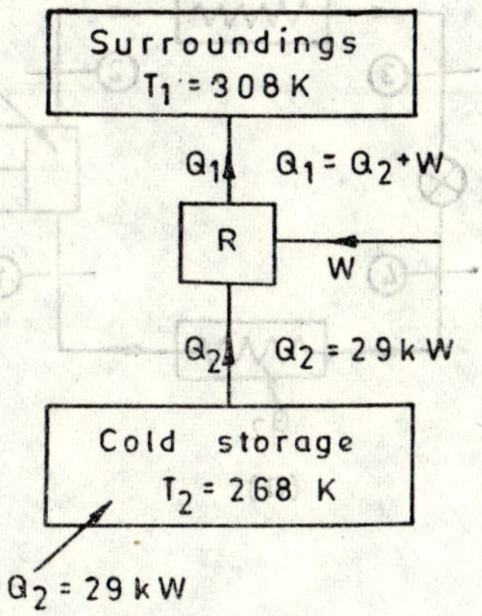

Fig. Ex. 14.1

EXAMPLE 14.2

A simple R-12 plant is to develop 5 tonnes of refrigeration. The condenser and evaporator temperatures are to be 40°C and −10°C respectively. Determine (a) the refrigerant flow rate in kg/s, (b) the volume flow rate handled by the compressor in m³/s, (c) the compressor discharge temperature, (d) the pressure ratio, (e) the heat rejected to the condenser in kW, (f) the flash gas percentage after throttling, (g) the COP, and (h) the power required to drive the compressor.

How does this COP compare with that of a Carnot refrigerator operating between 40°C and −10°C ?

Solution

From the table of the thermodynamic properties of a saturated Refrigerant-12 (Fig. Ex. 14.2), $p_1 = (p_{sat.})_{-10°C} = 2.1912$ bar, $h_1 = 183.19$ kJ/kg, $s_1 = 0.7019$ kJ/kg-K, $v_1 = 0.077$ m³/kg, $p_2 = (p_{sat.})_{40°C} = 9.6066$ bar, and $h_3 = 74.59$ kJ/kg $= h_4$. From the superheated table of R-12, when $p_2 = 9.6066$ bar, and $s_2 = s_1 = 0.7019$ kJ/kg-K, by interpolation,

$t_2 = 48°C$, and $h_2 = 209.41$ kJ/kg. The capacity of the plant $= 5 \times 14{,}000$ $= 70{,}000$ kJ/hr. If w is the refrigerant flow rate in kg/s

$$w(h_1 - h_4) = \frac{70.000}{3600} = 19.44 \text{ kW}$$

$$\therefore \quad w = \frac{19.44}{183.19 - 74.59} = 0.18 \text{ kg/s} \qquad \text{Ans. (a)}$$

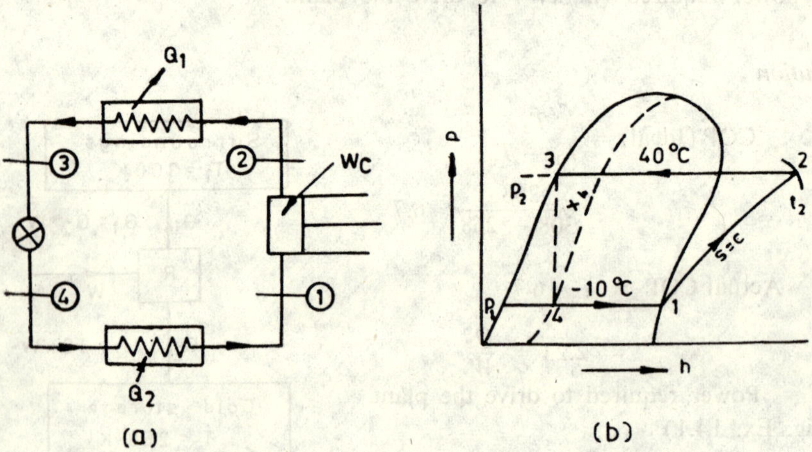

Fig. Ex. 14·2

$$\text{Volume flow rate} = w.v_1 = 0.18 \times 0.077$$
$$= 0.0139 \text{ m}^3\text{/s} \qquad \text{Ans. (c)}$$

Compressor discharge temperature = 48°C Ans. (b)

$$\text{Pressure ratio} = \frac{p_2}{p_1} = \frac{9.6066}{2.1912} = 4.39 \qquad \text{Ans. (d)}$$

$$\text{Heat rejected to the condenser} = w(h_2 - h_3)$$
$$= 0.18\ (209.41 - 74.59) \text{ units}$$
$$= 24.27 \text{ kW} \qquad \text{Ans. (e)}$$

$$h_4 = h_f + x_4 h_{fg} = 26.87 + x_4 \times 156.31 = 74.59$$

$$\therefore \quad x_4 = \frac{47.72}{156.31} = 0.305$$

$$\therefore \quad \text{Flash gas percentage} = 30.5\% \qquad \text{Ans. (f)}$$

$$\text{COP} = \frac{h_1 - h_4}{h_2 - h_1} = \frac{183.19 - 74.59}{209.41 - 183.19}$$

$$= \frac{108.60}{26.22} = 4.14 \qquad \text{Ans. (g)}$$

Power required to drive the compressor

$$= w(h_2 - h_1) = 0.18 \times 26.22 = 4.72 \text{ kW} \qquad \text{Ans. (h)}$$

$$\text{COP (Reversible)} = \frac{T_2}{T_1 - T_2} = \frac{263}{50} = 5.26$$

$$\therefore \quad \frac{\text{COP (Vap. Comp. cycle)}}{\text{COP (Carnot cycle)}} = \frac{4.14}{5.26} = 0.787 \qquad \text{Ans.}$$

EXAMPLE 14.3

A Refrigerant-12 vapour compression plant producing 10 tonnes of refrigeration operates with condensing and evaporating temperatures of 35°C and −10°C respectively. A suction line heat exchanger is used to subcool the saturated liquid leaving the condenser. Saturated vapour leaving the evaporator is superheated in the suction line heat exchanger to the extent that a discharge temperature of 60°C is obtained after isentropic compression. Determine (a) the subcooling achieved in the heat exchanger, (b) the refrigerant flow rate in kg/s, (c) the cylinder dimensions of the two-cylinder compressor, if the speed is 900 rpm, stroke-to-bore ratio is 1.1, and the volumetric efficiency is 80%, (d) the COP of the plant, and (e) the power required to drive the compressor in kW.

Solution

From the *p-h* chart of R-12, the property values at the states, as shown in Fig. Ex. 14.3,

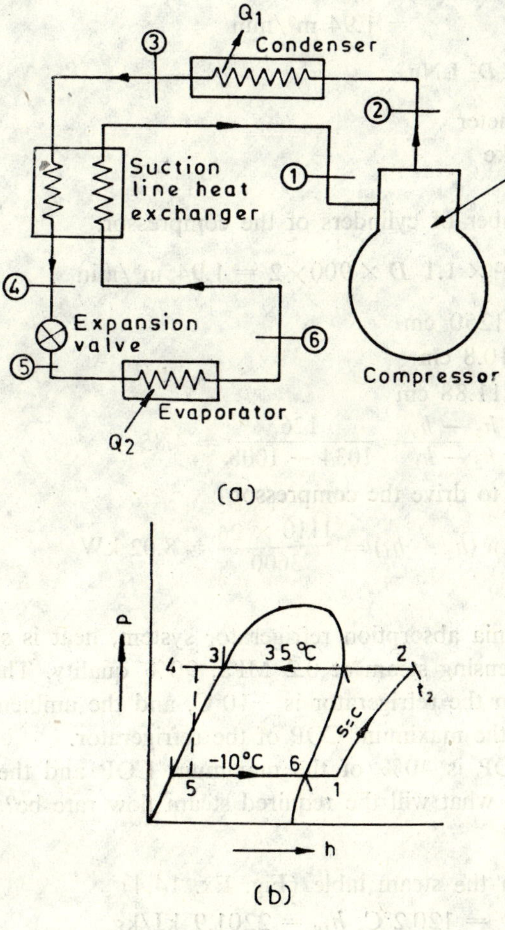

Fig. Ex. 14·3

$h_3 = 882,\ h_2 = 1034,$
$h_6 = 998,\ h_1 = 1008$ kJ/kg,
$v_1 = 0.084$ m³/kg,
$h_3 - h_4 = h_1 - h_6$
$882 - h_4 = 1008 - 998 = 10$
$\therefore\ h_4 = 872$ kJ/kg
$\therefore\ t_4 = 25°C$

So 10 C deg. subcooling is achieved in the heat exchanger. Refrigeration effect $= h_6 - h_5 = 998 - 872 = 126$ kJ/kg Ans. (a)

$$\therefore \text{ Refrigerant flow rate } \frac{10 \times 14000}{126} = 1110 \text{ kg/hr} \quad \text{Ans. (b)}$$

$$= 0.31 \text{ kg/s}$$

$$\text{Volume flow rate} = w.v_1 = 1110 \times 0.084 = 93 \text{ m}^3/\text{hr}$$

$$\text{Compressor displacement} = \frac{93}{8.0} = 116 \text{ m}^3/\text{hr}$$

$$= 1.94 \text{ m}^3/\text{min}$$

This is equal to $\frac{\pi}{4} D^2$ LNn

where D = diameter
L = stroke
N = rpm
n = number of cylinders of the compressor.

$$\frac{\pi}{4} D^2 \times 1.1\ D \times 900 \times 2 = 1.94 \text{ m}^3/\text{min}$$

$$D^3 = 1250 \text{ cm}^3$$

or $D = 10.8$ cm Ans. (c)

and $L = 11.88$ cm

$$\text{COP} = \frac{h_6 - h_5}{h_2 - h_1} = \frac{126}{1034 - 1008} = 4.85 \quad \text{Ans. (d)}$$

Power required to drive the compressor

$$= w(h_2 - h_1) = \frac{1110 \times 26}{3600} = 8.02 \text{ kW} \quad \text{Ans. (e)}$$

EXAMPLE 14.4

In an aqua-ammonia absorption refrigerator system, heat is supplied to the generator by condensing steam at 0.2 MPa, 90% quality. The temperature to be maintained in the refrigerator is −10°C, and the ambient temperature is 30°C. Estimate the maximum COP of the refrigerator.

If the actual COP is 40% of the maximum COP and the refrigeration load is 20 tonnes, what will the required steam flow rate be?

Solution

At 0.2 MPa, from the steam table (Fig. Ex. 14.4)

$$t_{sat.} = 120.2°C,\ h_{fg} = 2201.9 \text{ kJ/kg}$$

The maximum COP of the absorption refrigeration system is given by equation (14.4)

$$(COP)_{max} = \frac{(T_1 - T_2)\, T_R}{(T_2 - T_R)\, T_1}$$

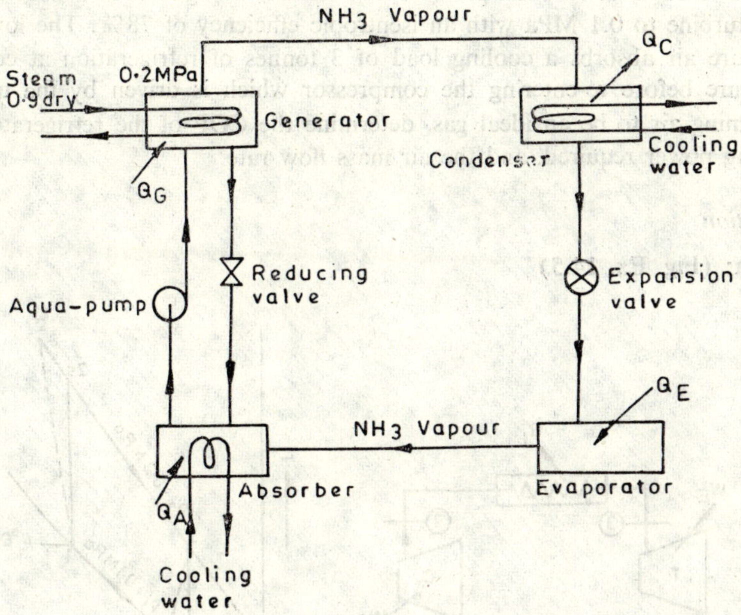

Fig. Ex 14.4

where T_1 = generator temperature

$= 119.62 + 273 = 392.62$ K

T_2 = condenser and absorber temperature

$= 30 + 273 = 303$ K

T_R = evaporator temperature

$= -10 + 273 = 263$ K

$$\therefore \quad (COP)_{max} = \frac{(392.62 - 303) \times 263}{(303 - 263) \times 392.62} = \frac{89.62 \times 263}{40 \times 392.62} = 1.5 \quad \text{Ans.}$$

$\therefore$ Actual COP $= 1.5 \times 0.4 = 0.60$

Since $\quad COP = \dfrac{Q_E}{Q_G}$

$$Q_G = \frac{Q_E}{COP} = \frac{20 \times 14000}{0.60 \times 3600} = 129.6 \text{ kW}$$

Heat transferred by 1 kg of steam on condensation

$= (h_f + xh_{fg}) - h_f = 0.9 \times 2201.9$

$= 1981.71$ kJ/kg

$\therefore$ Steam flow rate required

$$= \frac{129.6}{1981.71} = 0.0654 \text{ kg/s} \quad \text{Ans.}$$

EXAMPLE 14.5

In an aircraft cooling system, air enters the compressor at 0.1 MPa, 4°C, and is compressed to 0.3 MPa with an insentropic efficiency of 72%. After being cooled to 55°C at constant pressure in a heat exchanger the air then expands in a turbine to 0.1 MPa with an isentropic efficiency of 78%. The low temperature air absorbs a cooling load of 3 tonnes of refrigeration at constant pressure before re-entering the compressor which is driven by the turbine. Assuming air to be an ideal gas, determine the COP of the refrigerator, the driving power required, and the air mass flow rate.

Solution

Given: (Fig. Ex. 14.5)

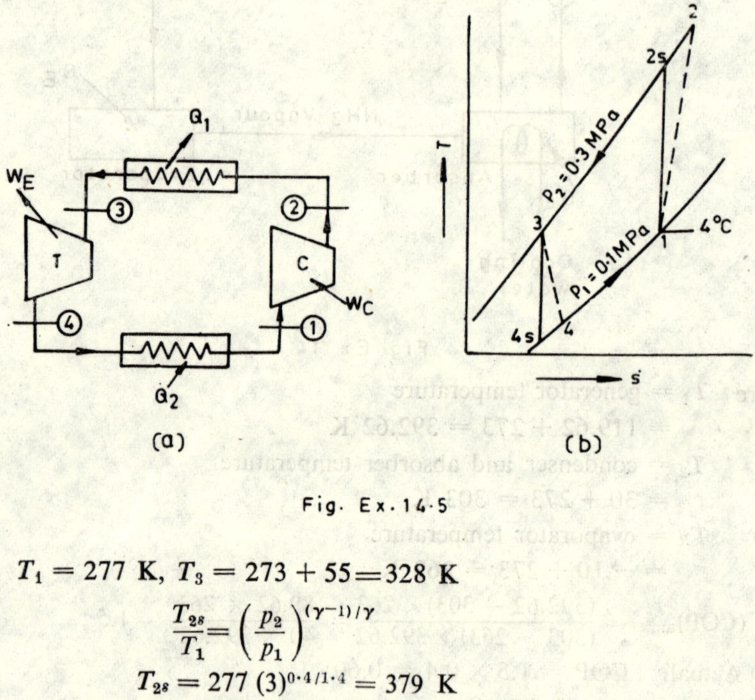

Fig. Ex. 14.5

$$T_1 = 277 \text{ K}, \; T_3 = 273 + 55 = 328 \text{ K}$$

$$\frac{T_{2s}}{T_1} = \left(\frac{p_2}{p_1}\right)^{(\gamma-1)/\gamma}$$

$$\therefore \quad T_{2s} = 277\,(3)^{0.4/1.4} = 379 \text{ K}$$

$$T_{2s} - T_1 = 102 \text{ K}$$

$$T_2 - T_1 = \frac{102}{0.72} = 141.8 \text{ K}$$

$$\frac{T_{4s}}{T_3} = \left(\frac{p_2}{p_1}\right)^{(\gamma-1)/\gamma}$$

$$\therefore \quad T_{4s} = 328\,(3)^{0.4/1.4} = \frac{328}{1.368} = 240 \text{ K}$$

$$T_3 - T_{4s} = 88 \text{ K}$$

$$\therefore \quad T_3 - T_4 = 0.78 \times 88 = 68.6 \text{ K}$$

$$\therefore \quad T_4 = 259.4 \text{ K}$$

Refrigerating effect $= c_p\,(T_1 - T_4) = 17.6\, c_p$ kJ/kg

Net work input $= c_p\,[(T_2 - T_1) - (T_3 - T_4)]$

$$= c_p\,[141.8 - 68.6] = 73.2\, c_p \text{ kJ/kg}$$

$$\therefore \quad \text{COP} = \frac{17.6\, c_p}{73.2\, c_p} = 0.24 \qquad \textit{Ans.}$$

Driving power required

$$= \frac{3 \times 14000}{0.24 \times 3600} = 48.6 \text{ kW} \qquad \textit{Ans.}$$

Mass flow rate of air

$$= \frac{3 \times 14000}{1.005 \times 17.6} = 2374.5 \text{ kg/hr}$$

$$= 0.66 \text{ kg/s} \qquad \textit{Ans.}$$

PROBLEMS

14.1 A Refrigerant-12 vapour compression cycle has a refrigeration load of 3 tonnes. The evaporator and condenser temperatures are —20°C and 40°C respectively. Find (a) the refrigerant flow rate in kg/s, (b) the volume flow rate handled by the compressor in m³/s, (c) the work input to the compressor in kW, (d) the heat rejected in the condenser in kW, and (e) the isentropic discharge temperature.

If there is 5 C deg. of superheating of vapour before it enters the compressor, and 5 C deg. subcooling of liquid before it flows through the expansion valve, determine the above quantities.

14.2 A 5 tonne R-12 plant maintains a cold store at —15°C. The refrigerant flow rate is 0.133 kg/s. The vapour leaves the evaporator with 5 C deg. superheat. Cooling water is available in plenty at 25°C. A suction line heat exchanger subcools the refrigerant before throttling. Find (a) the compressor discharge temperature, (b) the COP, (c) the amount of subcooling in C deg., and (d) the cylinder dimensions of the compressor, if the speed is 900 rpm, stroke-to-bore ratio is 1.2, and volumetric efficiency is 95%.

Allow approximately 5°C temperature difference in the evaporator and condenser.

Ans. (a) 66°C, (b) 4.1, (c) 125°C, (d) 104.5 mm, 125 mm.

14.3 A vapour compression refrigeration system uses R-12 and operates between pressure limits of 0.745 and 0.15 MPa. The vapour entering the compressor has a temperature of —10°C and the liquid leaving the condenser is at 28°C. A refrigerating load of 2 kW is required. Determine the COP and the swept volume of the compressor if it has a volumetric efficiency of 76% and runs at 600 rpm.

Ans. 4.15, 243 cm³.

14.4 A food-freezing system requires 20 tonnes of refrigeration at an evaporator temperature of —35°C and a condenser temperature of 25°C. The refrigerant, R-12, is subcooled 4°C before entering the expansion valve, and the vapour is superheated 5°C before leaving the evaporator. A six-cylinder single-acting compressor with

stroke equal to bore is to be used, operating at 1500 rpm. Determine (a) the refrigerating effect, (b) the refrigerant flow rate, (c) the theoretical piston displacement per sec, (d) the theoretical power required in kW, (e) the COP, (f) the heat removed in the condenser, and (g) the bore and stroke of the compressor.

14.5 A refrigeration plant produces 0.139 kg/s of the ice at —5°C from water at 30°C. If the power required to drive the plant is 22 kW, determine the capacity of the ice plant in tonnes and the actual COP (c_p of ice=2.1 kJ/kg-K).

14.6 An air conditioning unit using R-12 as a refrigerant is mounted in the window of a room. During steady operation 1.5 kW of heat is transferred from the air in the room to the evaporator coils of R-12. If this air is at 22°C and the temperature of R-12 in the evaporator is 15°C, determine (a) the refrigerant flow rate, and (b) the minimum power required to drive the compressor if the outside air is at 43°C and the temperature of the refrigerant during condensation is 50°C.

14.7 In a solar energy operated aqua-ammonia absorption refrigeration system, water is cooled at the rate of 10 kg/s from 38°C to 13°C. If the incident solar energy is 640 W/m^2 and the COP of the system is 0.32, estimate the area of the solar collector needed.

14.8 A gas refrigerating system using air as a refrigerant is to work between —12°C and 27°C using an ideal reversed Brayton cycle of pressure ratio 5 and minimum pressure 1 atm., and to maintain a load of 10 tonnes. Find (a) the COP, (b) the air flow rate in kg/s, (c) the volume flow rate entering the compressor in m^3/s, and (d) the maximum and minimum temperatures of the cycle.

14.9 An open cycle (Brayton) air craft cabin cooler expands air at 27°C through a turbine which is 30% efficient from 2 to 1 atm. The cabin temperature is not to exceed 24°C. Estimate the mass flow rate of air required (kg/s) for each tonne of cooling.

14.10 Determine the ideal COP of an absorption refrigerating system in which the heating, cooling, and refrigeration take place at 197°C, 7017°C, and —3°C respectively. **Ans. 5.16.**

14.11 A heat pump is to use an R-12 cycle to operate between outdoor air at —1°C and air in a domestic heating system at 40°C. The temperature difference in the evaporator and the condenser is 8°C. The compressor efficiency is 80%, and the compression begins with saturated vapour. The expansion begins with saturated liquid. The combined efficiency of the motor and belt drive is 75%. If the required heat supply to the warm air is 43.6 kW, what will be the electrical load in kW ?

14.12 An ideal (Carnot) refrigeration system operates between the temperature limits of —30°C and 25°C. Find the ideal COP and the horse power required from an external source to absorb 3.89 kW at low temperature.

If the system operates as a heat pump, determine the COP and the power required to discharge 3.89 kW at high temperature.

14.13 An ammonia-absorption system has an evaporator temperature of —12°C and a condenser temperature of 50°C. The generator temperature is 150°C. In this cycle, 0.42 kJ is transferred to the ammonia in the evaporator for each kJ transferred to the ammonia solution in the generator from the high temperature source.

It is desired to compare the performance of this cycle with the performance of a similar vapour compression cycle. For this, it is assumed that a reservoir is available at 150°C, and that heat is transferred from this reservoir to a reversible engine which rejects heat to the surroundings at 25°C. This work is then used to drive an ideal vapour compression system with ammonia as the refrigerant. Compare the amount of refrigeration that can be achieved per kJ from the high temperature source in this case with the 0.42 kJ that can be achieved in the absorption system.

14.14 An R-12 plant is to cool milk from 30°C to 1°C involving a refrigeration capacity of 10 tonnes. Cooling water for the condenser is available at 25°C and a 5 C deg. rise in its temperature is allowable. Determine the suitable condensing and evaporating temperatures, providing a minimum of 5 C deg. differential, and calculate the theoretical power required in kW and the cooling water requirement in kg/s. Also, find the percentage of flash gas at end of throttling. Assume a 2 C deg. subcooling in the liquid refrigerant leaving the condenser.

14.15 The following data pertain to an air cycle refrigeration system for an aircraft:

Capacity 5 tonnes

Cabin air inlet temperature 15°C and outlet temperature 25°C

Pressure ratio across the compressor 5

The aircraft is flying at 0.278 km/s where the ambient conditions are 0°C and 80 kPa. Find the COP and the cooling effectiveness of the heat exchanger. The cabin is at 0.1 MPa, and the cooling turbine powers the circulating fans.

14.16 A water cooler supplies chilled water at 7°C when water is supplied to it at 27°C at a rate of 0.7 litres/min., while the power consumed amounts to 200 watts. Compare the COP of this refrigeration plant with that of the ideal refrigeration cycle for a similar situation.

14.17 A refrigerating plant of 8 tonnes capacity has an evaporation temperature of −8°C and condenser temperature of 30°C. The refrigerant, R-12, is subcooled 5°C before entering the expansion valve and the vapour is superheated 6°C before leaving the evaporator coil. The compression of the refrigerant is isentropic. If there is a suction pressure drop of 20 kPa through the valves, and discharge pressure drop of 10 kPa through the valves, determine the COP of the plant, theoretical piston displacement per sec., and the heat removal rate in the condenser.

14.18 An ultra-low-temperature freezer system employs a coupling of two vapour compression cycles of R-12 and R-13, as shown in Fig. P. 14.18. The states and properties of both cycles are shown on the *T-s* plot. Determine the ratio of the circula-

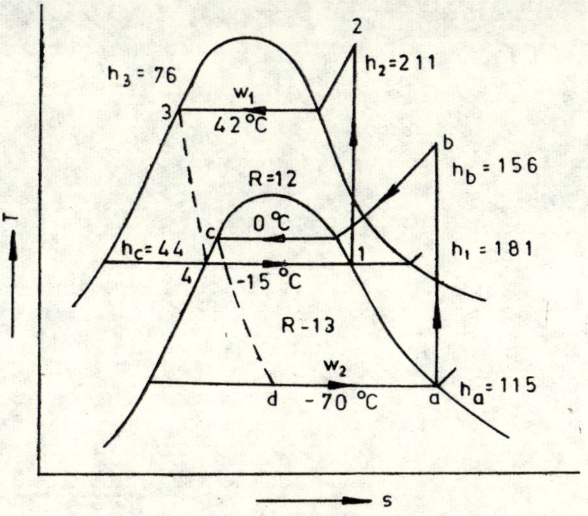

Fig. P. 14.18

tion rates of the two refrigerants, w_1/w_2, and the overall COP. How does this COP compare with the Carnot COP operating between 42°C and −70°C?

14.19 Derive an expression for the COP of an ideal gas refrigeration cycle with a regenerative heat exchanger. Express the result in terms of the minimum gas temperature during heat rejection (T_h) maximum gas temperature during heat absorption (T_1), and pressure ratio for the cycle (p_2/p_1)

$$\text{Ans. COP} = \frac{T_1}{T_h \, r_p^{(\gamma-1)/\gamma} \, {}_1 - T_1}$$

14.20 Large quantities of electrical power can be transmitted with relatively little loss when the transmission cable is cooled to a superconducting temperature. A regenerated gas refrigeration cycle operating with helium is used to maintain an electrical cable at 15 K. If the pressure ratio is 10 and heat is rejected directly to the atmosphere at 300 K, determine the COP and the performance ratio with respect to the Carnot cycle. **Ans.** 0.02, 0.38.

15

Psychrometrics

The properties of the mixtures of ideal gases were presented in Chapter 10. The name 'psychrometrics' is given to the study of the properties of air-water vapour mixtures. Atmospheric air is considered to be a mixture of dry air and water vapour. The control of moisture (or water vapour) content in the atmosphere is essential for the satisfactory operation of many processes involving *hygroscopic* materials like paper and textiles, and is important in comfort air conditioning.

15.1 PROPERTIES OF ATMOSPHERIC AIR

Dry air is a mechanical mixture of the gases: oxygen, nitrogen, carbon dioxide, hydrogen, argon, neon, krypton, helium, ozone, and xenon. However, oxygen and nitrogen make up the major part of the combination. Dry air is considered to consist of 21% oxygen and 79% nitrogen by volume, and 23% oxygen and 77% nitrogen by mass. Completely dry air does not exist in nature. Water vapour in varying amounts is diffused through it. If p_a and p_w are the partial pressures of dry air and water vapour respectively, then by Dalton's law of partial pressures

$$p_a + p_w = p$$

where p is the atmospheric pressure.

$\therefore$ Mole-fraction of dry air, x_a

$$= \frac{p_a}{p} = p_a \qquad (\because p = 1 \text{ atm.})$$

and mole-fraction of water vapour, x_w

$$= \frac{p_w}{p} = p_w$$

Since p_w is very small, the saturation temperature of water vapour at p_w is less than atmospheric temperature, t_{atm}. (Fig. 15.1). So the water vapour in air exists in the *superheated state,* and air is said to be *unsaturated.*

Relative humidity (R.H., ϕ) is defined as the ratio of partial pressure of water vapour, p_w, in a mixture to the saturation pressure, p_s, of pure water, at the same temperature of the mixture (Fig. 15.1)

$$\therefore \quad \text{R.H.}(\phi) = \frac{p_w}{p_s}$$

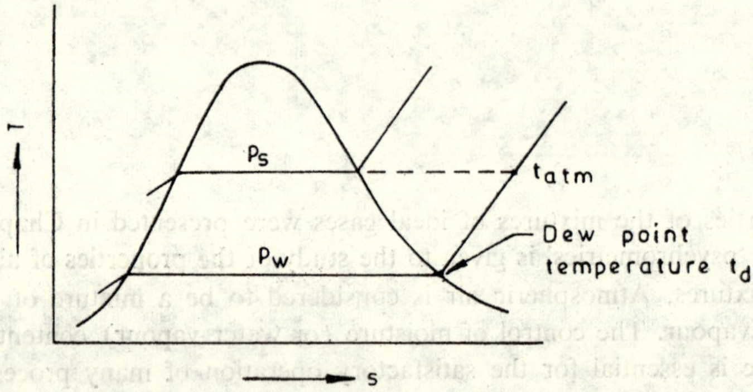

Fig. 15.1 States of water vapour in mixture

If water is injected into unsaturated air in a container, water will evaporate, which will increase the moisture content of the air, and p_w will increase. This will continue till air becomes saturated at that temperature, and there will be no more evaporation of water. For saturated air, the relative humidity is 100%. Assuming water vapour as an ideal gas

$$p_w V = m_w R_{H_2O} T = n_w \bar{R} T$$

and
$$p_s V = m_s R_{H_2O} T = n_s \bar{R} T$$

where V is the volume and T the temperature of air, the subscripts w and s indicating the unsaturated and saturated states of air respectively.

$$\phi = \frac{p_w}{p_s} = \frac{m_w}{m_s}$$

$$= \frac{\text{mass of water vapour in a given volume of air at temperature } T}{\text{mass of water vapour when the same volume of air is saturated at temperature } T}$$

$$= \frac{n_w}{n_s} = \frac{x_w}{x_s}$$

Specific humidity or *humidity ratio, W*, is defined as the mass of water vapour (or moisture) per unit mass of dry air in a mixture of air and water vapour.

If G = mass of dry air

m = mass of water vapour

$$W = \frac{m}{G}$$

Specific humidity is the maximum when air is saturated at temperature T, or

$$W_{\max} = W_s = \frac{m_s}{G}$$

If dry air and water vapour behave as ideal gases

$$p_w V = m R_w T$$

$$p_a V = G R_a T$$

$$\therefore \quad W = \frac{m}{G} = \frac{p_w}{p_a} \cdot \frac{R_a}{R_w} = \frac{p_w}{p - p_w} \cdot \frac{8.3143/28.96}{8.3143/18}$$

$$\therefore \quad W = 0.622 \frac{p_w}{p - p_w} \tag{15.1}$$

where p is the atmospheric pressure.

If p_w is constant, W remains constant.

If air is saturated at temperature T

$$W = W_s = 0.622 \frac{p_s}{p - p_s}$$

where p_s is the saturation pressure of water at temperature T.

The degree of saturation, μ, is the ratio of the actual specific humidity and the saturated specific humidity, both at the same temperature T.

$$\therefore \quad \mu = \frac{W}{W_s} = \frac{0.622 \dfrac{p_w}{p - p_w}}{0.622 \dfrac{p_s}{p - p_s}}$$

$$= \frac{p_w}{p_s} \cdot \frac{p - p_s}{p - p_w}$$

If $\phi = \dfrac{p_w}{p_s} = 0$, $p_w = 0$, $x_w = 0$, $W = 0$, i.e. for dry air, $\mu = 0$

If $\phi = 100\%$, $p_w = p_s$, $W = W_s$, $\mu = 1$

Therefore, μ varies between 0 and 1.

If a mixture of air and superheated (or unsaturated) water vapour is cooled at constant pressure, the partial pressure of each constituent remains constant until the water vapour reaches the saturated state. Further cooling causes condensation. The temperature at which water vapour starts condensing is called the *dew point temperature, t_{dp},* of the mixture (Fig. 15.1). It is equal to the saturation temperature at the partial pressure, p_w, of the water vapour in the mixture.

Dry bulb temperature (dbt) is the temperature recorded by the thermometer with a dry bulb.

Wet bulb temperature (wbt) is the temperature recorded by a thermometer when the bulb is enveloped by a cotton wick saturated with water. As the air stream flows past it, some water evaporates, taking the latent heat from the water-soaked wick, thus decreasing its temperature. Energy is then transferred to the wick from the air. When equilibrium condition is reached, there is a balance between energy removed from the water film by evaporation and energy supplied to the wick by heat transfer, and the temperature recorded is the wet bulb temperature.

A *psychrometer* is an instrument which measures both the dry bulb and the wet bulb temperatures of air. Figure 15.2 shows a continuous psychrometer with a fan for drawing air over the thermometer bulbs. A sling

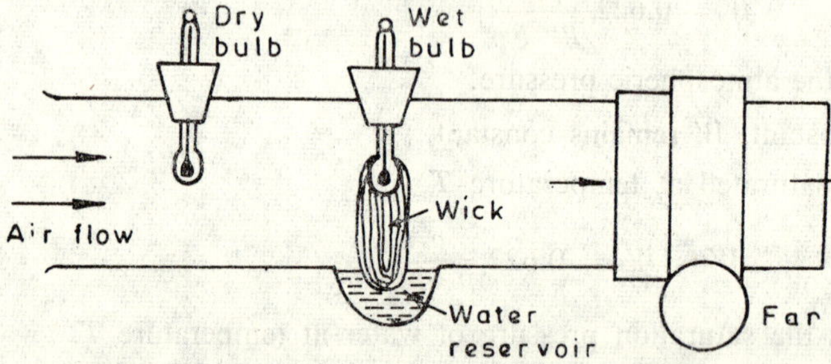

Fig. 15.2 Dry and wet bulb temperatures.

psychrometer has the two thermometers mounted on a frame with a handle. The handle is rotated so that there is good air motion. The wet bulb temperature is the lowest temperature recorded by the moistened bulb.

At any dbt, the greater the depression (difference) of the wbt reading below the dbt, the smaller is the amount of water vapour held in the mixture.

When unsaturated air flows over a long sheet of water (Fig. 15.3) in an insulated chamber, the water evaporates, and the specific humidity of the

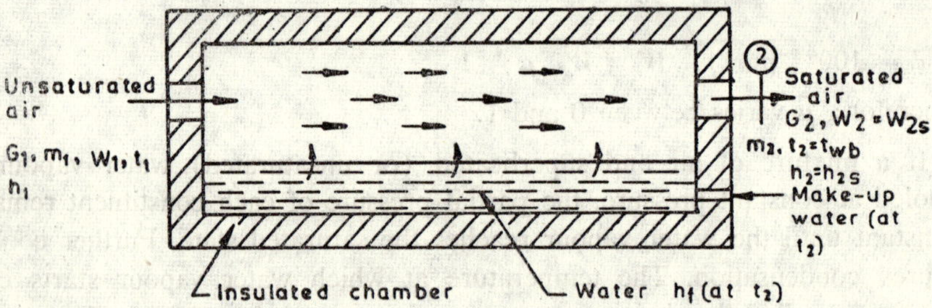

Fig. 15.3 Adiabatic saturation process

air increases. Both the air and water are cooled as evaporation takes places. The process continues until the energy transferred from the air to the water is equal to the energy required to vaporize the water. When this point is reached, thermal equilibrium exists with respect to the water, air and water vapour, and consequently the air is saturated. The equilibrium temperature is called the *adiabatic saturation temperature* or the *thermodynamic wet bulb temperature*. The make-up water is introduced at this temperature to make the water level constant.

The 'adiabatic' cooling process is shown in Fig. 15.4 for the vapour in the air-vapour mixture. Although the total pressure of the mixture is constant,

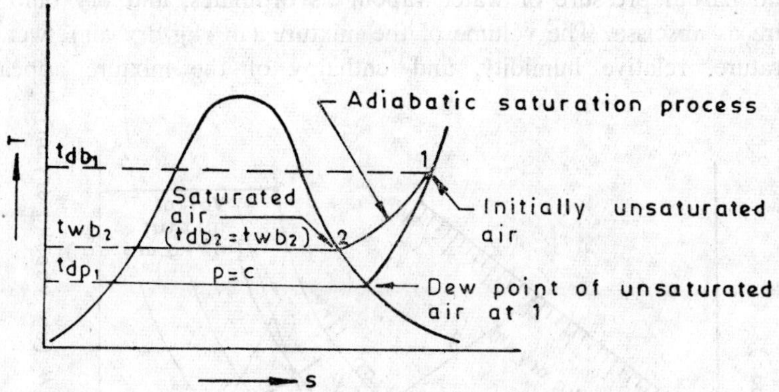

Fig. 15.4 Adiabatic cooling process on *T-s* plot

the partial pressure of the vapour increases, and in the saturated state corresponds to the adiabatic saturation temperature. The vapour is initially at the dbt t_{db_1} and is cooled adiabatically to the dbt t_{db_2} which is equal to the adiabatic saturation temperature t_{wb_2}. The adiabatic saturation temperature and the wet bulb temperature are taken to be equal for all practical purposes. The wbt lies between the dbt and dpt.

Since the system is insulated and no work is done, the first law yields

$$G\,h_{a_1} + m_1 h_{w_1} + (m_2 - m_1)\,h_{f_2} = G\,h_{a_2} + m_2\,h_{w_2}$$

where $(m_2 - m_1)$ is the mass of water added, h_{f_2} is the enthalpy of the liquid water at t_2 $(= t_{wb_2})$, h_a is the specific enthalpy of dry air, and h_w is the specific enthalpy of water vapour in air. Dividing by G, and since $h_{w_2} = h_{g_2}$

$$h_{a_1} + W_1 h_{w_1} + (W_2 - W_1)\,h_{f_2} = h_{a_2} + W_2\,h_{g_2} \qquad (15.2)$$

Solving for W_1

$$W_1 = \frac{(h_{a_2} - h_{a_1}) + W_2\,(h_{g_2} - h_{f_2})}{h_{w_1} - h_{f_2}}$$

$$= \frac{c_{p_a}\,(T_2 - T_1) + W_2 \cdot h_{fg_2}}{h_{w_1} - h_{f_2}} \qquad (15.3)$$

where $W_2 = \dfrac{m_2}{G} = \dfrac{m_s}{G} = 0.622\,\dfrac{p_s}{p - p_s}$

The *enthalpy* of the air-vapour mixture is given by

$$G\,h = G\,h_a + m\,h_w$$

where h is the enthalpy of the mixture per kg of dry air (it is not the specific enthalpy of the mixture)

$$\therefore \qquad h = h_a + W\,h_w \qquad (15.4)$$

15.2 PSYCHROMETRIC CHART

The psychrometric chart (Fig. 15.5) is a graphical plot with specific humidity and partial pressure of water vapour as ordinates, and dry bulb temperature as abscissa. The volume of the mixture (m^3/kg dry air), wet bulb temperature, relative humidity, and enthalpy of the mixture appear as

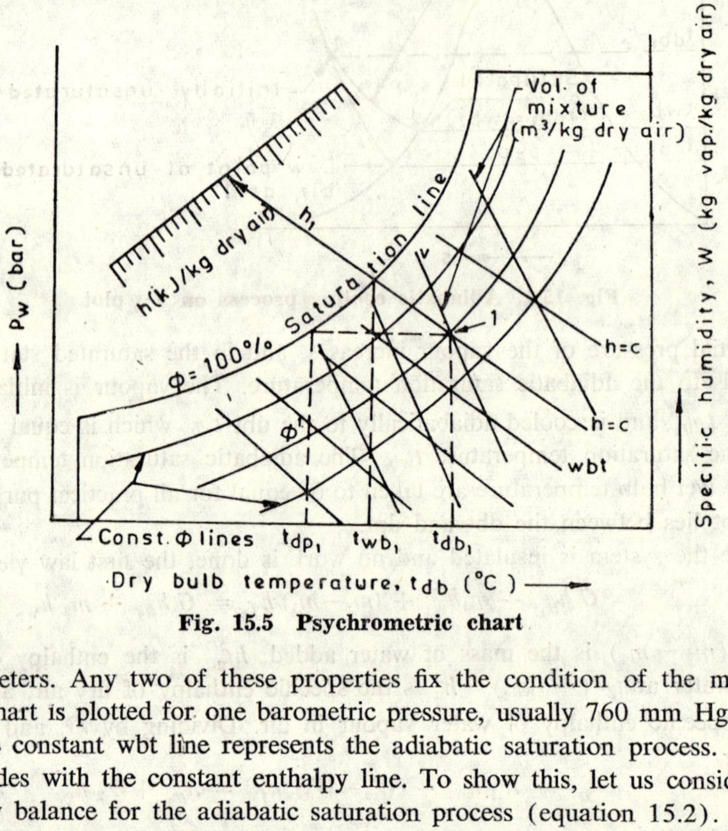

Fig. 15.5 Psychrometric chart

parameters. Any two of these properties fix the condition of the mixture. The chart is plotted for one barometric pressure, usually 760 mm Hg.

The constant wbt line represents the adiabatic saturation process. It also coincides with the constant enthalpy line. To show this, let us consider the energy balance for the adiabatic saturation process (equation 15.2).

$$h_{a_1} + W_1 h_{w_1} + (W_2 - W_1) h_{f_2} = h_{a_2} + W_2 h_{w_2}$$

Since $h_a + W h_w = h$ kJ/kg dry air (equation 15.4)

$$h_1 - W_1 h_{f_2} = h_2 - W_2 h_{f_2}$$

where subscript 2 refers to the saturation state, and subscript 1 denotes any state along the adiabatic saturation path. Therefore

$$h - W h_{f_2} = \text{constant.}$$

Since $W h_{f_2}$ is small compared to h (of the order of 1 or 2%)

$$h = \text{constant}$$

indicating that the enthalpy of the mixture remains constant during an adiabatic saturation process.

15.3 PSYCHROMETRIC PROCESSES

(a) Sensible Heating or Cooling (at W=Constant)

Only the dry bulb temperature of air changes. Let us consider sensible heating of air [Fig. 15.6(a), (b), (c)] Balance of

Dry air $\quad G_1 = G_2$

Moisture $\quad m_1 = m_2$

$$G_1 W_1 = G_2 W_2$$

Energy $\quad G_1 h_1 + Q_{1-2} = G_2 h_2$

$$Q_{1-2} = G\,(h_2 - h_1)$$

or

$$Q_{1-2} = G\,[h_{a_2} + W_2\, h_{w_2} - (h_{a_1} + W_1\, h_{w_1}\,)]$$

$$= G\,[1.005\,(t_2 - t_1) + W_2\,[h_{g_2} + 1.88\,(t_2 - t_{dp_2})] - W_1\,[h_{g_1} + 1.88\,(t_1 - t_{dp}\,)]\ \text{kJ}$$

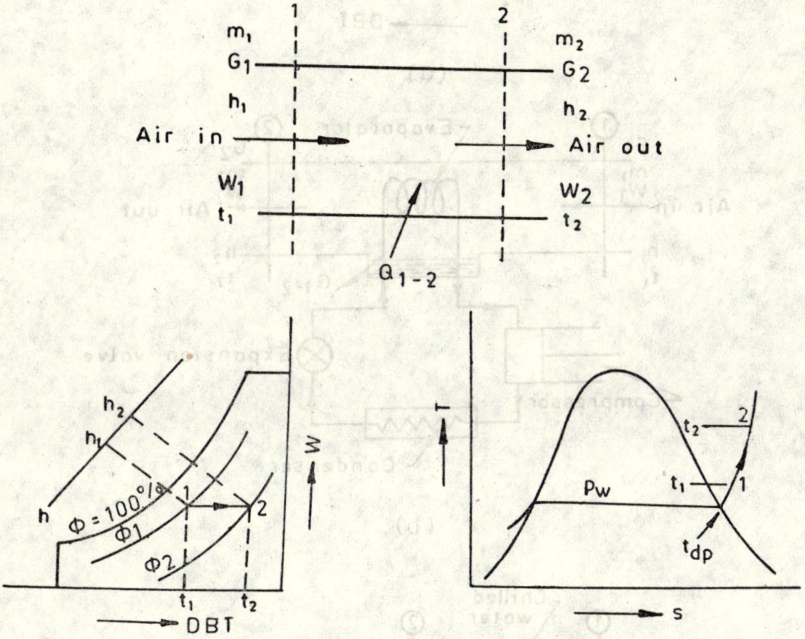

Fig. 15.6 Sensible heating

where the c_p of water vapour in a superheated state has been assumed to be 1.88 kJ/kg °C.

Since $\quad W_1 = W_2,\ h_{g_2} = h_{g_1}$

$$t_{dp_2} = t_{dp_1}$$

$$Q_{1-2} = G\,[1.005\,(t_2 - t_1) + 1.88\,W\,(t_2 - t_1)]$$

$$= G\ (1.005 + 1.88W)\ \ (t_2 - t_1)$$

Similar equations can be obtained for sensible cooling at constant humidity ratio.

(b) Cooling and Dehumidification

When the humidity ratio of air decreases, air is said to be dehumidified, and when it increases, air is humidified. Air may be cooled and dehumidified (a) by placing the evaporator coil across the air flow (Fig. 15.7b), (b) by

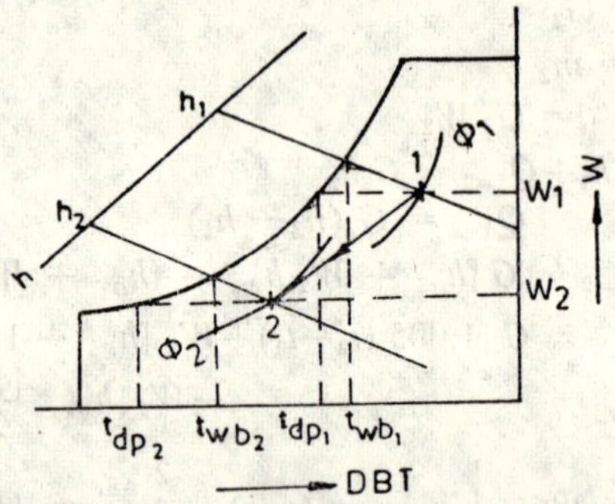

(a)

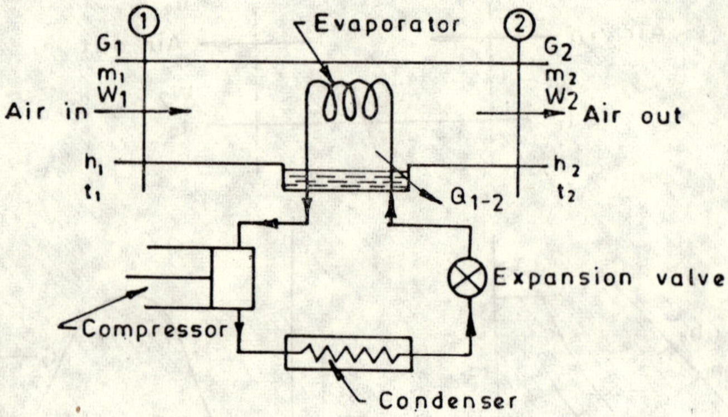

(b)

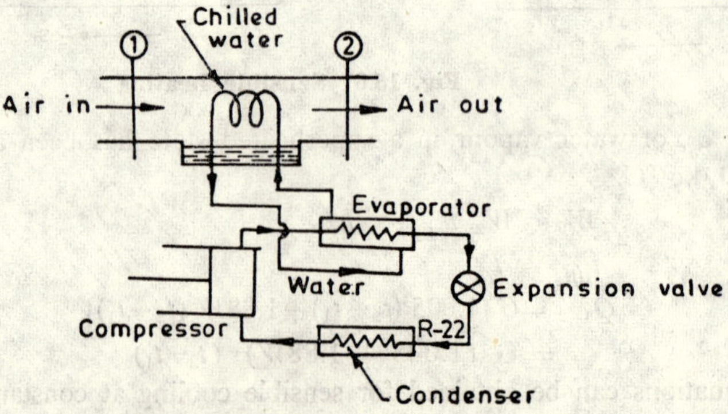

(c)

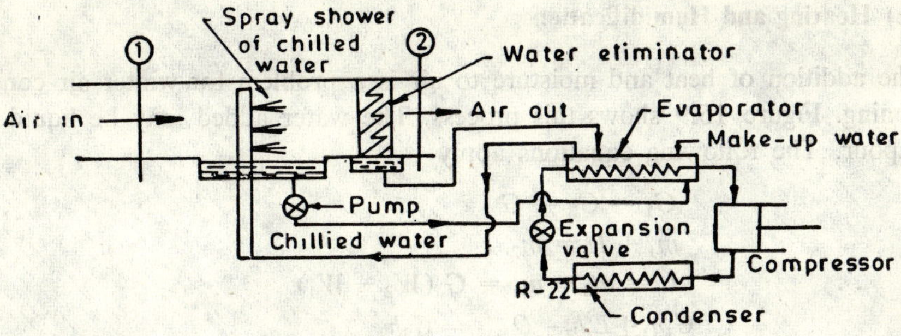

Fig. 15.7 Cooling and dehumidification

circulating chilled water (or brine) in a tube placed across the air flow (Fig. 15.7c), or (c) by spraying chilled water to air in the form of fine mist (Fig. 15.7d) to expose a large surface area. The temperature of the cooling surface or the spray water must be below the dew point at state 1 (Fig. 15.7a). If the cooling surface or the spray shower is of large magnitude, the air may come out at the saturation state, 2s, known as the *apparatus dew point* (adp) (Fig. 15.8).

$$G_1 = G_2 = G$$

If L is the amount of water vapour removed

$$m_1 = m_2 + L$$

or

$$L = G\,(W_1 - W_2)$$

The energy equation gives

$$G_1 h_1 = G_2 h_2 + Q_{1-2} + L \cdot hf_2$$

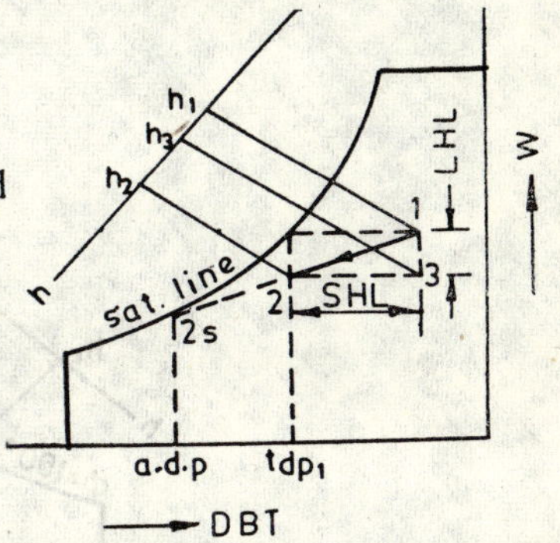

Fig. 15.8 Sensible and latent heat loads

where hf_2 is the specific enthalpy of water at temperature t_2.

$$\therefore \qquad Q_{1-2} = G\,[(h_1 - h_2) - (W_1 - W_2)\,h_{f_2}\,]$$

If $(W_1 - W_2)\;hf_2$ is small, the amount of heat removed is

$Q_{1-2} = G\,(h_1 - h_2)$ = Total heat load (THL) on the cooling coil (kJ/hr).

Now, as shown in Fig. 15.8, $h_1 - h_2 = (h_1 - h_3) + (h_3 - h_2)$

$$= \text{LHL} + \text{SHL} = \text{THL}$$

where $h_1 - h_3$ = enthalpy change at constant dbt, known as *latent heat load (LHL)*, and $(h_3 - h_2)$ is the enthalpy change at constant W, known as *sensible heat load (SHL)*.

Cooling and dehumidification of air is common in summer air conditioning.

(c) Heating and Humidification

The addition of heat and moisture to air is a problem for winter air conditioning. Figure 15.9 shows this process. The water added may be liquid or vapour. The following equations apply

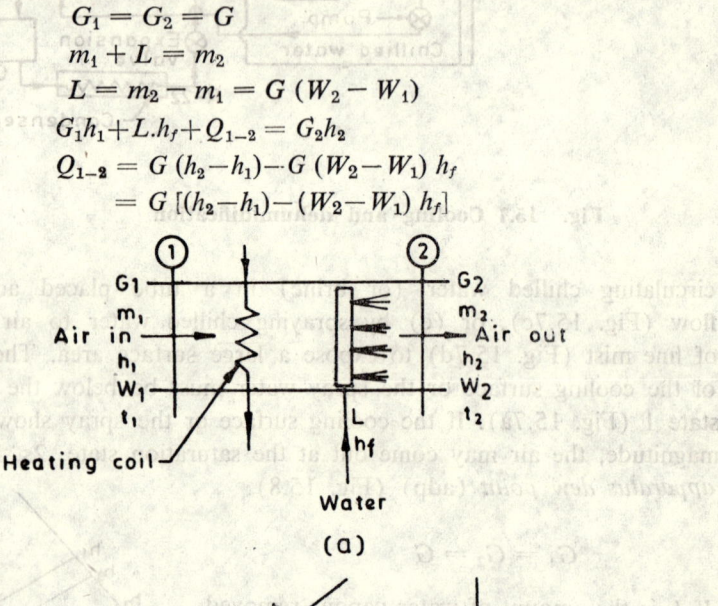

$$G_1 = G_2 = G$$
$$m_1 + L = m_2$$
$$\therefore \quad L = m_2 - m_1 = G\,(W_2 - W_1)$$
$$G_1 h_1 + L.h_f + Q_{1-2} = G_2 h_2$$
$$Q_{1-2} = G\,(h_2 - h_1) - G\,(W_2 - W_1)\,h_f$$
$$= G\,[(h_2 - h_1) - (W_2 - W_1)\,h_f]$$

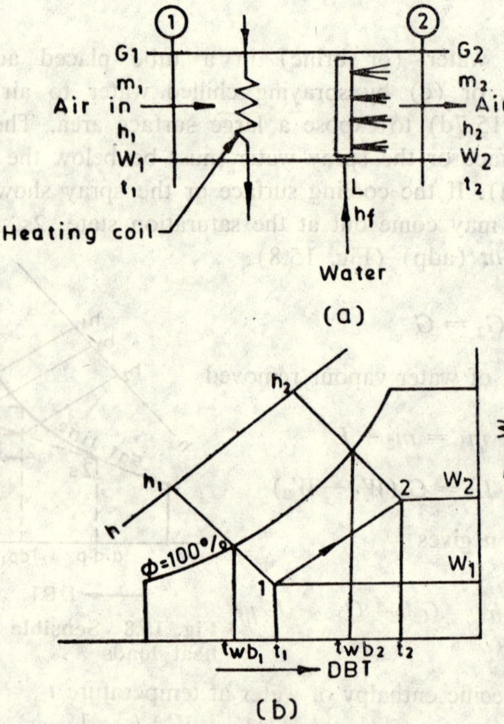

Fig. 15.9 Heating and humidification

(d) Adiabatic Mixing of Two Streams

This is a common problem in air conditioning, where ventilation air and some room air are mixed prior to processing it to the desired state (say, by cooling and dehumidification), and supplying it to the conditioned space. The process is shown in Fig. 15.10. The following equations hold good

$$G_1 + G_2 = G_3$$
$$G_1 W_1 + G_2 W_2 = G_3 W_3$$
$$G_1\,h_1 + G_2\,h_2 = G_3\,h_3$$

Combining these equations and rearranging

$$\frac{G_1}{G_2} = \frac{h_3 - h_2}{h_1 - h_3} = \frac{W_3 - W_2}{W_1 - W_3}$$

The points 1, 2, and 3 fall in a straight line, and the division of the line is inversely proportional to the ratio of the mass flow rates.

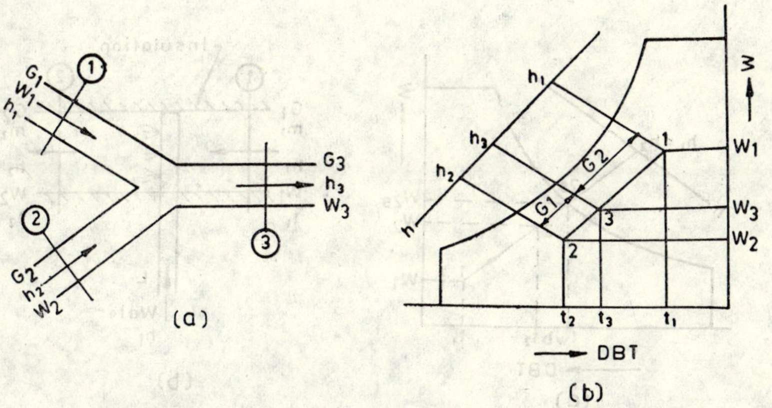

Fig. 15.10 Adiabatic mixing of two air streams

(e) Chemical Dehumidification

Some substances like silica gel (product of fused sodium silicate and sulphuric acid), and activated alumina have great affinity with water vapour. They are called *adsorbents*. When air passes through a bed of silica gel, water vapour molecules get adsorbed on its surface. Latent heat of condensation is released. So the dbt of air increases. The process is shown in Fig. 15.11.

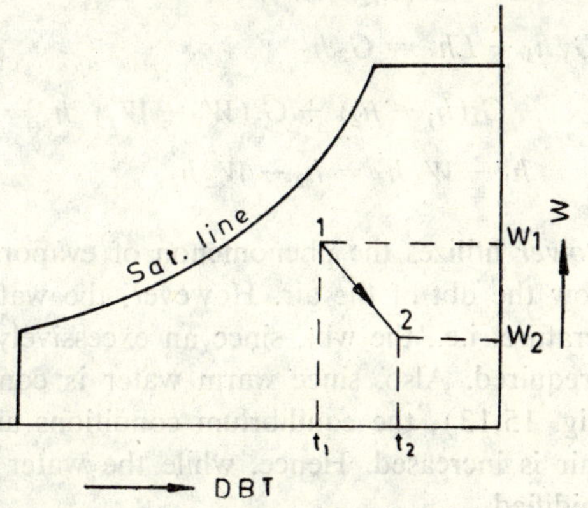

Fig 15.11 Chemical dehumidification

(f) Adiabatic Evaporative Cooling

A large quantity of water is constantly circulated through a spray chamber. The air-vapour mixture is passed through the spray and, in doing so, evaporates some of the circulating water. The air may leave at a certain humidity ratio or in a saturated state (Fig. 15.12). The increase in specific humidity

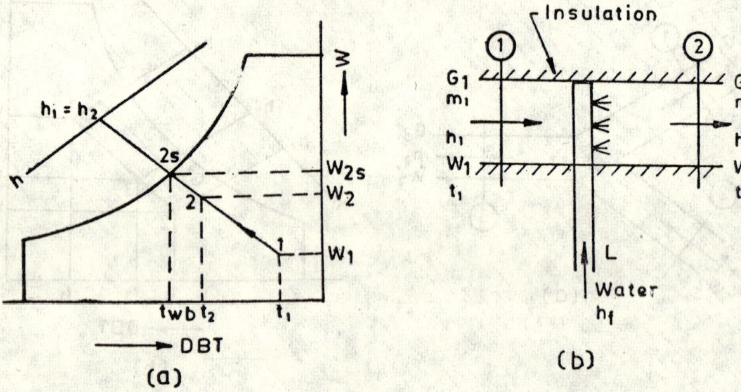

Fig. 15.12 Adiabatic evaporative cooling

is equal to the quantity of water evaporated per unit mass of dry air. No heat transfer takes place between the chamber and the surroundings. Therefore, the energy required for evaporation is supplied by the air, and consequently, the dbt is lowered. After the process has been in operation for a sufficient length of time, the circulating water approaches the wbt of air.

$$G_1 = G_2 = G$$

$$G_1 W_1 + L = G_2 W_2$$

$$\therefore \quad L = G (W_2 - W_1)$$

$$G_1 h_1 + L h_f = G_2 h_2$$

$$\therefore \quad G (h_1 - h_2) + G (W_2 - W_1) \; h_f = 0$$

$$\therefore \quad h_1 - W_1 h_f = h_2 - W_2 h_f$$

The *cooling tower* utilizes the phenomenon of evaporative cooling to cool warm water below the dbt of the air. However, the water never reaches the minimum temperature, i.e., the wbt, since an excessively large cooling tower would then be required. Also, since warm water is continuously introduced to the tower (Fig. 15.13), the equilibrium conditions are not achieved, and the dbt of the air is increased. Hence, while the water is cooled, the air is heated and humidified.

The warm water is introduced at the top of the tower in the form of spray to expose a large surface area for evaporation to take place. The

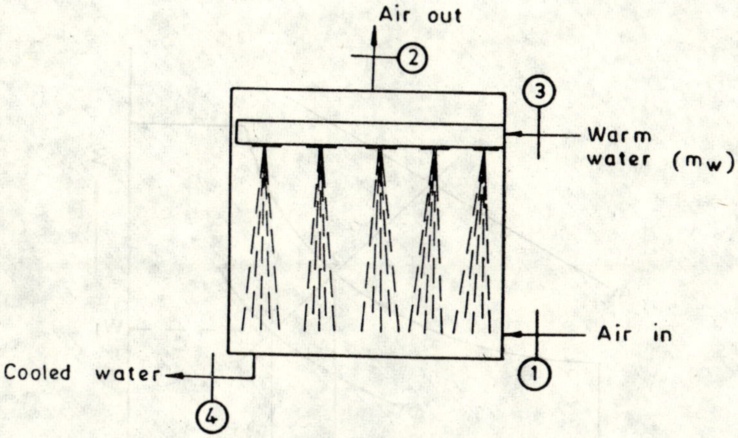

Fig. 15.13 Cooling tower

more the water evaporates, the more is the effect of coolng. Air leaves from the top very nearly saturated. The following equations apply

$$G_1 = G_2 = G$$

$$G_1 W_1 + m_{w3} = G_2 W_2 + m_{w4}$$

$$\therefore \quad m_{w3} - m_{w4} = G\,(W_2 - W_1)$$

$$G_1 h_1 + m_{w3}\, h_{w3} = G_2 h_2 + m_{w4}\, h_{w4}$$

$$\therefore \quad G(h_1 - h_2) + m_{w3}\, h_{w3} = m_{w4}\, h_{w4}$$

The difference in temperature of the cooled-water temperature and the wet bulb temperature of the entering air is known as the *approach*. The *range* is the temperature difference between the inlet and exit states of water. Cooling towers are rated in terms of approach and range.

EXAMPLE 15.1

Atmospheric air at 1.0132 bar has a dbt of 32°C and a wbt of 26°C. Compute (a) the partial pressure of water vapour, (b) the specific humidity, (c) the dew point temperature, (d) the relative humidity, (e) the degree of saturation, (f) the density of the dry air in the mixture, (g) the density of the vapour in the mixture, and (h) the enthalpy of the mixture.

Solution

The state of air is shown on the DBT-W plot in Fig. Ex. 15.1. The path 1-2 represents the constant wbt and enthalpy of the air, which also holds good approximately for an adiabatic saturation process. From equation

(15.1), the specific humidity at state 2 is given by

$$W_2 = 0.622 \frac{p_s}{p-p_s}$$

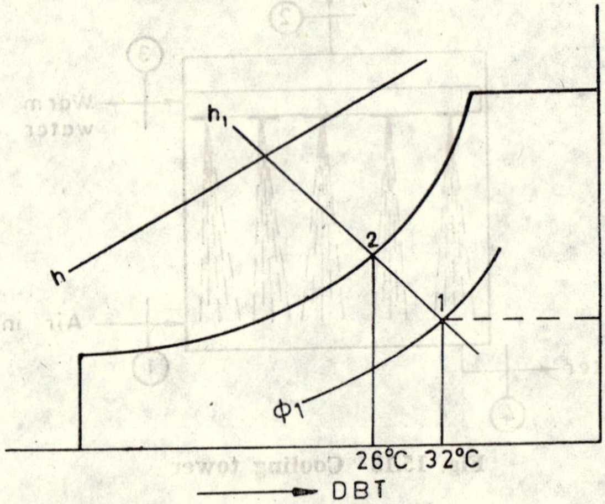

Fig. Ex. 15·1

The saturation pressure p_s at the wbt of 26°C is 0.03363 bar.

$$\therefore \quad W_2 = 0.622 \frac{0.03363}{1.0132-0.03363} = 0.021148 \text{ kg vap./kg dry air}$$

From equation (15.3), for adiabatic saturation

$$W_1 = \frac{c_{p_a}(T_2-T_1)+W_2.h_{fg_2}}{h_{w_1}-h_{f_2}}$$

From the steam tables, at 26°C

$$h_{fg2} = 2439.9 \text{ kJ/kg}, \; h_{f2} = 109.1 \text{ kJ/kg}$$

At $\quad$ 32°C, $h_{w1} = h_g = 2559.9$ kJ/kg

(a) $$W_1 = \frac{1.005\,(26-32)+0.021148\,(2439.9)}{2559.9-109.1}$$

$$= 0.0186 \text{ kg vap./kg dry air}$$ **Ans.**

(b) $$W_1 = 0.622 \frac{p_w}{p-p_w} = 0.0186$$

$$\frac{p-p_w}{p_w} = \frac{0.622}{0.0186} = 33.44$$

$\therefore \quad p_w = 0.03$ bar **Ans.**

(c) Saturation temperature at 0.03 bar, dpt = 24.1°C Ans.

(d) Relative humidity, $\phi = \frac{p_w}{p_{sat.}}$

At 32°C, $p_{sat.} = 0.048$ bar

$$\therefore \quad \phi = \frac{0.03}{0.048} = 0.625 \text{ or } 62.5\%$$ Ans.

(e) Deg. of saturation $\mu = \frac{W}{W_s} = \frac{p_w}{p_s} \cdot \frac{p-p_s}{p-p_w} = \frac{0.03\,(1.0132-0.048)}{0.048\,(1.0132-0.03)}$

$= 0.614$ Ans.

(f) Partial pressure of dry air

$$p_a = p - p_w = 1.0132 - 0.03 = 0.9832 \text{ bar}$$

∴ Density of dry air

$$\rho_a = \frac{p_a}{R_a T_{db}} = \frac{0.9832 \times 100}{0.287 \times (273+32)}$$

$= 1.12$ kg/m³ dry air Ans.

(g) Density of water vapour

$$\rho_w = 0.0186 \frac{\text{kg vap.}}{\text{kg dry air}} \times 1.12 \frac{\text{kg dry air}}{\text{m}^3 \text{ dry air}}$$

$= 0.021$ kg vap./m³ dry air Ans.

(h) Enthalpy of the mixture

$$h = h_a + W h_w$$
$$= c_p t_a + W\,[h_g + 1.88\,(t_{db} - t_{dp})]$$
$$= 1.005 \times 32 + 0.0186\,[2559.9 + 1.88\,(32 - 24.1)]$$

$= 80.55$ kJ/kg Ans.

EXAMPLE 15.2

An air-water vapour mixture enters an adiabatic saturator at 30°C and leaves at 20°C, which is the adiabatic saturation temperature. The pressure remains constant at 100 kPa. Determine the relative humidity and the humidity ratio of the inlet mixture.

Solution

The specific humidity at the exit

$$W_2 = 0.622 \frac{p_s}{p - p_s}$$
$$= 0.622 \left(\frac{2.339}{100 - 2.339}\right) = 0.0149 \frac{\text{kg vap.}}{\text{kg dry air}}$$

The specific humidity at the inlet (equation 15.3)

$$W_1 = \frac{c_{p_a}\,(T_2 - T_1) + W_2 h_{fg_2}}{h_{w_1} - h_{f_2}}$$
$$= \frac{1.005\,(20-30) + 0.0149 \times 2454.1}{2556.3 - 83.96}$$

$= 0.0107$ kg vap./kg dry air

$$W_1 = 0.622 \left(\frac{p_{w_1}}{100 - p_{w_1}} \right) = 0.0107$$

$$p_{w1} = 1.691 \text{ kPa}$$

$$\therefore \quad \phi_1 = \frac{p_{w_1}}{p_{s_1}} = \frac{1.691}{4.246} = 0.398 \text{ or } 39.8\% \qquad \textbf{Ans.}$$

EXAMPLE 15.3

Saturated air at 2°C is required to be supplied to a room where the temperature must be held at 20°C with a relative humidity of 50%. The air is heated and then water at 10°C is sprayed in to give the required humidity. Determine the temperature to which the air must be heated and the mass of spray water required per m³ of air at room conditions. Assume that the total pressure is constant at 1.013 bar and neglect the fan power.

Solution

The process is shown in Fig. Ex. 15.3. From the steam tables, at 20°C, $p_{sat.} = 2.339$ kPa

$$\phi_3 = \frac{p_{w_3}}{(p_{sat.})_{t_3}} = \frac{p_{w_3}}{2.339} = 0.50$$

$$\therefore \quad p_{w_3} = 1.17 \text{ kPa}$$

$$\therefore \quad p_{a_3} = 101.3 - 1.17 = 100.13 \text{ kPa}$$

$$W_3 = 0.622 \, \frac{p_{w_3}}{p_{a_3}} = 0.622 \times \frac{1.17}{100.13} = 0.00727$$

$$\phi_1 = \frac{p_{w_1}}{(p_{sat.})_{2°C}} = 1.00$$

At 2°C, $p_{sat.} = 0.7156$ kPa

$$\therefore \quad p_{w1} = 0.7156 \text{ kPa}$$

$$p_{a1} = 101.3 - 0.7156 = 100.5844 \text{ kPa}$$

$$W_1 = 0.622 \, \frac{0.7156}{100.5844} = 0.00442$$

$$W_3 - W_1 = 0.00727 - 0.00442 = 0.00285 \text{ kg vap./kg dry air}$$

$$v_{a_3} = \frac{R_a \, T_3}{p_{a_3}} = \frac{0.287 \times 293}{100.13} = 0.84 \text{ m}^3\text{/kg dry air}$$

$$\therefore \text{ Spray water} = 0.00285 \, \frac{\text{kg vap.}}{\text{kg dry air}} \times \frac{\text{kg dry air}}{0.84 \text{ m}^3}$$

$$= 0.00339 \text{ kg moisture/m}^3 \qquad \textbf{Ans.}$$

$$G_2 h_2 + m_{w_4} h_4 = G_3 h_3$$

$$\therefore \quad h_2 + (W_3 - W_2) \, h_4 = h_3$$

$$h_{a_2} + W_2 \, h_{w_2} + (W_3 - W_2) \, h_4 = h_{a_3} + W_3 \, h_{w_3}$$

$$\therefore \quad c_p \, (t_3 - t_2) + W_3 \, h_{w_3} - W_2 \, h_{w_2} - (W_3 - W_2) \, h_4 = 0$$

From the steam tables, at $p_w = 1.17$ kPa

$$h_g = 2518 \text{ kJ/kg and } t_{sat.} = 9.65°\text{C}$$

$$1.005\,(20 - t_2) + 0.00727\,[2518 + 1.884\,(20 - 9.65)] - 0.00442\,[2518 + 1.884\,(t_2 - 9.65)] - 0.00285 \times 10 = 0$$

$\therefore\ t_2 = 27.2°\text{C}$ Ans.

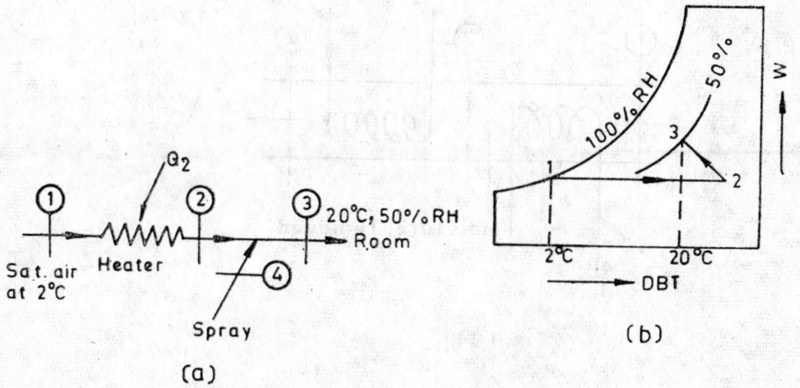

Fig. Ex. 15·3

EXAMPLE 15.4

An air conditioning system is designed under the following conditions:
Outdoor conditions—30°C dbt, 75% R.H.
Required indoor conditions—22°C dbt, 70% R.H.
Amount of free air circulated—3.33 m³/s
Coil dew point temperature—14°C

The required condition is achieved first by cooling and dehumidification, and then by heating. Estimate (a) the capacity of the cooling coil in tonnes, (b) the capacity of the heating coil in kW, and (c) the amount of water vapour removed in kg/s.

Solution

The processes are shown in Fig. Ex. 15.4. The property values, taken from the psychrometric chart, are

$$h_1 = 82,\ h_2 = 52,\ h_3 = 47,\ h_4 = 40 \text{ kJ/kg dry air}$$

$$W_1 = 0.020,\ W_2 = W_3 = 0.0115 \text{ kg vap./kg dry air}$$

$$v_1 = 0.887 \text{ m}^3\text{/kg dry air}$$

$$G = \frac{3.33}{0.887} = 3.754 \text{ kg dry air/sec}$$

$\therefore$ Cooling coil capacity $= G\,(h_1 - h_3) = 3.754\,(82 - 47)$ kJ/s

$$= \frac{3.754 \times 35 \times 3600}{14{,}000} = 33.79 \text{ tonnes}$$ Ans. (a)

Capacity of the heating coil = $G\ (h_2 - h_3)$ = 3.754 (52 − 47) kJ/s
= 3.754 × 5 = 18.77 kW Ans. (b)

Rate of water vapour removed = $G\ (W_1 - W_3)$
= 3.754 × (0.0200 − 0.0115)
= 0.319 kg/s Ans. (c)

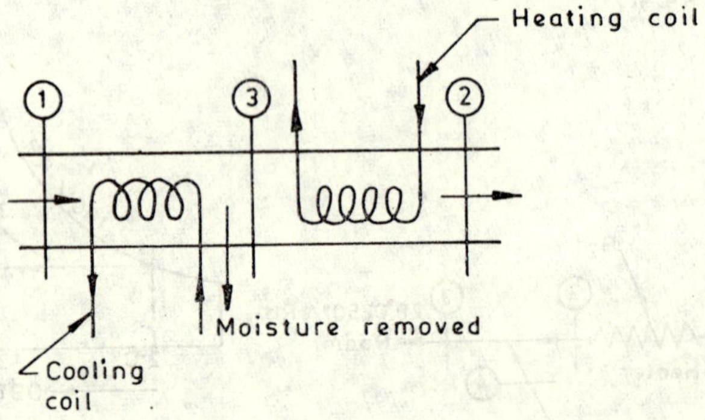

(a)

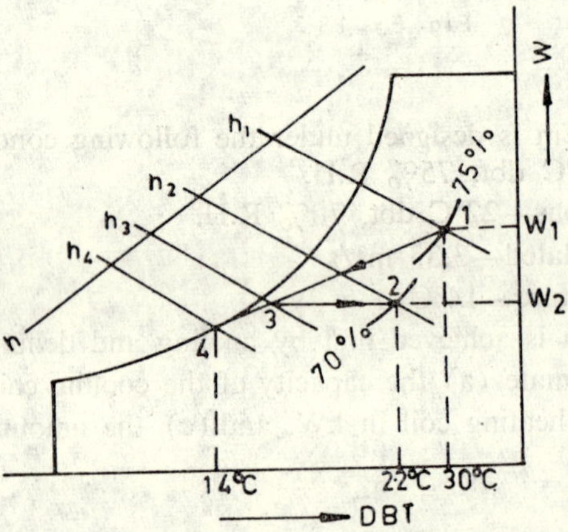

(b)

Fig. Ex. 15·4

Example 15.5

Air at 20°C, 40% RH is mixed adiabatically with air at 40°C, 40% RH in the ratio of 1 kg of the former with 2 kg of the later (on dry basis). Find the final condition of air.

Solution

Figure Ex. 15.5 shows the mixing process of two air streams. The equations

$$G_1 + G_2 = G_3$$

$$G_1W_1 + G_2W_2 = G_3W_3$$

$$G_1h_1 + G_2h_2 = G_3h_3$$

result in

$$\frac{W_2 - W_3}{W_3 - W_1} = \frac{h_2 - h_3}{h_3 - h_1} = \frac{G_1}{G_2}$$

From the psychrometric chart

$W_1 = 0.0058$, $W_2 = 0.0187$ kg vap./kg dry air

$h_1 = 35$, $h_2 = 90$ kJ/kg dry air

$$\therefore \quad \frac{0.0187 - W_3}{W_3 - 0.0058} = \frac{G_1}{G_2} = \frac{1}{2}$$

$$\therefore \quad W_3 = \tfrac{2}{3} \times 0.0187 + \tfrac{1}{3} \times 0.0058$$

$$= 0.0144 \text{ kg vap./kg dry air}$$

Again

$$\frac{h_2 - h_3}{h_3 - h_1} = \frac{G_1}{G_2} = \frac{1}{2}$$

$$\therefore \quad h_3 = \tfrac{2}{3}\, h_2 + \tfrac{1}{3}\, h_1 = \tfrac{2}{3} \times 90 + \tfrac{1}{3} \times 35$$

$$= 71.67 \text{ kJ/kg dry air}$$

$\therefore$ Final condition of air is given by

$W_3 = 0.0144$ kg vap./kg dry air

$h_3 = 71.67$ kJ/kg dry air **Ans.**

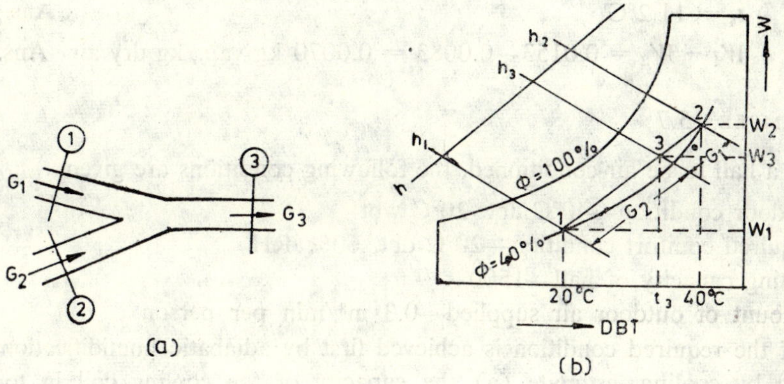

Fig. Ex. 15·5

EXAMPLE 15.6

Saturated air at 21°C is passed through a drier so that its final relative humidity is 20%. The drier uses silica gel adsorbent. The air is then passed through a cooler until its final temperature is 21°C without a change in specific humidity. Find out (a) the temperature of air at the end of the drying process, (b) the heat rejected in kJ/kg dry air during the cooling process, (c) the relative humidity at the end of the cooling process, (d) the dew point temperature at the end of the drying process, and (e) the moisture removed during the drying process in kg vap./kg dry air.

Solution

From the psychrometric chart (Fig. Ex. 15.6)

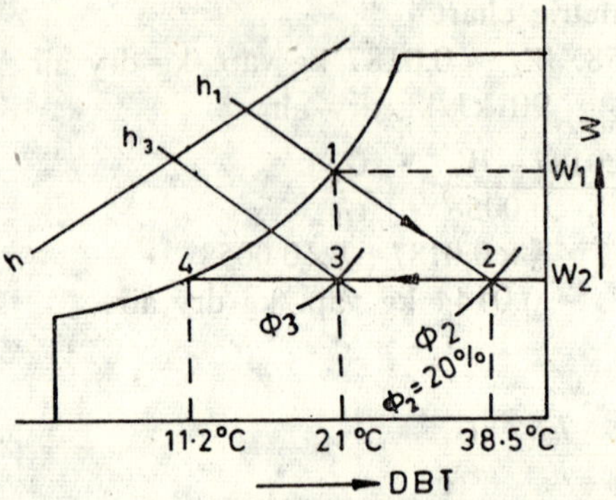

Fig. Ex. 15·6

$T_2 = 38.5°C$ Ans. (a)

$h_1 - h_3 = 60.5 - 42.0 = 18.5$ kJ/kg dry air Ans. (b)

$\phi = 53\%$ Ans. (c)

$t_4 = 11.2°C$ Ans. (d)

$W_1 - W_2 = 0.0153 - 0.0083 = 0.0070$ kg vap./kg dry air Ans. (e)

EXAMPLE 15.7

For a hall to be air-conditioned, the following conditions are given

Outdoor condition—40°C dbt, 20°C wbt

Required comfort condition—20°C dbt, 60% RH

Seating capacity of hall—1500

Amount of outdoor air supplied—0.3 m³/min per person

If the required condition is achieved first by adiabatic humidification and then by cooling, estimate (a) the capacity of the cooling coil in tonnes, and (b) the capacity of the humidifier in kg/hr.

Solution

From the psychrometric chart (Fig. Ex. 15.7)

$h_1 = h_2 = 57.0$, $h_3 = 42.0$ kJ/kg dry air

$W_1 = 0.0065$, $W_2 = W_3 = 0.0088$ kg vap./kg dry air

$t_2 = 34.5°C$, $v_1 = 0.896$ m³/kg dry air

Amount of dry air supplied

$$G = \frac{1500 \times 0.3}{0.896} = 502 \text{ kg/min}$$

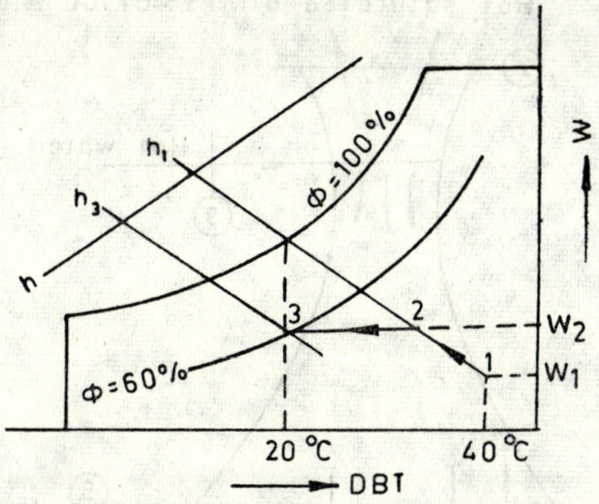

Fig. Ex. 15·7

∴ Capacity of the cooling coil

$$= G\,(h_2 - h_3) = 502\ \ (57 - 42)\ \ \text{kJ/min}$$

$$= \frac{502 \times 15 \times 60}{14{,}000} = 32.27 \text{ tonnes} \qquad \textbf{Ans. (a)}$$

Capacity of the humidifier

$$= G\,(W_2 - W_1) = 502\ \ (0.0088 - 0.0065)\ \ \text{kg/min}$$

$$= 502 \times 23 \times 10^{-4} \times 60 = 69.3\ \ \text{kg/hr} \qquad \text{Ans. (b)}$$

EXAMPLE 15.8

Water from a cooling system is itself to be cooled in a cooling tower at a rate of 2.78 kg/s. The water enters the tower at 65°C and leaves a collecting tank at the base at 30°C. Air flows through the tower, entering the base at 15°C, 0.1 MPa, 55% RH, and leaving the top at 35°C, 0.1 MPa, saturated. Make-up water enters the collecting tank at 14°C. Determine the air flow rate into the tower in m^3/s and the make-up water flow rate in kg/s.

Solution

Figure Ex. 15.8 shows the flow diagram of the cooling tower. From the steam tables

at 15°C, $p_{sat.} = 0.01705$ bar, $h_g = 2528.9$ kJ/kg
at 35°C, $p_{sat.} = 0.05628$ bar, $h_g = 2565.3$ kJ/kg

$$\phi = \frac{p_w}{(p_{sat.})_{15^\circ C}} = 0.55$$

$$\therefore \quad p_{w1} = 0.55 \times 0.01705 = 0.938 \times 10^{-2} \text{ bar}$$

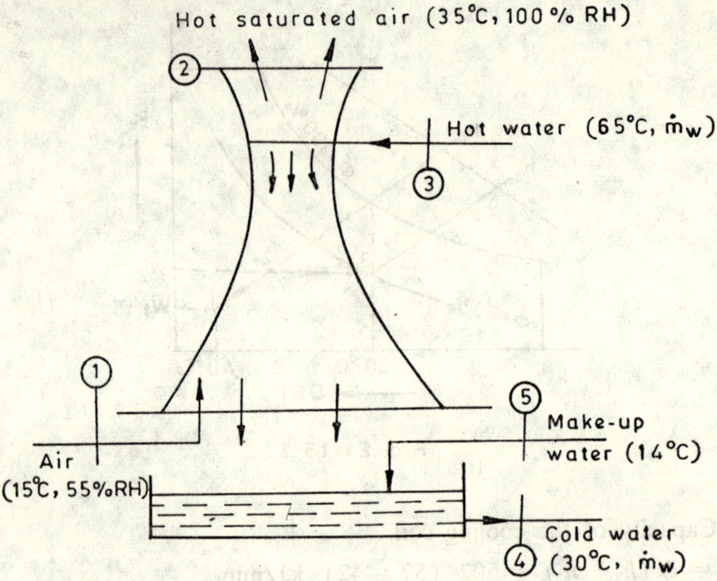

Fig. Ex. 15.8

$$\phi_2 = \frac{p_w}{(p_{sat.})_{35^\circ C}} = 1.00$$

$\therefore \quad p_{w_2} = 0.05628$ bar

$$W_1 = 0.622 \frac{p_w}{p - p_w} = 0.622 \times \frac{0.938 \times 10^{-2}}{1.00 - 0.00938}$$

$$= 0.00589 \text{ kg vap./kg dry air}$$

$$W_2 = 0.622 \times \frac{0.05628}{1.00 - 0.05628} = 0.0371 \text{ kg vap./dry air}$$

$\therefore$ Make-up water $= W_2 - W_1$

$$= 0.0371 - 0.00589$$

$$= 0.03121 \text{ kg vap./kg dry air}$$

Energy balance gives

$$H_2 + H_4 - H_1 - H_3 - H_5 = 0$$

For 1 kg of dry air

$$c_{p_a}(t_2 - t_1) + W_2 h_2 - W_1 h_1 + \dot{m}_w (h_4 - h_3) - (W_2 - W_1) h_5 = 0$$

$$\therefore \quad 1.005 \; (35 - 15) + 0.0371 \times 2565.3 - 0.00589 \times 2528.9$$

$$+ \dot{m}_w \; (-35) \; 4.187 - 0.03121 \times 4.187 \times 14 = 0$$

or $\quad 146.55 \; \dot{m}_w = 98.54$

$\therefore \quad \dot{m}_w = 0.672$ kg water/kg dry air

Since water flow rate is 2.78 kg/s

$$\therefore \quad \text{Rate of dry air flow} = \frac{2.78}{0.672} = 4.137 \text{ kg/s}$$

$\therefore$ Make-up water flow rate

$$= 0.03121 \times 4.137 = 0.129 \text{ kg/s} \qquad \text{Ans.}$$

Rate of dry air flow = 4.137 kg/s

$$\therefore \quad \text{Rate of wet air flow} = 4.137\,(1+W_1)$$

$$= 4.137 \times 1.00589$$

$$= 4.16 \text{ kg/s}$$

$\therefore$ Volume flow rate of air

$$= \frac{\dot{m}_a R T}{p} = \frac{4.16 \times 0.287 \times 288}{100} = 3.438 \text{ m}^3\text{/s} \qquad \text{Ans.}$$

PROBLEMS

15.1 An air-water vapour mixture at 0.1 MPa, 30°C, 80% RH has a volume of 50 m³. Calculate the specific humidity, dew point, wbt, mass of dry air, and mass of water vapour.

If the mixture is cooled at constant pressure to 5°C, calculate the amount of water vapour condensed.

15.2 A sling psychrometer reads 40°C dbt and 36°C wbt. Find the humidity ratio, relative humidity, dew point temperature, specific volume, and enthalpy of air.

15.3 Calculate the amount of heat removed per kg of dry air if the initial condition of air is 35°C, 70% RH, and the final condition is 25°C, 60% RH.

15.4 Two streams of air at 35°C, 50% RH and 25°C, 60% RH are mixed adiabatically to obtain 0.3 kg/s of dry air at 30°C. Calculate the amounts of air drawn from both the streams and the humidity ratio of the mixed air.

15.5 Air at 40°C dbt and 27°C wbt is to be cooled and dehumidified by passing it over a refrigerant-filled coil to give a final condition of 15°C and 90% RH. Find the amounts of heat and moisture removed per kg of dry air.

15.6 An air-water vapour mixture enters a heater-humidifier unit at 5°C, 100 kPa, 50% RH. The flow rate of dry air is 0.1 kg/s. Liquid water at 10°C is sprayed into the mixture at the rate of 0.002 kg/s. The mixture leaves the unit at 30°C, 100 kPa. Calculate (a) the relative humidity at the outlet, and (b) the rate of heat transfer to the unit.

15.7 A laboratory has a volume of 470 m³, and is to be maintained at 20°C, 52.5% RH. The air in the room is to be completely changed once every hour and is drawn from the atmosphere at 1.05 bar, 32°C, 86% RH, by a fan absorbing 0.45 kW. This air passes through a cooler which reduces its temperature and causes condensation, the condensate being drained off at 8°C. The resulting saturated air is heated to room condition. The total pressure is constant throughout. Determine (a) the temperature of the air leaving the cooler, (b) the rate of condensation, (c) the heat transfer in the cooler, and (d) the heat transfer in the heater.

Ans. (a) 10°C, (b) 10.35 kg/hr, (c) 11.33 kW, (d) 1.63 kW.

15.8 Air at 30°C, 80% RH is cooled by spraying in water at 12°C. This causes saturation, followed by condensation, the mixing being assumed to take place adiabatically and the condensate being drained off at 16.7°C. The resulting saturated mixture is then heated to produce the required conditions of 60% RH at 25°C. The total pressure is constant at 101 kPa. Determine the mass of water supplied to the sprays to provide 10 m^3/hr of conditioned air. What is the heater power required?

Ans. 2224 kg/hr, 2.75 kW.

15.9 An air-conditioned room requires 30 m^3/min of air at 1.013 bar, 20°C, 52.5% RH. The steady flow conditioner takes in air at 1.013 bar, 77% RH, which it cools to adjust the moisture content and reheats to room temperature. Find the temperature to which the air is cooled and the thermal loading on both the cooler and heater. Assume that a fan before the cooler absorbs 0.5 kW, and that the condensate is discharged at the temperature to which the air is cooled.

Ans. 10°C, 25 kW, 6.04 kW.

15.10 An industrial process requires an atmosphere having a RH of 88.4% at 22°C, and involves a flow rate of 2000 m^3/hr. The external conditions are 44.4% RH, 15°C. The air intake is heated and then humidified by water spray at 20°C. Determine the mass flow rate of spray water and the power required for heating, if the pressure throughout is 1 bar.

Ans. 23.4 kg/hr, 20.5 kW.

15.11 Cooling water enters a cooling tower at a rate of 1000 kg/hr and 70°C. Water is pumped from the base of the tower at 24°C and some make-up water is added afterwards. Air enters the tower at 15°C, 50% RH, 1.013 bar, and is drawn from the tower saturated at 34°C, 1 bar. Calculate the flow rate of the dry air in kg/hr and the make-up water required per hour.

Ans. 2088 kg/hr, 62.9 kg/hr.

15.12 A grain dryer consists of a vertical cylindrical hopper through which hot air is blown. The air enters the base at 1.38 bar, 65°C, 50% RH. At the top, saturated air is discharged into the atmosphere at 1.035 bar, 60°C.

Estimate the moisture picked up by 1 kg of dry air, and the total enthalpy change between the entering and leaving streams expressed per unit mass of dry air.

Ans. 0.0864 kJ/kg air, 220 kJ/kg air.

16

Reactive Systems

In this chapter we shall study the thermodynamics of mixtures that may be undergoing chemical reaction. With every chemical reaction is associated a chemical equation which is obtained by balancing the atoms of each of the atomic species involved in the reaction. The initial constituents which start the reaction are called the *reactants,* and the final constituents which are formed by chemical reaction with the rearrangement of the atoms and electrons are called the *products*. The reaction between the reactants, hydrogen and oxygen, to form the product water can be expressed as

$$H_2 + \tfrac{1}{2} O_2 \rightleftharpoons H_2O \qquad (16.1)$$

The equation indicates that one mole of hydrogen and half a mole of oxygen combine to form one mole of water. The reaction can also proceed in the reverse direction. The coefficients 1, $\frac{1}{2}$, 1 in the chemical equation (16.1) are called *stoichiometric coefficients.*

16.1 DEGREE OF REACTION

Let us suppose that we have a mixture of four substances, A_1, A_2, A_3 and A_4, capable of undergoing a reaction of the type

$$\nu_1 A_1 + \nu_2 A_2 \rightleftharpoons \nu_3 A_3 + \nu_4 A_4$$

where the ν's are the stoichiometric coefficients.

Starting with rabitrray amounts of both initial and final constituents, let us imagine that the reaction proceeds completely to the right with the disappearance of at least one of the initial constituents, say, A_1. Then the original number of moles of the initial constituents is given in the form

$$n_1 \text{ (original)} = n_0 \nu_1$$
$$n_2 \text{ (original)} = n_0 \nu_2 + N_2$$

where n_0 is an arbitrary positive number, and N_2 is the residue (or excess) of A_2, i.e., the number of moles of A_2 which cannot combine. If the reaction is assumed to proceed completely to the left with the disappearance of the final constituent A_3, then

$$n_3 \text{ (original)} = n_0' \nu_3$$

$$n_4 \text{ (original)} = n_0' \nu_4 + N_4$$

where n_0' is an arbitrary positive number and N_4 is the excess number of moles of A_4 left after the reaction is complete from right to left.

For a reaction that has occurred completely to the left, there is a maximum amount possible of each initial constituent, and a minimum amount possible of each final constituent, so that

$$n_1(\max) = n_0 \nu_1 \quad + \quad n_0' \nu_1$$

(Original number of moles of A_1) (Number of moles of A_1 formed by chemical reaction)

$$(n_0' \nu_3 A_3 + n_0' \nu_4 A_4 \rightarrow n_0' \nu_1 A_1 + n_0' \nu_2 A_2)$$

$$= (n_0 + n_0') \nu_1$$

$$n_2(\max) = (n_0 \nu_2 + N_2) \quad + \quad n_0' \nu_2$$

(Original number of moles of A_2) (Number of moles of A_2 formed by chemical reaction)

$$= (n_0 + n_0') \nu_2 + N_2$$

$n_3(\min) = 0$ (The constituent A_3 completely disappears by reaction)

$n_4(\min) = N_4$ (The excess number of moles of A_4 that are left after the reaction is complete to the left)

Similarly, if the reaction is imagined to proceed completely to the right, there is a minimum amount of each initial constituent, and a maximum amount of each final constituent, so that

$$n_1(\min) = 0$$

$$n_2(\min) = N_2$$

$$n_3(\max) = n_0' \nu_3 \quad + \quad n_0 \nu_3$$

(Original amount) (Amount formed by chemical reaction)

$$(n_0 \nu_1 A_1 + n_0 \nu_2 A_2 \rightarrow n_0 \nu_3 A_3 + n_0 \nu_4 A_4)$$

$$= (n_0 + n_0') \nu_3$$

$$n_4(\max) = (n_0 + n_0') \nu_4 + N_4$$

Let us suppose that the reaction proceeds partially either to the right or to the left to the extent that there are n_1 moles of A_1, n_2 moles of A_2, n_3 moles of A_3, and n_4 moles of A_4. The *degree (or advancement) of reaction* ε is defined in terms of any one of the initial constituents, say, A_1, as the fraction

$$\epsilon = \frac{n_1(\max) - n_1}{n_1(\max) - n_1(\min)}$$

It is seen that when $n_1 = n_1(\max)$, $\varepsilon = 0$, the reaction will start from left to right. When $n_1 = n_1(\min)$, $\varepsilon = 1$, reaction is complete from left to right.

The degree of reaction can thus be written in the form

$$\epsilon = \frac{(n_0 + n_0') \nu_1 - n_1}{(n_0 + n_0') \nu_1}$$

Therefore

$$n_1 = (n_0 + n_0') \nu_1 - (n_0 + n_0') \nu_1 \epsilon$$

$$= n \text{ (at start)} - n \text{ (consumed)}$$

$$= \text{Number of moles of } A_1 \text{ at start} - \text{number of moles of } A_1 \text{ consumed in the reaction}$$
$$= (n_0 + n_0')\,\nu_1\,(1-\epsilon)$$
$$n_2 = n \text{ (at start)} - n \text{ (consumed)}$$
$$= (n_0+n_0')\nu_2+N_2-(n_0+n_0')\,\nu_2\epsilon$$
$$= (n_0+n_0')\,\nu_2\,(1-\epsilon)+N_2$$
$$n_3 = n \text{ (at start)} + n \text{ (formed)}$$
$$= 0+(n_0+n_0')\,\nu_3\epsilon$$
$$= (n_0+n_0')\,\nu_3\epsilon$$
$$n_4 = n \text{ (at start)} + n \text{ (formed)}$$
$$= N_4+(n_0+n_0')\,\nu_4\epsilon$$
$$= (n_0+n_0')\,\nu_4\epsilon+N_4 \qquad (16.2)$$

The number of moles of the constituents change during a chemical reaction, not independently but restricted by the above relations. These equations are the *equations of constraint.* The n's are functions of ε only. In a homogeneous system, in a given reaction, the mole fractions x's are also functions of ε only, as illustrated below.

Let us take the reaction

$$H_2+\tfrac{1}{2}\,O_2 \rightleftharpoons H_2O$$

in which n_0 moles of hydrogen combine with $n_0/2$ moles of oxygen to form n_0 moles of water. The n's and x's as functions of ε are shown in the table given below.

A	ν	n	x
$A_1 = H_2$	$\nu_1 = 1$	$n_1 = n_0\,(1-\epsilon)$	$x_1 = \dfrac{n_1}{\Sigma n} = \dfrac{2\,(1-\epsilon)}{3-\epsilon}$
$A_2 = O_2$	$\nu_2 = \frac{1}{2}$	$n_2 = \dfrac{n_0}{2}(1-\epsilon)$	$x_2 = \dfrac{1-\epsilon}{3-\epsilon}$
$A_3 = H_2O$	$\nu_3 = 1$	$n_3 = n_0\epsilon$	$x_3 = \dfrac{2\,\epsilon}{3-\epsilon}$
		$\Sigma\, n = \dfrac{n_0}{2}\,(3-\epsilon)$	

If the reaction is imagined to advance to an infinitesimal extent, the degree of reaction changes from ε to $\varepsilon + d\varepsilon$, and the various n's will change by the amounts

$$dn_1 = -(n_0+n_0')\,\nu_1\,d\epsilon$$
$$dn_2 = -(n_0+n_0')\,\nu_2\,d\epsilon$$
$$dn_3 = (n_0+n_0')\,\nu_3\,d\epsilon$$
$$dn_4 = (n_0+n_0')\,\nu_4\,d\epsilon$$

or

$$\frac{dn_1}{-\nu_1} = \frac{dn_2}{-\nu_2} = \frac{dn_3}{\nu_3} = \frac{dn_4}{\nu_4} = (n_0+n_0')\,d\epsilon$$

which shows that the dn's are proportional to the ν's.

16.2 REACTION EQUILIBRIUM

Let us consider a homogeneous phase having arbitrary amounts of the constituents, A_1, A_2, A_3, and A_4, capable of undergoing the reaction

$$\nu_1 A_1 + \nu_2 A_2 \rightleftharpoons \nu_3 A_3 + \nu_4 A_4$$

The phase is at uniform temperature T and pressure p. The Gibbs function of the mixture is

$$G = \mu_1 n_1 + \mu_2 n_2 + \mu_3 n_3 + \mu_4 n_4$$

where the n's are the number of moles of the constituents at any moment, and the μ's are the chemical potentials.

Let us imagine that the reaction is allowed to take place at constant T and p. The degree of reaction changes by an infinitesimal amount from ε to $\varepsilon + d\,\varepsilon$. The change in the Gibbs function is

$$dG_{T,\,p} = \Sigma\ \mu_k\ dn_k$$
$$= \mu_1 dn_1 + \mu_2 dn_2 + \mu_3 dn_3 + \mu_4 dn_4$$

The equations of constraint in differential form are

$$dn_1 = -(n_0 + n_0')\ \nu_1\ d\epsilon, \qquad dn_3 = (n_0 + n_0')\ \nu_3\ d\epsilon$$
$$dn_2 = -(n_0 + n_0')\ \nu_2\ d\epsilon, \qquad dn_4 = (n_0 + n_0')\ \nu_4\ d\epsilon$$

On substitution

$$dG_{T,\,p} = (n_0 + n_0')\ (-\nu_1\mu_1 - \nu_2\mu_2 + \nu_3\mu_3 + \nu_4\mu_4)\ d\epsilon \qquad (16.3)$$

When the reaction proceeds *spontaneously* to the right, $d\varepsilon$ is positive, and since $dG_{T,\,p} < 0$

$$(\nu_1\mu_1 + \nu_2\mu_2) > (\nu_3\mu_3 + \nu_4\mu_4)$$

When the reaction proceeds spontaneously to the left, $d\varepsilon$ is negative

$$(\nu_1\mu_1 + \nu_2\mu_2) < (\nu_3\mu_3 + \nu_4\mu_4)$$

At equilibrium, the Gibbs function will be a minimum, and

$$\nu_1\mu_1 + \nu_2\mu_2 = \nu_3\mu_3 + \nu_4\mu_4 \qquad (16.4)$$

which is called the *equation of reaction equilibrium.*

16.3 LAW OF MASS ACTION

For a homogeneous phase chemical reaction at constant temperature and pressure, when the constituents are *ideal gases,* the chemical potentials are given by the expressions of the type

$$\mu_k = \bar{R}T\,(\phi_k + \ln\ p + \ln\ x_k)$$

where the ϕ's are functions of temperature only (Article 10.10).

Substituting in the equation of reaction equilibrium (16.4)

$$\nu_1\,(\phi_1 + \ln p + \ln\ x_1) + \nu_2\,(\phi_2 + \ln p + \ln\ x_2)$$
$$= \nu_3\,(\phi_3 + \ln p + \ln x_3) + \nu_4\,(\phi_4 + \ln p + \ln x_4)$$

On rearranging

$$\nu_3 \ln x_3 + \nu_4 \ln x_4 - \nu_1 \ln x_1 - \nu_2 \ln x_2 + (\nu_3 + \nu_4 - \nu_1 - \nu_2) \ln p$$
$$= -(\nu_3\phi_3 + \nu_4\phi_4 - \nu_1\phi_1 - \nu_2\phi_2)$$

$$\therefore \quad \ln \frac{x_3^{\nu_3} \cdot x_4^{\nu_4}}{x_1^{\nu_1} \cdot x_2^{\nu_2}} p^{\nu_3+\nu_4-\nu_1-\nu_2}$$
$$= -(\nu_3\phi_3 + \nu_4\phi_4 - \nu_1\phi_1 - \nu_2\phi_2)$$

Denoting

$$\ln K = -(\nu_3\phi_3 + \nu_4\phi_4 - \nu_1\phi_1 - \nu_2\phi_2)$$

where K, known as the *equilibrium constant,* is a function of temperature only

$$\left[\frac{x_3^{\nu_3} \cdot x_4^{\nu_4}}{x_1^{\nu_1} \cdot x_2^{\nu_2}}\right]_{\epsilon=\epsilon_e} p^{\nu_3+\nu_4-\nu_1-\nu_2} = K \tag{16.5}$$

This equation is called the *law of mass action.* K has the dimension of pressure raised to the $(\nu_3 + \nu_4 - \nu_1 - \nu_2)$ th power. Here the x's are the values of mole fractions at equilibrium when the degree of reaction is ε_e.

The law of mass action can also be written in this form

$$\frac{p_3^{\nu_3} \cdot p_4^{\nu_4}}{p_1^{\nu_1} \cdot p_2^{\nu_2}} = K$$

where the p's are the partial pressures.

16.4 HEAT OF REACTION

The equilibrium constant K is defined by the expression

$$\ln K = -(\nu_3\phi_3 + \nu_4\phi_4 - \nu_1\phi_1 - \nu_2\phi_2)$$

Differentiating ln K with respect to T

$$\frac{d}{dT} \ln K = -\left(\nu_3 \frac{d\phi_3}{dT} + \nu_4 \frac{d\phi_4}{dT} - \nu_1 \frac{d\phi_1}{dT} - \nu_2 \frac{d\phi_2}{dT} \right)$$

Now, from equation (10.73)

$$\phi = \frac{h_0}{\bar{R}T} - \frac{1}{\bar{R}} \int \frac{\int c_p \, dT}{T_2} \cdot dT - \frac{s_0}{\bar{R}}$$

Therefore

$$\frac{d\phi}{dT} = -\frac{h_0}{\bar{R}T^2} - \frac{\int c_p \, dT}{\bar{R}T^2}$$

$$= -\frac{1}{\bar{R}T^2}(h_0 + \int c_p dT)$$

$$= -\frac{h}{\bar{R}T^2}$$

Therefore

$$\frac{d}{dT} \ln K = -\frac{1}{\bar{R}T^2}(\nu_3 h_3 + \nu_4 h_4 - \nu_1 h_1 - \nu_2 h_2)$$

where the h's refer to the same temperature T and the same pressure p. If ν_1 moles of A_1 and ν_2 moles of A_2 combine to from ν_3 moles of A_3 and ν_4 moles of A_4 at constant temperature and pressure, the heat transferred would be, as shown in Fig. 16.1, equal to the final enthalpy $(\nu_3 h_3 + \nu_4 h_4)$ minus the initial enthalpy $(\nu_1 h_1 + \nu_2 h_2)$. This is known as the *heat o reaction,* and is denoted by ΔH.

$$\Delta H = \nu_3 h_3 + \nu_4 h_4 - \nu_1 h_1 - \nu_2 h_2 \quad (16.6)$$

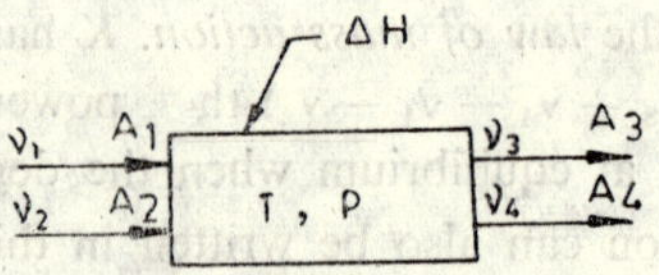

Fig. 16.1 Heat of reaction.

Therefore $$\frac{d}{dT} \ln K = \frac{\Delta H}{\bar{R}T^2} \quad (16.7)$$

This is known as the *van't Hoff equation.* The equation can be used to calculate the heat of reaction at any desired temperature or within a certain temperature range.

By rearranging equation (16.7)

$$\frac{d \ln K}{\frac{dT}{T^2}} = \frac{\Delta H}{\bar{R}}$$

$$\frac{d \ln K}{d(1/T)} = -\frac{\Delta H}{\bar{R}}$$

Therefore

$$\Delta H = -2.303\,\bar{R}\,\frac{d \log K}{d\left(\frac{1}{T}\right)} = -19.148\frac{d \log K}{d\left(\frac{1}{T}\right)} \text{ kJ/kg mol}$$

$$= -4.573\frac{d \log K}{d\left(\frac{1}{T}\right)} \text{ kcal/kgmol}$$

If K_1 and K_2 are the equilibrium constants evaluated at temperatures T_1 and T_2 respectively

$$\Delta H = -19.148 \frac{\log K_1 - \log K_2}{\frac{1}{T_1} - \frac{1}{T_2}}$$

or

$$\Delta H = 19.148 \frac{T_1 T_2}{T_1 - T_2} \log \frac{K_1}{K_2}$$

If ΔH is positive, the reaction is said to be *endothermic*. If ΔH is negative, the reaction is *exothermic*.

16.5 TEMPERATURE DEPENDENCE OF THE HEAT OF REACTION

$$\Delta H = \nu_3 h_3 + \nu_4 h_4 - \nu_1 h_1 - \nu_2 h_2$$

$$h = h_0 + \int c_p \, dT$$

Therefore

$$\Delta H = \nu_3 h_{03} + \nu_4 h_{04} - \nu_1 h_{01} - \nu_2 h_{02} + \int (\nu_3 c_{p_3} + \nu_4 c_{p_4} - \nu_1 c_{p_1} - \nu_2 c_{p_2}) \, dT$$

Denoting $\Delta H_0 = \nu_3 h_{03} + \nu_4 h_{04} - \nu_1 h_{01} - \nu_2 h_{02}$

$$\Delta H = \Delta H_0 + \int (\nu_3 c_{p_3} + \nu_4 c_{p_4} - \nu_1 c_{p_1} - \nu_2 c_{p_2}) \, dT$$

If c_p is known as a function of temperature and if at any temperature ΔH is known, then at any other temperature, ΔH can be determined for a certain chemical reaction from the above relation.

16.6 TEMPERATURE DEPENDENCE OF THE EQUILIBRIUM CONSTANT

$$\ln K = -(\nu_3 \phi_3 + \nu_4 \phi_4 - \nu_1 \phi_1 - \nu_2 \phi_2)$$

where $$\phi = \frac{h_0}{\bar{R}T} - \frac{1}{\bar{R}} \int \frac{\int c_p \, dT}{T_2} dT - \frac{s_0}{\bar{R}}$$

On substitution

$$\ln \mathrm{K} = -\frac{1}{\bar{R}T} (\nu_3 h_{03} + \nu_4 h_{04} - \nu_1 h_{01} - \nu_2 h_{02})$$

$$+ \frac{1}{\bar{R}} \int \frac{\int (\nu_3 c_{p_3} + \nu_4 c_{p_4} - \nu_1 c_{p_1} - \nu_2 c_{p_2}) \, dT}{T^2} \cdot dT$$

$$+ \frac{1}{\bar{R}} (\nu_3 s_{03} + \nu_4 s_{04} - \nu_1 s_{01} - \nu_2 s_{02})$$

If

$$\Delta H_0 = \nu_3 h_{03} + \nu_4 h_{04} - \nu_1 h_{01} - \nu_2 h_{02}$$

$$\Delta S_0 = \nu_3 s_{03} + \nu_4 s_{04} - \nu_1 s_{01} - \nu_2 s_{02}$$

$$\ln K = -\frac{\Delta H_0}{\bar{R}T} + \frac{1}{\bar{R}}\int \frac{\int (\nu_3 c_{p3} + \nu_4 c_{p4} - \nu_1 c_{p1} - \nu_2 c_{p2})\, dT}{T^2}\, dT + \frac{\Delta S_0}{\bar{R}} \quad (16.8)$$

This equation is sometimes known as the *Nernst's equation.*

16.7 THERMAL IONIZATION OF A MONATOMIC GAS

One interesting application of Nernst's equation was made by Dr. M.N. Saha to the thermal ionization of a monoatomic gas. If a monoatomic gas is heated to a high enough temperature (5000 K and above), some ionization occurs, with the electrons in the outermost orbit being shed off, and the atoms, ions, and electrons may be regarded as a mixture of three ideal monoatomic gases, undergoing the reaction

$$A \rightleftharpoons A^+ + e$$

Starting with n_0 moles of atoms

A	ν	n	x
$A_1 = A$	$\nu_1 = 1$	$n_1 = n_0 (1 - \epsilon_e)$	$x_1 = \dfrac{1-\epsilon_e}{1+\epsilon_e}$
$A_3 = A^+$	$\nu_3 = 1$	$n_3 = n_0 \epsilon_e$	$x_3 = \dfrac{\varepsilon_e}{1+\varepsilon_e}$
$A_4 = e$	$\nu_4 = 1$	$n_4 = n_0 \varepsilon_e$	$x_4 = \dfrac{\varepsilon_e}{1+\varepsilon_e}$
		$\Sigma n = n_0 (1 + \varepsilon_e)$	

The equilibrium constant is given by

$$\ln K = \ln \left\{ \frac{x_3^{\nu_3} \cdot x_4^{\nu_4}}{x_1^{\nu_1} \cdot x_2^{\nu_2}} \right\}_{\varepsilon_e} \cdot p^{\nu_3 + \nu_4 - \nu_1 - \nu_2}$$

$$\ln K = \ln \frac{\dfrac{\varepsilon_e}{1+\varepsilon_e} \cdot \dfrac{\varepsilon_e}{1+\varepsilon_e}}{\dfrac{1-\varepsilon_e}{1+\varepsilon_e}} \cdot p^{\nu_3 + \nu_4 - \nu_1}$$

$$= \ln \frac{\varepsilon_e^2}{1-\epsilon_e^2} \cdot p$$

Since the three gases are monoatomic, $c_p = \frac{5}{2}\bar{R}$ which, on being substituted in the Nernst's equation, gives

$$\ln \frac{\varepsilon_e^2}{1-\varepsilon_e^2} \cdot p = -\frac{\Delta H_0}{\bar{R}T} + \frac{5}{2} \ln T + \ln B$$

where

$$\frac{\Delta S_0}{\bar{R}} = \ln B$$

$$\ln \frac{\varepsilon_e^2}{1-\varepsilon_e^2} \cdot \frac{p}{T^{5/2} . B} = -\frac{\Delta H_0}{\bar{R}T}$$

$$\therefore \quad \frac{\varepsilon_e^2}{1-\varepsilon_e^2} = B\, e^{-\Delta H_0/\bar{R}T} \cdot \frac{T^{5/2}}{p} \tag{16.9}$$

where ε_e is the equilibrium value of the degree of ionization. This is known as the *Saha's equation.* For a particular gas the degree of ionization increases with an increase in temperature and a decrease in pressure.

16.8 GIBBS FUNCTION CHANGE

Molar Gibbs function of an ideal gas at temperature T and pressure p is equal to (Article 10.10)

$$g = \bar{R}T(\phi + \ln p)$$

For the reaction of the type

$$\nu_1 A_1 + \nu_2 A_2 \rightleftharpoons \nu_3 A_3 + \nu_4 A_4$$

the Gibbs function change of the reaction ΔG is defined by the expression

$$\Delta G = \nu_3 g_3 + \nu_4 g_4 - \nu_1 g_1 - \nu_2 g_2 \tag{16.10}$$

where the g's refer to the gases completely separated at T, p, ΔG is also known as the *free energy change*. Substituting the values of the g's

$$\Delta G = \bar{R}T(\nu_3\phi_3 + \nu_4\phi_4 - \nu_1\phi_1 - \nu_2\phi_2) + \bar{R}T \ln p^{\nu_3+\nu_4-\nu_1-\nu_2}$$

But

$$\ln K = -(\nu_3\phi_3 + \nu_4\phi_4 - \nu_1\phi_1 - \nu_2\phi_2)$$

$$\therefore \quad \Delta G = -\bar{R}T \ln K + \bar{R}T \ln p^{\nu_3+\nu_4-\nu_1-\nu_2}$$

If p is expressed in atmospheres and ΔG is calculated when each gas is at a pressure of 1 atm., the second term on the right drops out. Under these conditions ΔG is known as the *standard Gibbs function change* and is denoted by ΔG^0

$$\Delta G^0 = -\bar{R}T \ln K \tag{16.11}$$

This is an important equation which relates the standard Gibbs function change with temperature and the equilibrium constant. From this the equilibrium constant can be calculated from changes in the standard Gibbs function, or vice versa.

From equation (16.3), if $n_0 = 1$ and $n_0' = 0$

$$\left(\frac{\partial G}{\partial \varepsilon}\right)_{T,p} = (\nu_3\mu_3 + \nu_4\mu_4 - \nu_1\mu_1 - \nu_2\mu_2)$$

Since $\mu_k = \bar{R}T(\phi_k + \ln p + \ln x_k)$

and $g_k = \bar{R}T(\phi_k + \ln p)$

so $\mu_k = g_k + \bar{R}T \ln x_k$

Therefore

$$\left(\frac{\partial G}{\partial \epsilon}\right)_{T,p} = \nu_3 g_3 + \nu_4 g_4 - \nu_1 g_1 - \nu_2 g_2 + \bar{R}T \ln \frac{x_3^{\nu_3} \cdot x_4^{\nu_4}}{x_1^{\nu_1} \cdot x_2^{\nu_2}}$$

$$= \Delta G + \bar{R}T \ln \frac{x_3^{\nu_3}\, x_4^{\nu_4}}{x_1^{\nu_1}\, x_2^{\nu_2}}$$

At $\varepsilon = 0,\quad x_3 = 0,\quad x_4 = 0$

$$\left(\frac{\partial G}{\partial \varepsilon}\right)_{T,p} = -\infty$$

and at $\varepsilon = 1,\quad x_1 = 0,\quad x_2 = 0$

$$\left(\frac{\partial G}{\partial \varepsilon}\right)_{T,p} = +\infty$$

A't $\varepsilon = \frac{1}{2},\ x_1 = \frac{\nu_1}{\Sigma\nu},\ x_2 = \frac{\nu_2}{\Sigma\nu}$

$$x_3 = \frac{\nu_3}{\Sigma\nu},\ x_4 = \frac{\nu_4}{\Sigma\nu}$$

where $\Sigma\nu = \nu_3 + \nu_4 - \nu_1 - \nu_2$

$$\left(\frac{\partial G}{\partial \varepsilon}\right)_{T,p} \varepsilon\tfrac{1}{2} = \Delta G + \bar{R}T \ln \frac{\left(\frac{\nu_3}{\Sigma\nu}\right)^{\nu_3} \left(\frac{\nu_4}{\Sigma\nu}\right)^{\nu_4}}{\left(\frac{\nu_1}{\Sigma\nu}\right)^{\nu_1} \left(\frac{\nu_2}{\Sigma\nu}\right)^{\nu_2}}$$

If $p = 1$ atm., $T = 298$ K

$$\left(\frac{\partial G}{\partial \varepsilon}\right)_{\substack{p=1\text{ atm.}\\ T=298\text{ K}\\ \varepsilon=\frac{1}{2}}} = \Delta G^\circ_{298}$$

because the magnitude of the second term on the right hand side of the equation is very small compared to ΔG°. The slop $\left(\frac{\partial G}{\partial \varepsilon}\right)_{T,p}$ at $\varepsilon = \frac{1}{2}$ is called the 'affinity' of the reaction, and it is equal to ΔG^0 at the standard reference state. The magnitude of the slope at $\varepsilon = \frac{1}{2}$ (Fig. 16.2) indicates the direction in which the reaction will proceed. For water vapour reaction, ΔG^0_{298} is a large positive number, which indicates that the equilibrium point is far to the left of $\varepsilon = \frac{1}{2}$, and therefore, ε_e is very small. Again for the reaction NO $\rightleftharpoons$ $\frac{1}{2}$ N_2 + $\frac{1}{2}$ O_2, ΔG^0_{298} is a large negative value, which shows that the equilibrium point is far to the right of $\varepsilon = \frac{1}{2}$, and so ε_e is close to unity.

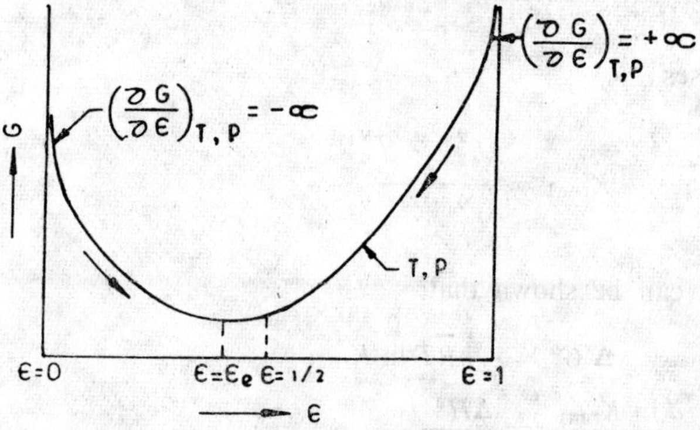

Fig. 16.2 **Plot of G against** ε **at constants** T and p

16.9 FUGACITY AND ACTIVITY

The differential of the Gibbs function of an ideal gas undergoing an isothermal process is

$$dG = Vdp = \frac{n\bar{R}T}{p}\,dp$$

$$= n\bar{R}T\,d(\ln p)$$

Analogously, the differential of the Gibbs function for a real gas is

$$dG = n\bar{R}T\,d(\ln f)$$

where f is called the *fugacity*, first used by Lewis. The value of fugacity approaches the value of pressure as the latter tends to zero, i.e., when ideal gas conditions apply. Therefore

$$\lim_{p\to 0} \frac{f}{p} = 1$$

For an ideal gas $f = p$. Fugacity has the same dimension as pressure. Integrating equation (16.12)

$$G - G^\circ = n\bar{R}T \ln \frac{f}{f^0}$$

where G^0 and f^0 refer to the reference state when $p_0 = 1$ atm. The ratio f/f^0 is called the *activity*.

Therefore

$$G - G^0 = n\bar{R}\ln a \tag{16.13}$$

For ideal gases, the equilibrium constant is given by

$$K = \frac{p_3^{\nu_3} \cdot p_4^{\nu_4}}{p_1^{\nu_1} \cdot p_2^{\nu_2}}$$

For real gases

$$K_{\text{real}} = \frac{f_3^{\nu_3} \cdot f_4^{\nu_4}}{f_1^{\nu_1} \cdot f_2^{\nu_2}}$$

Similarly, it can be shown that

$$\Delta G^0 = -n\bar{R}T \ln K_{\text{real}}$$

and $$\frac{d \ln K_{\text{real}}}{d T} = \frac{\Delta H^0}{\bar{R}T^2}$$

16.10 DISPLACEMENT OF EQUILIBRIUM DUE TO A CHANGE IN TEMPERATURE OR PRESSURE

The degree of reaction at equilibrium ε_e changes with temperature and also with pressure. From the law of mass action

$$\ln K = \ln \left[\frac{x_3^{\nu_3} \cdot x_4^{\nu_4}}{x_1^{\nu_1} \cdot x_2^{\nu_2}} \right]_{\varepsilon=\varepsilon_e} + (\nu_3+\nu_4-\nu_1-\nu_2) \ln p$$

where ln K is a function of temperature only and the first term on the right hand side is a function of ε_e only. Therefore

$$\left(\frac{\partial \varepsilon_e}{\partial T}\right)_p = \left(\frac{\partial \varepsilon_e}{\partial \ln K}\right)_p \left(\frac{\partial \ln K}{\partial T}\right)_p$$

$$= \frac{\left(\dfrac{\partial \ln K}{\partial T}\right)_p}{\left(\dfrac{\partial \ln K}{\partial \varepsilon_e}\right)_p} = \ln \frac{\Delta H}{\bar{R}\, T^2 \dfrac{d}{d\varepsilon_e} \ln \left[\dfrac{x_3^{\nu_3} \cdot x_4^{\nu_4}}{x_1^{\nu_1} \cdot x_2^{\nu_2}} \right]_{\varepsilon=\varepsilon_e}} \tag{16.14}$$

It can be shown (Problem 16.6) that

$$\frac{d}{d\,\varepsilon} \ln \frac{x_3^{\nu_3} \cdot x_4^{\nu_4}}{x_1^{\nu_1} \cdot x_2^{\nu_2}} = \frac{n_o}{\Sigma\, n_k} \cdot \frac{(\nu_1+\nu_2)(\nu_3+\nu_4)}{\epsilon\,(1-\varepsilon)}$$

which is always positive.

Therefore, for endothermic reaction, when ΔH is positive, $\left(\frac{\partial \varepsilon_e}{\partial T}\right)_p$ is positive, and for exothermic reaction, when ΔH is negative, $\left(\frac{\partial \varepsilon_e}{\partial T}\right)_p$ is negative.

Again,
$$\left(\frac{\partial \varepsilon_e}{\partial p}\right)_T = -\left(\frac{\partial \varepsilon_e}{\partial \ln K}\right)_p \left(\frac{\partial \ln K}{\partial p}\right)_{\varepsilon_e}$$
$$= \frac{-\left(\frac{\partial \ln K}{\partial p}\right)_{\varepsilon_e}}{\left(\frac{\partial \ln K}{\partial \varepsilon_e}\right)_p}$$

Using the law of mass action

$$\left(\frac{\partial \varepsilon_e}{\partial p}\right)_T = -\frac{\nu_3 + \nu_4 - \nu_1 - \nu_2}{p \frac{d}{d\varepsilon} \ln \left[\frac{x_3^{\nu_3} \cdot x_4^{\nu_4}}{x_1^{\nu_1} \cdot x_2^{\nu_2}}\right]_{\varepsilon = \varepsilon_e}} \tag{16.15}$$

If $(\nu_3+\nu_4) > (\nu_1+\nu_2)$, i.e., the number of moles increase or the volume increases due to reaction, $\left(\frac{\partial \varepsilon_e}{\partial p}\right)_T$ is negative. If $(\nu_3+\nu_4) < (\nu_1+\nu_2)$, i.e., volume decreases in an isothermal reaction, $\left(\frac{\partial \varepsilon_e}{\partial p}\right)_T$ is positive.

16.11 HEAT CAPACITY OF REACTING GASES IN EQUILIBRIUM

For a reaction of four ideal gases, such as

$$\nu_1 A_1 + \nu_2 A_2 \rightleftharpoons \nu_3 A_3 + \nu_4 A_4$$

the enthalpy of mixture at equilibrium is

$$H = \Sigma\, n_k\, h_k$$

where
$$n_1 = (n_0+n_0')\, \nu_1\, (1-\varepsilon_e),\quad n_2 = (n_0+n_0')\, \nu_2\, (1-\varepsilon_e)+N_2$$
$$n_3 = (n_0+n_0')\, \nu_3 \varepsilon_e, \text{ and } n_4 = (n_0+n_0')\, \nu_4 \varepsilon_e + N_4$$

Let us suppose that an infinitesimal change in temperature takes place at constant pressure in such a way that equilibrium is maintained.

ε_e will change to the value $\varepsilon + d\varepsilon_e$, and the enthalpy will change by the amount

$$dH_P = \Sigma\, n_k\, dh_k + \Sigma h_k\, dn_k$$

where
$$dh_k = c_{p_k}\, dT,\; dn_k = \pm\, (n_0+n_0')\, \nu_k\, d\varepsilon_e$$

Therefore

$$dH_p = \Sigma\, n_k\, c_{p_k}\, dT + (n_0 + n_0')\,(\nu_3 h_3 + \nu_4 h_4 - \nu_1 h_1 - \nu_2 h_2)\, d\,\varepsilon_e$$

The heat capacity of the reacting gas mixture is

$$C_p = \left(\frac{\partial H}{\partial T}\right)_p = \Sigma\, n_k\, c_{p_k} + (n_0 + n_0')\,\Delta H\,\left(\frac{\partial\,\varepsilon_e}{\partial T}\right)_P$$

Using equation (16.14)

$$C_p = \Sigma\, n_k\, c_{p_k} + (n_0 + n_0')\,\frac{(\Delta H)^2}{\bar{R}\,T^2\,\dfrac{d}{d\varepsilon}\,\ln\left[\dfrac{x_3^{\nu_3}\cdot x_4^{\nu_4}}{x_1^{\nu_1}\cdot x_2^{\nu_2}}\right]_{\varepsilon=\varepsilon_e}}$$

16.12 COMBUSTION

Combustion is a chemical reaction between a fuel and oxygen which proceeds at a fast rate with the release of energy in the form of heat. In the combustion of methane, e.g.

$$\underset{\text{Reactants}}{CH_4 + 2O_2} \rightarrow \underset{\text{Products}}{CO_2 + 2H_2O}$$

One mole of methane reacts with 2 moles of oxygen to form 1 mole of carbon dioxide and 2 moles of water. The water may be in the liquid or vapour state depending on the temperature and pressure of the products of combustion. Only the initial and final products are being considered without any concern for the intermediate products that usually occur in a reaction.

Atmospheric air contains 21% oxygen, 78% nitrogen, and 1% argon by volume. In combustion calculations, however, the argon is usually neglected, and air is assumed to consist of 21% oxygen and 79% nitrogen by volume (or molar basis). On a mass basis, air contains 23% oxygen and 77% nitrogen.

For each mole of oxygen taking part in a combustion reaction, there are 79.0/21.0 = 3.76 moles of nitrogen. So for the combustion of methane, the reaction can be written as

$$CH_4 + 2O_2 + 2\,(3.76)\,N_2 \rightarrow CO_2 + 2H_2O + 7.52\,N_2$$

The minimum amount of air which provides sufficient oxygen for the complete combustion of all the elements like carbon, hydrogen, etc., which may oxidize is called the *theoretical or stoichiometric air*. There is no oxygen in the products when complete combustion (oxidation) is achieved with this theoretical air. In practice, however, more air than this theoretical amount is required to be supplied for complete combustion. Actual air supplied is usually expressed in terms of percent theoretical air; 150%

theoretical air means that 1.5 times the theoretical air is supplied. Thus, with 150% theoretical air, the methane combustion reaction can be written as

$$CH_4 + 2(1.5)O_2 + 2(3.76)(1.5)N_2$$
$$\rightarrow CO_2 + 2H_2O + O_2 + 11.28\ N_2$$

Another way of expressing the actual air quantity supplied is in terms of excess air. Thus 150% theoretical air means 50% excess air.

With less than theoretical air supply, combustion will remain incomplete with some CO present in the products. Even with excess air supply also, there may be a small amount of CO present, depending on mixing and turbulence during combustion, e.g., with 115% theoretical air

$$CH_4 + 22(1.15)\ O_2 + 2(1.15)\ 3.76\ N_2 \rightarrow$$
$$\rightarrow 0.95\ CO_2 + 0.05 + CO + 2H_2O + 0.325\ O_2 + 8.65\ N_2$$

By analyzing the products of combustion, the actual amount of air supplied in a combustion process can be computed. Such analysis is often given on the 'dry' basis, i.e., the fractional analysis of all the components, except water vapour. Following the principle of the conservation of mass of each of the elements, it is possible to make a carbon balance, hydrogen balance, oxygen balance, and nitrogen balance from the combustion reaction equation, from which the actual air-fuel ratio can be determined.

16.13 ENTHALPY OF FORMATION

Let us consider the steady state steady flow combustion of carbon and oxygen to form CO_2 (Fig. 16.3). Let the carbon and oxygen each enter the control volume at 25°C and 1 atm. pressure, and the heat transfer be such that the product CO_2 leaves at 25°C, 1 atm. pressure. The measured value

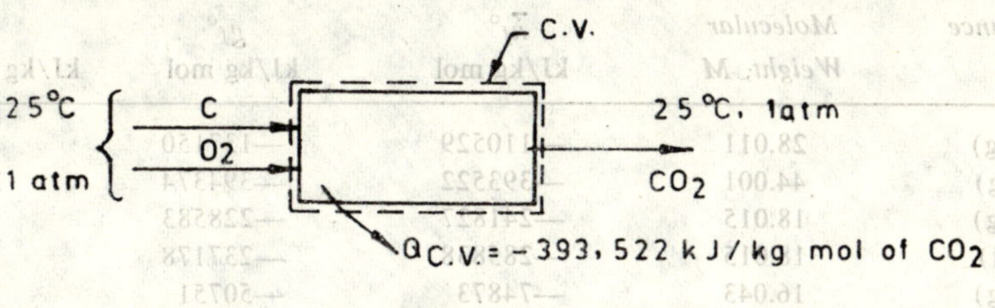

Fig. 16.3 Enthalpy of formation

of heat transfer is −393,522 kJ per kgmol of CO_2 formed. If H_R and H_P refer to the total enthalpy of the reactants and products respectively, then the first law applied to the reaction $C + O_2 \rightarrow CO_2$ gives

$$H_K + Q_{C.V.} = H_P$$

For all the reactants and products in a reaction, the equation may be written as

$$\sum_R n_i \bar{h}_i + Q_{c.v.} = \sum_P n_e \bar{h}_e$$

where R and P refer to the reactants and products respectively.

The enthalpy of all the elements at the standard reference state of 25°C, 1 atm. is assigned the value of zero. In the carbon-oxygen reaction, $H_R = 0$. So the energy equation gives

$$Q_{c,v.} = H_P = -393{,}522 \text{ kJ/kgmol}$$

This is what is known as the *enthalpy of formation of* CO_2 at 25°C, 1 atm., and designated by the symbol, $\bar{h}^0{}_f$. So

$$(\bar{h}_f{}^0)_{CO_2} = -393, 552 \text{ kJ/kg mol}$$

In most cases, however, the reactants and products are not at 25°C, 1 atm. Therefore, the change of enthalpy (in the case of constant pressure process or S.S.S.F. process) between 25°C, 1 atm. and the given state must be known. Thus the enthalpy at any temperature and pressure, $\bar{h}_{T,p}$ is

$$\bar{h}_{T,\,p} = (\bar{h}_f{}^0)_{298\text{K, 1 atm.}} + (\Delta \bar{h})_{298\text{K, 1 atm} \to T,\, p}$$

For convenience, the subscripts are usually dropped, and

$$\bar{h}_{T,\,p} = \bar{h}_f{}^0 + \Delta \bar{h}$$

where $\Delta \bar{h}$ represents the difference in enthalpy between any given state and the enthalpy at 298.15 K, 1 atm.

Table 16.1 gives the values of the enthalpy of formation of a number of substances in kJ/kgmol.

Table 16.1

Enthalpy of Formation, Gibbs Function of Formation, and Absolute Entropy of Various Substances at 25°C, 1 atm. pressure

Substance	*Molecular Weight, M*	$\bar{h}_f{}^\circ$ kJ/kg mol	$\bar{g}_f{}^\circ$ kJ/kg mol	s° kJ/kg mol K
CO (g)	28.011	—110529	—137150	197.653
CO_2(g)	44.001	—393522	—394374	213.795
H_2O(g)	18.015	—241827	—228583	188.833
H_2O(l)	18.015	—285838	—237178	70.049
CH_4(g)	16.043	—74873	—50751	186.256
C_2H_2(g)	26.038	+226731	+209234	200.958
C_2H_4(g)	28.054	+52283	+68207	219.548
C_2H_6(g)	30.070	—84667	—32777	229.602
C_3H_8(g)	44.097	—103847	—23316	270.019
C_4H_{10}(g)	58.124	—126148	—16914	310.227
C_8H_{18}(g)	114.23	—208447	+16859	466.835
C_8H_{18}(l)	114.23	—249952	+6940	360.896

Table C in the appendix gives the values of $\Delta\bar{h} = \bar{h}^0 - \bar{h}^0_{298}$ in kJ/kg mol for various substances at different temperatures.

16.14 FIRST LAW FOR REACTIVE SYSTEMS

For the S.S.S.F. process as shown in Fig. 16.4 the first law gives

$$H_R + Q_{\text{C.V.}} = H_P + W_{\text{C.V.}}$$

or

$$\sum_R n_i\bar{h}_i + Q_{\text{C.V.}} = \sum_P n_e\bar{h}_e + \dot{W}_{\text{C.V.}}$$

$$\sum_R n_i\,[\bar{h}_f{}^0 + \Delta\bar{h}]_i + Q_{\text{C.V.}} = \sum_P n_e\,[\bar{h}_f{}^0 + \Delta\bar{h}]_e + W_{\text{C.V.}}$$

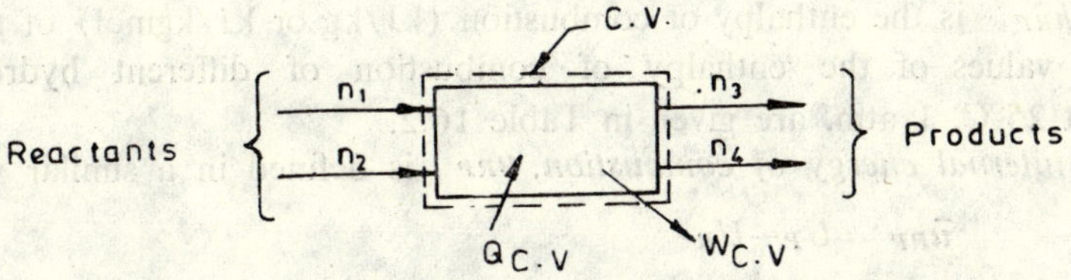

Fig. 16.4 First law for a reactive system

16.15 ADIABATIC FLAME TEMPERATURE

If a combustion process occurs adiabatically in the absence of work transfer or changes in K.E. and P.E., then the energy equation becomes

$$H_R = H_P$$

or

$$\sum_R n_i\bar{h}_i = \sum_P n_e\bar{h}_e$$

or

$$\sum_R n_i\,[\bar{h}_f{}^0 + \Delta\bar{h}]_i = \sum_P n_e\,[\bar{h}_f{}^0 + \Delta\bar{h}]_e$$

For such a process, the temperature of the products is called the *adiabatic flame temperature* which is the maximum temperature achieved for the given reactants. The adiabatic flame temperature can be controlled by the amount of excess air supplied; it is the maximum with a stoichiometric mixture. Since the maximum permissible temperature in a gas turbine is fixed from metallurgical considerations, close control of the temperature of the products is achieved by controlling the excess air.

For a given reaction the adiabatic flame temperature is computed by trial and error. The energy of the reactants H_R being known, a suitable temperature is chosen for the products so that the energy of the products at that temperature becomes equal to the energy of the reactants.

16.16 ENTHALPY AND INTERNAL ENERGY OF COMBUSTION: HEATING VALUE

The *enthalpy of combustion* is defined as the difference between the enthalpy of the products and the enthalpy of the reactants when complete combustion occurs at a given temperature and pressure.

Therefore

$$\bar{h}_{RP} = H_P - H_R$$

or
$$\bar{h}_{RP} = \sum_P n_e [\bar{h}_f^0 + \Delta\bar{h}]_e - \sum_R n_i [\bar{h}_f^0 + \Delta\bar{h}]_i$$

where $\bar{h}_{RP}$ is the enthalpy of combustion (kJ/kg or kJ/kgmol) of the fuel.

The values of the enthalpy of combustion of different hydrocarbon fuels at 25°C, 1 atm. are given in Table 16.2.

The *internal energy of combustion*, u_{RP}, is defined in a similar way.

$$\bar{u}_{RP} = U_P - U_R$$

$$= \sum_P n_e [\bar{h}_f^0 + \Delta\bar{h} - p\bar{v}]_e - \sum_R n_i [h_f^0 + \Delta\bar{h} - p\bar{v}]_i$$

If all the gaseous constituents are considered ideal gases and the volume of liquid and solid considered is assumed to be neglible compared to gaseous volume

$$\bar{u}_{RP} = \bar{h}_{RP} - \bar{R}\, T\, (n_{\text{gaseous products}} - n_{\text{gaseous reactants}})$$

In the case of a constant pressure or steady flow process, the negative of the enthalpy of combustion is frequently called the *heating value at constant pressure,* which represents the heat transferred from the chamber during combustion at constant temperature.

Similarly, the negative of the internal energy of combustion is sometimes designated as the *heating value at constant volume* in the case of combustion, because it represents the amount of heat transfer in the constant volume process.

The *higher heating value* (HHV) or higher calorific value (HCV) is the heat transferred when H_2O in the products is in the liquid state. The *lower heating value* (LHV) or lower calorific value (LCV) is the heat transferred in the reaction when H_2O in the products is in the vapour state.

Therefore

$$\text{LHV} = \text{HHV} - m_{H_2O} \cdot h_{fg}$$

where m_{H_2O} is the mass of water formed in the reaction.

Table 16.2

Enthalpy of Combustion of Some Hydrocarbons at 25°C

Hydrocarbon	Formula	Liquid H_2O in Products (Negative of Higher Heating Value): Liquid Hydrocarbon kJ/kg fuel	Liquid H_2O in Products (Negative of Higher Heating Value): Gaseous Hydrocarbon kJ/kg fuel	Vapour H_2O in Products (Negative of Lower Heating Value): Liquid Hydrocarbon, kJ/kg fuel	Vapour H_2O in Products (Negative of Lower Heating Value): Gaseous Hydrocarbon, kJ/kg fuel
Paraffin Family					
Methane	CH_4		—55496		—50010
Ethane	C_2H_6		—51875		—47484
Propane	C_3H_8	—49975	—50345	—45983	—46353
Butane	C_4H_{10}	—49130	—49500	—45344	—45714
Pentane	C_5H_{12}	—48643	—49011	—44983	—45351
Hexane	C_6H_{14}	—48308	—48676	—44733	—45101
Heptane	C_7H_{16}	—48071	—48436	—44557	—44922
Octane	C_8H_{18}	—47893	—48256	—44425	—44788
Decane	$C_{10}H_{22}$	—47641	—48000	—44239	—44598
Dodecane	$C_{12}H_{26}$	—47470	—47828	—44109	—44467
Olefin Family					
Ethene	C_2H_4		—50296		—47158
Propene	C_3H_6		—48917		—45780
Butene	C_4H_8		—48453		—45316
Pentene	C_5H_{10}		—48134		—44996
Hexene	C_6H_{12}		—47937		—44800
Heptene	C_7H_{14}		—47800		—44662
Octene	C_8H_{16}		—47693		—44556
Nonene	C_9H_{18}		—47612		—44475
Decene	$C_{10}H_{20}$		—47547		—44410
Alkylbenzene Family					
Benzene	C_6H_6	—41831	—42266	—40141	—40576
Methylbenzene	C_7H_8	—42473	—42847	—40527	—40937
Ethylbenzene	C_8H_{10}	—42997	—43395	—40924	—41322
Propylbenzene	C_9H_{12}	—43416	—43800	—41219	—41603
Butylbenzene	$C_{10}H_{14}$	—43748	—44123	—41453	—41828

16.17 ABSOLUTE ENTROPY AND THE THIRD LAW OF THERMODYNAMICS

So far only the first law aspects of chemical reactions have been discussed. The second law analysis of chemical reactions needs a base for the entropy of various substances. The entropy of substances at the absolute zero of temperature, called absolute entropy, is dealt with by the third law of thermodynamics formulated in the early twentieth century primarily by W.H. Nernst (1864-1941) and Max Planck (1858-1947). The hird law states that the entropy of a perfect crystal is zero at the absolute zero of temperature and it represents the maximum degree of order. A substance not having a perfect crystalline structure and possessing a degree of randomness such as a solid solution or a glassy solid, has a finite value of entropy at absolute zero. The third law provides an absolute base from which the entropy of each substance can be measured. The entropy relative to this base is referred to as the absolute entropy. Table 16.1 gives the absolute entropy of various substances at the standard state 25°C, 1 atm. For any other state

$$\bar{s}_{T,\,p} = \bar{s}_T{}^{o} + (\Delta\,\bar{s})_{T,\,1\text{ atm.},\,\rightarrow\,T,\,p}$$

where $\bar{s}_T{}^0$ refers to the absolute entropy at 1 atm. and temperature T, and $(\Delta\bar{s})_{T,\,1\text{ atm.}\,\rightarrow\,T,\,p}$ refers to the change of entropy for an isothermal change of pressure from 1 atm. to pressure p (Fig. 16.5). Table C in the

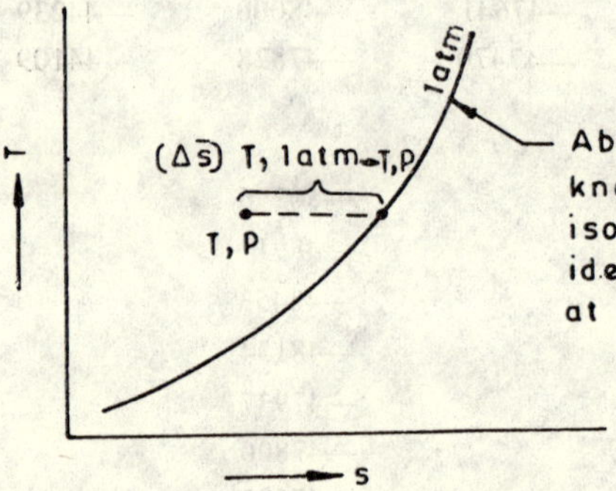

Fig. 16.5 Absolute entropy.

appendix gives the values of $\bar{s}^0$ for various substances at 1 atm. and at different temperatures. Assuming ideal gas behaviour $(\Delta\bar{s})_{T,\,1\text{ atm.}\rightarrow T,\,p}$ can be determined (Fig. 16.5)

$$\bar{s}_2 - \bar{s}_1 = -\bar{R}\,\ln\frac{p_2}{p_1}$$

or $$(\Delta\,\bar{s})_{T,\,1\text{ atm.}\,T,p\rightarrow} = -\,\bar{R}\ln p,$$

where p is in atm.

16.18 SECOND LAW ANALYSIS OF REACTIVE SYSTEMS

The reversible work for a steady state steady flow process, in the absence of changes in KE and PE, is given by

$$W_{\text{rev.}} = \Sigma\, n_i\,(h_i - T_o\, s_i) - \Sigma\, n_e\,(h_e - T_o\, s_e)$$

For an S.S.S.F. process involving a chemical reaction

$$W_{\text{rev.}} = \sum_R n_i\,[\bar{h}_f{}^o + \Delta\bar{h} - T_o\,\bar{s}]_i - \sum_P n_e\,[\bar{h}_f{}^o + \Delta\,\bar{h} - T_o\,\bar{s}]_e$$

The irreversibility for such a process is

$$I = \sum_P n_e\, T_o\, \bar{s}_e - \sum_R n_i\, T_o\, \bar{s}_i - Q_{\text{C.V.}}$$

The availability, ψ, in the absence of KE and PE changes, for an S.S.S.F. process is

$$\psi = (h - T_o\, s) - (h_o - T_o\, s_o)$$

When an S.S.S.F. chemical reaction takes place in such a way that both the reactants and products are in temperature equilibrium with the surroundings, the reversible work is given by

$$W_{\text{rev.}} = \sum_R n_i\, \bar{g}_i - \sum_P n_e\, \bar{g}_e$$

where the $\bar{g}$'s refer to the Gibbs function. The *Gibbs function of formation*, $\bar{g}_f{}^0$, is defined similar to enthalpy of formation, $\bar{h}_f{}^0$. The Gibbs function of each of the elements at 25°C and 1 atm. pressure is assumed to be zero, and the Gibbs function of each substance is found relative to the base. Table 16.1 gives $\bar{g}_f{}^0$ for some substances at 25°C, 1 atm.

EXAMPLE 16.1

Starting with n_0 moles of NH_3, which dissociates according to the equation

$$NH_3 \rightleftharpoons \frac{1}{2} N_2 + \frac{3}{2} H_2,$$

show that at equilibrium

$$K = \frac{27}{4}\, \frac{\varepsilon_e^2}{1 - \varepsilon_e^2} \,.\, p$$

Solution

$$NH_3 \rightleftharpoons \frac{1}{2} N_2 + \frac{3}{2} H_2$$

Comparing the equation with

$$\nu_1\, A_1 + \nu_2\, A_2 \rightleftharpoons \nu_3\, A_3 + \nu_4\, A_4$$

we have

$$A_1 = NH_3,\ \nu_1 = 1,\ n_1 = n_0\,(1 - \varepsilon),\ x_1 = \frac{1 - \varepsilon}{1 + \varepsilon}$$

$$A_3 = N_2,\ \nu_3 = \frac{1}{2},\ n_3 = \frac{n_o\,\varepsilon}{2},\ x_3 = \frac{\varepsilon}{2(1+\varepsilon)}$$

$$A_4 = H_2,\ \nu_4 = \frac{3}{2},\ n_4 = \frac{3\,n_o\,\varepsilon}{2},\ x_4 = \frac{3\,\varepsilon}{2(1+\varepsilon)}$$

$$\Sigma\, n = n_o\,(2+\varepsilon)$$

Substituting in the law of mass action

$$K = \left[\frac{x_3^{\nu_3}\,.\,x_4^{\nu_4}}{x_1^{\nu_1}\,.\,x_2^{\nu_2}}\right]_{\varepsilon=\varepsilon_e} p^{\nu_3+\nu_4-\nu_1-\nu_2}$$

$$= \frac{\left[\dfrac{\varepsilon_e}{2(1+\varepsilon_e)}\right]^{1/2}\left[\dfrac{3\,\varepsilon_e}{2(1+\varepsilon_e)}\right]^{2/3}}{\left(\dfrac{1-\varepsilon_e}{1+\varepsilon_e}\right)}\ .\ p^{3/2+1/2-1}$$

$$= \frac{\sqrt{27}\,\varepsilon^2}{4\,(1+\varepsilon_e)^2}\ .\ \frac{1+\varepsilon_e}{1-\varepsilon_e}\ .\ p$$

$$= \frac{\sqrt{27}}{4}\ .\ \frac{\varepsilon_e^2}{1-\varepsilon_e^2}\ .\ p$$

Proved.

EXAMPLE 16.2

At 35°C and 1 atm. the degree of dissociation of N_2O_4 at equilibrium is 0.27. (a) Calculate K. (b) Calculate ε_e at the same temperature when the pressure is 100 mm Hg. (c) The equilibrium constant for the dissociation of N_2O_4 has the values 0.664 and 0.141 at temperatures 318 and 298 K respectively. Calculate the average heat of reaction within this temperature range.

Solution

Nitrogen tetroxide dissociates according to the equation

$$N_2O_4 \rightleftharpoons 2NO_2$$

Comparing this equation with

$$\nu_1\,A_1 + \nu_2\,A_2 \rightleftharpoons \nu_3\,A_3 + \nu_4\,A_4$$

$$A_1 = N_2O_4,\ \nu_1 = 1,\ n_1 = n_o\,(1-\varepsilon),\ x_1 = \frac{1-\varepsilon}{1+\varepsilon}$$

$$A_3 = NO_2,\ \nu_3 = 2,\ n_3 = 2n_o\,\varepsilon,\ x_3 = \frac{2\,\varepsilon}{1+\varepsilon}$$

$$\Sigma\, n = n_o\,(1+\varepsilon)$$

$$\nu_2 = 0,\ \nu_4 = 0$$

Substituting in the law of mass action

$$K = \left[\frac{x_3^{\nu_3} \cdot x_3^{\nu_4}}{x_1^{\nu_1} \cdot x_2^{\nu_2}}\right] \cdot p^{\nu_3 + \nu_4 - \nu_1 - \nu_2}$$

$$= \frac{\left(\dfrac{2\,\varepsilon_e}{1+\varepsilon_e}\right)^2}{\dfrac{1-\varepsilon_e}{1+\varepsilon_e}} \cdot \quad p^{2+0-1+0} = \frac{4\,\varepsilon_e^2\,p}{1-\varepsilon_2^2}$$

(a) When $\varepsilon_e = 0.27$ and $p = 1$ atm.

$$K = \frac{4.\,(0.27)^2 \cdot 1}{1-(0.27)^2} = 0.3145 \quad \text{atm.}$$ Ans.

(b) Taking the value of K to be the same at the same temperature

$$0.3145 \text{ atm.} = \frac{4\,\varepsilon_e^2}{1-\varepsilon_e^2} \cdot \frac{100}{760}$$

$$\therefore \qquad \varepsilon_e = 0.612 \text{ Ans.}$$

(c) The heat of reaction is given by

$$\Delta H = -\,2.30\,\bar{R}\,\frac{d \log K}{d\dfrac{1}{T}}$$

$$= 2.30\,\bar{R}\,\frac{T_1\,T_2}{T_1 - T_2}\,\log\frac{K_1}{K_2}$$

$$= 2.30 \times 8.3143 \times \frac{318 \times 298}{318 - 298}\log\frac{0.664}{0.141}$$

$$= 90700 \log 4.7$$

$$= 61{,}000 \text{ kJ/kgmol}$$ Ans.

Example 16.3

Determine the heat transfer per kgmol of fuel for the following reaction

$$CH_4 + 2O_2 \rightarrow CO_2 + 2H_2O\ (1)$$

The reactants and products are each at a total pressure of 100 kPa and 25°C.

Solution

By the first law

$$Q_{C.V.} + \sum_R n_i\,\bar{h}_i = \sum_P n_e\,\bar{h}_e$$

From Table C in the appendix

$$\sum_R n_i \bar{h}_i = (\bar{h}_f{}^o)_{CH_4} = -74{,}874 \text{ kJ}$$

$$\sum n_e \bar{h}_e = (\bar{h}_f{}^o)_{CO_2} + 2(\bar{h}_f{}^o)_{H_2O(1)}$$

$$= -393{,}522 + 2(-285{,}838)$$

$$= -965{,}198 \text{ kJ} \qquad \text{Ans.}$$

$$\therefore \quad Q_{c.v.} = -965{,}198 = (-74{,}873)$$

$$= -890{,}325 \text{ kJ} \qquad \text{Ans.}$$

EXAMPLE 16.4

A gasoline engine delivers 150 kW. The fuel used is C_8H_{18} (1) and it enters the engine at 25°C. 150% theoretical air is used and it enters at 45°C. The products of combustion leave the engine at 750 K, and the heat transfer from the engine is 205 kW. Determine the fuel consumption per hour, if complete combustion is achieved.

Solution

The stoichiometric equation (Fig. Ex. 16.4) gives

$$C_8H_{18}(1) + 12.5\, O_2 + 3.76 \times 12.5\ N_2$$
$$\rightarrow 8CO_2 + 9H_2O + 3.76 \times 12.5\ N_2$$

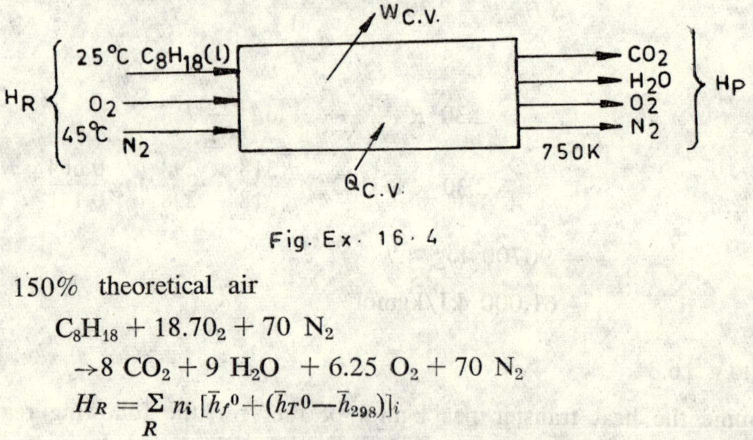

Fig. Ex. 16.4

With 150% theoretical air

$$C_8H_{18} + 18.7 O_2 + 70\ N_2$$
$$\rightarrow 8\ CO_2 + 9\ H_2O + 6.25\ O_2 + 70\ N_2$$

$$H_R = \sum_R n_i\,[\bar{h}_f{}^0 + (\bar{h}_T{}^0 - \bar{h}_{298})]_i$$

$$= 1\,(\bar{h}_f{}^0)_{C_8H_{18(1)}} + 18.7\,[\bar{h}_f{}^0 + (\bar{h}_T - \bar{h}_{298})]_{O_2}$$
$$+ 70\,[\bar{h}_f{}^0 + (\bar{h}_T - \bar{h}_{298})]_{N_2}$$

$$= (-249{,}952) + 18.7 \times 560 + 70 \times 540$$

$$= -201{,}652 \text{ kJ/kgmol fuel}$$

$$H_P = 8\,[-393522 + 20288] + 9\,[-241827 + 16087]$$
$$+ 6.25\,[14171] + 70\,[13491]$$

$$= 8\ (-373{,}234) + 9\ [-225740] + 6.25\,(14171) + 70\,[13491]$$

$$= -3{,}884{,}622 \text{ kJ/kgmol fuel}$$

Energy output from the engine, $W_{C.V.} = 150$ kW

$$Q_{C.V.} = -205 \text{ kW}$$

Let n kgmol of fuel be consumed per second.

By the first law

$$n\,(H_R - H_P) = W_{C.V.} - Q_{C.V.}$$

$$n\,[-201{,}652 + 3{,}884{,}662] = 150 - (-205) = 355 \text{ kW}$$

$$\therefore \quad n = \frac{355 \times 3600}{3683010} \text{ kgmol/hr}$$

$$= 0.346 \text{ kgmol/hr}$$

$\therefore$ Fuel consumption rate

$$= 0.346 \times 114 = 39.4 \text{ kg/hr}$$ Ans.

EXAMPLE 16.5

Determine the adiabatic flame temperature when liquid octane at 25°C is burned with 300% theoretical air at 25°C in a steady flow process.

Solution

The combustion equation with 300% theoretical air is given by

$$C_8H_{18}(1) + 3\,(12.5)\,O_2 + 3\,(12.5)(3.76)\;N_2 \rightarrow$$

$$\rightarrow 8\;CO_2 + 9\;H_2O + 25\;O_2 + 141\;N_2$$

By the first law

$$H_R = H_P$$

$$\sum_R n_i\,[\bar{h}_f{}^0 + \Delta\bar{h}]_i = \sum_P n_e\,[\bar{h}_f{}^0 + \Delta\bar{h}]_e$$

Now

$$H_R = (\bar{h}_f{}^0)_{C_8H_{18}\,(1)} = -249{,}952 \text{ kJ/kgmol fuel}$$

$$H_P = \sum_P n_e\,[\bar{h}_f{}^0 + \Delta\bar{h}]_e = 8\,[-393{,}522 + \Delta\bar{h}_{CO_2}]$$

$$+ 9\,[-241{,}827 + \Delta\bar{h}_{H_2O}] + 25\,\Delta\bar{h}_{O_2} + 141\,\Delta\bar{h}_{N_2}$$

The exit temperature, which is the adiabatic flame temperature, is to be computed by trial and error satisfying the above equation.

Let $T_e = 1000$ K, then

$$H_P = 8\,(-393{,}522 + 33{,}405) + 9\,(-241{,}827 + 25{,}978)$$

$$+ 25\,(22{,}707) + 141\,(21{,}460)$$

$$= -1{,}226{,}577 \text{ kJ/kgmol}$$

If $\quad T_e = 1200$ K

$$H_P = 8\ (-393{,}522 + 44{,}484) + 9\ (-241{,}827 + 34{,}476)$$
$$+ 25\ (29{,}765) + 141(28{,}108)$$
$$= +\ 46{,}537 \text{ kJ/kgmol}$$

If $T_e = 1100$ K,

$$H_P = 8\ (-393{,}522 + 38{,}894) + 9\ (-241{,}827 + 30{,}167)$$
$$+ 25\ (26{,}217) + 141\ (24{,}757)$$
$$= -595{,}964 \text{ kJ/kgmol}$$

Therefore, by interpolation, the adiabatic flame temperature is 1182 K. Ans.

EXAMPLE 16.6

(a) Propane (g) at 25°C and 100 kPa is burned with 400% theoretical air at 25°C and 100 kPa. Assume that the reaction occurs reversibly at 25°C, that the oxygen and nitrogen are separated before the reaction takes place (each at 100 kPa, 25°C), that the constituents in the products are separated, and that each is at 25°C, 100 kPa. Determine the reversible work for this process.

(b) If the above reaction occurs adiabilitically, and each constituent in the products is at 100 kPa pressure and at the adiabatic flame temperature, compute (a) the increase is entropy during combustion, (b) the irreversibility of the process, and (c) the availability of the products of combustion.

Solution

The combustion equation (Fig. Ex. 16.6) is

$$C_3H_8\ (g) + 5\ (4)\ O_2 + 5\ (4)\ (3.76)\ N_2 \rightarrow$$
$$3CO_2 + 4\ H_2O\ (g) \quad + 15\ O_2 + 75.2\ N_2$$

(a)
$$W_{rev.} = \sum_R n_i \bar{g}_i - \sum_P n_e \bar{g}_e$$

From Table 16.1

$$W_{rev.} = (\bar{g}_f{}^0)_{C_3H_8\ (g)} - 3\ (\bar{g}_f{}^0)_{CO_2} - 4\ (\bar{g}_f{}^0)_{H_2O\ (g)}$$
$$= -23{,}316 - 3\ (-394{,}374) - 4\ (-228{,}583)$$
$$= 2{,}074{,}128 \text{ kJ/kgmol} \qquad \text{Ans.}$$
$$= \frac{2{,}074{,}128}{44.097} = 47{,}035.6 \text{ kJ/kg} \qquad \text{Ans.}$$

(b)
$$H_R = H_P$$
$$(\bar{h}_f{}^0)_{C_3H_8\ (g)} = 3\ (\bar{h}_f{}^0 + \Delta\bar{h})_{CO_2} + 4\ (\bar{h}_f{}^0 + \Delta\bar{h})_{H_2O\ (g)}$$
$$+ 15\Delta\bar{h}_{O_2} + 75.2\ \Delta\bar{h}_{N_2}$$

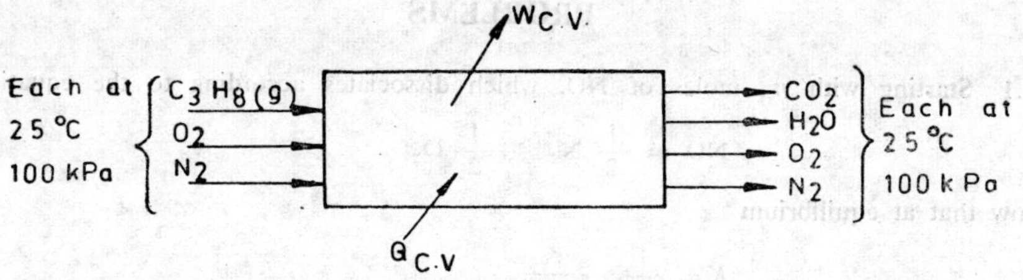

Fig. Ex 16.6

From Table 16.1

$$-103{,}847 = 3\,(-393{,}522+\Delta\bar{h})_{CO_2} + 4\,(-241{,}827+\Delta\bar{h})_{H_2O\,(g)} + 15\;\Delta\bar{h}_{O_2} + 75.2\;\Delta\bar{h}_{N_2}$$

Using Table C in the appendix, and by trial and error, the adiabatic flame temperature is found to be 980 K.

The entropy of the reactants

$$S_R = \sum_R (n_i\,\bar{s}_i^0)_{298} = (\bar{s}^0{}_{C_3H_8\,(g)} + 20\;\bar{s}^0{}_{O_2} + 75.2\;\bar{s}^0{}_{N_2})_{298}$$

$$= 270.019 + 20\;(205.142)\;+ 75.2\;(191.611)$$

$= 18{,}782.01$ kJ/kgmol-K

The entropy of the products

$$SP = \sum_P (n_e\,\bar{s}_e^0)_{980} = (3\;\bar{s}^0{}_{CO_2} + 4\;\bar{s}^0{}_{H_2O\,(g)} + 15\;\bar{s}^0{}_{O_2} + 75.2\;\bar{s}^0{}_{N_2})_{980}$$

$$= 3\;(268.194) + 4\;(231.849) + 15\;(242.855) + 75.2\;(227.485)$$

$$= 22{,}481.68\ \text{kJ/kgmol-K}$$

∴ The increase in entropy during combustion

$SP - S_R = 3{,}699.67$ kJ/kgmol-K   Ans.

The irreversibility of the process

$$I = T_0\,[\,\sum_P n_e\bar{s}_e - \sum_R n_i\bar{s}_i]$$

$$= 298 \times 3699.67$$

$$= 1{,}102{,}501.66\ \text{kJ/kgmol}$$

$$= \frac{1{,}102{,}501.66}{44.097} = 25{,}001\ \text{kJ/kg}$$

The availability of combustion products

$$\psi = W_{rev.} - I$$

$$= 47{,}035.6 - 25{,}001$$

$$= 22{,}034.6\ \text{kJ/kg}$$ Ans.

PROBLEMS

16.1 Starting with n_0 moles of NO, which dissociates according to the equation

$$NO \rightleftharpoons \frac{1}{2} N_2 + \frac{1}{2} O_2$$

show that at equilibrium

$$K = \frac{1}{2} \cdot \frac{\varepsilon_e}{1-\varepsilon_e}$$

16.2 A mixture of $n_0 \nu_1$ moles of A_1 and $n_0 \nu_2$ moles of A_2 at temperature T and pressure P occupies a volume V_0. When the reaction

$$\nu_1 A_1 + \nu_2 A_2 \rightleftharpoons \nu_3 A_3 + \nu_4 A_4$$

has come to equilibrium at the same T and p, the volume is V_e. Show that

$$\varepsilon_e = \frac{V_e - V_0}{V_0} \cdot \frac{\nu_1 + \nu_2}{\nu_3 + \nu_4 - \nu_1 - \nu_2}$$

16.3 The equilibrium constant of the reaction

$SO_3 \rightleftharpoons SO_2 + \frac{1}{2} O_2$ has the following values

T	800 K	900 K	1000 K	1105 K
K	0.0319	0.153	0.540	1.59

Determine the average heat of dissociation graphically.

16.4 (a) Show that

$$\Delta G = \Delta H + T \left(\frac{\partial \Delta G}{\partial T}\right)_p$$

(b) Show that

$$\Delta G = -\bar{R}\, T \ln \frac{x_3^{\nu_3} \cdot x_4^{\nu_4}}{x_1^{\nu_1} \cdot x_2^{\nu_2}}$$

where the x's are equilibrium values.

16.5 When 1 kgmol of HI dissociates according to the reaction

$$HI \rightleftharpoons \frac{1}{2} H_2 + \frac{1}{2} I_2$$

at $T = 675$ K, $K = 0.0174$, and $\Delta H = 5910$ kJ/kgmol.
Calculated $(\partial \varepsilon_e / \partial T)_p$ at this temperature.

16.6 (a) Prove that for a mixture of reacting ideal gases

$$\frac{d}{d\varepsilon} \ln \frac{x_3^{\nu_3} \cdot x_4^{\nu_4}}{x_1^{\nu_1} \cdot x_2^{\nu_2}} = \frac{n_0 + n_0'}{\Sigma\, n_K} \cdot \frac{1}{\psi}$$

where $$\frac{1}{\psi} = \frac{\nu_1^2}{x_1} + \frac{\nu_2^2}{x_2} + \frac{\nu_3^2}{x_3} + \frac{\nu_4^2}{x_4} - (\Delta\nu)^2$$

$$\nu = \nu_3 + \nu_4 - \nu_1 - \nu_2$$

(b) If we start with $n_0 \nu_1$ moles of A_1, $n_0 \nu_2$ moles of A_2, and no A_3 or A_4, show that

$$\psi = \frac{\varepsilon(1-\varepsilon)}{(\nu_1+\nu_2)(\nu_3+\nu_4)}$$

16.7 Prove that, for a mixture of reacting ideal gases in equilibrium

(a) $$\left(\frac{\partial V}{\partial p}\right)_T = -\frac{V}{p} - \frac{(n_0+n_0')\bar{R}T(\Delta\nu)^2}{p \cdot \frac{d}{d\varepsilon_e} \ln \frac{x_3^{\nu_3} \cdot x_4^{\nu_4}}{x_1^{\nu_1} \cdot x_2^{\nu_2}}}$$

(b) $$\left(\frac{\partial V}{\partial T}\right)_p = \frac{V}{T} + \frac{(n_0+n_0')\Delta\nu\,\Delta H}{pT \frac{d}{d\varepsilon_e} \ln \frac{x_3^{\nu_3} \cdot x_4^{\nu_4}}{x_1^{\nu_1} \cdot x_2^{\nu_2}}}$$

(c) $$\left(\frac{\partial p}{\partial T}\right)_{\varepsilon_e} = \frac{p\,\Delta H}{\bar{R}T^2\Delta\nu}$$

16.8 Liquid hydrazine (N_2H_4) and oxygen gas, both at 25°C, 0.1 MPa are fed to a rocket combustion chamber in the ratio of 0.5 kg O_2/kg N_2H_4. The heat transfer from the chamber to the surroundings is estimated to be 100 kJ/kg N_2H_4. Determine the temperature of the product, assuming only H_2O, H_2, and N_2 to be present. The enthalpy of the formation of $N_2H_4(1)$ is + 50, 417 kJ/kgmol. (2855 K)

If saturated liquid oxygen at 90K is used instead of 25°C oxygen gas in the combustion process, what will the temperature of the products be ?

16.9 Liquid ethanol (C_2H_5OH) is burned with 150% theoretical oxygen in a steady flow process. The reactants enter the combustion chamber at 25°C, and the products are cooled and leave at 65°C, 0.1 MPa. Calculate the heat transfer per kgmol of ethanol. The enthalpy of formation of C_2H_5OH (1) is −277,634 kJ/kgmol.

16.10 A small gas turbine uses $C_8H_{18}(1)$ for fuel and 400% theoretical air. The air and fuel enter at 25°C and the combustion products leave at 900 K. If the specific fuel consumption is 0.25 kg/s per MW output, determine the heat transfer from the engine per kgmol of fuel, assuming complete combustion. Ans. −48,830 kJ/kgmol.

16.11 Hydrogen peroxide (H_2O_2) enters a gas generator at the rate of 0.1 kg/s, and is decomposed to steam and oxygen. The resulting mixture is expanded through a turbine to atmospheric pressure, as shown in Fig. P. 16.11. Determine the power output of the turbine and the heat transfer rate in the gas generator. The enthalpy of formation of $H_2O_2(1)$ is −187,583 kJ/kgmol. Ans. 38.66 kW, −83.3 kW.

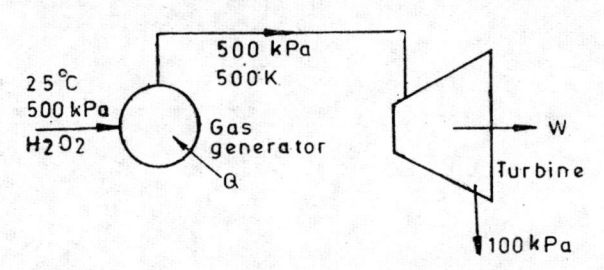

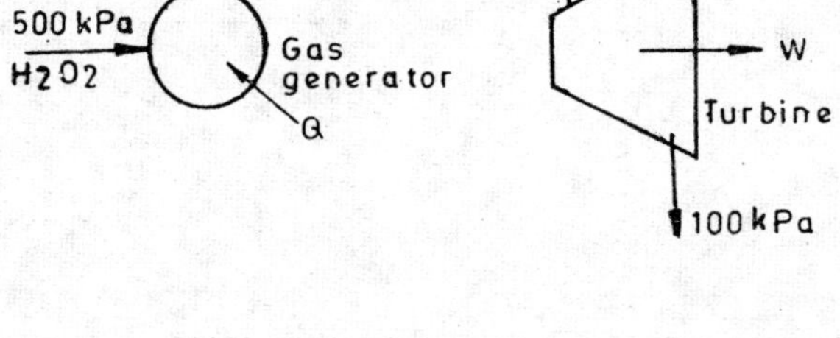

Fig. P. 16·11

16.12 An internal combustion engine burns liquid octane and uses 150% theoretical air. The air and fuel enter at 25°C, and the products leave the engine exhaust ports at 900 K. In the engine 80% of the carbon burns to CO_2 and the remainder burns to CO. The heat transfer from this engine is just equal to the work done by the engine. Determine (a) the power output of the engine if the engine burns 0.006 kg/s of fuel, and (b) the composition and the dew point of the products of combustion.

16.13 Gaseous butane at 25°C is mixed with air at 400 K and burned with 400% theoretical air. Determine the adiabatic flame temperature.

16.14 A mixture of butane and 150% theoretical air enters a combustion chamber at 25°C, 150 kPa, and the products of combustion leave at 1000K, 150 kPa. Determine the heat transfer from the combustion chamber and the irreversibility for the process.

16.15 The following data are taken from the test of a gas turbine:

Fuel—C_4H_{10}(g) 25°C, 0.1 MPa

Air—300% theoretical air at 25°C, 0.1 MPa

Velocity of inlet air—70 m/s

Velocity of products at exit—700 m/s

Temperature and pressure of products—900 K, 0.1 MPa

Assuming complete combustion, determine (a) the net heat transfer per kgmol of fuel, (b) the net increase of entropy per kgmol of fuel, and (c) the irreversibility for the process.

16.16 Calculate the equilibrium composition if argon gas is heated in an arc to 10,000 K, 1 kPa, assuming the plasma to consist of A, A^+, e^-. The equilibrium constant for the reaction

$$A \rightleftharpoons A^+ + e^-$$

at this temperature is 0.00042.

16.17 Methane is to be burned with oxygen in an adiabatic steady-flow process. Both enter the combustion chamber at 298K, and the products leave at 70 kPa. What percent of excess O_2 should be used if the flame temperature is to be 2800 K? Assume that some of the CO_2 formed dissociates to CO and O_2, such that the products leaving the chamber consist of CO_2, CO, H_2O, and O_2 at equilibrium.

17

Compressible Fluid Flow

A fluid is defined as a substance which continuously deforms under the action of shearing forces. Liquids and gases are termed as fluids. A fluid is said to be *incompressible* if its density (or specific volume) does not change (or changes very little) with a change in pressure (or temperature or velocity). Liquids are incompressible. A fluid is said to be *compressible* if its density changes with a change in pressure or temperature or velocity. Gases are compressible. The effect of compressibility must be considered in flow problems of gases. Thermodynamics is an essential tool in studying compressible flows, because of which Theodore von Karman suggested the name 'Aerothermodynamics' for the subject which studies the dynamics of compressible fluids.

The basic principles in compressible flow are :

(a) Conservation of mass (continuity equation)
(b) Newton's second law of motion (momentum principle)
(c) Conservation of energy (first law of thermodynamics)
(d) Second law of thermodynamics (entropy principle)
(e) Equation of state.

For the first two principles, the student is advised to consult a book on fluid mechanics, and the last three principles have been discussed in the earlier chapters of this book.

17.1 VELOCITY OF PRESSURE PULSE IN A FLUID

Let us consider an infinitesimal pressure wave initiated by a slight movement of a piston to the right (Fig. 17.1) in a pipe of uniform cross-section. The pressure wave front propagates steadily with a velocity *c,* which is known as the velocity of sound, sonic velocity or acoustic velocity. The fluid near the piston will have a slightly increased pressure and will be slightly more dense, than the fluid away from the piston.

To simplify the analysis, let the observer be assumed to travel with the wave front to the right with the velocity *c*. Fluid flows steadily from right to left and as it passes through the wave front, the velocity is reduced from c to $c\text{-}d\mathbf{V}$. At the same time, the pressure rises from p to $p + dp$ and the density from ρ to $\rho + d\rho$.

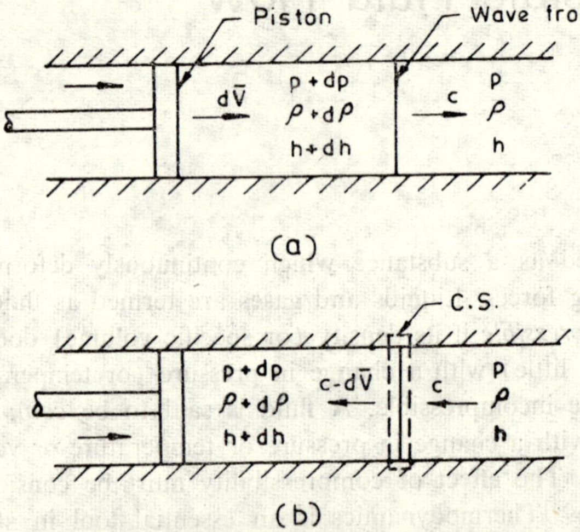

Fig. 17.1 Diagram illustrating sonic velocity. (a) Stationary observer, (b) Observer travelling with the wave front

The continuity equation for the control volume gives

$$\rho Ac = (\rho + d\rho)\, A\, (c - d\mathbf{V})$$

$$\rho c = \rho c - \rho\, d\mathbf{V} + c d\rho - d\rho \,.\, d\mathbf{V}$$

Neglecting the product $d\rho . d\mathbf{V}$, both being very small

$$\rho\, d\mathbf{V} = c d\rho \tag{17.1}$$

The momentum equation for the control volume gives

$$[p - (p + dp)]\ A = w\ [(c - d\mathbf{V}) - c]$$

$$- dp\ \mathrm{A} = \rho Ac\ (c - d\mathbf{V} - c)$$

$$dp = \rho\, c d\mathbf{V} \tag{17.2}$$

From equations (17.1) and (17.2)

$$\frac{dp}{c} = c d\rho$$

$$\therefore \qquad c = \sqrt{\frac{dp}{d\rho}}$$

Since the variations in pressure and temperature are negligibly small and the change of state is so fast as to be essentially adiabatic, and in the absence of any internal friction or viscosity, the process is reversible and isentropic. Hence, the sonic velocity is given by

$$c = \sqrt{\left(\frac{\partial p}{\partial \rho}\right)_s} \tag{17.3}$$

No fluid is truly incompressible, although liquids show little change in density. The velocity of sound in common liquids is of the order of 1650 m/s.

17.1.1 Velocity of Sound in an Ideal Gas

For an ideal gas, in an isentropic process

$$pv^{\gamma} = \text{constant}$$

or
$$\frac{p}{\rho^{\gamma}} = \text{constant}$$

By logarithmic differentiation (i.e., first taking logarithm and then differentiating)

$$\frac{dp}{p} = \gamma \frac{d\rho}{\rho} = 0$$

$$\therefore \quad \frac{dp}{d\rho} = \gamma \frac{p}{\rho}$$

Since
$$c^2 = \frac{dp}{d\rho} \text{ and } p = \rho RT$$

$$c^2 = \gamma RT$$

or
$$c = \sqrt{\gamma RT} \tag{17.4}$$

where R = characteristic gas constant

$$= \frac{\text{Universal gas constant}}{\text{Molecular weight}}$$

The lower the molecular weight of the fluid and higher the value of γ, the higher the sonic velocity at the same temperature. c is a thermodynamic property of the fluid.

17.1.2 Mach Number

The Mach number, M, is defined as the ratio of the actual velocity $\mathbf{V}$ to the sonic velocity c.

$$M = \frac{\mathbf{V}}{c}$$

When $M > 1$, the flow is supersonic, when $M < 1$, the flow is subsonic, and when $M = 1$, the flow is sonic.

17.2 STAGNATION PROPERTIES

The isentropic *stagnation state* is defined as the state a fluid in motion would reach if it were brought to rest isentropically in a steady-flow, adiabatic, zero work output device. This is a reference state in compressible fluid flow and is commonly designated with the subscript zero. The stagnation enthalpy h_0 (Fig. 17.2) is related to the enthalpy and velocity of the moving fluid by

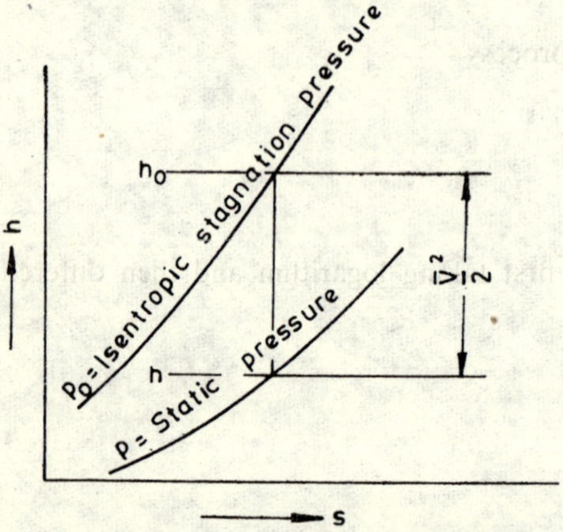

Fig. 17.2 Stagnation state

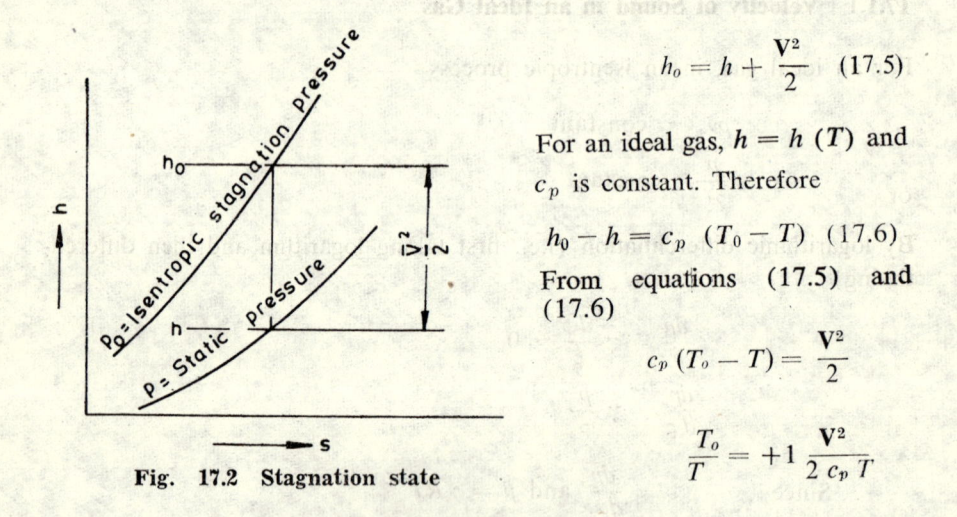

$$h_0 = h + \frac{\mathbf{V}^2}{2} \quad (17.5)$$

For an ideal gas, $h = h(T)$ and c_p is constant. Therefore

$$h_0 - h = c_p \ (T_0 - T) \quad (17.6)$$

From equations (17.5) and (17.6)

$$c_p \, (T_0 - T) = \frac{\mathbf{V}^2}{2}$$

$$\frac{T_0}{T} = +1 \, \frac{\mathbf{V}^2}{2 \, c_p \, T}$$

The properties without any subscript denote static properties.

Since $$c_p = \frac{\gamma R}{\gamma - 1},$$

$$\frac{T_0}{T} = 1 + \frac{\mathbf{V}^2 \, (\gamma - 1)}{2 \, \gamma \, RT}$$

Using equation (17.4) and the Mach number

$$\frac{T_0}{T} = 1 + \frac{\gamma - 1}{2} M^2 \quad (17.7)$$

The stagnation pressure p_0 is related to the Mach number and static pressure in the case of an ideal gas by the following equation

$$\frac{p_0}{p} = \left(\frac{T_0}{T}\right)^{\gamma/(\gamma-1)} = \left(1 + \frac{\gamma - 1}{2} M^2\right)^{\gamma/(\gamma-1)} \quad (17.8)$$

Similarly $$\frac{\rho_0}{\rho} = \left(1 + \frac{\gamma - 1}{2} M^2\right)^{1/(\gamma-1)} \quad (17.8a)$$

17.3 ONE DIMENSIONAL STEADY ISENTROPIC FLOW

From a one dimensional point of view, the three most common factors which tend to produce continuous changes in the state of a flowing stream are

(a) Changes in cross-sectional area
(b) Wall friction
(c) Energy effects, such as external heat exchange, combustion, etc.

A study will first be made of the effects of area change in the absence of friction and energy effects. The process, which has been called isentropic flow, might aptly be termed as simple area change.

A nozzle is any duct which increases the kinetic energy of a fluid at the expense of its pressure. A diffuser is a passage through which a fluid loses kinetic energy and gains pressure. The same duct or passage may be either a nozzle or diffuser depending upon the end conditions across it. A nozzle or diffuser with both a converging and a diverging section is shown in Fig. 17.3. The minimum section is known as the throat. For the control volume shown in Fig. 17.3, since stagnation enthalpy and stagnation temperature do not change in adiabatic flow

$$h_o = h + \frac{\mathbf{V}^2}{2}$$

$$dh = -\mathbf{V}\,d\mathbf{V} \tag{17.9}$$

From the property relation

$$Tds = dh - vdp$$

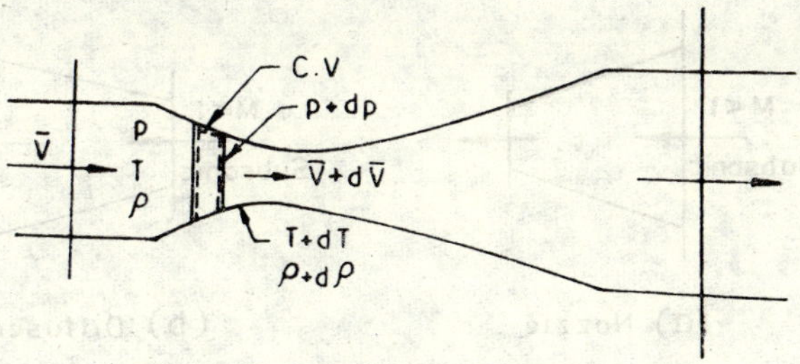

Fig. 17.3 Reversible adiabatic flow through a nozzle

For isentropic flow

$$dh = \frac{dp}{\rho} \tag{17.10}$$

From equations (17.9) and (17.10)

$$dp = -\rho\mathbf{V}\,d\mathbf{V} \tag{17.11}$$

or

$$\frac{dp}{d\mathbf{V}} < 0 \tag{17.12}$$

As pressure decreases, velocity increases, and vice versa.

The continuity equation gives

$$w = \rho A \mathbf{V}$$

By logarithmic differentiation

$$\frac{d\rho}{\rho} + \frac{dA}{A} + \frac{d\mathbf{V}}{\mathbf{V}} = 0$$

$$\therefore \quad \frac{dA}{A} = -\frac{d\mathbf{V}}{\mathbf{V}} - \frac{d\rho}{\rho}$$

Substituting from equation (17.11)

$$\frac{dA}{A} = \frac{dp}{\rho \mathbf{V}^2} - \frac{d\rho}{\rho} = \frac{dp}{\rho \mathbf{V}^2}\left[1 - \mathbf{V}^2 \frac{d\rho}{dp}\right]$$

or
$$\frac{dA}{A} = \frac{dp}{\rho \mathbf{V}^2}(1 - M^2) \quad (17.13)$$

Also
$$\frac{dA}{A} = (M^2 - 1)\frac{d\mathbf{V}}{\mathbf{V}} \quad (17.14)$$

When $M < 1$, i.e., the inlet velocity is subsonic, as flow area A decreases, the pressure decreases and velocity increases, and when flow area A increases, pressure increases and velocity decreases. So for subsonic flow, a convergent passage becomes a nozzle (Fig. 17.4a) and a divergent passage becomes a diffuser (Fig. 17.4b).

When $M > 1$, i.e., when the inlet velocity is supersonic, as flow area A decreases, pressure increases and velocity decreases, and as flow area A increases, pressure decreases and velocity increases. So for supersonic flow, a convergent passage is a diffuser (Fig. 17.4c) and a divergent passage is a nozzle (Fig. 17.4d).

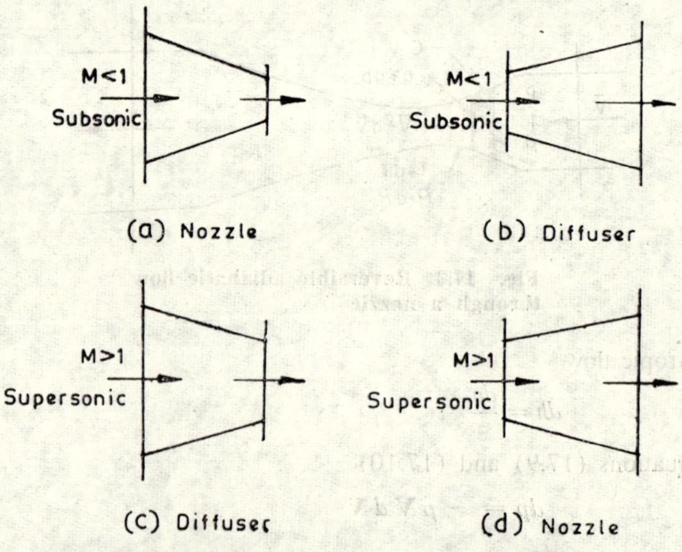

Fig. 17.4 Effect of area change in subsonic and supersonic flow

17.4 CRITICAL PROPERTIES—CHOKING IN ISENTROPIC FLOW

Let us consider the mass rate of flow of an ideal gas through a nozzle. The flow is isentropic.

$$w = \rho A \mathbf{V}$$

or

$$\frac{w}{A} = \frac{p}{RT} \cdot cM = \frac{p}{RT}\sqrt{\gamma RT} \, . \, M$$

$$= \frac{p}{p_o} \cdot p_0 \sqrt{\frac{T_o}{T}} \cdot \sqrt{\frac{1}{T_o}} \sqrt{\frac{\gamma}{R}} \cdot M$$

$$= \left(\frac{T_o}{T}\right)^{-\gamma/\gamma-1} \left(\frac{T_o}{T}\right)^{\frac{1}{2}} \cdot \frac{p_o}{\sqrt{T_o}} \sqrt{\frac{\gamma}{R}} \cdot M$$

$$= \sqrt{\frac{\gamma}{R}} \frac{p_o M}{\sqrt{T_o}} \cdot \frac{1}{\left(1 + \frac{\gamma - 1}{2} M^2\right)^{(\gamma+1)/2(\gamma-1)}} \tag{17.15}$$

Since p_0, T_0, γ and R are constant, the discharge per unit area w/A is a function of M only. There is a particular value of M when w/A is a maximum. Differentiating equation (17.15) with respect to M and equating it to zero,

$$\frac{d(w/A)}{dM} = \sqrt{\frac{\gamma}{R}} \cdot \frac{p_o}{\sqrt{T_o}} \frac{1}{\left(1 + \frac{\gamma-1}{2} M^2\right)^{(\gamma+1)/2(\gamma-1)}}$$

$$+ \sqrt{\frac{\gamma}{R}} \cdot \frac{p_o M}{\sqrt{T_o}} \left[-\frac{\gamma+1}{2(\gamma-1)}\right] \left(1 + \frac{\gamma-1}{2} M^2\right)^{-(\gamma+1)/2(\gamma-1)-1} \left(\frac{\gamma-1}{2} \cdot 2M\right) = 0$$

or

$$1 - \frac{M^2(\gamma+1)}{2\left(1 + \frac{\gamma-1}{2} M^2\right)} = 0$$

$$M^2(\gamma + 1) = 2 + (\gamma - 1) M^2$$

$$M^2 = 1$$

or

$$M = 1$$

So, the discharge w/A is maximum when $M = 1$.

Since $\mathbf{V} = c\,M = \sqrt{\gamma RT}.\ M$, by logarithmic differentiation

$$\therefore \qquad \frac{d\mathbf{V}}{\mathbf{V}} = \frac{dM}{M} + \frac{1}{2}\frac{dT}{T} \tag{17.16}$$

and

$$\frac{T}{T_o} = \left(1 + \frac{\gamma-1}{2} M^2\right)^{-1}$$

by logarithmic differentiation

$$\therefore \quad \frac{dT}{T} = -\frac{(\gamma-1)\,M^2}{1+\frac{\gamma-1}{2}M^2}\,\frac{dM}{M} \tag{17.17}$$

From equations (17.16) and (17.17)

$$\therefore \quad \frac{d\mathbf{V}}{\mathbf{V}} = \frac{1}{1+\frac{\gamma-1}{1}M^2}\cdot\frac{dM}{M} \tag{17.18}$$

From equations (17.14) and (17.18)

$$\frac{dA}{A}\left(\frac{1}{M-1}\right) = \frac{1}{1+\frac{\gamma-1}{2}M^2}\cdot\frac{aM}{M}$$

$$\therefore \quad \frac{dA}{A} = \frac{(M^2-1)\,dM}{M\left(1+\frac{\gamma-1}{2}M^2\right)} \tag{17.19}$$

By substituting $M = 1$ in any one of the equations (17.13), (17.14) or (17.19), $dA = 0$ or A = constant. *So $M=1$ occurs only at the throat and nowhere else, and this happens only when the discharge is the maximum.*

If the convergent-divergent duct acts as a nozzle, in the divergent part also, the pressure will fall continuously to yield a continuous rise in velocity. The velocity of the gas is subsonic before the throat, becomes sonic at the throat, and then becomes supersonic till its exit in isentropic flow, provided the exhaust pressure is low enoguh. The reverse situation prevails when the inlet velocity is supersonic. The whole duct then becomes a diffuser. The transition from subsonic flow to supersonic flow and vice versa can occur only when the compressible fluid flows through a throat and the exit pressure is maintained at the appropriate value.

When $M = 1$, the discharge is maximum and the nozzle is said to be choked. The properties at the throat are then termed as critical properties and these are designed by a superscript asterisk (*). Substituting $M = 1$ in equation (17.7)

$$\frac{T_0}{T^*} = 1 + \frac{\gamma-1}{2}M^2 = 1 + \frac{\gamma-1}{2} = \frac{\gamma+1}{2}$$

$$\therefore \quad \frac{T^*}{T_0} = \frac{2}{\gamma+1} \tag{17.20}$$

The *critical pressure ratio* p^*/p_0 is then given by

$$\frac{p^*}{p_0} = \left(\frac{2}{\gamma+1}\right)^{\gamma/\gamma-1} \tag{17.20a}$$

For diatomic gases, like air, $\gamma = 1.4$

$$\therefore \quad \frac{p^*}{p_0} = \left(\frac{2}{2.4}\right)^{1.4/1.4} = 0.528$$

The critical pressure ratio for air is 0.528.
For superheated steam, $\gamma = 1.3$ and p^*/p_0 is 0.546.

$$\text{For air, } \frac{T^*}{T_0} = 0.833$$

and

$$\frac{\rho^*}{\rho_0} = \left(\frac{2}{\gamma+1}\right)^{1/(\gamma-1)} = 0.634$$

By substituting $M = 1$ in equation (17.15)

$$\frac{w}{A^*} = \sqrt{\frac{\gamma}{R}} \cdot \frac{p_0}{\sqrt{T_0}} \cdot \frac{1}{\left(\frac{\gamma+1}{2}\right)^{(\gamma+1)/2(\gamma-1)}} \tag{17.21}$$

Dividing equation (17.21) by equation (17.15)

$$\frac{A}{A^*} = \left[\left(\frac{2}{\gamma+1}\right)\left(1+\frac{\gamma-1}{2}M^2\right)\right]^{(\gamma+1)/2(\gamma-1)} \frac{1}{M} \tag{17.22}$$

The area ratio A/A^* is the ratio of the area at the point where the Mach number is M to the throat area A^*. Figure 17.5 shows a plot of A/A^* vs. M, which shows that a subsonic nozzle is converging and a supersonic nozzle is diverging.

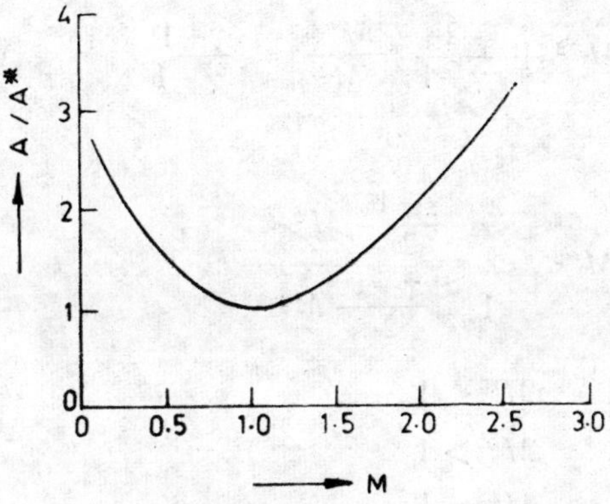

Fig. 17.5 Area ratio as a function of Mach number in an isentropic nozzle

17.4.1 Dimensionless Velocity, M*

Since the Mach number M is not proportional to the velocity alone and it tends towards infinity at high speeds, one more dimensionless parameter M^* is often used, which is defined as

$$M^* = \frac{\mathbf{V}}{c^*} = \frac{\mathbf{V}}{\mathbf{V}^*} \tag{17.23}$$

where $c^* = \sqrt{\gamma RT^*} = \mathbf{V}^*$

$$\therefore \quad M^{*2} = \frac{V^2}{c^{*2}} = \frac{V^2}{c^2} \cdot \frac{c^2}{c^{*2}}$$

$$= M^2 \cdot \frac{c^2}{c^{*2}}$$

For the adiabatic flow of an ideal gas

$$\frac{V^2}{2} + c_p T = \text{constant} = h_0 = c_p T_0$$

$$\frac{V^2}{2} + \frac{\gamma RT}{\gamma+1} = \frac{\gamma R T_0}{\gamma-1}$$

$$\frac{V^2}{2} + \frac{c^2}{\gamma-1} = \frac{c_0^2}{\gamma-1}$$

Since $\frac{c_0}{c^*} = \frac{\sqrt{\gamma R T_0}}{\sqrt{\gamma R T^*}} = \sqrt{\frac{T_0}{T^*}} = \sqrt{\left(\frac{\gamma+1}{2}\right)}$

$$\therefore \quad \frac{V^2}{2} + \frac{c^2}{\gamma-1} = \frac{\gamma+1}{2} c^{*2} \frac{1}{\gamma-1}$$

$$\frac{V^2}{c^{*2}} + \frac{2}{\gamma-1} \frac{c^2}{c^{*2}} = \frac{\gamma+1}{\gamma-1}$$

$$M^{*2} + \frac{2}{\gamma-1} \frac{M^{*2}}{M^2} = \frac{\gamma+1}{\gamma-1}$$

On simplification

$$M^{*2} = \frac{\frac{\gamma+1}{2} M^2}{1 + \frac{\gamma-1}{2} M^2} \qquad (17.24)$$

When $M < 1$, $M^* < 1$

When $M > 1$, $M^* > 1$

When $M = 1$, $M^* = 1$

When $M = 0$, $M^* = 0$

When $M = \infty$, $M^* = \sqrt{\frac{\gamma+1}{\gamma-1}}$

17.4.2 Pressure Distribution and Chocking in a Nozzle

Let us first consider a convergent nozzle as shown in Fig. 17.6, which also shows the pressure ratio p/p_0 along the length of the nozzle. The inlet condition of the fluid is the stagnation state at p_0, T_0, which is assumed to be constant. The pressure at the exit plane of the nozzle is denoted by p_E and the back pressure is p_B which can be varied by the valve. As the back pressure p_B is decreased, the mass flow rate w and the exit plane pressure p_E/p_0 vary, as shown in Fig. 17.7.

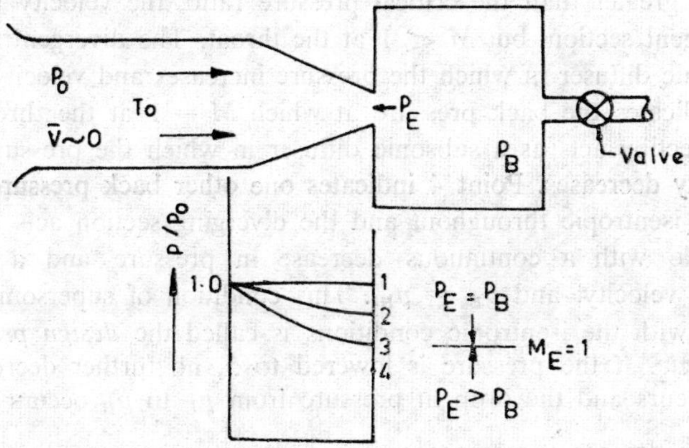

Fig. 17.6 Compressible flow through a converging nozzle

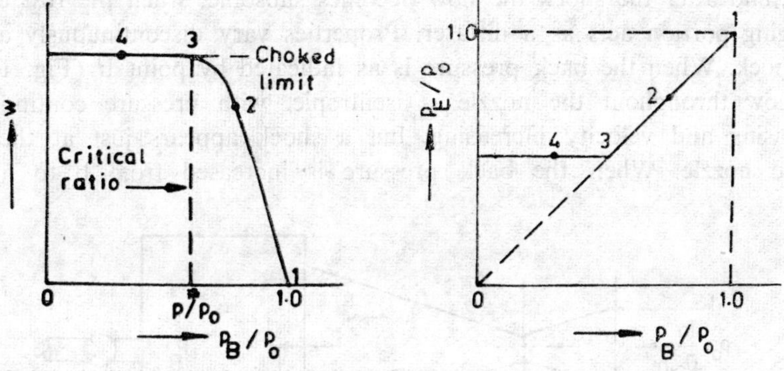

Fig. 17.7 Mass flow rate and exit pressure as a function of back pressure in a converging nozzle

When $p_B/p_0 = 1$, there is no flow, and $p_E/p_0 = 1$, as designated by point 1. If the back pressure p_B is now decreased to a value as designated by point 2, such that p_B/p_0 is greater than the critical pressure ratio, the mass flow rate has a certain value, and $p_E = p_B$. The exit Mach number M_E is less than 1. Next the back pressure p_B is lowered to the critical pressure, denoted by point 3. The exit Mach number M_E is now unity, and $p_E = p_B$. When p_B is increased below the critical pressure, indicated by point 4, there is no increase in the mass flow rate, and p_E remains constant at a value equal to critical pressure, and $M_E = 1$. The drop in pressure from P_E to p_B occurs outside the nozzle exit. This is the choking limit which means that for given stagnation conditions the nozzle is passing the maximum possible mass flow.

Let us next consider a convergent-divergent nozzle, as shown in Fig. 17.8. Point 1 designates the condition when $p_B = p_0$, and there is no flow. When p_B is lowered to the pressure denoted by point 2, so that p_B/p_0 is less than 1 but greater than the critical pressure ratio, the velocity increases in the convergent section, but $M < 1$ at the throat. The divergent section acts as a subsonic diffuser in which the pressure increases and velocity decreases. Point 3 indicates the back pressure at which $M = 1$ at the throat, but the diverging section acts as a subsonic diffuser in which the pressure increases and velocity decreases. Point 4 indicates one other back pressure for which the flow is isentropic throughout and the diverging section acts as a supersonic nozzle with a continuous decrease in pressure and a continuous increase in velocity, and $p_{E4} = p_{B4}$. This condition of supersonic flow past the throat with the isentropic conditions is called the *design pressure ratio of the nozzle*. If the pressure is lowered to 5, no further decrease in exit pressure occurs and the drop in pressure from p_E to p_B occurs outside the nozzle.

Between the back pressures designated by points 3 and 4, flow is not isentropic in the diverging part, and it is accompanied by a highly irreversible phenomenon, known as *shocks*. Shocks occur only when the flow is supersonic, and after the shock the flow becomes subsonic, when the rest of the diverging portion acts as a diffuser. Properties vary discontinuously across the shock. When the back pressure is as indicated by point b (Fig. 17.8), the flow throughout the nozzle is isentropic, with pressure continuously decreasing and velocity increasing, but a shock appears just at the exit of the nozzle. When the back pressure is increased from b to a, the

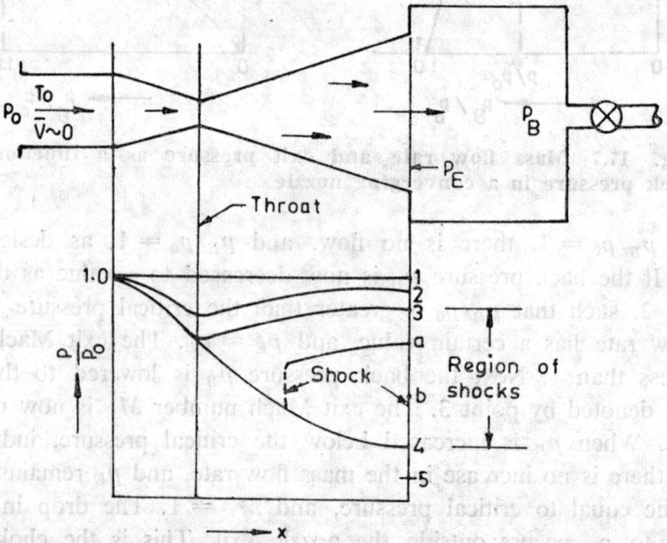

Fig. 17.8 Compressible flow through a convergent-divergent nozzle

shock moves upstream, as indicated. When the back pressure is further increased, the shock moves further upstream and disappears at the nozzle throat where the back pressure corresponds to 3. Since flow throughout is subsonic, no shock is possible.

17.4.3 Gas Tables for Isentropic Flow

The values of M^*, A/A^*, p/p_0, ρ/ρ_0, and T/T_0 computed for an ideal gas having $\gamma = 1.4$ for various values of Mach number M from the equations (17.24), (17.22), (17.8), and (17.7) respectively are given in Table D.1 in the appendix. These may be used with advantage for computations of problems of isentropic flow.

17.5 NORMAL SHOCKS

Shock waves are highly localized irreversibilities in the flow. Within the distance of a mean free path of a molecule, the flow passes from a supersonic to a subsonic state, the velocity decreases abruptly, and the pressure rises sharply. Figure 17.9 shows a control surface that includes such a normal

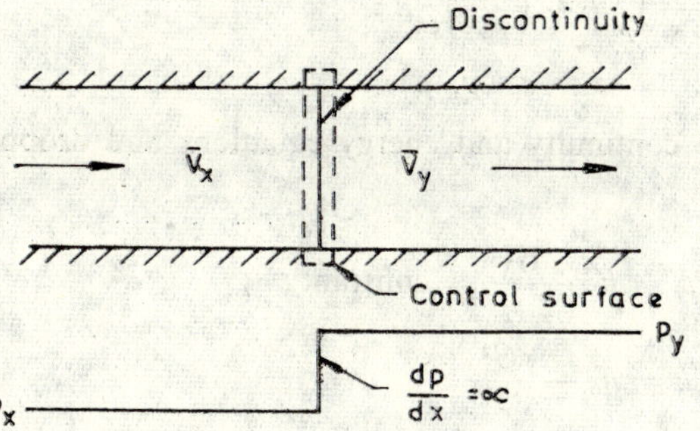

Fig. 17.9 One dimensional normal shock

shock. Normal shocks may be treated as shock waves perpendicular to the flow. The fluid is assumed to be in thermodynamic equilibrium upstream and downstream of the shock, the properties of which are designated by the subscripts x and y respectively. For the control surface

Continuity equation

$$\frac{w}{A} = \rho_x \mathbf{V}_x = \rho_y \mathbf{V}_y = G \tag{17.25}$$

where G is the mass velocity (kg/m²s).

Momentum equation

$$p_x - p_y = \frac{w}{A} (\mathbf{V}_y - \mathbf{V}_x)$$

$$= \rho_y \mathbf{V}_y^2 - \rho_x \mathbf{V}_x^2$$

$$\therefore \quad p_x + \rho_x \mathbf{V}_x^2 = p_y + \rho_y \mathbf{V}_y^2 \tag{17.26}$$

or $$F_x = F_y \tag{17.26a}$$

where $F = pA + \rho A \mathbf{V}^2$ is known as the *impulse function.*

Energy equation

$$h_x + \frac{\mathbf{V}_x^2}{2} = h_y + \frac{\mathbf{V}_y^2}{2} = h_{0x} = h_{0y} = h_0 \tag{17.27}$$

where h_0 is the stagnation enthalpy on both sides of the shock.

Second law

$$s_y - s_x \geqslant 0 \tag{17.28}$$

The equation of state of the fluid may be written implicity in the form

$$h = h(s, \rho) \tag{17.29a}$$

$$dh = \left(\frac{\partial h}{\partial s}\right)_\rho ds + \left(\frac{\partial h}{\partial \rho}\right)_s d\rho$$

or $$s = s(p, \rho) \tag{17.29b}$$

Combining the continuity and energy equations and dropping the subscripts x and y

$$h_o = h + \frac{G^2}{2\rho^2} = \text{constant} \tag{17.30}$$

or $$h = h_0 - \frac{G^2}{2\rho^2}$$

Given the values of G and h_0, this equation relates h and ρ (Fig. 17.10). The line representing the locus of points with the same mass velocity and stagnation enthalpy is called a *Fanno line*. The end states of the normal shock must lie on the Fanno line.

The Fanno line may also be represented in the h-s diagram. The upstream properties h_x, ρ_x, p_x, and $\mathbf{V}_x$ are known. Let a particular value of $\mathbf{V}_y$ be chosen, then ρ_y may be computed from the continuity equation (17.25), h_y from the energy equation (17.27), and s_y from equation (17.29a). By repeating the calculation for various values of $\mathbf{V}_y$, the Fanno line may easily be constructed (Fig. 17.11). Since the momentum equation has not been introduced, the Fanno line represents states with the same mass velocity and stagnation enthalpy, but not the same value of the impulse function.

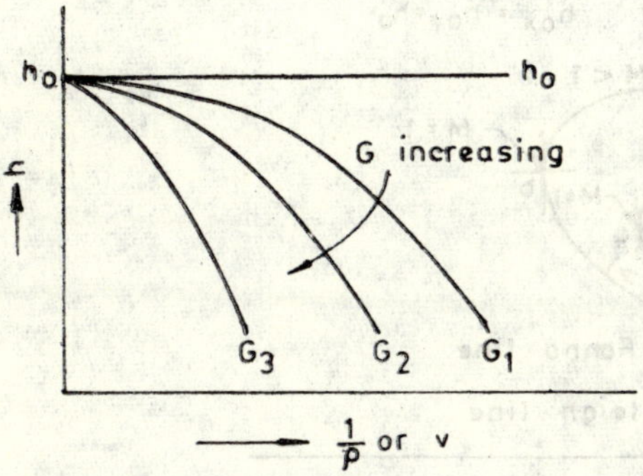

Fig. 17.10 Fanno line on h-$1/\rho$ coordinate

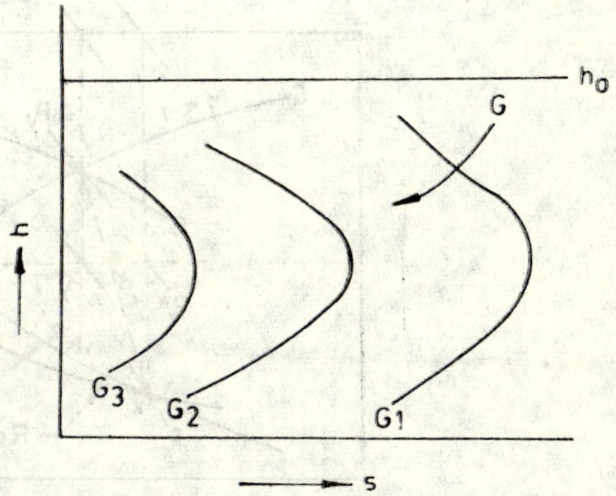

Fig. 17.11 Fanno lines on h-s diagram

Adiabatic flow in a constant area duct with friction, in a one dimensional model, has both constant G and constant h_0, and hence must follow a Fanno line.

Let us next consider the locus of states which are defined by the continuity equation (17.25), the momentum equation (17.26) and the equation of state (17.29). The impulse function in this case becomes

$$F = pA + \rho A \mathbf{V}^2$$

or the *impulse pressure* I is given by

$$I = \frac{F}{A} = p + \rho \mathbf{V}^2 = p + \frac{G^2}{\rho} \tag{17.31}$$

Given the values for I and G, the equation relates p and ρ. The line representing the locus of states with the same impulse pressure and mass velocity is called the *Rayleigh line*. The end states of the normal shock must lie on the Rayleigh line, since $I_x = I_y$, and $G_x = G_y$.

The Rayleigh line may also be drawn on the h-s plot. The properties upstream of the shock are all known. The downstream properties are to be known. Let a particular value of $\mathbf{V}_y$ be chosen. Then ρ_y may be computed from the continuity equation (17.25) and p_y from the momentum equation (17.26), and s_y from equation (17.29b) may be found. By repeating the calculations for various values of $\mathbf{V}_y$, the locus of possible states reachable from, say, state x may be plotted, and this is the Rayleigh line (Fig. 17.12).

Since the normal shock must satisfy equations (17.25), (17.26), (17.27), and (17.29) simultaneously, the end states x and y of the shock must lie at the intersections of the Fanno line and the Rayleigh line for the same G (Fig. 17.12).

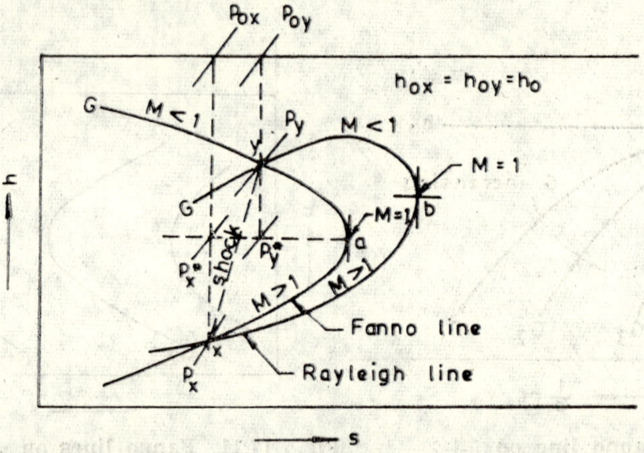

Fig. 17.12 End states of a normal shock on h-s diagram

The Rayleigh line is also a model for flow in a constant area duct with heat transfer, but without friction.

For an infinitesimal process in the neighbourhood of the point of maximum entropy (point a) on the Fanno line, from the energy equation

$$dh + \mathbf{V}\, d\mathbf{V} = 0 \tag{17.32}$$

and from the continuity equation

$$\rho\, d\mathbf{V} + \mathbf{V}\, d\rho = 0 \tag{17.33}$$

From the thermodynamic relation

$$Tds = dh - vdp$$

or

$$dh = \frac{dp}{\rho} \tag{17.34}$$

By combining equations (17.32), (17.33), and (17.34)

$$\frac{dp}{\rho} + \mathbf{V}\left(-\frac{\mathbf{V}d\rho}{\rho}\right) = 0$$

$$\therefore \quad \frac{dp}{d\rho} = \mathbf{V}^2$$

or $\mathbf{V} = \sqrt{\left(\frac{\partial p}{\partial \rho}\right)_s}$, since the flow is isentropic.

This is the local sound velocity.

So the Mach number is unity at point a. Similarly, it can be shown that at point b on the Rayleigh line, $M = 1$. It may also be shown that the upper branches of the Fanno and Rayleigh lines represent subsonic speeds ($M < 1$) and the lower branches represent supersonic speeds ($M > 1$).

The normal shock always involves a change from supersonic to subsonic speed with a consequent pressure rise, and never the reverse. By the second law, entropy always increases during irreversible adiabatic change.

17.5.1 Normal Shock in an Ideal Gas

The energy equation for an ideal gas across the shock becomes

$$c_p T_x + \frac{\mathbf{V}_x^2}{2} = c_p T_y + \frac{\mathbf{V}_y^2}{2} = c_p T_0$$

Now $h_{0x} = h_{0y} = h_0$, and $T_{0x} = T_{0y} = T_0$

ubstituting $c_p = \frac{\gamma R}{\gamma - 1}$, $c_x = \sqrt{\gamma R T_x}$, and $c_y = \sqrt{\gamma R T_0}$

$$\frac{T_0}{T_x} = 1 + \frac{\gamma-1}{2} M_x^2, \text{ and } \frac{T_0}{T_y} = 1 + \frac{\gamma-1}{2} M_y^2$$

$$\therefore \quad \frac{T_y}{T_x} = \frac{1 + \frac{\gamma-1}{2} M_x^2}{1 + \frac{\gamma-1}{2} M_y^2} \tag{17.35}$$

Again $\rho_x \mathbf{V}_x = \rho_y \mathbf{V}_y$

$$\therefore \quad \frac{p_x}{R T_x} \mathbf{V}_x = \frac{p_y}{R T_y} \cdot \mathbf{V}_y$$

or $$\frac{T_y}{T_x} = \frac{p_y}{p_x} \cdot \frac{\mathbf{V}_y}{\mathbf{V}_x} = \frac{p_y}{p_x} \cdot \frac{M_y c_y}{M_x c_x} = \frac{p_y}{p_x} \cdot \frac{M_y}{M_x} \sqrt{\frac{T_y}{T_x}}$$

or $$\frac{T_y}{T_x} = \left(\frac{p_y}{p_x}\right)^2 \left(\frac{M_y}{M_x}\right)^2 \tag{17.36}$$

From equations (17.35) and (17.36)

$$\frac{p_y}{p_x} = \frac{M_x}{M_y} \sqrt{\frac{1 + \frac{\gamma-1}{2} M_x^2}{1 + \frac{\gamma-1}{2} M_y^2}} \tag{17.37}$$

Also

$$p_x + \rho_x \mathbf{V}_x^2 = p_y + \rho_y \mathbf{V}_y^2$$

or $$p_x + \frac{\gamma p_x}{\gamma R T_x} \mathbf{V}_x^2 = p_y + \frac{\gamma p_y}{R T_y \gamma} \mathbf{V}_y^2$$

$$p_x (1 + \gamma M_x^2) = p_y (1 + \gamma M_y^2)$$

$$\therefore \quad \frac{p_y}{p_x} = \frac{1 + \gamma M_x^2}{1 + \gamma M_y^2} \tag{17.38}$$

From equations (17.37) and (17.38), upon rearrangement

$$M_y^2 = \frac{M_x^2 + \frac{2}{\gamma-1}}{\frac{2\gamma}{\gamma-1}M_x^2 - 1} \tag{17.39}$$

Then from equations (17.37), (17.38), and (17.39)

$$\frac{p_y}{p_x} = \frac{2\gamma}{\gamma+1} M_x^2 - \frac{\gamma-1}{\gamma+1} \tag{17.40}$$

and from equations (17.35) and (17.39)

$$\frac{T_y}{T_x} = \frac{\left(1 + \frac{\gamma-1}{2} M_x^2\right)\left(\frac{2\gamma}{\gamma-1} M_x^2 - 1\right)}{\frac{(\gamma+1)^2}{2(\gamma-1)} M_x^2} \tag{17.41}$$

Then

$$\frac{\rho_y}{\rho_x} = \frac{p_y}{T_y} \cdot \frac{T_x}{P_x}$$

$$= \frac{\left(\frac{2\gamma}{\gamma+1} M_x^2 - \frac{\gamma-1}{\gamma+1}\right)\left(\frac{(\gamma+1)^2}{2(\gamma-1)} M_x^2\right)}{\left(1 + \frac{\gamma-1}{2} M_x^2\right)\left(\frac{2\gamma}{\gamma-1} M_p^2 - 1\right)} \tag{17.42}$$

The ratio of the stagnation pressures is a measure of the irreversibility of the shock process. Now

$$\frac{p_{oy}}{p_{ox}} = \frac{p_{oy}}{p_y} \cdot \frac{p_y}{p_x} \cdot \frac{p_x}{p_{ox}}$$

and

$$\frac{p_{oy}}{p_y} = \left(1 + \frac{\gamma-1}{2} M_y^2\right)^{\gamma/(\gamma-1)}$$

$$\frac{p_{ox}}{p_x} = \left(1 + \frac{\gamma-1}{2} M_x^2\right)^{\gamma/(\gamma-1)}$$

$$\therefore \quad \frac{p_{oy}}{p_{ox}} = \left\{\frac{\frac{\gamma+1}{2} M_x^2}{1 + \frac{\gamma-1}{2} M_x^2}\right\}^{\gamma/(\gamma-1)} \Big/ \left[\frac{2\gamma}{\gamma+1} M_x^2 - \frac{\gamma-1}{\gamma+1}\right]^{1/(\gamma-1)} \tag{17.43}$$

$$\frac{p_{oy}}{p_x} = \frac{p_{oy}}{p_y} \cdot \frac{p_y}{p_p}$$

$$= \left(1 + \frac{\gamma-1}{2} M_y^2\right)^{\gamma/(\gamma-1)} \left[\frac{2\gamma}{\gamma+1} M_x^2 - \frac{\gamma-1}{\gamma+1}\right] \tag{17.44}$$

For different values of M_x, and for $\gamma = 1.4$, the values of M_y, p_y/p_x, T_y/T_x ρ_y/ρ_x, p_{oy}/p_{xx}, and p_{oy}/p_x, computed from equations (17.39), (17.40),

(17.42), (17.43), and (17.44) respectively, are given in Table D.2 in the appendix.

To evaluate the entropy change across the shock, for an ideal gas

$$ds = c_p \frac{dT}{T} - R \frac{dp}{p}$$

$$\therefore \quad s_y - s_x = c_p \ln \frac{T_y}{T_x} - R \ln \frac{p_y}{p_x}$$

$$= c_p \left[\ln \frac{T_y}{T_x} - \ln \left(\frac{p_y}{p_x} \right)^{(\gamma-1)/\gamma} \right]$$

since $$R = \frac{c_p (\gamma - 1)}{\gamma}$$

$$\therefore \quad s_y - s_x = c_p \ln \frac{T_y/T_x}{(p_y/p_x)^{(\gamma-1)/\gamma}}$$

$$= c_p \ln \frac{T_{oy}/T_{ox}}{(p_{oy}/p_{ox})^{(\gamma-1)/\gamma}} = - R \ln \frac{p_{oy}}{p_{ox}} \qquad (17.45)$$

The strength of a shock wave, P, is defined as the ratio of the pressure increase to the initial pressure, i.e.

$$P = \frac{p_y - p_x}{p_x} = \frac{p_y}{p_x} - 1$$

Substituting from equation (17.40)

$$P = \frac{2\gamma}{\gamma+1} M_x^2 - \frac{\gamma-1}{\gamma+1} - 1$$

$$= \frac{2\gamma}{\gamma+1} (M_x^2 - 1) \qquad (17.46)$$

17.6 ADIABATIC FLOW WITH FRICTION AND DIABATIC FLOW WITHOUT FRICTION

It was stated that the Fanno line representing the states of constant mass velocity and constant stagnation enthalpy also holds for adiabatic flow in a constant area duct with friction. For adiabatic flow the entropy must increase in the flow direction. Hence a Fanno process must follow its Fanno line to the right, as shown in Fig. 17.13. Since friction will tend to move the state of the fluid to the right on the Fanno line, the Mach number of subsonic flows increases in the downstream section (Fig. 17.13), and in supersonic flows friction acts to decrease the Mach number. Hence, friction tends to drive the flow to the sonic point.

Let us consider a short duct with a given h_0 and G, i.e., a given Fanno line, with a given subsonic exit Mach number represented by point 1 in Fig. 17.13. If some more length is added to the duct, the new exit Mach

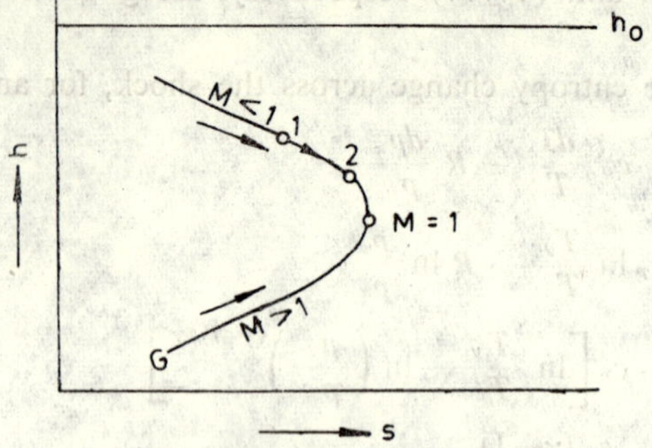

Fig. 17.13 A Fanno line on h-s **plot**

number will be increased due to friction, as represented by, say, point 2. The length of the duct may be further increased till the exit Mach number is unity. Any further increase in duct length is not possible without incurring a reduction in the mass flow rate. Hence subsonic flows can be choked by friction. There is a maximum flow rate that can be passed by a pipe with given stagnation conditions. Choking also occurs in supersonic flow with friction, usually in a very short length. It is thus difficult to use such flows in applications.

Diabatic flows, i.e., flows with heating or cooling, in a constant area duct, in the absence of friction, can be treated by the Rayleigh process (Fig. 17.14).

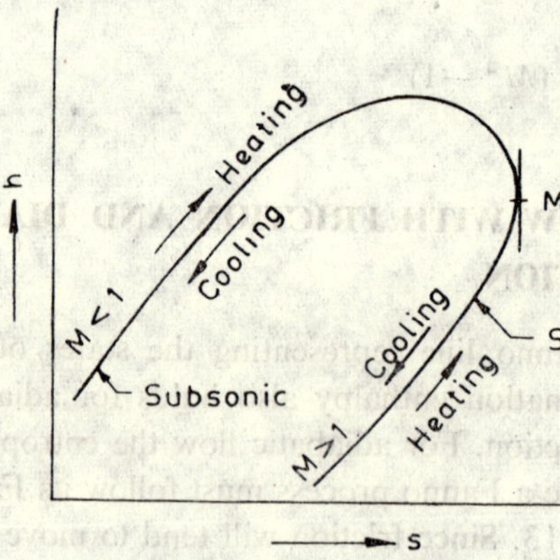

Fig. 17.14 A Rayleigh line on h-s **plot**

The process is reversible, and the direction of entropy change is determined by the sign of the heat transfer. Heating a compressible flow has the same effect as friction, and the Mach number goes towards unity. Therefore, there is a maximum heat input for a given flow rate which can be passed by the duct, which is then choked. Although the cooling of the fluid increases the flow stagnation pressure with a decrease in entropy, a nonmechanical pump is not feasible by cooling a compressible flow, because of the predominating effect of friction.

EXAMPLE 17.1

A stream of air flows in a duct of 100 mm diameter at a rate of 1 kg/s. The stagnation temperature is 37°C. At one section of the duct the static pressure is 40 kPa. Calculate the Mach number, velocity, and stagnation pressure at this section.

Solution

$$T_0 = 37 + 273 = 310 \text{ K}, \quad p = 40 \text{ kPa}, \gamma = 1.4$$

The mass flow rate per unit area is

$$\frac{w}{A} = \rho \mathbf{V} = \frac{p}{RT} \sqrt{\gamma RT}.\, M$$

$$= \sqrt{\frac{\gamma}{R}} \cdot \frac{pM}{\sqrt{T_0}} \sqrt{\frac{T_0}{T}} = \sqrt{\frac{\gamma}{R}} \cdot \frac{p\, M}{\sqrt{T_0}} \left(1 + \frac{\gamma - 1}{2} M^2\right)^{\frac{1}{2}}$$

$$\frac{1}{\pi/4\,(0.1)^2}\ \frac{\text{kg}}{\text{sm}^2} = \sqrt{\frac{1.4}{0.287 \text{ kJ/kg-K}}} \cdot \frac{40 \text{ kN/m}^2 \times M}{\sqrt{310 \text{ K}}} (1 + 0.2\, M^2)^{1/2}$$

$$127.39 = \sqrt{\frac{1.4}{0.287 \times 10^3 \times 310}} \cdot 40 \times 10^3 M\, (1 + 0.2\, M^2)^{1/2}$$

$$\therefore \quad M\,(1 + 0.2\, M^2)^{1/2} = 0.803$$

$$M^2\,(1 + 0.2M^2) = 0.645$$

$$M^4 + 5M^2 - 3.225 = 0$$

$$M^2 = \frac{-5 \pm \sqrt{25 + 12.9}}{2} = \frac{-5 + 6.16}{2} = 0.58$$

$$\therefore \quad M = 0.76$$ Ans.

$$\frac{T_0}{T} = 1 + \frac{\gamma - 1}{2} M^2 = 1 + 0.2 \times (0.76)^2 = 1.116$$

$$\therefore \quad T = \frac{310}{1.116} = 277.78 \text{ K}$$

$$c = \sqrt{\gamma RT} = \sqrt{1.4 \times 0.287 \times 277.78 \times 10^3}$$

$$= 334.08 \text{ m/s}$$

$$\therefore \quad \mathbf{V} = c\, M = 334.08 \times 0.76 = 253.9 \text{ m/s}$$ Ans.

$$\frac{p_0}{p} = \left(\frac{T_0}{T}\right)^{\gamma/(\gamma-1)} = (1.116)^{1.4/0.4} = 1.468$$

$$\therefore \quad p_0 = 40 \times 1.468 = 58.72 \text{ kPa}$$ Ans.

EXAMPLE 17.2

A conical air diffuser has an intake area of 0.11 m² and an exit area of 0.44 m². Air enters the diffuser with a static pressure of 0.18 MPa, static temperature of 37°C, and velocity of 267 m/s. Calculate (a) the mass flow rate of air through the diffuser, (b) the Mach number, static temperature, and static pressure of air leaving the diffuser, and (c) the net thrust acting upon the diffuser, assuming that its outer surfaces are wetted by atmospheric pressure at 0.1 MPa.

Solution

The conical diffuser is shown in Fig. Ex. 17.2. The mass flow rate of air through it

$$w = \rho A \mathbf{V}$$

$$= \frac{p_1}{RT_1} \cdot A_1 \mathbf{V}_1 = \frac{0.18 \times 10^3}{0.287 \times 310} \times 0.11 \times 267$$

$$= 59.42 \text{ kg/s} \qquad \text{Ans.(a)}$$

$$c_1 = \sqrt{\gamma R T_1}$$

$$= 20.045 \sqrt{T_1} = 20.045 \sqrt{310}$$

$$= 352 \text{ m/s}$$

$$\therefore \quad M_1 = \frac{\mathbf{V}_1}{c_1} = \frac{267}{352} = 0.76$$

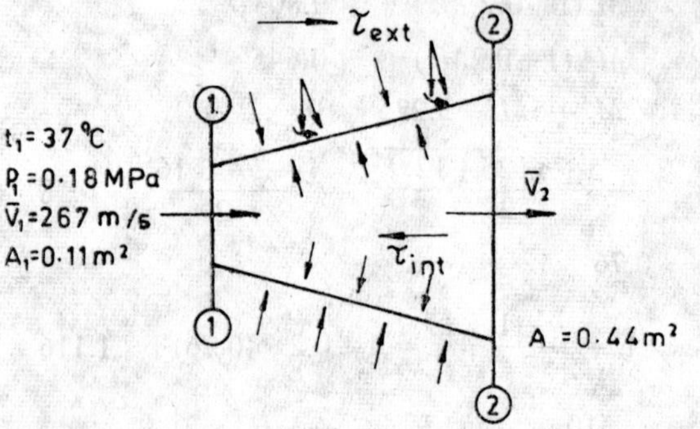

Fig. Ex. 17·2

From the gas tables for the isentropic flow of air ($\gamma = 1.4$), given in the appendix, when $M_1 = 0.76$

$$\frac{A_1}{A^*} = 1.0570, \quad \frac{p_1}{p_{01}} = 0.68207, \quad \frac{T_1}{T_{01}} = .89644$$

$$\frac{F_1}{F^*} = 1.0284$$

Now $\quad \dfrac{A_2}{A_1} = \dfrac{0.44 \text{ m}^2}{0.11 \text{ m}^2} = 4 = \dfrac{A_2/A^*}{A_1/A^*}$

$\therefore \quad \dfrac{A_2}{A^*} = 4 \times 1.0570 = 4.228$

Again, from the isentropic flow tables, when $\dfrac{A_2}{A^*} = 4.228$

$$M_2 = 0.135, \quad \frac{p_2}{p_{02}} = 0.987, \quad \frac{T_2}{T_{02}} = 0.996, \quad \frac{F_2}{F^*} = 3.46$$

$$\frac{p_2}{p_1} = \frac{p_2/p_{02}}{p_1/p_{01}} = \frac{0.987}{0.682} = 1.447$$

$$\frac{T_2}{T_1} = \frac{T_2/T_{02}}{T_1/T_{01}} = \frac{0.996}{0.89644} = 1.111$$

$$\frac{F_2}{F_1} = \frac{F_2/F^*}{F_1/F^*} = \frac{3.46}{1.0284} = 3.364$$

$\therefore \quad p_2 = 1.447 \times 0.18 = 0.26$ MPa **Ans.**

$T_2 = 1.111 \times 310 = 344.4 \text{ K} = 71.3°\text{C}$ **Ans.**

Impulse function at inlet

$$F_1 = p_1 A_1 + \rho_1 A_1 V_1^2$$

$$= p_1 A_1 \left(1 + \frac{1}{RT_1} V_1^2\right)$$

$$= p_1 A_1 (1 + \gamma M_1^2)$$

$$= 0.18 \times 10^3 \times 0.11 \ (1 + 1.4 \times 0.76^2)$$

$$= 35.82 \text{ kN}$$

Internal thrust τ will be from right to left, as shown in Fig. Ex. 17.2

$$\tau_{int.} = F_2 - F_1 = 2.364 \ F_1 - F_1$$

$$= 2.364 \times 35.82$$

$$= 84.68 \text{ kN}$$

External thrust is from left to right

$$\tau_{ext.} = p_0 (A_2 - A_1)$$

$$= 0.1 \times 10^3 \ (0.44 - 0.11)$$

$$= 33 \text{ kN}$$

Net thrust $= \tau_{int.} - \tau_{ext.}$

$= 84.68 - 33 = 51.68$ kN **Ans.** (c)

EXAMPLE 17.3

A convergent-divergent nozzle has a throat area 500 mm² and an exit area of 1000 mm². Air enters the nozzle with a stagnation temperature of

360 K and a stagnation pressure of 1 MPa. Determine the maximum flow rate of air that the nozzle can pass, and the static pressure, static temperature, Mach number, and velocity at the exit from the nozzle, if (a) the divergent section acts as a nozzle, and (b) the divergent section acts as a diffuser.

Solution

$$\frac{_2}{*} = \frac{1000}{500} = 2$$

From the isentropic flow tables, when $A_2/A^* = 2$ there are two values of the Mach number, one for supersonic flow when the divergent section acts as a nozzle, and the other for subsonic flow when the divergent section acts as a diffuser, which are $M_2 = 2.197$, 0.308 (Fig. Ex. 17.3).

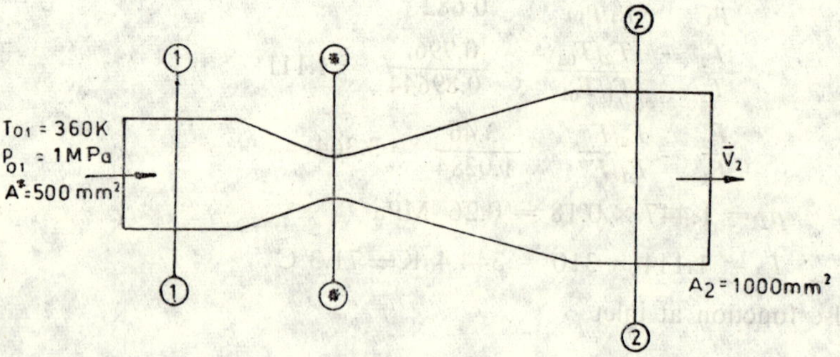

Fig. Ex. 17·3

(a) When $M_2 = 2.197$, $\frac{p_2}{p_0} = 0.0939$, $\frac{T_2}{T_0} = 0.5089$

$$\therefore \quad p_2 = 0.0939 \times 1000 = 93.9 \text{ kPa}$$

$$T_2 = 0.5089 \times 360 = 183.2 \text{ K}$$

$$c_2 = \sqrt{\gamma R T_2} = 20.045\sqrt{183.2}$$

$$= 271.2 \text{ m/s}$$

$$\therefore \quad V_2 = 271.2 \times 2.197 = 596 \text{ m/s} \qquad \textbf{Ans.}$$

Mass flow rate

$$w = A^* \rho^* V^* = \rho_2 A_2 V_2 = \rho_1 A_1 V_1$$

For air $\frac{p^*}{p_0} = 0.528$ and $\frac{T^*}{T_0} = 0.833$

$$\rho^* = \frac{p^*}{R\,T^*} = \frac{0.528 \times 1000}{0.287 \times 0.833 \times 360} = 6.13 \text{ kg/m}^3$$

$$T^* = 360 \times 0.833 = 300 \text{ K}$$

$$V^* = \sqrt{\gamma R\,T^*} = 20.045\sqrt{300}$$

$$= 347.2 \text{ m/s}$$

$$\therefore \quad w = (500 \times 10^{-6}) \times 6.13 \times 347.2$$

$$= 1.065 \text{ kg/s} \qquad \textbf{Ans.}$$

(b) When $M = 0.308$, $\frac{p_2}{p_0} = 0.936$, $\frac{T_2}{T_0} = 0.9812$

$$p_2 = 0.936 \times 1000 = 93.6 \text{ kPa}$$

$$T_2 = 0.9812 \times 360 = 353.2 \text{ K}$$

$$c_2 = \sqrt{\gamma R T_2} = 20.045 \sqrt{353.2} = 376.8 \text{ m/s}$$

$$\mathbf{V}_2 = 376.8 \times 0.308 = 116 \text{ m/s}$$

$$w = 1.065 \text{ kg/s}$$ Ans.

EXAMPLE 17.4

When a Pitot tube is immersed in a supersonic stream, a curved shock wave is formed ahead of the Pitot tube mouth. Since the radius of the curvature of the shock is large, the shock may be assumed to be a normal shock. After the normal shock, the fluid stream decelerates isentropically to the total pressure p_{oy} at the entrance to the Pitot tube.

A Pitot tube travelling in a supersonic wind-tunnel gives values of 16 kPa and 70 kPa for the static pressure upstream of the shock and the pressure at the mouth of the tube respectively. Estimate the Mach number of the tunnel. If the stagnation temperature is 300°C, calculate the static temperature and the total (stagnation) pressure upstream and downstream of the tube.

Solution

With reference to the Fig. Ex. 17.4

$$p_x = 16 \text{ kPa}, \; p_{oy} = 70 \text{ kPa}$$

$$\therefore \quad \frac{p_{oy}}{p_x} = \frac{70}{16} = 4.375$$

From the gas tables for normal shock

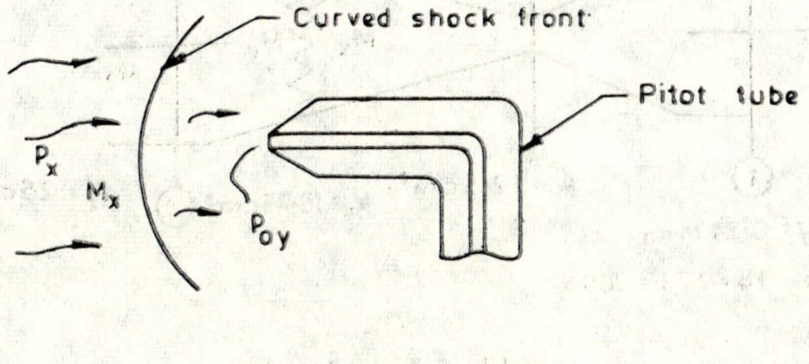

Fig. Ex. 17.4

when $\frac{p_{oy}}{p_x} = 4.375,\ M_x = 1.735,\ \frac{p_y}{p_x} = 3.34,\ \frac{\rho_y}{\rho_x} = 2.25$

$$\frac{T_y}{T_x} = 1.483,\ \frac{p_{oy}}{p_{ox}} = 0.84,\ M_y = 0.631$$

$$T_{ox} = T_{oy} = 573 \text{ K}$$

$$T_{ox} = \left(1 + \frac{\gamma - 1}{2} M_x^2\right) T_x$$

$$= (1 + 0.2 \times 3)\ T_x = 1.6\ T_x$$

$$\therefore \quad T_x = \frac{573}{1.6} = 358 \text{ K}$$

$$\therefore \quad T_y = 358 \times 1.483 = 530 \text{ K} = 257°\text{C}$$

$$p_{ox} = \frac{p_{oy}}{0.84} = \frac{70}{0.84} = 83.3 \text{ kPa}$$

$$M_x = 1.735$$

Ans.

EXAMPLE 17.5

A convergent-divergent nozzle operates at off-design condition while conducting air from a high pressure tank to a large container. A normal shock occurs in the divergent part of the nozzle at a section where the cross-sectional area is 18.75 cm². If the stagnation pressure and stagnation temperature at the inlet of the nozzle are 0.21 MPa and 36°C respectively, and the throat area is 12.50 cm² and the exit area is 25 cm² estimate the exit Mach number, exit pressure, loss in stagnation pressure, and entropy increase, during the flow between the tanks.

Solution

With reference to Fig. Ex. 17.5, at shock section

$$\frac{A_x}{A^*} = \frac{18.75}{12.50} = 1.5$$

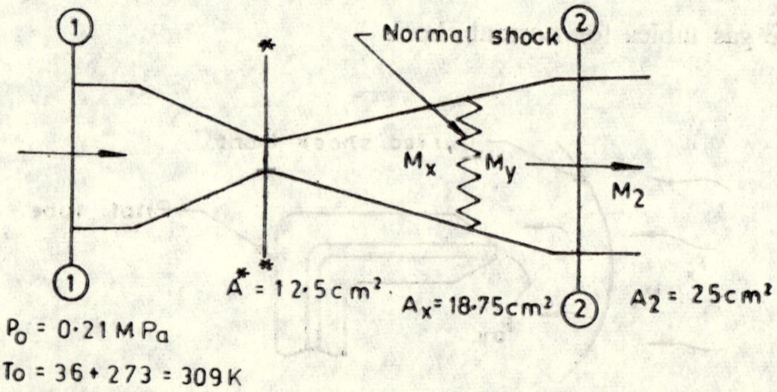

Fig. Ex. 17.5

Upto the shock, the flow is isentropic. From the isentropic flow tables, when

$$\frac{A}{A^*} = 1.5, \qquad M_x = 1.86$$

$$\frac{p_x}{p_{ox}} = 0.159$$

$\therefore$ Static pressure upstream of the shock

$$p_x = 0.159 \times 0.21 \times 10^3 = 33.4 \text{ kPa}$$

From the gas tables on normal shocks

when $M_x = 1.86$, $M_y = 0.604$, $\frac{p_y}{p_x} = 3.87$, $\frac{p_{oy}}{p_x} = 4.95$, $\frac{p_{oy}}{p_{ox}} = 0.0786$

$$\therefore \quad p_y = 3.87 \times 33.4 = 129.3 \text{ kPa}$$

$$p_{oy} = 4.95 \times 33.4 = 165.3 \text{ kPa}$$

From the shock section to the exit of the nozzle, the flow is again isentropic.

When $M_y = 0.604$, from the isentropic flow tables

$$\frac{A_y}{A^*} = 1.183$$

$$\therefore \quad \frac{A_2}{A^*} = \frac{A_2}{A_y}\frac{A_y}{A^*} = \frac{25}{18.75} \times 1.183 = 1.582$$

When $A_2/A^* = 1.582$, from the isentropic flow tables, $M_2 = 0.402$ Ans.

$$\frac{p_2}{p_{oy}} = 0.895$$

$$\therefore \quad p_2 = 0.895 \times 165.3 = 147.94 \text{ kPa}$$ Ans.

Loss in stagnation pressure occurs only across the shock.

$$p_{ox} - p_{oy} = 210 - 165.3 = 44.7 \text{ kPa}$$ Ans.

Entropy increase, $s_y - s_x$

$$= -R \ln \frac{p_{oy}}{p_{ox}}$$

$$= 0.287 \ln \frac{210}{165.3}$$

$$= 0.287 \times 0.239 = 0.0686 \text{ kJ/kg-K}$$ Ans.

PROBLEMS

17.1 Air in a reservoir has a temperature of 27°C and a pressure of 0.8 MPa. The air is allowed to escape through a channel at a rate of 2.5 kg/s. Assuming that the air velocity in the reservoir is negligible and that the flow through the channel is isentropic, find the Mach number, the velocity, and the area at a section in the channel where the static pressure is 0.6 MPa.

17.2 A supersonic wind tunnel nozzle is to be designed for $M=2$, with a throat section, 0.11 m^2 in area. The supply pressure and temperature at the nozzle inlet, where the velocity is negligible, are 70 kPa and 37°C respectively. Compute the mass flow rate, the exit area, and the fluid properties at the throat and exit. Take γ $=1.4$.

17.3 An ideal gas flows into a convergent nozzle at a pressure of 0.565 MPa, a temperature of 280°C, and negligible velocity. After reversible adiabatic expansion in the nozzle the gas flows directly into a large vessel. The gas in the vessel may be maintained at any specified state while the nozzle supply state is held constant. The exit area of the nozzle is 500 mm^2. For this gas $\gamma=1.3$ and c_p $=1.172$ kJ/kg-K. Determine (a) the pressure of the gas leaving the nozzle when its temperature is 225°C, and (b) the gas mass flow rate when the pressure in the vessel is 0.21 MPa.

Ans. 0.36 MPa, 0.48 kg/s.

17.4 Air flows adiabatically through a pipe with a constant area. At point 1, the stagnation pressure is 0.35 MPa and the Mach number is 0.4. Further downstream the stagnation pressure is found to be 0.25 MPa. What is the Mach number at the second point for subsonic flow ?

17.5 The intake duct to an axial flow air compressor has a diameter of 0.3 m and compresses air at 10 kg/s. The static pressure inside the duct is 67 kPa and the stagnation temperature is 40°C. Calculate the Mach number in the duct. Ans. 0.526.

17.6 Show that for an ideal gas the fractional change in pressure across a small pressure pulse is given by

$$\frac{dp}{p} = \gamma \frac{dV}{c}$$

and that the fractional change in absolute temperature is given by

$$\frac{dT}{T} = (\gamma - 1) \frac{dV}{c}$$

17.7 An airplane flies at an altitude of 13,000 m (temperature −55°C, pressure 18.5 kPa) with a speed of 180 m/s. Neglecting frictional effects, calculate (a) the critical velocity of the air relative to the aircraft, and (b) the maximum possible velocity of the air relative to the aircraft.

17.8 A stream of air flowing in a duct is at a pressure of 150 kPa, has a Mach number of 0.6, and flows at a rate of 0.25 kg/s. The cross-sectional area of the duct is 625 mm^2. (a) Compute the stagnation temperature of the stream in °C. (b) What is the maximum percentage reduction in area which could be introduced without reducing the flow rate of the stream? (c) For the maximum area reduction of part (b), find the velocity and pressure at the minimum area, assuming no friction and heat transfer.

17.9 A rocket combustion chamber is supplied with 12 kg/s of hydrogen and 38 kg/s of oxygen. Before entering the nozzle all the oxygen is consumed, the pressure is 2.3 MPa, and the temperature is 2800°C. Neglecting dissociation and friction, find the throat area of the nozzle required. Assume $=1.25$.

17.10 A gas with specific heat ratio $\gamma=1.4$ passes through a plane normal shock. Immediately before the shock the Mach number is 2, and the stagnation pressure and temperature are 600 kPa and 400 K respectively. Find the static pressure and static temperature immediately downstream of the shock. Ans. 345 kPa, 376 K.

17.11 Air flowing through a nozzle encounters a shock. The Mach number upstream of the shock is 1.8, and the static temperature downstream of the shock is 450 K. How much has the velocity changed across the shock ? Ans. 378 m/s.

17.12 The stagnation temperature and stagnation pressure of air in a reservoir supplying a convergent-divergent nozzle are 450 K and 400 kPa respectively. The nozzle throat area is 625 mm^2 and the nozzle exit area is 1875 mm^2. A shock is noted at a position in the divergent section where the area is 1250 mm². (a) What is the exit pressure, temperature, and velocity? (b) What value of the back pressure would cause the flow through the nozzle to be completely supersonic? (c) What value of back pressure would result in completely isentropic flow interior and exterior to the nozzle? Ans. (a) 234 kPa, 440 K, 141 m/s, (b) 150 kPa, (c) 18.8 kPa.

17.13 Assume that a flow through the nozzle has been established by setting the back pressure at 200 kPa. Because a back pressure greater than 150 kPa leads to a shock in the nozzle, find the area ratio where the shock occurs.

17.4 Air flows through a frictionless convergent-divergent nozzle. The area of the exit section is three times the area of the throat section, and the ratio of stagnation pressure at the entrance to the static pressure in the exit section is 2.5. Calculate (a) the Mach number M_x where the shock occurs, (b) the area A_x where the shock occurs, and (c) the entropy increase.

17.15 A nozzle is designed assuming reversible adiabatic flow with an exit Mach number of 2.6. Air flows through it with a stagnation pressure and temperature of 2 MPa and 150°C respectively. The mass rate of flow is 5 kg/s. (a) Determine the exit pressure, temperature, area, and throat area. (b) If the back pressure at the nozzle exit is raised to 1.35 MPa, and the flow remains isentropic except for a normal shock wave, determine the exit Mach number and temperature, and the mass flow rate through the nozzle.

17.16 A jet plane travels through the air with a speed of 1000 km per hour at an altitude of 6000 m, where the pressure is 40 kPa and the temperature is −12°C. Consider the diffuser of the engine. The air leaves the diffuser with a velocity of 100 m/s. Determine the pressure and temperature leaving the diffuser, and the ratio of the inlet to the exit area of the diffuser, assuming isentropic flow.

Ans. 61 kPa, 295 K, 0.487.

18

Elements of Heat Transfer

18.1 BASIC CONCEPTS

The subject of heat transfer is concerned with the determination of the *rate* at which energy is transferred as heat by virtue of temperature difference between two bodies. An elementary knowledge of heat transfer is a prerequisite to the study of any branch of thermal engineering.

There are three modes in which heat may be transferred: (a) conduction, (b) convection, and (c) radiation.

Heat transfer theory rests on certain *rate equations* as illustrated below. It is a surface phenomenon, i.e., the more the surface area provided, the unidirectional heat conduction,

18.2 CONDUCTION HEAT TRANSFER

Conduction refers to the transfer of heat between two bodies or two parts of the same body through molecules which are more or less stationary.

Fourier's law of heat conduction states that the rate of heat flux is linearly proportional to the temperature gradient. For one dimensional or unindirectional heat conduction.

$$\mathbf{q} \propto \frac{dt}{dx}$$

or

$$\mathbf{q} = -K\frac{dt}{dx} \qquad (18.1)$$

where $\mathbf{q}$ is he rate of heat flux in W/m^2, dt/dx is the temperature gradient in x-direction, and K is the constant of proportionality which is a property of the material through which heat is being conducted and is known as *thermal conductivity*. $\mathbf{q}$ is a vector quantity. The negative sign is being used because heat flows from a high to a low temperature region and the slope dt/dx is negative.

For a finite temperature difference $(t_1 - t_2)$ across a wall of thickness x (Fig. 18.1)

$$\mathbf{q} = -K\frac{t_2 - t_1}{x} = K\frac{t_1 - t_2}{x} \text{ W/m}^2$$

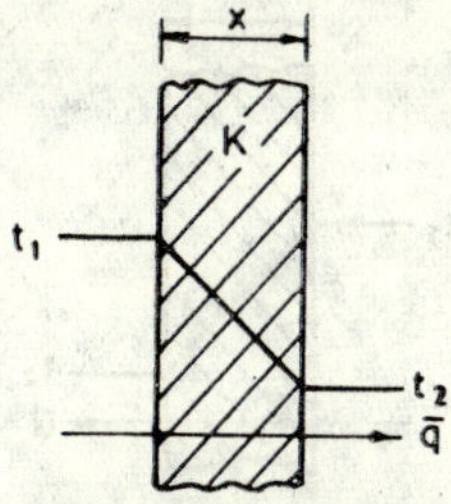

Fig. 18.1 Heat conduction through a wall.

If A is the surface area normal to heat flow, then the rate of heat transfer

$$Q = \mathbf{q} \cdot A = -KA\frac{t_2 - t_1}{x}$$

or $$Q = KA\frac{t_1 - t_2}{x} \text{ Watts}$$

The dimension of thermal conductivity in MKS units is kcal/mhr°C, and in S.I. units it is W/mK. Since $dt/dx = \mathbf{q}/K$ for the same $\mathbf{q}$, if K is low (i.e., for an insulator), dt/dx will be large, i.e., there will be a large temperature difference across the wall, and if K is high (i.e., for a conductor), dt/dx will be small, or there will be a small temperature difference across the wall.

18.2.1 Resistance Concept

Heat flow has an analogy in the flow of electricity. Ohm's law states that the current I flowing through a wire (Fig. 18.2) is proportional to the potential difference E, or

$$I = \frac{E}{R}$$

where $1/R$ is the constant of proportionality, and R is known as the resistance of the wire, which is a property of the material. Since the temperature difference and heat flux in conduction are similar to the potential difference and electric current respectively, the heat conduction rate through the wall may be written as

$$Q = -KA\frac{t_2 - t_1}{x} = \frac{t_1 - t_2}{x/KA} = \frac{t_1 - t_2}{R}$$

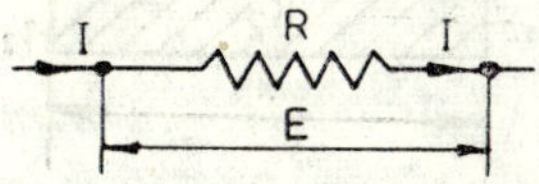

Fig. 18.2 Electrical resistance concept.

where $R = x/KA$ is the thermal resistance to heat flow offered by the wall (Fig. 18.3). For a composite wall, as shown in Fig. 18.4, there are two resistances in series. The slope of the temperature profile depends on

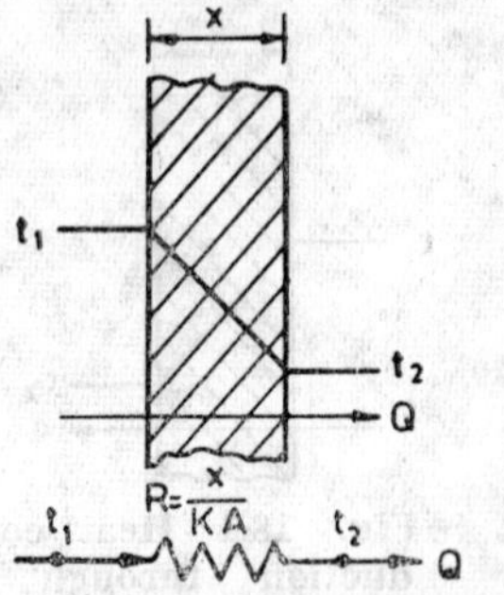

Fig. 18.3 Thermal resistance offered by a wall

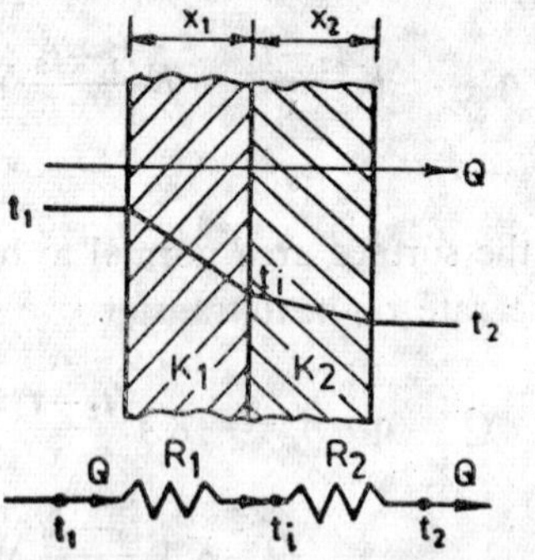

Fig. 18.4 Heat conduction through resistances in series

the thermal conductivity of the material. t_i is the interface temperature. The total thermal resistance

$$R = R_1 + R_2 = \frac{x_1}{K_1\, A} + \frac{x_2}{K_2\, A}$$

and the rate of heat flow

$$Q = \frac{t_1 - t_2}{R}$$

Again, $t_1 - t_i = Q \cdot R_1 = Q \cdot \frac{x_1}{K_1 A}$, from which t_i can be evaluated.

For two resistances in parallel (Fig. 18.5), the total resistance R is given by

$$\frac{1}{R} = \frac{1}{R_1} + \frac{1}{R_2} \quad \text{or} \quad R = \frac{R_1\, R_2}{R_1 + R_2}$$

where $\quad R_1 = \frac{x}{K_1\, A_1} \quad$ and $\quad R_2 = \frac{x}{K_2\, A_2}$

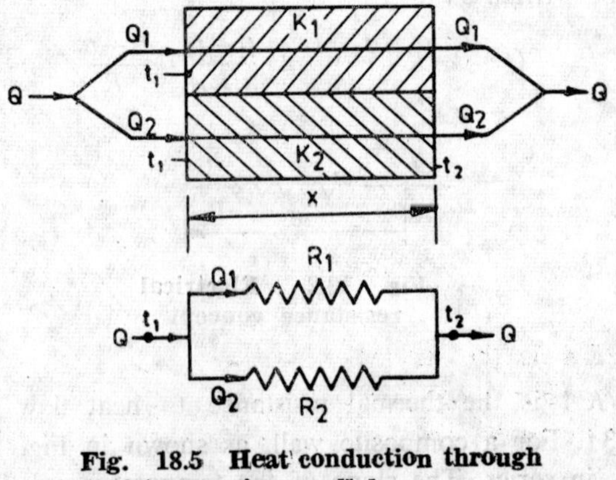

Fig. 18.5 Heat conduction through resistances in parallel

and the rate of heat flow

$$Q = \frac{t_1 - t_2}{R}$$

18.2.2 Heat Conduction through a Cylinder

Let us assume that the inside and outside surfaces of the cylinder (Fig. 18.6) are maintained at temperatures t_1 and t_2 respectively, and t_1 is greater than t_2. We will also assume that heat is flowing, under steady state, only in the radial direction, and there is no heat conduction along the length or the periphery of the cylinder. The rate of heat transfer through the thin cylinder of thickness dr is given by

$$Q = -KA \frac{dt}{dr}$$

$$= -K\, 2\pi r L \frac{dt}{dr}, \qquad (18.3)$$

where L is the length of the cylinder

or $$\int_{t=t_1}^{t=t_2} dt = \int_{r=r_1}^{r=r_2} -\frac{Q}{2\pi KL} \frac{dr}{r}$$

$$\therefore \quad t_2 - t_1 = -\frac{Q}{2\pi KL} \ln \frac{r_2}{r_1}$$

$$\therefore Q = \frac{2\pi KL (t_1 - t_2)}{\ln \dfrac{r_2}{r_1}} \qquad (18.4)$$

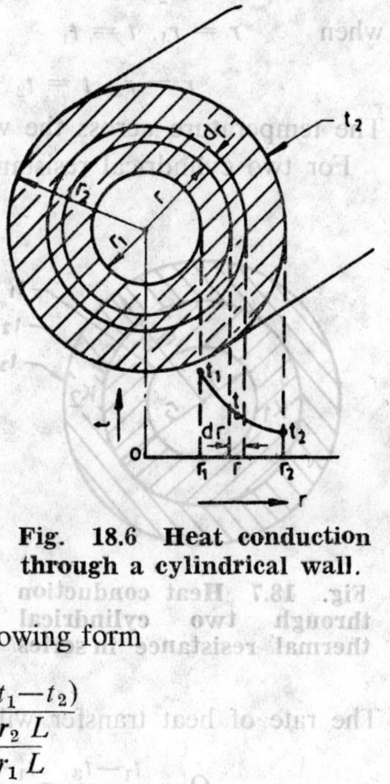

Fig. 18.6 Heat conduction through a cylindrical wall.

Equation (18.4) can also be written in the following form

$$Q = \frac{2\pi L (r_2 - r_1) K (t_1 - t_2)}{(r_2 - r_1) \ln \dfrac{2\pi r_2 L}{2\pi r_1 L}}$$

$$= \frac{K (A_2 - A_1)(t_1 - t_2)}{(r_2 - r_1) \ln \dfrac{A_2}{A_1}}$$

where A_1 and A_2 are the inside and outside surface areas of the cylinder.

$$\therefore \quad Q = -KA_{\text{l.m.}} \frac{t_2 - t_1}{r_2 - r_1} \text{ or } Q = -KA_{\text{l.m.}} \frac{t_2 - t_1}{x_w} \qquad (18.5)$$

where $A_{\text{l.m.}}$ = log-mean area = $\dfrac{A_2 - A_1}{\ln \dfrac{A_2}{A_1}}$

and x_w = wall thickness of the cylinder

$= r_2 - r_1$

Here the thermal resistance offered by the cylinder wall to radial heat conduction is

$$R = \frac{x_w}{K.\, A_{l \cdot m.}}$$

From equation (18.3)

$$dt = -\frac{Q}{2\pi KL}\frac{dr}{r} = C_1 \frac{dr}{r}$$

or $$t = C_1 \ln r + C_2 \qquad (18.6)$$

where C_1 and C_2 are the constants to be evaluated from the conditions:

when $r = r_1,\ t = t_1$

$r = r_2,\ t = t_2$

The temperature across the wall of the cylinder thus varies logarithmically.

For two cylindrical resistances in series (Fig. 18.7)

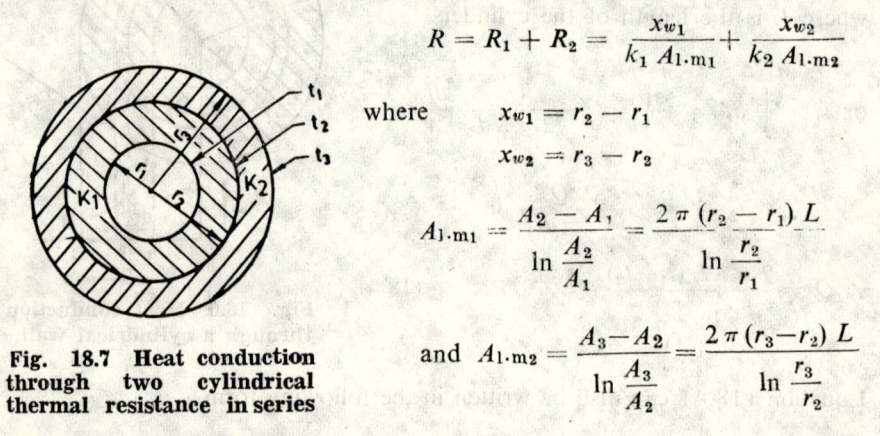

$$R = R_1 + R_2 = \frac{x_{w1}}{k_1 A_{l \cdot m1}} + \frac{x_{w2}}{k_2 A_{l \cdot m2}}$$

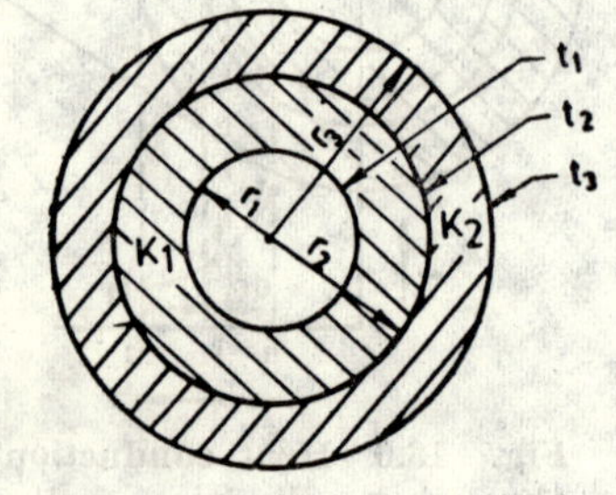

Fig. 18.7 Heat conduction through two cylindrical thermal resistance in series

where $x_{w1} = r_2 - r_1$

$x_{w2} = r_3 - r_2$

$$A_{l \cdot m1} = \frac{A_2 - A_1}{\ln \dfrac{A_2}{A_1}} = \frac{2\pi (r_2 - r_1) L}{\ln \dfrac{r_2}{r_1}}$$

and $$A_{l \cdot m2} = \frac{A_3 - A_2}{\ln \dfrac{A_3}{A_2}} = \frac{2\pi (r_3 - r_2) L}{\ln \dfrac{r_3}{r_2}}$$

The rate of heat transfer will be

$$Q = \frac{t_1 - t_3}{R} = \frac{t_1 - t_2}{R_1} = \frac{t_2 - t_3}{R_2}$$

from which the interface temperature t_2 can be evaluated.

18.2.3 Heat Conduction through a Sphere

Heat flowing through the thin spherical strip (Fig. 18.8) at radius r, of thickness dr, is

$$Q = -KA\frac{dt}{dr}$$

where A is the spherical surface at radius r normal to heat flow,

$$\therefore \quad Q = -K\,4\,\pi r^2 \frac{dt}{dr}$$

$$\int_{t=t_1}^{t=t_2} dt = \int_{r=r_1}^{r=r_2} - \frac{Q}{4\,\pi\,K}\frac{dr}{r^2}$$

$$t_2 - t_1 = -\frac{Q}{4\,\pi\,K}\left(\frac{1}{r_1} - \frac{1}{r_2}\right)$$

$$\therefore \quad Q = \frac{4\,\pi\,K\,(t_1 - t_2)\,r_1 r_2}{r_2 - r_1}$$

$$\text{or } Q = -K\,A_{g.m}\frac{t_2 - t_1}{x_w} \qquad (18.7)$$

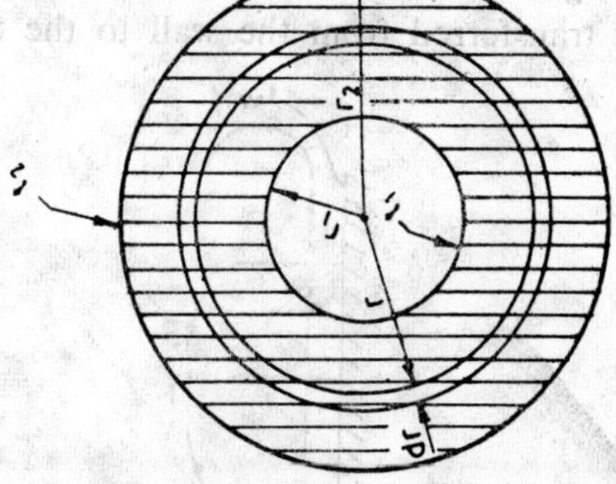

Fig. 18.8 Heat conduction through a sphere

where $A_{g.m.}$ = geometrical mean area

$$= \sqrt{A_1 A_2} = \sqrt{4\,\pi\,r_1^2\,4\,\pi\,r_2^2}$$

$$= 4\pi\,r_1\,r_2$$

and x_w = wall thickness of the sphere

$$= r_2 - r_1$$

Here the thermal resistance offered by the wall to heat conduction is

$$R = \frac{x_w}{K\,A_{g.m}}$$

Thus similar expressions of thermal resistance hold good for flat plate, cylinder, and sphere, which are

$$R_{plate} = \frac{x_w}{KA}, \quad R_{cylinder} = \frac{x_w}{K\,A_{l.m.}} \text{ and}$$

$$R_{sphere} = \frac{x_w}{K\,A_{g.m.}}$$

where K is the thermal conductivity of the wall material.

18.3 CONVECTIVE HEAT TRANSFER

Convection is a process involving the mass movement of fluids. When a temperature difference produces a density difference which results in mass movement (Fig. 18.9), the process is called *free or natural convection.* Here the plate is maintained isothermal at temperature t_w, which is higher than the surrounding fluid temperature t_f. The fluid near the wall, on getting heated, moves up due to the effect of buoyancy, and is replaced by the cold fluid moving towards the wall. Thus a circulation current is set up.

When the mass motion of the fluid is caused by an external device like a pump, compressor, blower or fan, the process is called *forced convection* (Fig. 18.10). Here the fluid is made to flow along the hot surface and heat is transferred from the wall to the fluid.

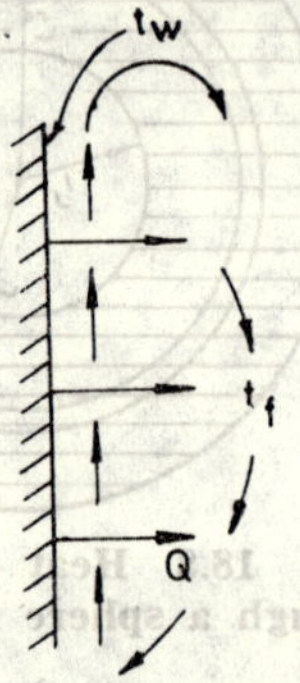

Fig. 18.9 Natural or free convection

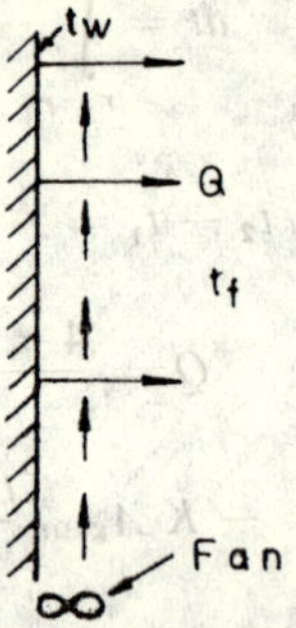

Fig. 18.10 Forced convection

Whether the convection process is natural or forced, there is always a fluid film immediately adjacent to the wall where the temperature varies from t_w to t_f (Fig. 18.11). Heat is first conducted through this fluid film and then it is transported by fluid motion.

Rate of heat transfer through the film

$$Q = -K_f \cdot A \frac{t_f - t_w}{\delta}$$

where δ is the film thickness and K_f is the thermal conductivity of the film.

The film coefficient of heat transfer, or heat transfer coefficient h, is defined as

$$h = \frac{K_f}{\delta} \text{ (kcal/m}^2\text{ hr °C or W/m}^2\text{ °C)}$$

$$\therefore \quad Q = hA\,(t_w - t_f) \qquad (18.8)$$

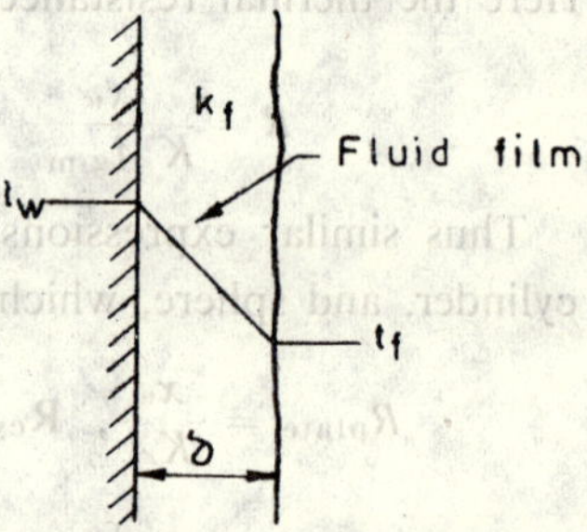

Fig. 18.11 Heat conduction through fluid film at the wall

The rate of heat transfer Q increases with the increase in the value of heat transfer coefficient h. The higher the value of K_f and the lower the value of film thickness δ, the higher will be the value of h. As the velocity of the fluid increases, the film thickness decreases. For gases, K_f is low, so the rate of heat transfer from the solid wall to a gas is small compared to a liquid.

The above equation is known as Newton's law of cooling. Strictly speaking, convection applies only to fluid motion. The mechanism of heat transfer is by conduction.

The thermal resistance offered by the fluid film

$$R = \frac{t_w - t_f}{Q} = \frac{1}{hA}$$

For heat transfer from a hot fluid to a cold fluid through a wall (Fig. 18.12), there are three resistances in series

$$R = R_1 + R_2 + R_3$$

$$= \frac{1}{h_1 A} + \frac{x}{KA} + \frac{1}{h_2 A}$$

$$Q = \frac{t_h - t_c}{R}$$

or $$Q = \frac{t_h - t_c}{\frac{1}{h_1 A} + \frac{x}{KA} + \frac{1}{h_2 A}} = UA\,(t_h - t_c)$$

where U is known as the overall heat transfer coefficient (W/m²K or kcal/m²hr °C), and is given by

$$\frac{1}{UA} = \frac{1}{h_1 A} + \frac{x}{KA} + \frac{1}{h_2 A} = R$$

or $$\frac{1}{U} = \frac{1}{h_1} + \frac{x}{K} + \frac{1}{h_2}$$

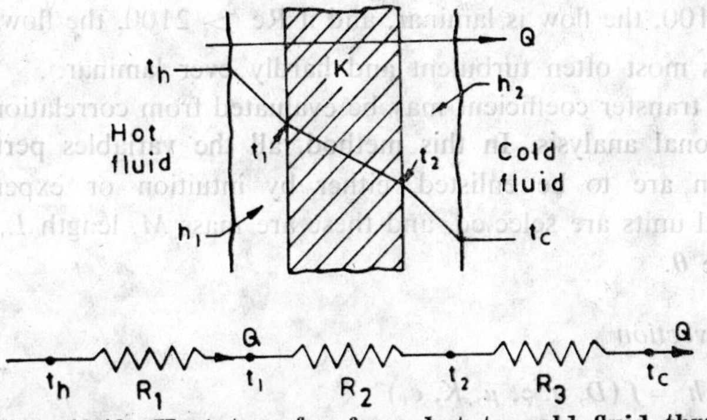

Fig. 18.12 Heat transfer from hot to cold fluid through a plane wall

For heat transfer from a hot fluid inside a cylinder to the cold fluid outside (Fig. 18.13)

$$Q = \frac{t_h - t_c}{R_1 + R_2 + R_3} = \frac{t_h - t_c}{\frac{1}{h_i A_i} + \frac{x_w}{K_w A_{l.m.}} + \frac{1}{h_o A_o}}$$

or $$Q = U_o A_o\,(t_h - t_c)$$

where $$\frac{1}{U_o A} = \frac{1}{h_i A_i} + \frac{1}{K_w A_{l.m.}} + \frac{1}{h_o A_o} \qquad (18.9)$$

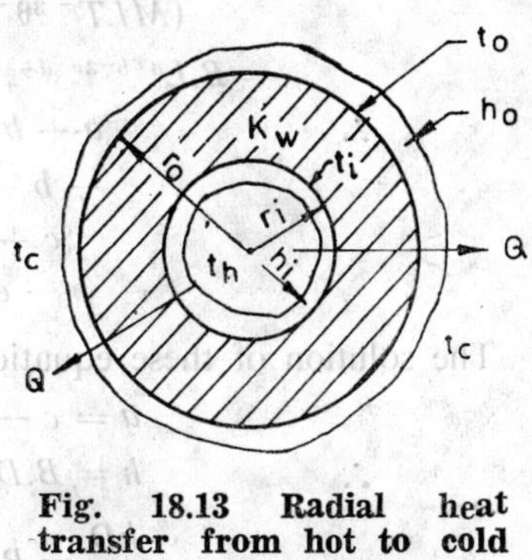

U_o being the overall heat transfer coefficient based on outsdie area A_o, h_i the inside heat transfer coefficient, and h_o the outside heat transfer coefficient.

Fig. 18.13 Radial heat transfer from hot to cold fluid through a cylinder

When the wall thickness x_w is small, $A_o \cong A_{lm} \cong A_i$, and

$$\frac{1}{U_o} = \frac{1}{h_i} + \frac{x_w}{k_w} + \frac{1}{h_o}$$

18.3.1 Correlations in Convective Heat Transfer

Imagine there to be a curve in a fluid system in motion. If the tangent at every point of this curve indicates the direction of the velocity of the fluid particle, then the curve is known as a *streamline*. When one steamline slides over the other, the flow is laminar. When the streamlines are interwoven with one other and there is transverse flow of fluid particles, the flow is *turbulent*. The transition from laminar to turbulent flow in a tube depends on the mean velocity u_m, the diameter of the tube D, the density ρ, and the viscosity μ of the fluid. These variables are grouped together in a dimensionless parameter, called Reynolds number (Re), which is

$$\text{Re} = \frac{u_m D \rho}{\mu} = \frac{u_m D}{\nu}$$

where ν = kinematic viscosity

$$= \frac{\mu}{\rho} \text{ (m}^2\text{/sec.)}$$

If Re < 2100, the flow is laminar, and if Re > 2100, the flow is turbulent. The flow is most often turbulent and hardly ever laminar.

The heat transfer coefficient may be evaluated from correlations developed by dimensional analysis. In this method, all the variables pertinent to the phenomenon are to be enlisted, either by intuition or experience. Four fundamental units are selec'ed, and these are mass M, length L, time T, and temperature θ.

Forced convection

$$h = f(D, u, \rho, \mu, K, c_p)$$

$$\text{Let } h = B \cdot D^a \cdot u^b \cdot \rho^c \cdot \mu^d \cdot K^e \cdot c_p{}^f$$

Expressing the variables in terms of their dimensions

$$MT^{-3}\,\theta^{-1} = B \cdot L^a \cdot (LT^{-1})^b \cdot (ML^{-3})^c \cdot (ML^{-1}\,T^{-1})^d$$

$$(MLT^{-3}\theta^{-1})^e \cdot (L^2 \cdot T^{-2}\,\theta^{-1})^f$$

$$= B.L^{a+b-3c-d+e+2f} \cdot T^{-b-d-3e-2f} \cdot M^{c+d+e} \cdot \theta^{-e-f}$$

$$\therefore \quad a + b - 3c - d + e + 2f = 0$$

$$-b - d - 3e - 2f = -3$$

$$c + d + e = 1$$

$$-e - f = -1$$

The solution of these equations gives

$$a = c - 1,\; b = c,\; d = -c + f,\; e = 1 - f$$

$$\therefore \quad h = B.D^{c-1} \cdot u^c \cdot \rho^c \cdot \mu^{-c+f} \cdot k^{1-f} \cdot c_p{}^f$$

$$\frac{hD}{K} = B\left(\frac{uD\rho}{\mu}\right)^c \cdot \left(\frac{\mu c_p}{K}\right)^f$$

$$\text{Nu} = B\,(\text{Re})^c \cdot (\text{Pr})^f \qquad (18.10)$$

where

$$\text{Nu} = \text{Nusselt number} = \frac{hD}{K}$$

$$\text{Re} = \text{Reynolds number} = \frac{uD\rho}{\mu}$$

$$\text{Pr} = \text{Prandtl number} = \frac{\mu c_p}{K}$$

The constants B, c, and f are evaluated from experimental data. For turbulent flow inside tubes, the following equation, attributed to Dittus and Boelter, may be used

$$\text{Nu} = 0.023\ \text{Re}^{0.8}\ \text{Pr}^n$$

where

$n = 0.4$, when the fluid is heated

$n = 0.3$, when the fluid is cooled

Free convection

Let a fluid at T_o, with density ρ_o, change to temperature T with density ρ.

Then the buoyancy force, $F = \dfrac{(\rho_0 - \rho)\, g}{\rho}$

Now, let β = coefficient of volume expansion

then
$$\frac{1}{\rho} = \frac{1}{\rho_0} + \beta\,(T_0 - T)$$

or
$$\rho_0 = \rho\,(1 + \beta \cdot \Delta T)$$

$\therefore$
$$F = \beta \cdot g \cdot \Delta T$$

where
$$\Delta T = T_0 - T$$

For an ideal gas

$$\beta \frac{1}{v}\left(\frac{\partial v}{\partial T}\right)_p = \frac{1}{v} \cdot \frac{R}{p} = \frac{1}{T}\ (K^{-1})$$

The heat transfer coefficient in free convection may be assumed to be a function of the variables as given below

$$h = f\,(L, K, c_p, \rho, \mu, g\,\beta\,\Delta T)$$

By dimensional analysis, the above variables may be arranged in three non-dimensional groups

$$\frac{hL}{K} = B\left(\frac{g\,\beta\Delta T \cdot L^3 \cdot \rho^2}{\mu^2}\right)^a \cdot \left(\frac{\mu\, c_p}{K}\right)^b$$

or
$$\text{Nu} = B.\ \text{Gr}^a.\ \text{Pr}^b$$

where

$$\text{Nu} = \text{Nusselt number} = \frac{hL}{K}$$

$$\text{Gr} = \text{Grashof number} = \frac{g\,\beta\Delta T L^3 \rho^2}{\mu^2}$$

$$\text{Pr} = \text{Prandtl number} = \frac{\mu c_p}{K}$$

For a large number of experiments made on fluids it has been found that exponents a and b are of the same value. So the expression reduces to

$$\text{Nu} = B.\ (\text{Gr. Pr})^{a} \tag{18.11}$$

L is the characteristic length, which is the length in the case of a vertical plate and cylinder, diameter in the case of a horizontal cylinder, and radius in the case of a sphere.

For Gr. Pr $< 10^9$, the flow is laminar, and

$$\text{Nu} = 0.59\ (\text{Gr. Pr})^{1/4}$$

and for Gr. Pr $> 10^9$, the flow is turbulent, and

$$\text{Nu} = 0.13\ (\text{Gr. Pr})^{1/3}$$

18.4 HEAT EXCHANGERS

A heat exchanger is a device in which heat is transferred between two moving fluids.

Heat exchangers may be parallel flow, counterflow or crossflow, depending upon the direction of the motion of the two fluids. If both the fluids move in the same direction, it is a parallel flow heat exchanger. If the fluids flow in the opposite directions, it is a counterflow heat exchanger. If they flow normal to each other, it is a crossflow heat exchanger.

18.4.1 Parallel Fow Heat Exchanger

Let us assume that the cold fluid (subscript c) is flowing through the inner tube and the hot fluid (subscript h) is flowing through the annulus. The hot fluid enters at t_{h1} and leaves at t_{h2}, while the cold fluid enters at t_{c1} and leaves at t_{c2}. Let us consider a differential length dL of the heat exchanger, as shown in Fig. 18.14, where the hot fluid is at t_h and the cold fluid is at t_c, and the temperature difference between the two fluids is $\Delta t\ (= t_h - t_c)$. Δt varies from Δt_i at the inlet to Δt_e at the exit of the heat exchanger. Let dQ be the rate of heat transfer in that differential length. Then by energy balance

$$dQ = U_o .\ dA_o .\ \Delta t$$

$$= -\dot{m}_h\, c_h\, dt_h$$

$$= \dot{m}_c\, c_c\, dt_c$$

Here dt_h is negative, because there is a decrease in t_h along the flow, where dt_c is positive. Symbols $\dot{m}$ and c represent the mass flow rate and specific heat respectively.

$$\Delta t = t_h - t_c$$

$$\therefore \quad d(\Delta t) = dt_h - dt_c$$

$$= -\frac{dQ}{\dot{m}_h c_h} - \frac{dQ}{\dot{m}_c c_c} = -dQ\left(\frac{1}{\dot{m}_h c_h} + \frac{1}{\dot{m}_c c_c}\right)$$

$$= -\mu_p \cdot dQ$$

where μ_p is used for the expression in parenthesis.

$$\int_{\Delta t_i}^{\Delta t_e} d(\Delta t) = -\int_i^e \mu_p \, dQ \text{ or } \mu_p = \frac{\Delta t_i - \Delta t_e}{Q}$$

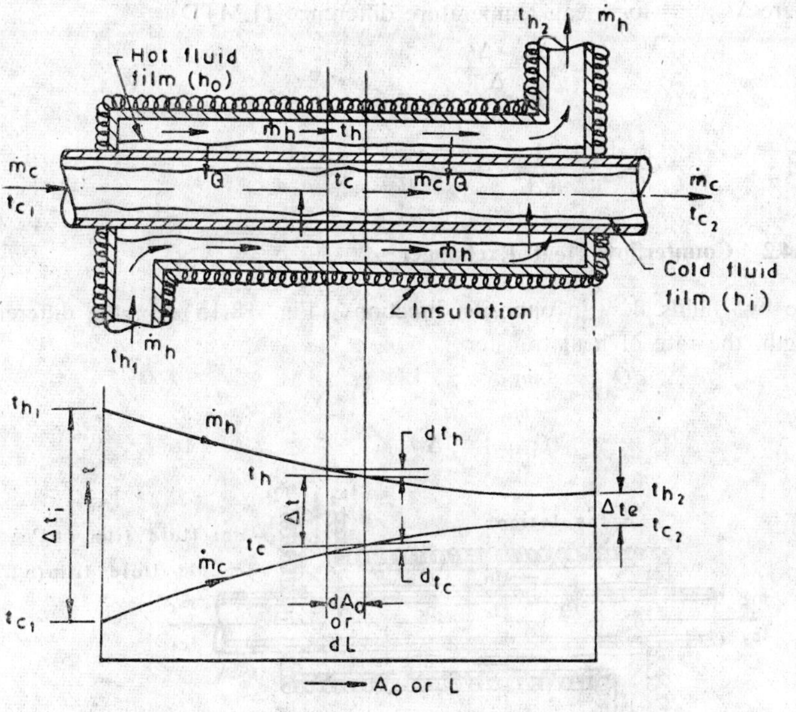

Fig. 18.14 **Parallel flow heat exchanger**

Also $\quad dQ = U_0 dA_0 \Delta t$

or
$$-\frac{d(\Delta t)}{\mu_p} = U_0 \, dA_0 \, \Delta t$$

$$\therefore \quad -\int_{\Delta t_i}^{\Delta t} \frac{d(\Delta t)}{\Delta t} = \int_i^e \mu_p U_0 \, dA_{60}$$

$$\therefore \quad \ln \frac{\Delta t_i}{\Delta t_e} = \mu_p U_0 A_0$$

Substituting the expression for μ_p

$$\ln \frac{\Delta t_i}{\Delta t_e} = \frac{\Delta t_i - \Delta t_e}{Q} \cdot U_0 A_0$$

$$\therefore \qquad Q = \frac{\Delta t_i - \Delta t_e}{\ln \dfrac{\Delta t_i}{\Delta t_e}} U_0 A_0$$

where $\Delta t_i = t_{h_1} - t_{c_1}$ and $\Delta t_e = t_{h_2} - t_{c_2}$

$$\therefore \qquad Q = U_o A_o \, \Delta t_{1.m.}$$

$$= \dot{m}_h c_h (t_{h_1} - t_{h_2}) \tag{18.12}$$

$$= \dot{m}_c c_c (t_{c_2} - t_{c1})$$

where $\Delta t_{l.m.}$ = log-mean temperature difference (LMTD)

$$= \frac{\Delta t_i - \Delta t_e}{\ln \dfrac{\Delta t_i}{\Delta t_e}}$$

and $$\frac{1}{U_o A_o} = \frac{1}{h_i A_i} + \frac{x_w}{k_m A_{1.m.}} + \frac{1}{h_o A_o}$$

18.4.2 Counterflow Heat Exchanger

The two fluids flow in opposite directions (Fig. 18.15). In the differential length, the rate of heat transfer

$$dQ = - \dot{m}_h c_h dt_h$$

$$= - \dot{m}_c c_c dt_c$$

$$= U_0 \cdot d A_0 \cdot \Delta t$$

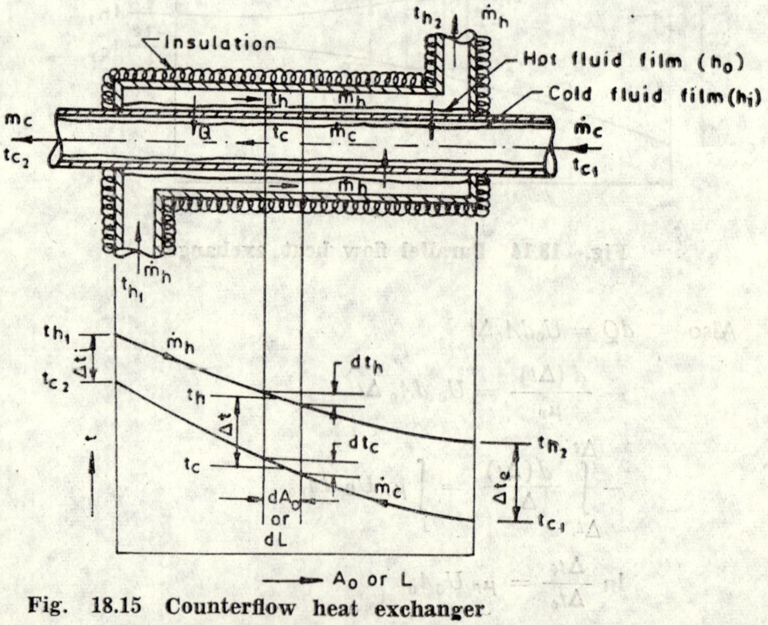

Fig. 18.15 Counterflow heat exchanger

where both dt_h and dt_c are negative for positive x direction (towards the right). Now

$$\Delta t = t_h - t_c$$

$$d(\Delta t) = dt_h - dt_c$$

$$= -\frac{dQ}{\dot{m}_h c_h} + \frac{dQ}{\dot{m}_c c_c}$$

$$= -dQ\left(\frac{1}{\dot{m}_h c_h} - \frac{1}{\dot{m}_c c_c}\right)$$

$$= -dQ\,\mu_c$$

where

$$\mu_c = \frac{1}{\dot{m}_h c_h} - \frac{1}{\dot{m}_c c_c}$$

$$\int_{\Delta t_i}^{\Delta t_e} d(\Delta t) = \int_i^e -dQ \cdot \mu_c$$

$$\Delta t_i - \Delta t_e = \mu_c \cdot Q$$

or

$$\mu_c = \frac{\Delta t_i - \Delta t_e}{Q}$$

Again

$$dQ = U_0\, dA_0\, \Delta t$$

$$-\frac{d(\Delta t)}{\mu_c} = U_0 \cdot dA_0 \cdot \Delta t$$

$$\int_{\Delta t_i}^{\Delta t_e} -\frac{d(\Delta t)}{\Delta t} = \int_i^e U_0\, dA_0 \cdot \mu_c$$

$$\ln\frac{\Delta t_i}{\Delta t_e} = U_0 \cdot A_0 \cdot \mu_c = U_0 A_0 \cdot \frac{\Delta t_i - \Delta t_e}{Q}$$

$\therefore$

$$Q = U_0 A_0\, \Delta t_{l.m.}$$

where

$$\Delta t_{l.m.} \frac{\Delta t_i - \Delta t_e}{\ln\frac{\Delta t_i}{\Delta t_e}}$$

$$\Delta t_i = t_{h1} - t_{c2}, \quad \Delta t_e = t_{h2} - t_{c1}$$

$$\frac{1}{U_0 A_0} = \frac{1}{h_i A_i} + \frac{x_w}{K_w A_{l.m}} + \frac{1}{h_0 A_0}$$

and when x_w is small,

$$\frac{1}{U_0} \simeq \frac{1}{h_i} + \frac{x_w}{K_w} + \frac{1}{h_0}$$

$\therefore$

$$Q = U_0 A_0\, \Delta t_{l.m.}$$

$$= \dot{m}\, c_h (t_{h1} - t_{h2}) = \dot{m}_c\, c_c (t_{c2} - t_{c1}) \qquad (18.13)$$

For the same rate of heat transfer, and inlet and exit temperatures, ($\Delta t_{l.m.}$) counterflow is greater than ($\Delta t_{l.m.}$) parallel flow. So the surface area required is less for counterflow operation. For parallel flow, $t_{h_2} - t_{c_2}$, i.e., the hot fluid cannot be cooled below t_{c_2} or the cold fluid cannot be heated above t_{h_2}. But for counterflow operation, t_{h_2} may be less than t_{c_2} which means that the hot fluid can be cooled below t_{c_2} or the cold fluid heated above t_{h_2}. For these reasons, counterflow heat exchangers are much more common in practice.

When one of the two fluids undergoes phase change (at constant temperature and pressure), e.g., condensation and evaporation, $\Delta t_{l \cdot m.}$ is the same for parallel flow and counterflow (Fig. 18.16), and the heating surface required is also the same.

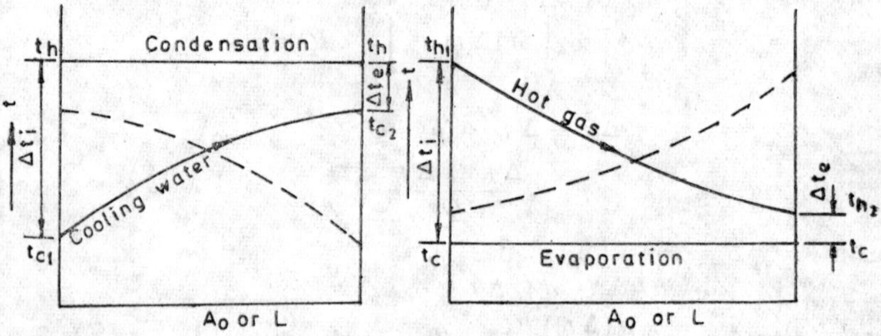

Fig. 18.16 Heat transfer with phase change

18.5 RADIATION HEAT TRANSFER

All bodies radiate heat. The phenomenon is identical to the emission of light. Two similar bodies isolated together in a vacuum radiate heat to each other, but the colder body will receive more heat than the hot body and thus become heated.

If Q is the total radiant energy incident upon the surface of a body, some part of it (Q_a) will be absorbed, some (Q_r) will be reflected, and some (Q_t) will be transmitted through the body. Therefore

$$Q = Q_a + Q_r + Q_t$$

or

$$\frac{Q_a}{Q} + \frac{Q_r}{Q} + \frac{Q_t}{Q} = 1$$

or

$$\alpha + \rho + \tau = 1$$

where α is known as *absorptivity*, ρ as *reflectivity*, and τ as transmissivity.

For an *opaque* body, $\tau = 0$ and $\alpha + \rho = 1$. Most solids are opaque.

A body which absorbs all the incident radiation is called a *black body*. A black body is also the best radiator. Most radiating surfaces are gray and have an emissivity factor ε less than unity, where

$$\varepsilon = \frac{\text{Actual radiation of gray body at } T°K}{\text{Radiation of a black body at } T °K}$$

It can be shown that the *emissivity* or ability to radiate heat is equal to the *absorptivity* or ability to absorb heat (Kirchhoff's law), which jsutifies the statement that good absorbers are also good emitters. A brightly polished surface will have a low absorptivity and low emissivity.

The rate at which energy is radiated by a black body at temperature T (K) is given by the *Stefan-Boltzmann law*

$$Q = \sigma A T^4$$

where Q = rate of energy radiation, W

A = surface area radiating heat, m^2

and σ = Stefan-Boltzmann constant = 5.67×10^{-8} W/m².K⁴

$= 4.88 \times 10^{-8}$ kcal/m² hr K⁴

The radiant heat exchange between two gray bodies at temperatures T_1 and T_2 depends on how the two bodies view each other and their emissivities, and it is given by

$$Q_{1-2} = \sigma A_1 \mathcal{J}_{1-2} (T_1^4 - T_2^4) \qquad (18.15)$$

where $\mathcal{J}_{1-2}$ = view factor or configuration factor for gray bodies

= fraction of total radiant energy leaving gray surface 1 and reaching gray surface 2.

It can be shown that (See Reference 4),

$$\mathcal{J}_{1-2} = \frac{1}{\left(\frac{1}{\varepsilon_1} - 1\right) + \frac{1}{F_{1-2}} + \frac{A_1}{A_2}\left(\frac{1}{\varepsilon_2} - 1\right)}$$

where ε_1 and ε_2 are the emissivities of the two bodies of surface areas A_1 and A_2, and F_{12} is the configuration factor of two similar black bodies, or the fraction of energy that leaves the black surface 1 and is incident on the black surface 2. It can be shown that

$$A_1 F_{12} = A_2 F_{21}$$

which is known as the reciprocity theorem.

Below are given the expressions of the view factor for some simple situations.

(a) For two infinite parallel gray planes, $A_1 = A_2$, $F_{12} = 1$

($\because$ All the energy leaving surface 1 strikes surface 2)

$$\therefore \quad \mathcal{J}_{1-2} = \frac{1}{\frac{1}{\varepsilon_1} + \frac{1}{\varepsilon_2} - 1}$$

(b) For two concentric cylinders or spheres, $F_{12} = 1$, A_1 = surface area of the inner cylinder or sphere

$$\mathcal{J}_{1-2} = \frac{1}{\dfrac{1}{\varepsilon_1} + \dfrac{A_1}{A_2}\left(\dfrac{1}{\varepsilon_2} - 1\right)}$$

(c) When the enclosed body (area A_1) is very small compared to the enclosure surface, $A_2 \gg A_1$, then

$$\mathcal{J}_{1-2} = \varepsilon_1$$

18.5.1 Combined Convection and Radiation

Heat is transferred from a hot body both by natural convection and radiation.
Rate of heat transfer by natural convection

$$Q_c = h_c\, A\, (t_w - t_f)$$

where h_c is the convective heat transfer coefficient.
Rate of heat transfer by radiation

$$Q_r = \sigma A_1 \mathcal{J}_{1-2}\, (T_w^4 - T_f^4) = h_r\, A_1\, (t_w - t_f)$$

where h_r is known as the radiation heat transfer coefficient.

$$\therefore \quad h_r = \sigma \mathcal{J}_{1-2}\, (T_w + T_f)\, (T_w^2 + T_f^2)$$

$\therefore$ Total rate of heat transfer

$$Q = Q_c + Q_r = (h_c + h_r)\, A_1\, (t_w - t_f)$$

Illustrative Examples

EXAMPLE 18.1

A cold storage room has walls made of 0.23 m of brick on the outside, 0.08 m of plastic foam, and finally 1.5 cm of wood on the inside. The outside and inside air temperatures are 22°C and −2°C respectively. If the inside and outside heat transfer coefficients are respectively 29 and 12 W/m²K, and the thermal conductivities of brick, foam, and wood are 0.98, 0.02, and 0.17 W/mK respectively, determine (a) the rate of heat removal by refrigeration if the total wall area is 90 m², and (b) the temperature of the inside surface of the brick.

Solution

Figure Ex. 18.1 shows the wall of the cold storage

$$\frac{1}{U} = \frac{1}{h_o} + \frac{x_1}{k_1} + \frac{x_2}{k_2} + \frac{x_3}{k_3} + \frac{1}{h_i}$$

$$= \frac{1}{12} + \frac{0.23}{0.98} + \frac{0.08}{0.02} + \frac{1.5}{100 \times 0.17} + \frac{1}{29}$$

$$\frac{1}{U} = 0.0833 + 0.2347 + 4.0 + 0.0892 + 0.0345$$

$$= 4.4407 \text{ m}^2 \text{ k/W}$$

$$\therefore \quad U = 0.2252 \text{ W/m}^2\text{K}$$

$\because$ Rate of heat transfer

$$Q = UA \ (t_o - t_i) = 0.2252 \times 90 \, [22 - (-2)]$$

$$= 486.4 \text{ W} \qquad \text{Ans.}$$

Now

$$R_1 + R_2 = \frac{1}{h_0} + \frac{x_1}{k_1} = \frac{1}{12} + \frac{0.23}{0.98} = 0.318$$

$$\therefore \quad t_0 - t_2 = \frac{486.4 \times 0.318}{90} = 1.72 \text{ °C}$$

$$\therefore \quad t_2 = 22 - 1.72 = 20.28\text{°C} \qquad \text{Ans.}$$

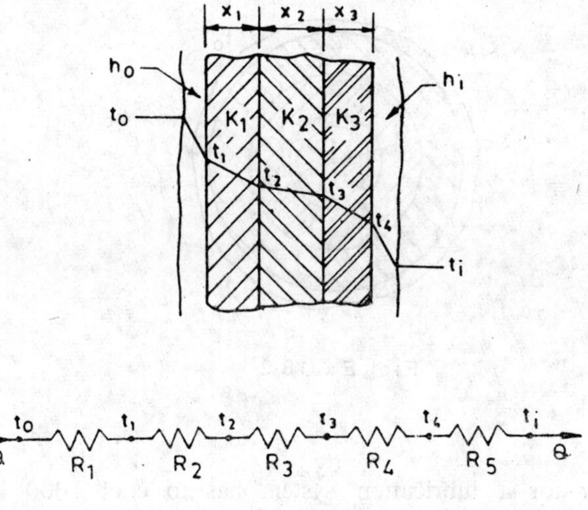

Fig. Ex. 18·1

EXAMPLE 18.2

Hot air at a temperature of 60°C is flowing through a steel pipe of 10 cm diameter. The pipe is covered with two layers of different insulating materials of thicknesses 5 cm and 3 cm, and their corresponding thermal conductivities are 0.23 and 0.37 W/mK. The inside and outside heat transfer coefficients are 58 and 12 W/m²K. The atmosphere is at 25°C. Find the rate of heat loss from a 50 m length of pipe. Neglect the resistance of the steel pipe.

Solution

As shown in Fig. Ex. 18.2

$$r_1 = 5 \text{ cm}, \; r_2 = 10 \text{ cm}, \; r_3 = 13 \text{ cm}$$

$$k_1 = 0.23 \text{ and } k_2 = 0.37 \text{ W/mK}$$

$$h_i = 58 \text{ and } h_o = 12 \text{ W/m}^2\text{K}$$

$$t_i = 60°\text{C}, \; t_o = 25°\text{C}$$

Rate of heat transfer

$$Q = \frac{2\pi L(t_i - t_o)}{\dfrac{1}{h_i r_1} + \dfrac{1}{k_1}\ln\dfrac{r_2}{r_1} + \dfrac{1}{k_2}\ln\dfrac{r_3}{r_2} - \dfrac{1}{h_o r_3}}$$

$$= \frac{2\pi \times 50\,(60 - 25)}{\dfrac{1}{58 \times 0.05} - \dfrac{1}{0.23}\ln\dfrac{10}{5} + \dfrac{1}{0.37}\ln\dfrac{13}{10} + \dfrac{1}{12 \times 0.13}}$$

$$= \frac{6.28 \times 50 \times 35}{0.3448 + 0.7092 + 0.3448 + 0.6410} = 2334 \text{ W}$$

$$= 2.334 \text{ kW} \qquad \text{Ans.}$$

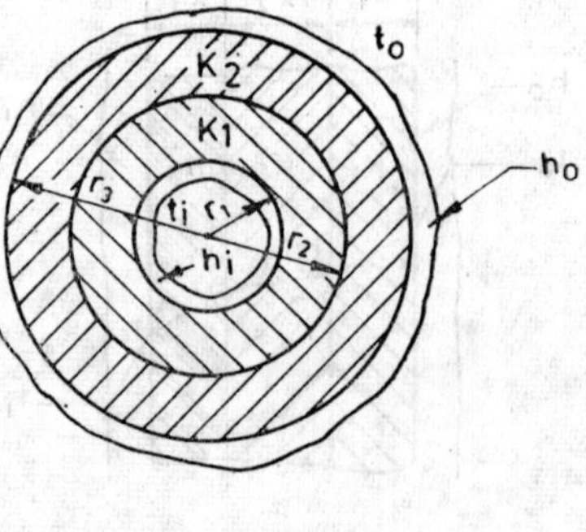

Fig. Ex. 18·2

EXAMPLE 18.3

An oil cooler for a lubrication system has to cool 1000 kg/hr of oil (c_p = 2.09 kJ/kg-K) from 80°C to 40°C by using a cooling water flow of 1000 kg/hr available at 30°C. Give your choice for a parallel flow or counter-flow heat exchanger, with reasons. Estimate the surface area of the heat exchanger, if the overall heat transfer coefficient is 24 W/m²K (c_p of water = 4.18 kJ/kg-K).

Solution

Rate of heat transfer

$$Q = \dot{m}_h c_h (t_{h2} - t_{h1}) = \dot{m}_c c_c (t_{c2} - t_{c1})$$

$$1000 \times 2.09 \; (80 - 40) = 1000 \times 4.18 \quad (t_{c2} - 30)$$

$$\therefore \; t_{c2} = 50°C$$

Since $t_{c2} > t_{h2}$, counterflow arrangement must be used (Fig. Ex. 18.3).

$$(\Delta t)_{l.m.} = \frac{30-10}{\ln \frac{30}{10}} = 18.2°C$$

$$Q = 1000 \times 2.09 \times 40 = 83.6 \times 10^3 = U_o \; A_o \; \Delta t_{l.m.}$$

$$\therefore \quad A_o = \frac{83,600}{24 \times 18.2 \times 3.6}$$

$$= 53.16 \text{ m}^2$$ Ans.

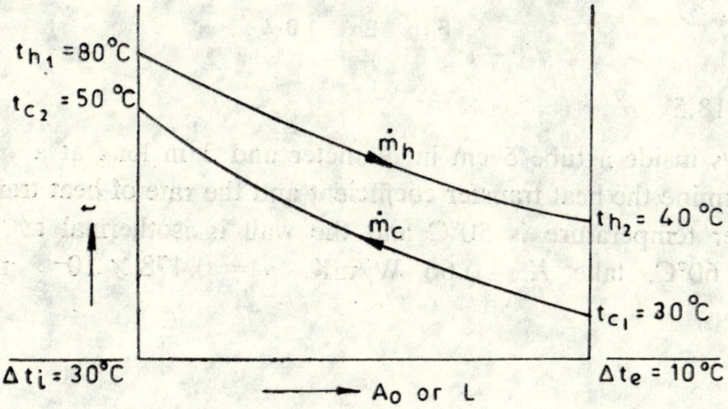

Fig. Ex. 18·3

EXAMPLE 18.4

Water is evaporated continuously at 100°C in an evaporator by cooling 500 kg/hr of air from 260°C to 150°C. Calculate the heat transfer surface area required and the steam evaporation rate per hour if the liquid enters at 100°C. Take U_o = 46 W/m²K and c_p of air = 1.005 kJ/kg-K.

Solution

At 100°C, h_{fg} = 2257 kJ/kg. If $\dot{m}$ is the rate of evaporation

$$Q = \dot{m}\, h_{fg} = \dot{m}_a \, c_{pa} \, (t_{a1} - t_{a2})$$

$$\therefore \quad \dot{m} \times 2257 = 500 \times 1.005 \; (260 - 150)$$

$$\therefore \quad \dot{m} = 24.49 \text{ kg/hr}$$ Ans.

As shown in Fig. Ex. 18.4

$$\Delta t_{l.m.} \; \frac{160-50}{\ln \frac{160}{50}} = 94.5°C$$

$$\therefore \quad Q = U_o \; A_o \; \Delta t_{l.m.} = 500 \times 1.005 \times 110$$

$$\therefore \quad A_o = \frac{500 \times 1.005 \times 110}{46 \times 94.5 \times 3.6} = 3.53 \text{ m}^2$$ Ans.

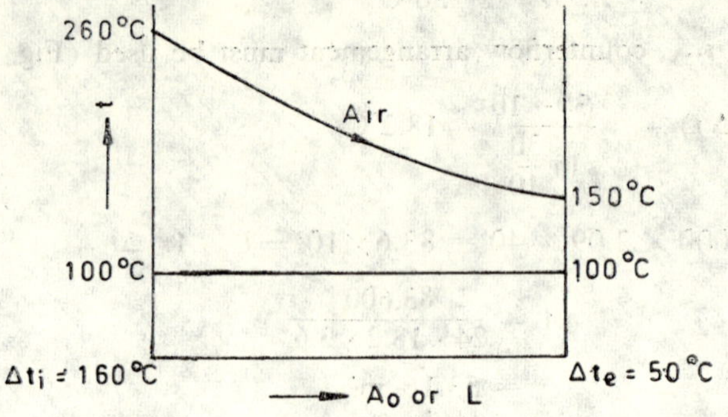

Fig. Ex. 18·4

EXAMPLE 18.5

Water flows inside a tube 5 cm in diameter and 3 m long at a velocity 0.8 m/s. Determine the heat transfer coefficient and the rate of heat transfer if the mean water temperature is 50°C and the wall is isothermal at 70°C. For water at 60°C, take $K = 0.66$ W/mK, $\nu = 0.478 \times 10^{-6}$ m²/s, and $Pr = 2.98$.

Solution

$$\text{Reynolds number, Re} = \frac{u_m D}{\nu} = \frac{0.8 \times 0.05}{0.478 \times 10^{-6}}$$
$$= 83{,}700$$

Prandtl number, $Pr = 2.98$

From the Dittus and Boelter equation

$$Nu = 0.023\,(Re)^{0.8}\,(Pr)^{0.4}$$

$$\frac{hD}{K} = 0.023\,(83{,}700)^{0.8}\,(2.98)^{0.4}$$

$$\frac{h \times 0.05}{0.66} = 0.023 \times 8673.6 \times 1.5478$$

$$\therefore \quad h = 4075 \text{ W/m}^2\text{K}$$ Ans.

$$\therefore \quad Q = hA\,(t_w - t_f) = 4075 \times \pi \times 0.05 \times 3 \times 20$$
$$= 38387 \text{ W} = 38.39 \text{ kW}$$ Ans.

EXAMPLE 18.6

An electrically heated plate 15 cm high and 10 cm wide is maintained at 140°C. Estimate the rate of heat dissipation from both sides of the plate in an atmosphere at 20°C, if the radiation heat transfer coefficient is 8.72 W/m²K. For air at the mean temperature of 80°C, take $\nu = 21.09 \times 10^{-6}$ m²/s, $Pr = 0.692$, and $K = 0.0305$ W/mK.

Solution

$$Q = (h_c + h_r)\ A\ (t_w - t_f)$$

where $A = 0.1 \times 0.15 \times 2 = 0.03\ \text{m}^2$

$$h_r = 8.72\ \text{W/m}^2\text{K},\ t_w - t_f = 140 - 20 = 120°\text{C}$$

$$\beta = \frac{1}{v}\left(\frac{\partial v}{\partial T}\right)_p = \frac{1}{T_{\text{mean}}} = \frac{1}{273+80} = \frac{1}{353}\ (K)^{-1}$$

$$L = 0.15\ \text{m}$$

$$\therefore \quad \text{Grashoff number, Gr} = \frac{g\ \beta \Delta T_3}{\nu^2}$$

$$= \frac{9.81 \times 120 \times (0.15)^3}{353 \times (21.09 \times 10^{-6})^2} = 25.4 \times 10^6$$

Gr. Pr $= 25.4 \times 10^6 \times 0.692 = 17.6 \times 10^6$

Since the product Gr. Pr is less than 10^9

$$\text{Nu} = 0.59\ (\text{Gr. Pr})^{1/4} = 0.59\ (17.6 \times 10^6)^{1/4} = 38.2$$

$$\therefore \quad h_c = \frac{38.2 \times 0.0305}{0.15} = 7.77\ \text{W/m}^2\ \text{K}$$

$$\therefore \quad Q = (8.72 + 7.77) \times 0.03 \times 120$$

$$= 59.36\ W$$

Ans.

EXAMPLE 18.7

A long steel rod, 2 cm in diameter, is to be heated from 427°C to 538°C. It is placed concentrically in a long cylindrical furnace which has an inside diameter of 16 cm. The inner surface of the furnace is at a temperature of 1093°C, and has an emissivity of 0.85. If the surface of the rod has an emissivity of 0.6, find the time required for the heating operation. Take for steel, $\rho = 7845\ \text{kg/m}^3$ and $c = 0.67\ \text{kJ/kg-K}$.

Solution

The surface area of the rod (Fig. Ex. 18.7), $A_1 = \pi \times 2 \times L\ \text{cm}^2$, and the surface area of the furnace, $A_2 = \pi \times 16 \times L\ \text{cm}^2$

$$\therefore \quad \frac{A_1}{A_2} = \frac{1}{8}$$

Now

$$F_{1-2} = \frac{1}{\dfrac{1}{\varepsilon_1} + \dfrac{A_1}{A_2}\left(\dfrac{1}{\varepsilon_2} - 1\right)} = \frac{1}{\dfrac{1}{0.6} + \dfrac{1}{8}\left(\dfrac{1}{0.85} - 1\right)}$$

$$= \frac{1}{1.684} = 0.592$$

Initial rate of heat absorption by radiation, when the rod is at 427°C or 700 K

$$Q_i = \sigma\ A_1\ F_{1-2}\ (T_1^4 - T_2^4)$$

$$= 5.67 \times 10^{-8} \times \pi \times 2 \times 10^{-2} \times 1 \times 0.592\ (700^4 - 1366^4)$$

$$= -6332\ \text{W/m}$$

Rate of heat absorption at the end of the heating process, when the rod is at 538°C

$$Q_e = 5.67\times10^{-8}\times p\times2\times10^{-2}\times1\times0.592\,(811^4-1366^4)$$

$$= -\,5937\ \text{W/m}$$

$\therefore$ Average rate of heat absorption during the heating process

$$Q_{av.} = \frac{6322+5937}{2} = 6130\ \text{W/m}$$

Time required for heating, τ is obtained from the equation

$$m\,c_p\,\Delta T = Q_{av},\ \tau = 6130\ \tau$$

$$\therefore\qquad \tau = \frac{7845\times\pi/4\times10^{-4}\times1\times0.67\times111}{6130\times10^{-3}}\ \text{seconds}$$

$$= 29.88\ \text{seconds}$$ Ans.

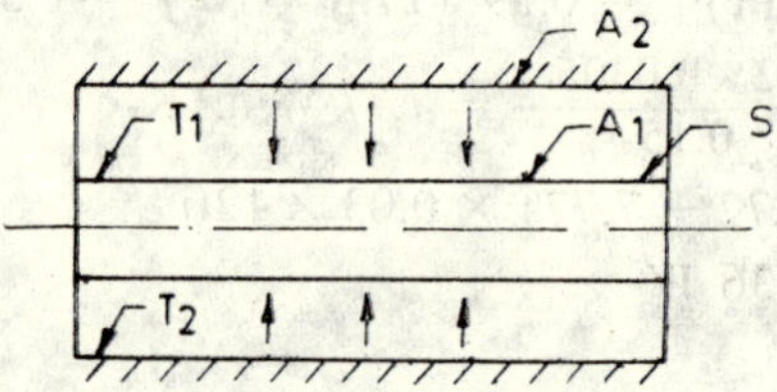

Fig. Ex. 18·7

PROBLEMS

18.1 A room has a brick wall 25 cm in thickness. The inside air is at 25°C and the outside air is at −15°C. The heat transfer coefficients on the inside and outside are 8.72 and 28 W/m²K respectively. The thermal conductivity of brick is 0.7 W/mK. Find the rate of heat transfer through the wall and the inside surface temperature.

18.2 For the wall in the above problem, it is proposed to reduce the heat transfer by fixing an insulating board (K = 0.05 W/mK), 2.5 cm in thickness, to the inside surface. Find the rate of heat transfer and the inside surface temperature.

18.3 Sheets of brass and steel, each 1 cm thick, are placed in contact. The outer surface of brass is kept at 100°C and the outer surface of steel is kept at 0°C. What is the temperature of the common interface ? The thermal conductivities of brass and steel are in the ratio of 2 : 1. Ans. 66.7°C.

18.4 In a pipe carrying steam, the outside surface (15 cm OD) is at 300°C. The pipe is to be covered with insulation (K=0.07 W/mK) such that the outside surface temperature does not exceed 60°C. The atmosphere is at 25°C and the heat transfer coefficient is 11.6 W/m²K. Find the thickness of insulation required and the rate of heat loss per m length of pipe.

18.5 The passenger compartment of a jet transport is essentially a cylindrical tube of diameter 3 m and length 20 m. It is lined with 3 cm of insulating material (K=0.04 W/mK), and must be maintained at 20°C for passenger comfort, although the average outside temperature is −30°C at its operating height. What rate of heating is required in the compartment, neglecting the end effects ?

18.6 A hollow sphere ($K=35$ W/mK), the inner and outer diameters of which are 28 cm and 32 cm respectively, is heated by means of a 20 ohm coil placed inside the sphere. Calculate the current required to keep the two surfaces at a constant temperature difference of 50°C, and calculate the rate of heat supply.

18.7 (a) Develop an expression for the steady state heat transfer rate through the walls of a spherical container of inner radius r_0 and outer radius r_1. The temperatures are t_0 and t_1 at radii r_0 and r_1 respectively. Assume that the thermal conductivity of the wall varies as

$$K = K_o + (K_1 - K_o)\frac{t - t_o}{t_1 - t_o}$$

(b) Estimate the rate of evaporation of liquid oxygen from a spherical container, 1.8 m ID, covered with 30 cm of asbestos insulation. The temperatures at the inner and outer surfaces of the insulation are −183°C and 0°C respectively. The boiling point of oxygen is −183°C and the latent heat of vaporization is 215 kJ/kg. The thermal conductivities of the insulation are 0.16 and 0.13 W/mK at 0°C and −183°C respectively.

Ans. 19.8 kg/hr.

18.8 A counterflow double-pipe heat exchanger using superheated steam is used to heat water at the rate of 10,500 kg/hr. The steam enters the heat exchanger at 180°C and leaves at 130°C. The inlet and exit temperatures of water are 30°C and 80°C respectively. If the overall heat transfer coefficient from steam to water is 814 W/m^{2}K, calculate the heat transfer area. What would the increase in area be if the fluid flows were parallel?

18.9 An oil cooler consists of a straight tube of ID 1.25 cm, wall thickness 0.125 cm, enclosed within a pipe and concentric with it. The external surface of the pipe is well insulated. Oil flows through the tube at the rate of 250 kg/hr and cooling water flows in the annulus at the rate of 300 kg/hr in the direction opposite to that of oil. The oil enters the tube at 180°C and is cooled to 66°C. The cooling water enters at 10°C. Estimate the length of tube required, given that the heat transfer coefficient from the oil to the tube surface is 1628 and that from the tube surface to the water is 3722 W/m^{2}K. Neglect the temperature drop across the tube wall. c_p of oil = 1.675 kJ/kg-K.

Ans. 2.81 m.

18.10 A marine steam turbine has a condenser flow (of steam) at a full load of 12,000 kg/hr. The quality of steam at the turbine exhaust is 0.85. The condenser pressure is 0.07 bar. Sea water used for circulation is at 21°C. The terminal temperature difference between the steam and outgoing water is 5.7°C. The circulating water velocity is maintained at 1.7 m/s. The condenser tubes are of 1.3 cm ID and 0.122 cm thickness. Determine (a) the rate of flow of cooling water, (b) the length of tubes, and (c) the number of tubes, in the condenser. Take $U_0 = 3256$ W/m^{2}K.

Ans. 490 t/hr, 6.8 m, 604.

18.11 Fifty kg of water per min is heated from 30°C to 50°C by passing through a pipe of 2 cm in diameter. The pipe is heated by condensing steam on its surface maintained at 100°C. Find the length of the pipe required. Take the following properties of water at the mean bulk temperature of 40°C:

$\rho = 992.2$ kg/m^3, $\nu = 0.659 \times 10^{-6}$ m^2/s, $K = 0.63$ W/mK, and Pr–4.31.

18.12 Air flows at a velocity of 0.3 m/s through the annular space between two concentric tubes. The outer tube is 5 cm ID and the inner tube is 3.125 cm OD. The air enters at 16°C and leaves at 32°C. The temperature of the outside surface of the inner tube is 50°C. Determine the heat transfer coefficient between the air and inner tube. Properties of air at 24°C : $\rho = 1.18$ kg/m^3, $K = 0.03$ W/mK, Pr=0.7, and $\nu = 15.5 \times 10^{-6}$ m^2/s.

18.13 A rectangular duct, 30 cm × 20 cm in cross-section, carries cold air. The temperature of the outer surface of the duct is 5°C and the surrounding air temperature is 25°C. Estimate the rate of heat gain by the duct, assuming that the duct, one metre in length, is exposed to the air in the vertical position. Properties of air at 15°C : $\rho = 1.22$ kg/m.³, $\nu = 14.6 \times 16^{-6}$ m²/s, $K = 0.03$ W/mK, and Pr = 0.7.

Ans. 0.186 W.

18.14 A hot square plate, 50 cm×50 cm at 100°C, is exposed to atmospheric air at 20°C. Compute the rate of heat loss from both the surfaces of the plate if the plate is kept in the vertical plane. Properties of air at 60°C :

$\rho = 1.06$ kg/m³, $\nu = 18.97 \times 10^{-6}$ m²/s, $K = 0.03$ W/mk, and Pr = 0.696.

18.15 Determine the heat lost by radiation per metre length of 7.5 cm diameter oxidized steel pipe at 300°C if (a) located in a large room with red brick walls at a temperature of 25°C, and (b) enclosed in a 25 cm×25 cm red brick conduit at a temperature of 25°C. The emissivity of oxidized steel is 0.79 and that of red brick is 0.93.

Ans. 1.052, 1.035 kW.

18.16 Two concentric spheres, 21 cm and 30 cm in diameter, with the space between them evacuated, are to be used to store liquid air (−153°C) in a room at 27°C. The surfaces of the spheres are flushed with aluminium ($\varepsilon = 0.03$) and the latent heat of vaporization of liquid air is 209.35 kJ/kg. Calculate the rate of evaporation of liquid air.

Ans. 0.0216 kg/hr.

18.17 Estimate the net radiant heat exchange per square metre for two very large parallel planes at temperatures 560°C and 300°C respectively. Assume that the emissivities of the hot and cold planes are 0.8 and 0.6 respectively.

Ans. 11.28 kW.

APPENDICES

Appendix A—Steam Tables*

Table A.1(a) Saturated Steam: Temperature Table

Temperature °C t	*Pressure* kPa p	*Specific Volume* *Sat. Liquid* v_f cm³/g	*Sat. Vapour* v_g m³/kg	*Enthalpy* kJ/kg *Sat. Liquid* h_f	*Evap.* h_{fg}	*Sat. Vapour* h_g	*Entropy* kJ/kg-K *Sat. Liquid* s_f	*Sat. Vapour* s_g
0.01	0.6113	1.000	206.14	0.01	2501.3	2501.4	0.0000	9.1562
5	0.8721	1.000	147.12	20.98	2489.6	2510.6	0.0761	9.0257
10	1.2276	1.000	106.38	42.01	2477.7	2519.8	0.1510	8.9008
15	1.7051	1.001	77.93	62.99	2465.9	2528.9	0.2245	8.7814
20	2.339	1.002	57.79	83.96	2454.1	2538.1	0.2966	8.6672
25	3.169	1.003	43.36	104.89	2442.3	2547.2	0.3674	8.5580
30	4.246	1.004	32.89	125.79	2430.5	2556.3	0.4369	8.4533
35	5.628	1.006	25.22	146.68	2418.6	2565.3	0.5053	8.3531
40	7.384	1.008	19.52	167.57	2406.7	2574.3	0.5725	8.2570
45	9.593	1.010	15.26	188.45	2394.8	2583.2	0.6387	8.1648
50	12.349	1.012	12.03	209.33	2382.7	2592.1	0.7038	8.0763
55	15.758	1.015	9.568	230.23	2370.7	2600.9	0.7679	7.9913

*Abridged and adapted from Joseph H. Keenan, Frederick G Keyes, Philip G. Hill, and Joan G. Moore, *Steam Tables*, John Wiley, New York, 1969, (with the kind permission of the publishers, John Wiley & Sons, Inc., New York).

t	p	v_f	v_g	h_f	h_{fg}	h_g	s_f	s_g
60	19.940	1.017	7.671	251.13	2358.5	2609.6	0.8312	7.9096
65	25.03	1.020	6.197	272.06	2346.2	2618.3	0.8935	7.8310
70	31.19	1.023	5.042	292.98	2333.8	2626.8	0.9549	7.7553
75	38.58	1.026	4.131	313.93	2321.4	2635.3	1.0155	7.6824
80	47.39	1.029	3.407	334.91	2308.8	2643.7	1.0753	7.6122
85	57.83	1.033	2.828	355.90	2296.0	2651.9	1.1343	7.5445
90	70.14	1.036	2.361	376.92	2283.2	2660.1	1.1925	7.4791
95	84.55	1.040	1.982	397.96	2270.2	2668.1	1.2500	7.4159
100	0.10135	1.044	1.6729	419.04	2257.0	2676.1	1.3069	7.3549
105	0.12082	1.048	1.4194	440.15	2243.7	2683.8	1.3630	7.2958
110	0.14327	1.052	1.2102	461.30	2230.2	2691.5	1.4185	7.2387
115	0.16906	1.056	1.0366	482.48	2216.5	2699.0	1.4734	7.1833
120	0.19853	1.060	0.8919	503.71	2202.6	2706.3	1.5276	7.1296
125	0.2321	1.065	0.7706	524.99	2188.5	2713.5	1.5813	7.0775
130	0.2701	1.070	0.6685	546.31	2174.2	2720.5	1.6344	7.0269
135	0.3130	1.075	0.5822	567.69	2159.6	2727.3	1.6870	6.9777
140	0.3613	1.080	0.5089	589.13	2144.7	2733.9	1.7391	6.9299
145	0.4154	1.085	0.4463	610.63	2129.6	2740.3	1.7907	6.8833
150	0.4758	1.091	0.3928	632.20	2114.3	2746.5	1.8418	6.8379
160	0.6178	1.102	0.3071	675.55	2082.6	2758.1	1.9427	6.7502

t	p	v_f	v_g	h_f	h_{fg}	h_g	s_f	s_g
170	0.7917	1.114	0.2428	719.21	2049.5	2768.7	2.0419	6.6663
180	1.0021	1.127	0.19406	763.22	2015.0	2778.2	2.1396	6.5857
190	1.2544	1.141	0.15654	807.62	1978.8	2786.4	2.2359	6.5079
200	1.5538	1.157	0.12736	852.45	1940.7	2793.2	2.3309	6.4323
210	1.9062	1.173	0.10441	897.76	1900.7	2798.5	2.4248	6.3585
220	2.318	1.190	0.08619	943.62	1858.5	2802.1	2.5178	6.2861
230	2.795	1.209	0.07158	990.12	1813.8	2804.0	2.6099	6.2146
240	3.344	1.229	0.05976	1037.32	1766.5	2803.8	2.7015	6.1437
250	3.973	1.251	0.05013	1085.36	1716.2	2801.5	2.7927	6.0730
260	4.688	1.276	0.04221	1134.37	1662.5	2796.9	2.8838	6.0019
270	5.499	1.302	0.03564	1184.51	1605.2	2789.7	2.9751	5.9301
280	6.412	1.332	0.03017	1235.99	1543.6	2779.6	3.0668	5.8571
290	7.436	1.366	0.02557	1289.07	1477.1	2766.2	3.1594	5.7821
300	8.581	1.404	0.02167	1344.0	1404.9	2749.0	3.2534	5.7045
310	9.856	1.447	0.018350	1401.3	1326.0	2727.3	3.3493	5.6230
320	11.274	1.499	0.015488	1461.5	1238.6	2700.1	3.4480	5.5362
340	14.586	1.638	0.010797	1594.2	1027.9	2622.0	3.6594	5.3357
360	18.651	1.893	0.006945	1760.5	720.5	2481.0	3.9147	5.0526
374.14	22.09	3.155	0.003155	2099.3	0	2099.3	4.4298	4.4298

Table A.1 (b) Saturated Steam: Pressure Table

Pressure kPa p	Temperature °C t	Specific Volume: Sat. Liquid v_f cm³/kg	Specific Volume: Sat. Vapour v_g m³/kg	Enthalpy kJ/kg: Sat Liquid h_f	Enthalpy kJ/kg: Evap. h_{fg}	Enthalpy kJ/kg: Sat. Vapour h_g	Entropy kJ/kgK: Sat. Liquid s_f	Entropy kJ/kgK: Sat. Vapour s_g
0.6113	0.01	1.000	206.14	.01	2501.3	2501.4	0.0000	9.1562
1.0	6.98	1.000	129.21	29.30	2484.9	2514.2	0.1059	8.9756
1.5	13.03	1.001	87.98	54.71	2470.6	2525.3	0.1957	8.8279
2.0	17.50	1.001	67.00	73.48	2460.0	2533.5	0.2607	8.7237
2.5	21.08	1.002	54.25	88.49	2451.6	2540.0	0.3120	8.6432
3.0	24.08	1.003	45.67	101.05	2444.5	2545.5	0.3545	8.5776
4.0	28.96	1.004	34.80	121.46	2432.9	2554.4	0.4226	8.4746
5.0	32.88	1.005	28.19	137.82	2423.7	2561.5	0.4764	8.3951
7.5	40.29	1.008	19.24	168.79	2406.0	2574.8	0.5764	8.2515
10	45.81	1.010	14.67	191.83	2392.8	2584.7	0.6493	8.1502
15	53.97	1.04	10.02	225.94	2373.1	2599.1	0.7549	8.0085
20	60.06	1.017	7.649	251.40	2358.3	2609.7	0.8320	7.9085
25	64.97	1.020	6.204	271.93	2346.3	2618.2	0.8931	7.8314
30	69.10	1.022	5.229	289.23	2336.1	2625.3	0.9439	7.7686
40	75.87	1.027	3.993	317.58	2319.2	2636.8	1.0259	7.6700

p	t	v_f	v_g	h_f	h_{fg}	h_g	s_f	s_g
50	81.33	1.030	3.240	340.49	2305.4	2645.9	1.0910	7.5936
75	91.78	1.037	2.217	384.39	2278.6	2663.0	1.2130	7.4564
MPa								
0.10	99.63	1.043	1.6940	417.46	2258.0	2675.5	1.3026	7.3594
0.15	111.37	1.053	1.1593	467.11	2226.5	2693.6	1.4336	7.2233
0.20	120.23	1.061	0.8857	504.70	2201.9	2706.7	1.5301	7.1271
0.25	127.44	1.067	0.7187	535.37	2181.5	2716.9	1.6072	7.0527
0.30	133.55	1.073	0.6058	561.47	2163.8	2725.3	1.6718	6.9919
0.35	138.88	1.079	0.5243	584.33	2148.1	2732.4	1.7275	6.9405
0.40	143.63	1.084	0.4625	604.74	2133.8	2738.6	1.7766	6.8959
0.45	147.93	1.088	0.4140	623.25	2120.7	2743.9	1.8207	6.8565
0.50	151.86	1.093	0.3749	640.23	2108.5	2748.7	1.8607	6.8213
0.60	158.85	1.101	0.3157	670.56	2086.3	2756.8	1.9312	6.7600
0.70	164.97	1.108	0.2729	697.22	2066.3	2763.5	1.9922	6.7080
0.80	170.43	1.115	0.2404	721.11	2048.0	2769.1	2.0462	6.6628
0.90	175.38	1.121	0.2150	742.83	2031.1	2773.9	2.0946	6.6226
1.00	179.91	1.127	0.19444	762.	2015.3	2778.1	2.1387	6.5865
1.20	187.99	1.139	0.16333	798.65	1986.2	2784.8	2.2166	6.5233
1.40	195.07	1.149	0.14084	830.30	1959.7	2790.0	2.2842	6.4693

p	t	v_f	v_g	h_f	h_{fg}	h_g	s_f	s_g
1.5	198.32	1.154	0.13177	844.89	1947.3	2792.2	2.3150	6.4448
2.0	212.42	1.177	0.09963	908.79	1890.7	2799.5	2.4474	6.3409
2.5	223.99	1.197	0.07998	962.11	1841.0	2803.1	2.5547	6.2575
3.0	233.90	1.217	0.06668	1008.42	1795.7	2804.2	2.6457	6.1869
3.5	242.60	1.235	0.05707	1049.75	1753.7	2803.4	2.7253	6.1253
4	250.40	1.252	0.04978	1087.31	1714.1	2801.4	2.7964	6.0701
5	263.99	1.286	0.03944	1154.23	1640.1	2794.3	2.9202	5.9734
6	275.64	1.319	0.03244	1213.35	1571.0	2784.3	3.0267	5.8892
7	285.88	1.351	0.02737	1267.00	1505.1	2772.1	3.1211	5.8133
8.	295.06	1.384	0.02352	1316.64	1441.3	2758.0	3.2068	5.7432
9	303.40	1.418	0.02048	1363.26	1378.9	2742.1	3.2858	5.6772
10	311.06	1.452	0.018026	1407.65	1317.1	2724.7	3.3596	5.6141
11	318.15	1.489	0.015987	1450.1	1255.5	2705.6	3.4295	5.5527
12	324.75	1.527	0.014263	1491.3	1193.6	2684.9	3.4962	5.4924
13	330.93	1.567	0.012780	1531.5	1130.7	2662.2	3.5606	5.4323
14	336.75	1.611	0.011485	1571.1	1066.5	2637.6	3.6232	5.3717
15	342.24	1.658	0.010337	1610.5	1000.0	2610.5	3.6848	5.3098
16	347.44	1.711	0.009306	1650.1	930.6	2580.6	3.7461	5.2455
18	357.06	1.840	0.007489	1732.0	777.1	2509.1	3.8715	5.1044
20	365.81	2.036	0.005834	1826.3	583.4	2409.7	4.0139	4.9269
22.09	374.14	3.155	0.003155	2099.3	0	2099.3	4.4298	4.4298

Table A.2 Superheated Vapour

	p = .010 MPa (45.81)			p = .050 MPa (81.33)			p = .10 MPa (99.63)		
t °C	v m³-kg	h kJ/kg	s kJ/kg-K	v m³/kg	h kJ/kg	s kJ/kg-K	v m³/kg	h kJ/kg	s kJ/kg-K
Sat.	14.674	2584.7	8.1502	3.240	2645.9	7.5939	1.6940	2675.5	7.3594
50	14.869	2592.6	8.1749						
100	17.196	2687.5	8.4479	3.418	2682.5	7.6947	1.6958	2676.2	7.3614
150	19.512	2783.0	8.6882	3.889	2780.1	7.9401	1.9364	2776.4	7.6134
200	21.825	2879.5	8.9038	4.356	2877.7	8.1580	2.172	2875.3	7.8343
250	24.136	2977.3	9.1002	4.820	2976.0	8.3556	2.406	2974.3	8.0333
300	26.445	3076.5	9.2813	5.284	3075.5	8.5373	2.639	3074.3	8.2158
400	31.063	3279.6	9.6077	6.209	3278.9	8.8642	3.103	3278.2	8.5435
500	35.679	3489.1	9.8978	7.134	3488.7	9.1546	3.565	3488.1	8.8342
600	40.259	3705.4	10.1608	8.057	3705.1	9.4178	4.028	3704.7	9.0976
700	44.911	3928.7	10.4028	8.981	3928.5	9.6599	4.490	3928.2	9.3398
800	49.526	4159.0	10.6281	9.904	4158.9	9.8852	4.952	4158.6	9.5652
	p = .20 MPa (120.23)			p = .30 MPa (133.55)			p = .40 MPa (143.63)		
Sat.	0.8857	2706.7	7.1272	0.6058	2725.3	6.9919	0.4625	2738.6	6.8959
150	0.9596	2768.8	7.2795	0.6339	2761.0	7.0778	0.4708	2752.8	6.9299
200	1.0803	2870.5	7.5066	0.7163	2865.6	7.3115	0.5342	2860.5	7.1706
250	1.1988	2971.0	7.7086	0.7964	2967.6	7.5166	0.5951	2964.2	7.3789
300	1.3162	3071.8	7.8926	0.8753	3069.3	7.7022	0.6548	3066.8	7.5662
400	1.5493	3276.6	8.2218	1.0315	3275.0	8.0330	0.7726	3273.4	7.8985

t	v	h	s	v	h	s	v	h	s
500	1.7814	3487.1	8.5133	1.1867	3486.0	8.3251	0.8893	3484.9	8.1913
600	2.013	3704.0	8.770	1.3414	3703.2	8.5892	1.0055	3702.4	8.4558
700	2.244	3927.6	9.0194	1.4957	3927.1	8.8319	1.1215	3926.5	8.6987
800	2.475	4158.2	9.2449	1.6499	4157.8	9.0576	1.2372	4157.3	8.9244
	$p = .50$ MPa (151.86)			$p = .60$ MPa (158.85)			$p = .08$ MPa (170.43)		
Sat.	0.3749	2748.7	6.8213	0.3157	2756.8	6.7600	0.2404	2769.1	6.6628
200	0.4249	2855.4	7.0592	0.3520	2850.1	6.9665	0.2608	2839.3	6.8158
250	0.4744	2960.7	7.2709	0.3938	2957.2	7.1816	0.2931	2950.0	7.0384
300	0.5226	3064.2	7.4599	0.4344	3061.6	7.3724	0.3241	3056.5	7.2328
350	0.5701	3167.7	7.6329	0.4742	3165.7	7.5464	0.3544	3161.7	7.4089
400	0.6173	3271.9	7.7938	0.5137	3270.3	7.7079	0.3843	3267.1	7.5716
500	0.7109	3483.9	8.0873	0.5920	3482.8	8.0021	0.4433	3480.6	7.8673
600	0.8041	3701.7	7.3522	0.6697	3700.9	8.2674	0.5018	3699.4	8.1333
700	0.8969	3925.9	8.5952	0.7472	3925.3	8.5107	0.5601	3924.2	8.3770
800	0.9896	4156.9	8.8211	0.8245	4156.5	8.7367	0.6181	4155.6	8.6033
	$p = 1.00$ MPa (179.91)			$p = 1.20$ MPa (187.99)			$p = 1.40$ MPa (195.07)		
Sat.	0.19444	2778.1	6.5865	0.16333	2784.8	6.5233	0.14084	2790.0	6.4693
200	0.2060	2827.9	6.6940	0.16930	2815.9	6.5898	0.14302	2803.3	6.4975
250	0.2327	2942.6	6.9247	0.19234	2935.0	6.8294	0.16350	2927.2	6.7467

t	v	h	s	v	h	s	v	h	s
300	0.2579	3051.2	7.1229	0.2138	3045.8	7.0317	0.18228	3040	6.9534
350	0.2825	3157.7	7.3011	0.2345	3153.6	7.2121	0.2003	3149.5	7.1360
400	0.3066	3263.9	7.4651	0.2548	3260.7	7.3774	0.2178	3257.5	7.3026
500	0.3541	3478.5	7.7622	0.2946	3476.3	7.6759	0.2521	3474.1	7.6027
600	0.4011	3697.9	8.0290	0.3339	3696.3	7.9435	0.2860	3694.8	7.8710
700	0.4478	3923.1	8.2731	0.3729	3922.0	8.1881	0.3195	3920.8	8.1160
800	0.4943	4154.7	8.4996	0.4118	4153.8	8.4148	0.3528	4153.0	8.3431
	p = 1.60 MPa (201.41)			p = 1.80 MPa (207.15)			p = 2.00 MPa (212.42)		
Sat.	0.12380	2794.0	6.4218	0.11042	2797.1	6.3794	0.09963	2799.5	6.3409
225	0.13287	2857.3	6.5518	0.11673	2846.7	6.4808	0.10377	2835.8	6.4147
250	0.14184	2919.2	6.6732	0.12497	2911.0	6.6066	0.11144	2902.5	6.5453
300	0.15862	3034.8	6.8844	0.14021	3029.2	6.8226	0.12547	3023.5	6.7664
350	0.17456	3145.4	7.0694	0.15457	3141.2	7.0100	0.13857	3137.0	6.9563
400	0.19005	3254.2	7.2374	0.16847	3250.9	7.1794	0.15120	3247.6	7.1271
500	0.2203	3472.0	7.5390	0.19550	3469.8	7.4825	0.17568	3467.6	7.4317
600	0.2500	3693.2	7.8080	0.2220	3691.7	7.7523	0.19960	3690.1	7.7024
700	0.2794	3919.7	8.0535	0.2482	3918.5	7.9983	0.2232	3917.4	7.9487
800	0.3086	4152.1	8.2808	0.2742	4151.2	8.2258	0.2467	4150.3	8.1765

t	v	h	s	v	h	s	v	h	s
	p = 2.50 MPa (223.99)			p = 3.00 MPa (233.90)			p = 3.50 MPa (242.60)		
Sat.	0.07998	2803.1	6.2575	0.06668	2804.2	6.1869	0.05707	2803.4	6.1253
225	0.08027	2806.3	6.2639						
250	0.08700	2880.1	6.4085	0.07058	2855.8	6.2872	0.05872	2829.2	6.1749
300	0.09890	3008.8	6.6438	0.08114	2993.5	6.5390	0.06842	2977.5	6.4461
350	0.10976	3126.3	6.8403	0.09053	3115.3	6.7428	0.07678	3104.0	6.6579
400	0.12010	3239.3	7.0148	0.09936	3230.9	6.9212	0.08453	3222.3	6.8405
450	0.13014	3350.8	7.1746	0.10787	3344.0	7.0834	0.09196	3337.2	7.0052
500	0.13998	3462.1	7.3234	0.11619	3456.5	7.2338	0.09918	3450.9	7.1572
600	0.15930	3686.3	7.5960	0.13243	3682.3	7.5085	0.11324	3678.4	7.4339
700	0.17832	3914.5	7.8435	0.14838	3911.7	7.7571	0.12699	3908.8	7.6837
800	0.19716	4148.2	8.0720	0.16414	4145.9	7.9862	0.14056	4143.7	7.9134
	p = 4.0 MPa (250.40)			p = 4.5 MPa (257.49)			p = 5.0 MPa (263.99)		
Sat.	0.04978	2801.4	6.0701	0.04406	2798.3	6.0198	0.03944	2794.3	5.9734
275	0.05457	2886.2	6.2285	0.04730	2863.2	6.1401	0.04141	2838.3	6.0544
300	0.05884	2960.7	6.3615	0.05135	2843.1	6.2828	0.04532	2924.5	6.2084
350	0.06645	3092.5	6.5821	0.05840	3080.6	6.5131	0.05194	3068.4	6.4493
400	0.07341	3213.6	6.7690	0.06475	3204.7	6.7047	0.05781	3195.7	6.6459

t	v	h	s	v	h	s	v	h	s
450	0.08002	3330.3	6.9363	0.07074	3323.3	6.8746	0.06330	3316.2	6.8186
500	0.08643	3445.3	7.0901	0.07651	3439.6	7.0301	0.06857	3433.8	6.9759
600	0.09885	3674.4	7.3688	0.08765	3670.5	7.3110	0.07869	3666.5	7.2589
700	0.11095	3905.9	7.6198	0.09847	3903.0	7.5631	0.08849	3900.1	7.5122
800	0.12287	4141.5	7.8502	0.10911	4139.3	7.7942	0.09811	4137.1	7.7440
	p = 6.0 MPa (275.64)			p = 7.0 MPa (285.88)			p = 8.0 MPa (295.06)		
Sat.	0.03244	2784.3	5.8892	0.02737	2772.1	5.8133	0.02352	2758.0	5.7432
300	0.03616	2884.2	6.0674	0.02947	2838.4	5.9305	0.02426	2785.0	5.7906
350	0.04223	3043.0	6.3335	0.03524	3016.0	6.2283	0.02995	2987.3	6.1301
400	0.04739	3177.2	6.5408	0.03993	3158.1	6.4487	0.03432	3138.3	6.3634
450	0.05214	3301.8	6.7193	0.04416	3287.1	6.6327	0.03817	3272.0	6.5551
500	0.05665	3422.2	6.8803	0.04814	3410.3	6.7975	0.04175	3398.3	6.7240
550	0.06101	3540.6	7.0288	0.05195	3530.9	6.9486	0.04516	3521.0	6.8778
600	0.06525	3658.4	7.1677	0.05565	3650.3	7.0894	0.04845	3642.0	7.0206
700	0.07352	3894.2	7.4234	0.06283	3888.3	7.3476	0.05481	3882.4	7.2812
800	0.8160	4132.7	7.6566	0.06981	4128.2	7.5822	0.06097	4123.8	7.5173
	p = 9.0 MPa (303.40)			p = 10.0 MPa (311.06)			p = 12.5 MPa (327.89)		
Sat.	0.02048	2742.1	5.6772	0.018026	2724.7	5.6141	0.013495	2673.8	5.4624
325	0.02327	2856.0	5.8712	0.019861	2809.1	5.7568			
350	0.02580	2956.6	6.0361	0.02242	2923.4	5.9443	0.016126	2826.2	5.7118

t	*v*	*h*	*s*	*v*	*h*	*s*	*v*	*h*	*s*
400	0.02993	3117.8	6.2853	0.02641	3096.5	6.2120	0.02000	3039.3	6.0417
450	0.03350	3256.6	6.4844	0.02975	3240.9	6.4190	0.02299	3199.8	6.2719
500	0.03677	3386.1	6.6576	0.03279	3373.7	6.5966	0.02560	3341.8	6.4618
550	0.03987	3511.0	6.8142	0.03564	3500.9	6.7561	0.02801	3475.2	6.6290
600	0.04285	3633.7	6.9589	0.03837	3625.3	6.9029	0.03029	3604.0	6.7810
650	0.04574	3755.3	7.0943	0.04101	3748.2	7.0398	0.03248	3730.4	6.9218
700	0.04857	3876.5	7.2221	0.04358	3870.5	7.1687	0.03460	3855.3	7.0536
800	0.05409	4119.3	7.4536	0.04859	4114.8	7.4077	0.03869	4103.6	7.2965
	$p = 15.0$ MPa (342.24)			$p = 17.5$ MPa (354.75)			$p = 20.0$ MPa (365.81)		
Sat.	0.010337	2610.5	5.3098	0.007920	2528.8	5.1419	0.005834	2409.7	4.9269
350	0.011470	2692.4	5.4421						
400	0.015649	2975.5	5.8811	0.012447	2902.9	5.7213	0.009942	2818.1	5.5540
450	0.018445	3156.2	6.1404	0.015174	3109.7	6.0184	0.012695	3060.1	5.9017
500	0.02080	3308.6	6.3443	0.017358	3274.1	6.2383	0.014768	3238.2	6.1401
550	0.02293	3448.6	6.5199	0.019288	3421.4	6.4230	0.016555	3393.5	6.3348
600	0.02491	3582.3	6.6776	0.02106	3560.1	6.5866	0.018178	3537.6	6.5048
650	0.02680	3712.3	6.8224	0.02274	3693.9	6.7357	0.019693	3675.3	6.6582
700	0.02861	3840.1	6.9572	0.02434	3824.6	6.8736	0.02113	3809.0	6.7993
800	0.03210	4092.4	7.2040	0.02738	4081.1	7.1244	0.02385	4069.7	7.0544

Appendix B—Thermodynamic Properties of Refrigerant-12* (Dichlorodifluoromethane)

Table B.1 Saturated Refrigerant-12

		Specific Volume		*Enthalpy*			*Entropy*	
Temperature °C	*Abs. Pressure* MPa	*Sat. Liquid*	*Sat. Vapour*	*Sat. Liquid*	*Evap.*	*Sat. Vapour*	*Sat. Liquid*	*Sat. Vapour*
t	p	v_f cm³/kg	v_g m³/kg	h_f	h_{fg} kJ/kg	h_g	s_f kJ/kg-K	s_g
—90	0.0028	0.608	4.415545	—43.243	189.618	146.375	—0.2084	0.8268
—85	0.0042	0.612	3.037316	—38.968	187.608	148.640	—0.1854	0.8116
—80	0.0062	0.617	2.138345	—34.688	185.612	150.924	—0.1630	0.7979
—75	0.0088	0.622	1.537651	—30.401	183.625	153.224	—0.1411	0.7855
—70	0.0123	0.627	1.127280	—26.103	181.640	155.536	—0.1197	0.7744
—65	0.0168	0.632	0.841166	—21.793	179.651	157.857	—0.0987	0.7643
—60	0.0226	0.637	0.637910	—17.469	177.653	160.184	—0.0782	0.7552
—55	0.0300	0.642	0.491000	—13.129	175.641	162.512	—0.0581	0.7470
—50	0.0391	0.648	0.383105	— 8.772	173.611	164.840	—0.0384	0.7396

* Adapted from *Fundamentals of Classical Thermodynamics* by G. J. Van Wylen and R. Sonntag, John Wiley, New York, 1976, P. 667-673 (with the kind permission of the publishers, John Wiley & Sons, Inc., New York).

t	p	v_f	v_g	h_f	h_{fg}	h_g	s_f	s_g
—45	0.0504	0.654	0.302682	— 4.396	171.558	167.163	—0.0190	0.7329
—40	0.0642	0.659	0.241910	— 0.000	169.479	169.479	—0.0000	0.7269
—35	0.0807	0.666	0.195398	4.416	167.368	171.784	0.0187	0.7214
—30	0.1004	0.672	0.159375	8.854	165.222	174.076	0.0371	0.7165
—25	0.1237	0.679	0.131166	13.315	163.037	176.352	0.0552	0.7121
—20	0.1509	0.685	0.108847	17.800	160.810	178.610	0.0730	0.7082
—15	0.1826	0.693	0.091018	22.312	158.534	180.846	0.0906	0.7046
—10	0.2191	0.700	0.076646	26.851	156.207	183.058	0.1079	0.7014
— 5	0.2610	0.708	0.064963	31.420	153.823	185.243	0.1250	0.6986
0	0.3086	0.716	0.055389	36.022	151.376	187.397	0.1418	0.6960
5	0.3626	0.724	0.047485	40.659	148.859	189.518	0.1585	0.6937
10	0.4233	0.733	0.040914	45.337	146.265	191.602	0.1750	0.6916
15	0.4914	0.743	0.035413	50.058	143.586	193.644	0.1914	0.6897
20	0.5673	0.752	0.030780	54.828	140.812	195.641	0.2076	0.6879
25	0.6516	0.763	0.026854	59.653	137.933	197.586	0.2237	0.6863
30	0.7449	0.774	0.023508	64.539	134.936	199.475	0.2397	0.6848
35	0.8477	0.786	0.020641	69.494	131.805	201.299	0.2557	0.6834
40	0.9607	0.798	0.018171	74.527	128.525	203.051	0.2716	0.6820
45	1.0843	0.811	0.016032	79.647	125.074	204.722	0.2875	0.6806
50	1.2193	0.826	0.014170	84.868	121.430	206.298	0.3034	0.6792

t	p	v_f	v_g	h_f	h_{fg}	h_g	s_f	s_g
55	1.3663	0.841	0.012542	90.201	117.565	207.766	0.3194	0.6777
60	1.5259	0.858	0.011111	95.665	113.443	209.109	0.3355	0.6760
65	1.6988	0.877	0.009847	101.279	109.024	210.303	0.3518	0.6742
70	1.8858	0.897	0.008725	107.067	104.255	211.321	0.3683	0.6721
75	2.0874	0.920	0.007723	113.058	99.068	212.126	0.3851	0.6697
80	2.3046	0.946	0.006821	119.291	93.373	212.665	0.4023	0.6667
85	2.5380	0.976	0.006005	125.818	87.047	212.865	0.4201	0.6631
90	2.7885	1.012	0.005258	132.708	79.907	212.614	0.4385	0.6585
95	3.0569	1.056	0.004563	140.068	71.658	211.726	0.4579	0.6526
100	3.3440	1.113	0.003903	148.076	61.768	209.843	0.4788	0.6444
105	3.6509	1.197	0.003242	157.085	49.014	206.099	0.5023	0.6319
110	3.9784	1.364	0.002462	168.059	28.425	196.484	0.5322	0.6064

Table B.2 Superheated Refrigerant-12

t °C	v m³/kg	h kJ/kg	s kJ/kg-K	v m³/kg	h kJ/kg	s kJ/kg-K	v m³/kg	h kJ/kg	s kJ/kg-K
		0.05 MPa			0.10 MPa			0.15 MPa	
— 20.0	0.341857	181.042	0.7912	0.167701	179.861	0.7401			
— 10.0	0.356227	186.757	0.8133	0.175222	185.707	0.7628	0.114716	184.619	0.7318
0.0	0.370508	192.567	0.8350	0.182647	191.628	0.7849	0.119866	190.660	0.7543
10.0	0.384716	198.471	0.8562	0.189994	197.628	0.8064	0.124932	196.762	0.7763
20.0	0.398863	204.469	0.8770	0.197277	203.707	0.8275	0.129930	202.927	0.7977
30.0	0.412959	210.557	0.8974	0.204506	209.866	0.8482	0.134873	209.160	0.8186
40.0	0.427012	216.733	0.9175	0.211691	216.104	0.8684	0.139768	215.463	0.8390
50.0	0.441030	222.997	0.9372	0.218839	222.421	0.8883	0.144625	221.835	0.8591
60.0	0.455017	229.344	0.9565	0.225955	228.815	0.9078	0.149450	228.277	0.8787
70.0	0.468978	235.774	0.9755	0.233044	235.285	0.9269	0.154247	234.789	0.8980
80.0	0.482917	242.282	0.9942	0.240111	241.829	0.9457	0.159020	241.371	0.9169
90.0	0.496838	248.868	1.0126	0.247159	248.446	0.9642	0.163774	248.020	0.9354
		0.20 MPa			0.25 MPa			0.30 MPa	
0.0	0.088608	189.669	0.7320	0.069752	188.644	0.7139	0.057150	187.583	0.6984
10.0	0.092550	195.878	0.7543	0.073024	194.969	0.7366	0.059984	194.034	0.7216
20.0	0.096418	202.135	0.7760	0.076218	201.322	0.7587	0.062734	200.490	0.7440
30.0	0.100228	208.446	0.7972	0.079350	207.715	0.7801	0.065418	206.969	0.7658
40.0	0.103989	214.814	0.8178	0.082431	214.153	0.8010	0.068049	213.480	0.7869

t	v	h	s	v	h	s	v	h	s
50.0	0.107710	221.243	0.8381	0.085470	220.642	0.8214	0.070635	220.030	0.8075
60.0	0.111397	227.735	0.8578	0.088474	227.185	0.8413	0.073185	226.627	0.8276
70.0	0.115055	234.291	0.8772	0.091449	233.785	0.8608	0.075750	233.273	0.8473
80.0	0.118690	240.910	0.8962	0.094398	240.443	0.8800	0.078200	239.971	0.8665
90.0	0.122304	247.593	0.9149	0.097327	247.160	0.8987	0.080673	246.723	0.8853
100.0	0.125901	254.339	0.9332	0.100238	253.936	0.9171	0.083127	253.530	0.9038
110.0	0.129483	261.147	0.9512	0.103134	260.770	0.9352	0.085566	260.391	0.9220
		0.40 MPa			0.50 MPa			0.60 MPa	
20.0	0.045836	198.762	0.7199	0.035646	196.935	0.6999			
30.0	0.047971	205.428	0.7423	0.037464	203.814	0.7230	0.030422	202.116	0.7063
40.0	0.050046	212.095	0.7639	0.039214	210.656	0.7452	0.031966	209.154	0.7291
50.0	0.052072	218.779	0.7849	0.040911	217.484	0.7667	0.033450	216.141	0.7511
60.0	0.054059	225.488	0.8054	0.042565	224.315	0.7875	0.034887	223.104	0.7723
70.0	0.056014	232.230	0.8253	0.044184	232.161	0.8077	0.036285	230.062	0.7929
80.0	0.057941	239.012	0.8448	0.045774	238.031	0.8275	0.037653	237.027	0.8129
90.0	0.059846	245.837	0.8638	0.047340	244.932	0.8467	0.038995	244.009	0.8324
100.0	0.061731	252.707	0.8825	0.048886	251.869	0.8656	0.040316	251.016	0.8514
110.0	0.063600	259.624	0.9008	0.050415	258.845	0.8840	0.041619	258.053	0.8700

t	v	h	s	v	h	s	v	h	s
		0.70 MPa			0.80 MPa			0.90 MPa	
40.0	0.026761	207.580	0.7148	0.022830	205.924	0.7016	0.019744	204.170	0.6982
50.0	0.028100	214.745	0.7373	0.024068	213.290	0.7248	0.020912	211.765	0.7131
60.0	0.029387	221.854	0.7590	0.025247	220.558	0.7469	0.022012	218.212	0.7358
70.0	0.030632	228.931	0.7799	0.026380	227.766	0.7682	0.023062	226.564	0.7575
80.0	0.031843	235.997	0.8002	0.027477	234.941	0.7888	0.024072	233.856	0.7785
90.0	0.033027	243.066	0.8199	0.028545	242.101	0.8088	0.025051	141.113	0.7987
100.0	0.034189	250.146	0.8392	0.029588	249.260	0.8283	0.026005	248.355	0.8184
110.0	0.035332	257.247	0.8579	0.030612	256.428	0.8472	0.026937	255.593	0.8376
		1.00 MPa			1.20 MPa			1.40 MPa	
50.0	0.018366	210.162	0.7021	0.014483	206.661	0.6812			
60.0	0.019410	217.810	0.7254	0.015463	214.805	0.7060	0.012579	211.457	0.6876
70.0	0.020397	225.319	0.7476	0.016368	222.687	0.7293	0.013448	219.822	0.7123
80.0	0.021341	232.739	0.7689	0.017221	230.398	0.7514	0.014247	227.891	0.7355
90.0	0.022251	240.101	0.7895	0.018032	237.995	0.7727	0.014997	235.766	0.7575
100.0	0.023133	247.430	0.8094	0.018812	245.518	0.7931	0.015710	243.512	0.7785
110.0	0.023993	254.743	0.8287	0.019567	252.993	0.8129	0.016393	251.170	0.7988
		1.60 MPa			1.80 MPa			2.00 MPa	
70.0	0.011208	216.650	0.6959	0.009406	213.049	0.6794			
80.0	0.011984	225.177	0.7204	0.010187	222.198	0.7057	0.008704	218.859	0.6909
90.0	0.012698	233.390	0.7433	0.010884	230.835	0.7298	0.009406	228.056	0.7166
100.0	0.013366	241.397	0.7651	0.011526	239.155	0.7524	0.010035	236.760	0.7402
110.0	0.014000	249.264	0.7859	0.012126	2467.264	0.7739	0.010615	245.154	0.7624

Appendix C*—Enthalpy of Formation at 25°C, Ideal Gas Enthalpy, and Absolute Entropy at 0.1 MPa (1 bar) Pressure

	Nitrogen, Diatomic (N_2) $(h^\circ_f)_{298} = 0$ kJ/kmol M=28.013		*Oxygen, Diatomic* (O_2) $(\bar{h}^\circ_f)_{298} = 0$ kJ/kmol M=31.999	
Temp. K	$(\bar{h}^\circ - \bar{h}^\circ_{298})$ kJ/kmol	s° kJ/kmol K	$(\bar{h}^\circ - \bar{h}^\circ_{298})$ kJ/kmol K	$\bar{s}^\circ$ kJ/kmol K
0	—8669	0	—8682	0
100	—5770	159.813	—5778	173.306
200	—2858	179.988	—2866	193.486
298	0	191.611	0	205.142
300	54	191.791	54	205.322
400	2971	200.180	3029	213.874
500	5912	206.740	6088	220.698
600	8891	212.175	9247	226.455
700	11937	216.866	12502	231.272
800	15046	221.016	15841	235.924
900	18221	224.757	19246	239.936
1000	21406	288.167	22707	243.585
1100	24757	231.309	26217	246.928
1200	28108	234.225	29765	250.016
1300	31501	236.941	33351	252.886
1400	34936	239.484	36966	255.564
1500	38405	241.878	40610	258.078
1600	41903	244.137	44279	260.446
1700	45430	246.275	47970	262.685
1800	48982	248.304	51689	264.810
1900	52551	250.237	55434	266.835
2000	56141	252.078	59199	268.764

* Adapted from Tables A.11 (Pages 687 to 696) in *Fundamentals of Classical Thermodynamics* by G.J. Van Wylen and R. Sonntag, John Wiley, New York, 1976 (with the kind permission of the publishers, John Wiley & Sons, Inc., New York).

	Carbon Dioxide (CO_2) $(\bar{h}^\circ_f)_{298} = -393522$ kJ/kmol M = 44.01		Carbon Monoxide (CO) $(\bar{h}^\circ_f)_{298} = -110529$ kJ/kmol M = 28.01	
Temp. K	$(\bar{h}^\circ - \bar{h}^\circ_{298})$ kJ/kmol	$\bar{s}^\circ$ kJ/kmol K	$(\bar{h}^\circ - \bar{h}^\circ_{298})$ kJ/kmol	$\bar{s}^\circ$ kJ/kmol K
0	—9364	0	—8669	0
100	—6456	179.109	—5770	165.850
200	—3414	199.975	—2858	186.025
298	0	213.795	0	197.653
300	67	214.025	54	197.833
400	4008	225.334	2975	206.234
500	8314	234.924	5929	212.828
600	12916	243.309	8941	218.313
700	17765	250.773	12021	223.062
800	22815	257.517	15175	227.271
900	28041	263.668	18397	231.006
1000	33405	269.325	21686	234.531
1100	38894	274.555	25033	237.719
1200	44484	279.417	28426	240.673
1300	50158	283.956	31865	243.426
1400	55907	288.216	35338	245.999
1500	61714	292.224	38848	248.421
1600	67580	296.010	42384	250.702
1700	73492	299.592	45940	252.861
1800	79442	302.993	49522	254.907
1900	85429	306.232	53124	256.852
2000	91450	309.320	56739	258.710

	Water (H_2O) $(\bar{h}^\circ_f)_{298} = -241827$ kJ/kmol M = 18.015		*Hydrogen, Diatomic* (H_2) $(\bar{h}^\circ_f)_{298} = 0$ kJ/kmol M = 2.016	
Temp. K	$(\bar{h}^\circ - \bar{h}^\circ_{298})$ kJ/kmol	$\bar{s}^\circ$ kJ/kmol K	$(\bar{h}^\circ - \bar{h}^\circ_{298})$ kJ/kmol	$\bar{s}^\circ$ kJ/kmol K
0	—9904	0	—8468	0
100	—6615	152.390	—5293	102.145
200	—3280	175.486	—2770	119.437
298	0	188.833	0	130.684
300	63	189.038	54	130.864
400	3452	198.783	2958	139.215
500	6920	206.523	5883	145.738
600	10498	213.037	8812	151.077
700	14184	218.719	11749	155.608
800	17991	223.803	14703	159.549
900	21924	228.430	17682	163.060
1000	25978	232.706	20686	166.223
1100	30167	236.694	23723	169.118
1200	34476	240.443	26794	171.792
1300	38903	243.986	29907	174.281
1400	43447	247.350	33062	176.620
1500	48095	250.560	36267	178.833
1600	52844	253.622	39522	180.929
1700	57685	256.559	42815	182.929
1800	62609	259.372	46150	184.833
1900	67613	262.078	49522	186.657
2000	72689	264.681	52932	188.406

Appendix D—Gas Tables

Table D.1 Isentropic Flow

M	M^*	$\frac{A}{A^*}$	$\frac{p}{p_0}$	$\frac{\rho}{\rho_0}$	$\frac{T}{T_o}$
0	0	∞	1.00000	1.00000	1.00000
0.10	0.10943	5.8218	0.99303	0.99502	0.99800
0.20	0.21822	2.9635	0.97250	0.98027	0.99206
0.30	0.32572	2.0351	0.93947	0.95638	0.98232
0.40	0.43133	1.5901	0.89562	0.92428	0.96899
0.50	0.53452	1.3398	0.84302	0.88517	0.95238
0.60	0.63480	1.1882	0.78400	0.84045	0.93284
0.70	0.73179	1.09437	0.72092	0.79158	0.91075
0.80	0.82514	1.03823	0.65602	0.74000	0.88652
0.90	0.91460	1.00886	0.59126	0.68704	0.86058
1.00	1.00000	1.00000	0.52828	0.63394	0.83333
1.10	1.08124	1.00793	0.46835	0.58169	0.80515
1.20	1.1583	1.03044	0.41238	0.53114	0.77640
1.30	1.2311	1.06631	0.36092	0.48291	0.74738
1.40	1.2999	1.1149	0.31424	0.43742	0.71839
1.50	1.3646	1.1762	0.27240	0.39498	0.68965
1.60	1.4254	1.2502	0.23527	0.35573	0.66138
1.70	1.4825	1.3376	0.20259	0.31969	0.63372
1.80	1.5360	1.4390	0.17404	0.28682	0.60680
1.90	1.5861	1.5552	0.14924	0.25699	0.58072
2.00	1.6330	1.6875	0.12780	0.23005	0.55556
2.10	1.6769	1.8369	0.10935	0.20580	0.53135
2.20	1.7179	2.0050	0.09352	0.18405	0.50813
2.30	1.7563	2.1931	0.07997	0.16458	0.48591
2.40	1.7922	2.4031	0.06840	0.14720	0.46468
2.50	1.8258	2.6367	0.05853	0.13169	0.44444
2.60	1.8572	2.8960	0.05012	0.11787	0.42517
2.70	1.8865	3.1830	0.04295	0.10557	0.40684
2.80	1.9140	3.5001	0.03685	0.09462	0.38941
2.90	1.9398	3.8498	0.03165	0.08489	0.37286
3.00	1.9640	4.2346	0.02722	0.07623	0.35714
3.50	2.0642	6.7896	0.01311	0.04523	0.28986
4.00	2.1381	10.719	0.00658	0.02766	0.23810
4.50	2.1936	16.562	0.00346	0.01745	0.19802
5.00	2.2361	25.000	$189(10)^{-5}$	0.01134	0.16667
6.00	2.2953	53.180	$633(10)^{-6}$	0.00519	0.12195
7.00	2.3333	104.143	$242(10)^{-6}$	0.00261	0.09259
9.00	2.3772	327.189	$474(10)^{-7}$	0.000815	0.05814
10.00	2.3904	535.938	$236(10)^{-7}$	0.000495	0.04762
∞	2.4495	∞	0	0	0

Table D. 2 Normal Shocks

M_x	M_y	$\frac{p_y}{p_x}$	$\frac{\rho_y}{\rho_y}$	$\frac{T_y}{T_x}$	$\frac{p_{oy}}{p_{ox}}$	$\frac{p_{oy}}{p_x}$
1.00	1.00000	1.00000	1.0000	1.0000	1.00000	1.8929
1.10	0.91177	1.2450	1.1691	1.0649	0.99892	2.1328
1.20	0.84217	1.5133	1.3416	1.1280	0.99280	2.4075
1.30	0.78596	1.8050	1.5157	1.1909	0.97935	2.7135
1.40	0.73971	2.1200	1.6896	1.2547	0.95819	3.0493
1.50	0.70109	2.4583	1.8621	1.3202	0.92978	3.4133
1.60	0.66844	2.8201	2.0317	1.3880	0.89520	3.8049
1.70	0.64055	3.2050	2.1977	0.4583	0.85573	4.2238
1.80	0.61650	3.6133	2.3592	1.5316	0.81268	4.6695
1.90	0.59562	4.0450	2.5157	1.6079	0.76735	5.1417
2.00	0.57735	4.5000	2.6666	1.6875	0.72088	5.6405
2.10	0.56128	4.9784	2.8119	1.7704	0.67422	6.1655
2.20	0.54706	5.4800	2.9512	1.8569	0.62812	6.7163
2.30	0.53441	6.0050	3.0846	1.9968	0.58331	7.2937
2.40	0.52312	6.5533	3.2119	2.0403	0.54015	7.8969
2.50	0.51299	7.1250	3.3333	2.1375	0.49902	8.5262
2.60	0.50387	7.7200	3.4489	2.2383	0.46012	9.1813
2.70	0.49563	8.3383	3.5590	2.3429	0.42359	9.8625
2.80	0.48817	8.9800	3.6635	2.4512	0.38946	10.569
2.90	0.48138	9.6450	3.7629	2.5632	0.35773	11.302
3.00	0.47519	10.333	3.8571	2.6790	0.32834	12.061
4.00	0.43496	18.500	4.5714	4.0469	0.13876	21.068
5.00	0.41523	29.000	5.0000	5.8000	0.06172	32.654
10.00	0.38757	116.50	5.7143	20.388	0.00304	129.217
∞	0.37796	∞	6.000	∞	0	∞

Appendix E—Critical Point Data*

Substance	*Formula*	*Molecular Weight*	*Temperature* K	*Pressure* MPa	*Volume* m^3/kmol
Ammonia	NH_3	17.03	405.5	11.28	.0724
Argon	Ar	39.948	151	4.86	.0749
Bromine	Br_2	159.808	584	10.34	.1355
Carbon Dioxide	CO_2	44.01	304.2	7.39	.0943
Carbon Monoxide	CO	28.001	133	3.50	.0930
Chlorine	Cl_2	70.906	417	7.71	.1242
Deuterium (Normal)	D_2	4.00	38.4	1.66	—
Helium	He	4.003	5.3	0.23	.0578
Helium3	He	3.00	3.3	0.12	—
Hydrogen (Normal)	H_2	2.016	33.3	1.30	.0649
Krypton	Kr	83.80	209.4	5.50	.0924
Neon	Ne	20.183	44.5	2.73	.0417
Nitrogen	N_2	28.013	126.2	3.39	.0899
Nitrous Oxide	N_2O	44.013	309.7	7.27	.0961
Oxygen	O_2	31.999	154.8	5.08	.0780
Sulfur Dioxide	SO_2	64.063	430.7	7.88	.1217
Water	H_2O	18.015	647.3	22.09	.0568
Xenon	Xe	131.30	289.8	5.88	.1186
Benzene	C_6H_6	78.115	562	4.92	.2603
n-Butane	C_4H_{10}	58.124	425.5	3.80	.2547
Carbon Tetrachloride	CCl_4	153.82	556.4	4.56	.2759
Chloroform	$CHCl_3$	119.38	536.6	5.47	.2403
Dichlorodifluoromethane	CCl_2F_2	120.91	384.7	4.01	.2179
Dichloroflouromethane	$CHCl_2F$	102.92	451.7	5.17	.1973
Ethane	C_2H_6	30.070	305.5	4.88	.1480
Ethyl Alcohol	C_2H_5OH	46.07	516	6.38	.1673
Ethylene	C_2H_4	28.054	282.4	5.12	.1242
n-Hexane	C_6H_{14}	86.178	507.9	3.03	.3677
Methane	CH_4	16.043	191.1	4.64	.0993
Methyl Alcohol	CH_3OH	32.042	513.2	7.95	.1180
Methyl Chloride	CH_3Cl	50.488	416.3	6.68	.1430
Propane	C_3H_8	44.097	370	4.26	.1998
Propene	C_3H_6	42.081	365	4.62	.1810
Propyne	C_3H_4	40.065	401	5.35	—
Trichlorofluoromethane	CCl_3F	137.37	471.2	4.38	.2478

* K.A. Kobe and R.E. Lynn, Jr., *Chem. Rev.*, 52: 117-236 (1953).

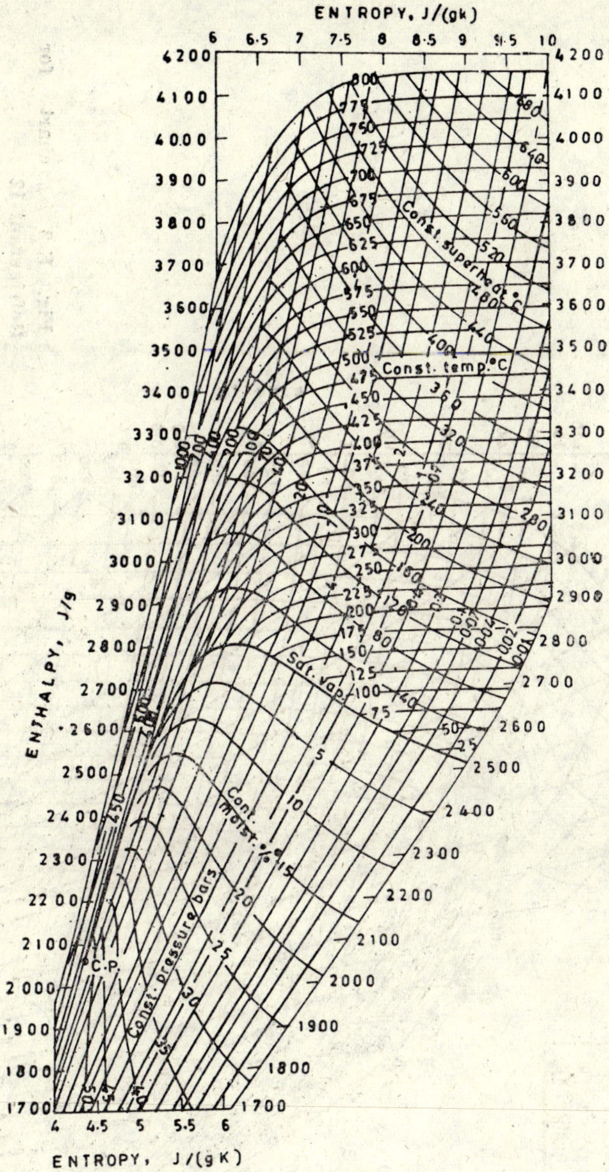

Fig. E-1 Mother diagram for steam
(From Joseph H. Keenan an Joseph Keyes, *Thermodynamic Properties of Steam,* John Wiley, New York, 1969, by permission.)

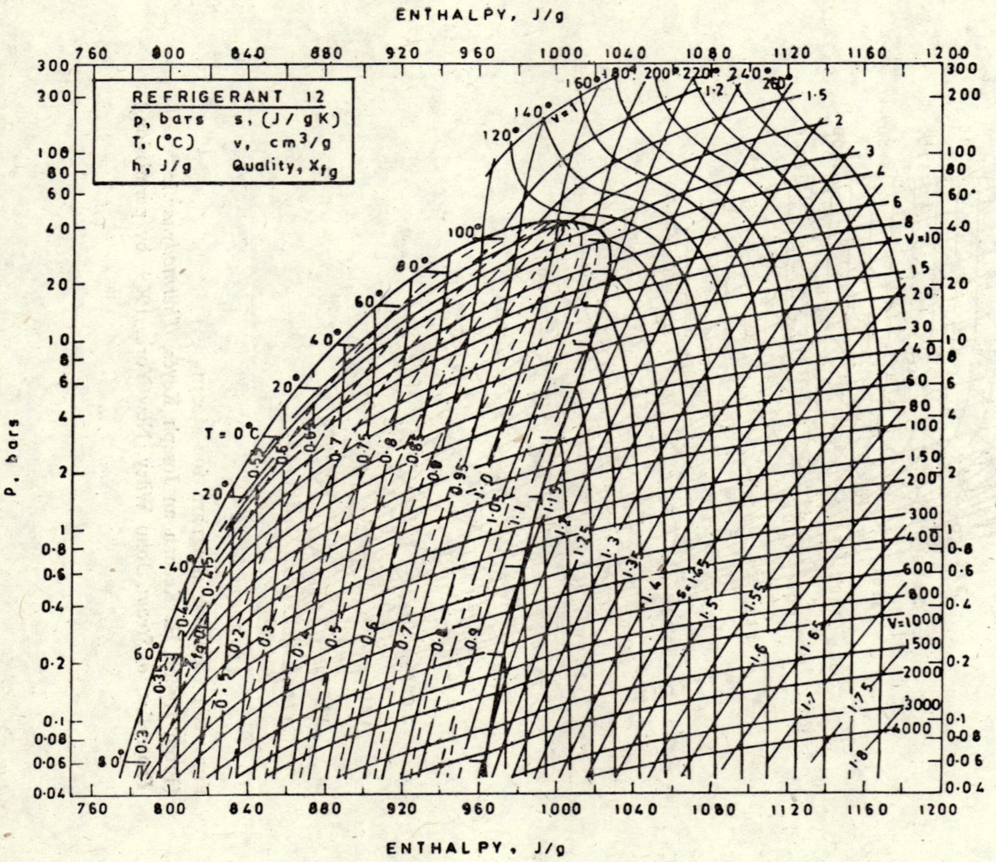

Fig. E-2 *p-h* chart for Refrigerant-12.

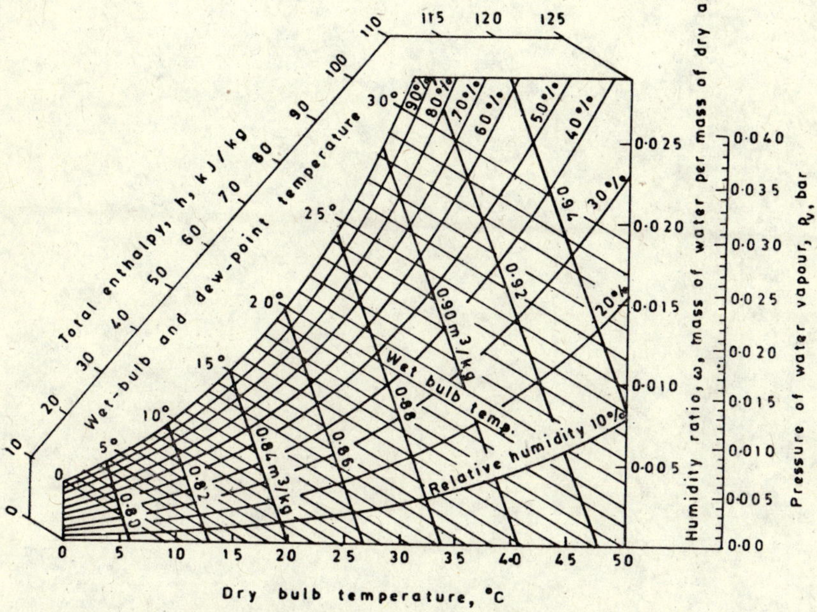

Fig. E-3 Psychometric chart
(Barometric pressure 760 mm Hg)

Bibliography

1. R.W. Benson, *Advanced Engineering Thermodynamics,* Pergamon Press, London, 1969.
2. H.B. Callen, *Thermodynamics,* John Wiley, New York, 1960.
3. G.N. Hatsopoulos and J.H. Keenan, *Principles of General Thermodynamics,* John Wiley, New York, 1965.
4. Jack P. Holman, *Heat Transfer,* McGraw-Hill, New York, 1978.
5. J.B. Jones and G.A. Hawkins, *Engineering Thermodynamics,* John Wiley, New York, 1960.
6. J.H. Keenan, *Thermodynamics,* John Wiley, New York, 1941.
7. J.F. Lee and F.W. Sears, *Thermodynamics,* Addison Wesely, Cambridge, Mass, 1962.
8. David A. Mooney, *Mechanical Engineering Thermodynamics,* Prentice-Hall, N.J., 1953.
9. E.F. Obert, *Concepts of Thermodynamics,* McGraw-Hill, New York, 1960.
10. W.C. Reynolds and H.C. Perkins, *Engineering Thermodynamics,* McGraw-Hill, New York, 1977.
11. Michel A. Saad, *Thermodynamics for Engineers,* Prentice Hall, New Delhi, 1969.
12. A.H. Shapiro, The Dynamics and *Thermodynamics of Compressible Fluid Flow,* The Ronald Press, New York, 1953.
13. D.B. Spalding and E.H. Cole, *Engineering Thermodynamics,* Edward Arnold, London, 1967.
14. G.J. Van Wylen and R.E. Sonntag, *Fundamentals of Classical Thermodynamics,* John Wiley, New York, 1976.
15. F.J. Wallace and W.A. Linning, *Basic Engineering Thermodynamics,* Pitman, London, 1970.
16. Kenneth Wark, *Thermodynamics,* McGraw-Hill, New York, 1977.
17. M.W. Zemansky, *Heat and Thermodynamics,* McGraw-Hill, New York, 1957.

Index